Stress Biology in Photosynthetic Organisms

Arun Kumar Mishra
Editor

Stress Biology in Photosynthetic Organisms

Molecular Insights and Cellular Responses

Springer

Editor
Arun Kumar Mishra
Laboratory of Microbial Genetics, Department of Botany
Institute of Science, Banaras Hindu University
Varanasi, India

ISBN 978-981-97-1882-5 ISBN 978-981-97-1883-2 (eBook)
https://doi.org/10.1007/978-981-97-1883-2

This Springer imprint is published by the registered company Springer Nature Singapore Pte Ltd.
The registered company address is: 152 Beach Road, #21-01/04 Gateway East, Singapore 189721, Singapore

Preface

The origin of life on Earth is challenging to predict, but it is estimated that photoautotrophs appeared approximately 3.5 billion years ago. These early photoautotrophs were simple and single-celled microorganisms known as cyanobacteria, playing a crucial role in shaping today's atmosphere. They are attributed to the Great Oxidation Event, which is believed to have occurred around 2.4 billion years ago. This event marked a significant turning point in the history of planet Earth, as it paved the way for the evolutionary origin of other organisms, including higher photoautotrophs.

Stress Biology in Photosynthetic Organisms: Molecular Insights and Cellular Responses delves into the intriguing realm of how photosynthetic life adapts to environmental challenges, aiming to present a comprehensive view of the adaptive strategies and molecular intricacies of these organisms in response to stress. Commencing with an exploration of their evolution, this book delves into the historical context of photosynthetic life, revealing the adaptive strategies that have allowed these organisms to thrive over time. Subsequently, the book explores the molecular machinery governing stress responses, with emphasis on signaling systems, reactive oxygen species, and the roles of epigenetic modifications and microRNAs. Further collaborative strategies are emphasized with the importance of synergistic efforts within populations and symbiotic relationships, contributing to stress resilience. Additionally, this book unravels the interconnected signaling pathways that govern stress responses, highlighting the role of phytohormones.

Extending beyond the prokaryotic realm, we explored recent advances in stress biology within plant tissue culture. Our journey encompasses diverse environments inhabited by photosynthetic organisms, ranging from extreme conditions with an emphasis on thermophilic photoautotrophs along with underwater photosynthesis by photoautotrophs and potential adaptations of the photoautotrophs in extraterrestrial environments. Further specialized topics have been discussed deepening exploration in the molecular aspects of the photoautotrophs, covering stress granules, nanobionics for bioenergy and crop productivity, the role of nitric oxide in photosynthetic stress response, and case studies on stress responses in various phytoplankton. Each topic provides a unique perspective, showcasing the multidimensional nature of stress biology, covering molecular aspects in the photoautotrophs. Implications of cell death in photoautotrophs, with its potential impact on nutrient cycling, biogeochemical processes, and the intricate mechanism of microbial interactions within diverse ecosystems, are profoundly discussed.

As editor, I extend sincere gratitude to all the contributors who have enriched this collaborative effort. Our hope is that this work becomes a valuable resource for researchers, educators, and students, offering a comprehensive overview of the dynamic field of stress biology in photosynthetic organisms. This book is more than a compilation; it is a shared journey through the captivating intricacies of how photosynthetic life responds to stress.

Varanasi, India Arun Kumar Mishra

Acknowledgments

It is with immense gratitude that I acknowledge the collaborative efforts and support from all the contributors in shaping this book entitled ***Stress Biology in Photosynthetic Organisms: Molecular Insights and Cellular Responses***. Further, I would like to extend my heartfelt thanks to the Head of the Department of Botany, Banaras Hindu University, for ensuring a conducive environment for research and collaboration. The authors of this book deserve special recognition for their invaluable contributions. Their dedication, expertise, and passion have enriched the pages with knowledge, offering readers a diverse and insightful exploration of the subject. I am deeply grateful for their commitment to this book and their role in making it a reality. The book is owed to the support received from various funding agencies, including Indian Council of Agricultural Research-National Bureau of Agriculturally Important Microorganisms, Department of Science and Technology and Institute of Eminence-Banaras Hindu University. The support from these esteemed agencies has not only facilitated the research and compilation processes but also substantially enriched the overall quality of this work.

Thank you all for your unwavering support and contributions.

Laboratory of Microbial Genetics, Department of Botany,
Institute of Science, Banaras Hindu University,
Varanasi, India

Arun Kumar Mishra

Contents

Editor and Contributors

About the Editor

Arun Kumar Mishra is a Professor at the Laboratory of Microbial Genetics at Banaras Hindu University, India. His group has been working on several aspects of cyanobacterial metabolisms and stress-signaling attributes. However, in recent times his lab is dedicated to understanding certain key metabolisms and signaling prospects in heterocytous cyanobacteria like regulation of sulfur homeostasis, mechanistic signal transduction of circadian cycle, regulation of nitric oxide signaling mediated by NOS enzyme as well as the pertinence of different cell demise modalities. He and his team confer expertise in proteome and metabolome profiling in cyanobacteria, polyphase-based species identification, microscopy, and genetic manipulation techniques. He has published more than 100 high-quality peer-reviewed research articles in various reputed international journals. Further, he has also edited a book entitled *Cyanobacteria: From Basic Science to Application* dealing with the current research advancement in cyanobacteria.

Contributors

Altaf Ahmad Department of Botany, Aligarh Muslim University, Aligarh, India

Bandaru Aishwarya Department of Microbiology and FST, GITAM School of Science, GITAM Deemed to be University, Visakhapatnam, India

Afreen Akhtar Department of Botany, Aligarh Muslim University, Aligarh, India

Alka Bhardwaj Laboratory of Microbial Genetics, Department of Botany, Institute of Science, Banaras Hindu University, Varanasi, India

Samujjal Bhattacharjee Laboratory of Microbial Genetics, Department of Botany, Institute of Science, Banaras Hindu University, Varanasi, India

Sindhunath Chakraborty Laboratory of Microbial Genetics, Department of Botany, Banaras Hindu University, Varanasi, India

Vineeta Dixit Department of Botany, Nilamber-Pitamber University, Garhwa, Jharkhand, India

Neha Gupta Laboratory of Microbial Genetics, Department of Botany, Institute of Science, Banaras Hindu University, Varanasi, India

Satyam Jaiswal Department of Botany, University of Lucknow, Lucknow, Uttar Pradesh, India

Tameshwar Jaiswal Laboratory of Cyanobacterial Systematics and Stress Biology, Department of Botany, Institute of Science, Banaras Hindu University, Varanasi, India

Saad Bin Javed Department of Botany, Aligarh Muslim University, Aligarh, India

Veda Sai Kalavalapalli Department of Microbiology and FST, GITAM School of Science, GITAM Deemed to be University, Visakhapatnam, India

Vaishnavi Kedia Department of Botany, University of Lucknow, Lucknow, Uttar Pradesh, India

Surbhi Kharwar Department of Botany, University of Lucknow, Lucknow, Uttar Pradesh, India

Deepak Kumar Department of Botany, Institute of Science, Banaras Hindu University, Varanasi, Uttar Pradesh, India

Ravinsh Kumar Department of Life Science, Central University of South Bihar, Gaya, Bihar, India

Vinod Kumar Department of Botany, Pandit Prithi Nath College, Kanpur, Uttar Pradesh, India

Arun Kumar Mishra Laboratory of Microbial Genetics, Department of Botany, Institute of Science, Banaras Hindu University, Varanasi, India

Awadhesh Kumar Mishra Department of Botany, Ramjas College, University of Delhi, Delhi, India

Deepankar Mondal Department of Biotechnology, Guru Ghasidas Vishwavidyalaya, Bilaspur, Chhattisgarh, India

Prashasti Pandey Department of Botany, Institute of Science, Banaras Hindu University, Varanasi, Uttar Pradesh, India

Deepanshu Pant Genetics and Tree Improvement Division, ICFRE-Himalayan Forest Research Institute, Conifer Campus, Shimla, India

Anirbana Parida Laboratory of Microbial Genetics, Department of Botany, Institute of Science, Banaras Hindu University, Varanasi, India

Preeti Patel Department of Plant Pathology, Ohio State University, Columbus, OH, USA

Rajani Prabha Department of Botany, Balrampur Girls Degree College, Balrampur, Uttar Pradesh, India

Archana Prasad Department of Fruit Tree Sciences, Institute of Plant Sciences, Volcani Center, Agricultural Research Organization, Rishon LeZion, Israel

R. Gyana Prasuna Department of Microbiology and FST, GITAM School of Science, GITAM Deemed to be University, Visakhapatnam, India

Rajesh Saini Department of Botany, MMV, Banaras Hindu University, Varanasi, India

P. K. Satheeshkumar Department of Botany, Banaras Hindu University, Varanasi, Uttar Pradesh, India

Himani Sharma Genetics and Tree Improvement Division, ICFRE-Himalayan Forest Research Institute, Conifer Campus, Shimla, India

Sanjay Sharma Laboratory of Cyanobacterial Systematics and Stress Biology, Department of Botany, Institute of Science, Banaras Hindu University, Varanasi, India

Dhananjay Shukla Department of Biotechnology, Guru Ghasidas Vishwavidyalaya, Bilaspur, Chhattisgarh, India

Praveen Kumar Shukla Department of Botany, MMV, Banaras Hindu University, Varanasi, India

Vandana Sindhu Laboratory of Cyanobacterial Systematics and Stress Biology, Department of Botany, Institute of Science, Banaras Hindu University, Varanasi, India

Aakanksha Singh Department of Botany, Banaras Hindu University, Varanasi, Uttar Pradesh, India

Ashutosh Singh Department of Life Science, Central University of South Bihar, Gaya, Bihar, India

Prashansa Singh Laboratory of Microbial Genetics, Department of Botany, Institute of Science, Banaras Hindu University, Varanasi, India

Satya Shila Singh Laboratory of Cyanobacterial Systematics and Stress Biology, Department of Botany, Institute of Science, Banaras Hindu University, Varanasi, Uttar Pradesh, India

Laboratory of Cyanobacterial Systematics and Stress Biology, Institute of Science, Banaras Hindu University, Varanasi, India

Savita Singh Babu Shivnath Agrawal (PG) College, Mathura, India

Ankit Srivasatava Laboratory of Microbial Genetics, Department of Botany, Institute of Science, Banaras Hindu University, Varanasi, India

Amrita Srivastava Department of Life Science, Central University of South Bihar, Gaya, Bihar, India

Taruna Genetics and Tree Improvement Division, ICFRE-Himalayan Forest Research Institute, Conifer Campus, Shimla, India

Balkrishna Tiwari Genetics and Tree Improvement Division, ICFRE-Himalayan Forest Research Institute, Conifer Campus, Shimla, India

Kavindra Nath Tiwari Department of Botany, MMV, Banaras Hindu University, Varanasi, India

Madhu Tiwari Aix-Marseille University, Saint-Paul-lez-Durance, France

Ranjan Kumar Tiwari Laboratory of Cyanobacterial Systematics and Stress Biology, Department of Botany, Banaras Hindu University, Varanasi, India

Jyoti Vishwakarma Department of Botany, MMV, Banaras Hindu University, Varanasi, India

Naveen K. Vishwakarma Department of Biotechnology, Guru Ghasidas Vishwavidyalaya, Bilaspur, Chhattisgarh, India

Photosynthetic Organisms: Their Existence in Evolutionary Prospective

1

Rajesh Saini, Praveen Kumar Shukla, Jyoti Vishwakarma, Awadhesh Kumar Mishra, and Kavindra Nath Tiwari

Abstract

Photosynthesis is a very old process on this Earth. Based on fossil discoveries and chemical evidence, cyanobacteria first appeared 2.5–2.6 billion years ago (bya). Their evolution was undoubtedly continued by a number of anaerobic, photosynthetic bacterial life forms. Carbon isotope data revealed that autotrophic carbon fixation may have begun at least a bya. It is unclear, nevertheless, what the earliest photosynthetic organisms were like. The primary elements of the photosynthetic system are the carbon fixation mechanism, electron transport complexes, antenna complexes, and reaction centers. It is most likely true that these components have not all evolved at the same point in time. Consequently, it is better to think of the photosynthetic apparatus as a mosaic made up of numerous structural components, each with its own unique evolutionary background. One early instance of a cyanobacterium's endosymbiotic absorption by a heterotrophic organism appears to have been the source of the chloroplasts seen in yellow-green algae, glaucophytes, brown algae, cryptophytes, red algae, and other algae in the "red" line of development. The variety of species present in the algae's "red line" is the outcome of a single secondary endo-symbiotic occurrence in which an organism resembling red algae was ingested by another eukaryote. This "red line" is further expanded by tertiary (third-level) endosymbiotic events. Photosynthetic units are found in reaction centers involving complexes for gathering light. Two of these units are necessary for oxygenic photosynthesis, which currently accounts for the majority of biological transfer

R. Saini · P. K. Shukla · J. Vishwakarma · K. N. Tiwari (✉)
Department of Botany, MMV, Banaras Hindu University, Varanasi, India

A. K. Mishra
Laboratory of Microbial Genetics, Department of Botany, Institute of Science, Banaras Hindu University, Varanasi, India

A. K. Mishra (ed.), *Stress Biology in Photosynthetic Organisms*, https://doi.org/10.1007/978-981-97-1883-2_1

of energy in the various trophic levels of the biosphere. The emergence of photosynthesis utilizing oxygen among cyanobacteria, which paved the path for the formation of complex life forms with multicellular levels of organization, had a profound influence on the biology, geology, and environment of Earth. In this review, we have discussed the early evidence of photosynthesis, the origin of reaction centers, antenna, pigments, and how oxygenic photosynthesis came into existence. The origin of the chloroplasts is a necessary event that occurred earlier and was added to the history of photosynthetic origin in this review.

Keywords

Blue-green algae · Endosymbiosis · Evolution · Photosynthesis · Photosynthetic organism

1.1 Introduction

Photosynthesis is a mechanism that plants employ to transform solar energy into biochemical energy, which is then used to sustain almost all life on Earth. The growth rate of a photosynthetic organism is directly dependent on the photosynthetic rate, and the production of food, fiber, wood, grain for cattle feed, and fuel for human use is made possible by photosynthesis. Early photosynthesis most likely evolved within extinct bacterial lineages during the Paleoarchean (3.6–3.2 bya) or Eoarchean (4.0–3.6 bya) periods of time (Schad et al. 2019). The antenna and the core are the two components that make up a photosystem. The antenna proteins are the key proteins for light energy absorption and transferring it to the reaction center. Chlorophyll molecules assist in a separation of charge that occurs within the reaction center. The proteins encoded in nuclear matter are sometimes referred to as the light-harvesting complex (LHC) (Giovagnetti and Ruban 2018). The process of oxygenic photosynthesis involves four electrons and is linked to both the water-oxidizing complex (WOC) and chlorophyll-a (He et al. 2021). Anaerobic autotrophic metabolisms like methanogenesis, along with anoxygenic photosynthesis, maintained the pre-oxygen biosphere (Ward et al. 2019). The crucial protein complexes known as Photosystems (PS) I and II, which contain chlorophyll and are situated in the membranes of the thylakoid of cyanobacteria and chloroplasts, work in sequence and are responsible for light-induced charge separation (Ben-Shem et al. 2004). PSI and PSII present in higher plants and cyanobacteria share some notable structural characteristics between the type I RC of green sulfur bacteria and the type II RC of purple bacteria. These relationships provide support for the well-established theory for all reaction center complexes share a similar evolutionary ancestry (Nelson and Ben-Shem 2005). There is no longer much evidence supporting the hypothesis that the earliest chloroplasts developed through symbiosis between a free-living photosynthetic organism (the precursor of current cyanobacteria) and a non-photosynthetic host (Howe et al. 2008). Oxygenic photosynthesis is among the most notable metabolic processes that have developed on Earth which led to the evolution of complex life forms. The first oxygenic event occurred between 2.32

and 2.4 bya, and the second oxygenic event, commonly named the Neoproterozoic Oxidation Event (NOE), happened between 600 and 800 million years ago. These two events were responsible for the present O_2 concentration in the atmosphere (Sánchez-Baracaldo and Cardona 2020). Furthermore, the development of aerobic respiration and new biosynthetic pathways was directly linked with the availability of biological O_2 (Fischer et al. 2016).

1.2 Early Evolution of Photosynthesis

The ancient history of photosynthesis is mostly unknown to all of us. The origin of the process was suggested by different workers but none of the ideas was based on evidence (Olson and Blankenship 2004). Four bya, which was the beginning of life forms on Earth, the environment and the surface of our planet underwent major transformations (Blankenship 1992; Blank 2004). Although life first depended on a severely reduced environment with hardly any oxygen throughout the atmosphere, it at present appears to have around 21% oxygen within the atmospheric environment as well as highly oxidative conditions in the majority of ecological niches. Based on carbon isotope studies, ribulose-1,5-bisphosphate, an enzyme, favoured the deployment of carbon in the biosphere from a rich atmospheric pool, regulating the worldwide carbon equilibrium throughout not less than 3.5 billion years across the atmosphere-ocean biosphere (Blankenship 1992; Schidlowski 1984). According to Des Marais (2000), there is proof that photosynthetic organisms called stromatolites existed 3.2 to 3.5 bya. These organisms had layered structures similar to those found in some contemporary cyanobacteria, along with several microfossils that are thought to have originated from the phototrophs. The most primitive Archean stromatolite fossil remains exhibiting supporting proof of microbial carbon fixation was 3.8 billion years old (Schidlowski 1988). The initial cyanobacteria-like cellular microfossils were found at 3.5 bya (Schopf 1983; Schopf and Packer 1987) but after 2.7 billion years, when oxygen production first started to emerge on Earth, cyanobacteria took over (Des Marais 2000).

Cyanobacteria contribute to increasing the O_2 concentration to around one-fourth of its current concentration. The most primitive known eukaryotic organism emerged 1.8 bya. The integration of photosynthetic cyanobacteria as chloroplasts into eukaryotes required an additional 0.6–0.8 billion years. Descended from algae, terrestrial plants first appeared on Earth 0.5 bya, when the planet's current topography was formed.

The term "Pre-Phanerozoic" (Precambrian) describes an eon of geological time originating 3.96 bya, when the earliest known terrestrial rocks evolved. The fossil record of the pre-Phanerozoic eon differs significantly from that of the Phanerozoic eon. Microbial life dominates the Earth's ecosystem. Cyanobacteria are prokaryotic phototrophs, and they have existed for 3.5 billion years.

In the geological evidence, photosynthetic organisms, such as macroscopic algae, have been discovered in rocks dating back as far as 1.4–1.5 billion years, but they are rather rare. Despite this, there was no evidence of plant life. Pre-Phanerozoic fossils are mostly divided into three main categories: Stromatolites, chemical fossils, and microbial fossils.

1.2.1 Stromatolites

Microbes capture, tethered, and/or condensed sediment to form macroscopic organic sedimentary structures called stromatolites. Stromatolite formation is dependent on photosynthetic microorganisms, specifically cyanobacteria. According to the geological evidence, stromatolites are highly stratified rock formations comprised mostly of mineral carbonate (such as calcite, $CaCO_3$) that are created when laminae from the outermost layer of a prehistoric lake or seafloor accumulate layer by layer due to microbial deposition. The mat-building microorganism's photosynthetic metabolic activity is reflected in its stratified arrangement. Through diagenesis, the sequence of processes that contribute to the lithification and preservation of these formations, the mineral silica (quartz, SiO_2) may substitute for the originally deposited carbonate matrix. The dimension, form, and laminar organization of modern calcareous microbiological stromatolites have appeared quite analogous to the characteristics of Precambrian fossils. Nowadays, those contemporary stromatolites usually occur exclusively in refugia-hypersaline lagoons, for example—where burrowing metazoans and grazing animals have no impact on the slow-growing microbial mats. As a result of this, stromatolites weren't prevalent in Phanerozoic sediments that were accumulated in surroundings where a variety of metazoan species predominated.

1.2.1.1 Archean Stromatolites

Stromatolites developed through microbial processes have been found in an unprecedented range of Archean age geological units, which are of particular significance due to their desirable influence on the evolution of aerobic photosynthesis.

1.2.1.2 Filamentous Cyanobacteria

The Oscillatoriaceae, which produce stromatolites, are one of the taxonomic families of filamentous cyanobacteria with the largest geological history, exhibiting a variety of fossils reported in hundreds of historical communities of microbes. Two prominent examples from the Precambrian time period, dating around 850 Mya, are *Helioconema funiculum*, a spirally coiled cellular structure is likely as an organism that belongs to the recent oscillatoriacean genus *Spirulina*, and *Cephalophytarion laticellulosum*, a tapered cell trichome that is comparable to the *Oscillatoria amoenum* cyanobacterium (Schopf et al. 2006).

1.2.1.3 Archean Microbes

The evidence makes it difficult to prove that ancient Archean stromatolites were biogenic but several identified abiogenic, laminated, stromatolite-like frameworks, including geyserites (Walter 1976), laminating calcretes (Read 1976), and spelothems (Thrailkill 1976) are available. Stromatolites having a broad variety of forms, including stratiform, enclosed, dome-shaped, pseudo-columnar, columnar-layer, and oncolytic, were recently identified at the two distinct Swaziland and Pilbara sedimentary sequential patterns.

1.2.2 Chemical Fossils

Chemical fossils are defined as traces of previous life that have been preserved through chemical processes. These consist of organic compounds and stable isotopes that are physiologically distinct, including carbon. These crystals are frequently laminated in domical to columnar patterns. Many chemical compounds that are produced through specific biological constituents can be employed as biological markers for distinct classes of organisms. Photosynthesis, which produces oxygen, permits the biosynthesis of organic compounds that are oxygen-dependent. It was believed that methyl-substituted hopanoids (Summons et al. 1999) represented the primary biological signal for a newly developing oxygen-dependent synthesis pathway that emerged 2.7 bya. These compounds are still not considered acceptable indicators for oxygenated photosynthesis because they were recently discovered in anaerobic purple bacteria (Rashby et al. 2007; Welander et al. 2010). It has been proclaimed that molecules revealing the existence of purple bacteria such as okenone (Brocks et al. 2005) along with isorenieratene and chlorobactene which reveals the presence of green sulfur bacteria may have evolved in ancient 1.64 bya.

1.2.3 Micro Fossils

The microorganisms preserved through permineralization represent the earliest identified microbial fossil evidence. This petrification method of preserving fossils produces a highly accurate description of the natural form of any mode of fossil preservation. This type of preservation, most frequent in organisms such as fungi, plants, and preserved prokaryotic cells, is caused by the penetration of charged mineral fluids entering the cells corresponding to the initial phases of diagenesis before their degradation and fragmentation.

Penetrating liquids fill microscopic spaces, replacing cellular components' aqueous environment, creating a mineral-infused inorganic–organic composite with strong architecture, retaining miniature details of cellular architecture and organismal morphology (William Schopf 2011).

Sequences from Swaziland and the Pilbara contain microbiological fossils, with four types of filaments identified in Warrawoona Group feldspar. The filamentous Warrawoona microfossils' potential syngenetic or biogenic origin is debated. The chert fragment found at Locality A contains microbial fossils that have never been identified (Awramik et al. 1983; Buick 1984, 1990; Schopf and Packer 1987).

Further microfaunal fossils representing two new Warrawoona locations have been reported by Schopf and Packer (1987). There are two forms of pluricellular spheroidal cell aggregates, in combination with filaments around 3 cm in diameter. One of these has four cells enclosed by a multi-lammelated envelope (17–24 cm in diameter). These organisms are most similar to chroococcalean cyanobacteria. The discoveries of Schopf and Packer support preceding proof of genuine microfossils found at Warrawoona. According to Pflug (1967), The Swaziland Supergroup has been looking for microbiological fossilized remains as proof of life for a long time, ever since Pflug announced in 1967 that filamentous and coccoidal microbial fossils had been found in the 3.3 Billion-year-old Fig Tree Group.

The current state of purported microbiological fossil remains in published articles reveals that each of them, with the exception of tiny spheroids claimed by various researchers (Walsh 1992; Pflug 1967), constituted non-fossils. The carbonaceous spheroids found in old rocks are referred to as "dubio microfossils" despite their lack of biogenic character. Walsh's (1992) study explores the biogenicity of various morphotypes, ranging from true microfossils to Dubai microfossils, with potential candidates including tiny coccoids, tubular filaments, and thread-like filaments.

Morphological evidence indicates the occurrence of phototrophs and cyanobacteria 3500 million years ago. The partitioned threads resembling trichomes and empty, tubular structures are found in Warrawoona rocks, suggesting that many of the microorganisms are motile and photoresponsive (Awramik et al. 1983; Schopf and Packer 1987). The conservation of filaments inside and similar to dark laminae which have similarities to stromatoliticlaminae provides evidence for this. Schopf and Packer (1987) found coccoids in Warrawoona rocks, indicating the presence of cyanobacteria through their morphological evidence.

1.3 The Proof of Photosynthesis from 3500 Million Years Ago: An Argument in Favour of Cyanobacteria

It was earlier believed that oxygen-releasing photosynthetic microorganisms evolved during the Archean and early Proterozoic eras based on banded iron forms (Cloud 1973). The earliest known examples of banded iron form have been from rocks estimated around 3.8 bya (Appel 1980). The oxygenation of reduced iron to form banded iron formations was a result of the oxygen produced in the photosynthetic process. It would offer unambiguous proof of cyanobacterial existence. Nevertheless, as was already indicated, there are two abiogenic pieces of evidence that point to the beginnings of iron production.

As previously explained, gypsum implies the existence of sulfur-oxidizing bacteria along with other oxygen-releasing cyanobacteria. Reduced sulfur is directly oxidized into sulfate by oxygen in the absence of sulfur-oxidizing bacteria. In the early Earth's history, oxygen-releasing phototrophy is considered the most believable origin of geochemically valuable oxygen (Cloud 1976). Many sulfur-oxidizing bacteria are phototrophs, except for *Beggiatoa*. As a result, the gypsum provides very confined support for the photosynthesis process.

As was previously mentioned, the stromatolite evidence, as well as some microfossils, suggest the origin of cyanobacteria 3.5 bya. The entire body of evidence supports this concept. Nonetheless, less plausible explanations (such as the formation of stromatolites by non-phototrophic bacteria or by phototrophic bacteria other than cyanobacteria, particularly *Chloroflexus*) must be taken into consideration. While some scholars believe that the ancient record of bacterial photosynthesis shows the existence of the phototrophs around 3.5 bya, others are unwilling to accept the inference that both oxygenated photosynthesis and cyanobacteria existed by then (Olson and Pierson 1986). In contrast to Photosystem I, which is involved in anoxygenic photosynthesis (Olson and Pierson 1987a, b), according to Vermaas (1989), oxygenic photosynthesis (Photosystem II) is thought to be a highly intricate bio-energetic process.

1.4 Molecular Evidence of Early Evolution of Photosynthesis

Xiong et al. (2000) suggested that concentrated on the sequence analysis as well as the molecular advancement of a variety of such pigment-producing enzymes. The variety of enzymes associated with the biosynthesis of chlorophyll, present in all five classes of photosynthesis-dependent prokaryotes as well as a number of algal and plant species, are being compared with each other. These findings significantly support the previous assumption that anoxygenic photosynthesis originated earlier compared to oxygenic photosynthesis. The purple photosynthetic bacteria represented the oldest branching genera in all situations, while all oxygenic genera clustered among the more recently evolved. Observations clearly reject an intense analysis given to the Granick hypothesis 15, which suggested that the pigment synthesis pathway resembles the pigment's evolutionary architecture (Granick 1965). The results are imprecise because they do not show whether the older microorganisms consisted of chlorophyll or bacteriochlorophyll; instead, they only indicate that the organisms that contain chlorophyll are currently more developed through the bacteriochlorophyll-containing microorganisms. Further investigation into this topic might be possible with the evaluation of new genes (Lazcano and Miller 1999; Blankenship 2001, 2021).

Furthermore, during the study of pigment-synthesizing enzymes, both green sulfur and green non-sulfur bacteria were found to be constantly clustered together. These species have a homologous antenna complex, termed the chlorosome; however, they are generally quite separated in terms of photosynthetic activity along with almost every other component of cellular metabolism. This could indicate that these organisms have laterally transferred genes involved in pigment production and structural proteins involved in chlorosome formation. The hypothesis that is gradually developing from these and other studies is that photosynthesis's evolutionary growth is an intricate process that can't be characterized by an easy, branching, linear evolutionary approach. Photosynthesis evolved through the recruitment and modification of genes encoding components of numerous additional currently present metabolic pathways, as well as a few crucial processes maybe multiple lateral gene-transfer mechanism.

The origins of photosynthesis reaction centers originated from a gene that transcribed a significant protein that functioned as a homodimer, identical to PscA from the Chlorobiaceae. All of the monomers had a tendency to bind around 20 and 40 bacterial-type chlorophyll molecules; two of them existed in adjacent vicinity under the optimum conditions to generate stimulation traps that could be utilized to initiate oxidation–reduction events (Heathcote et al. 2003). There were two analogous binding sites for quinone on the reverse end of the membrane (one for each monomer).

Each monomer had a cysteine residue that was specifically placed to serve as a framework for the production of the acceptor of electrons in the non-heme iron core. When cyclic electron transport facilitates the synthesis of ATP, this specific type of reaction center may function as both the type I reaction center, like the PSI of today, and the type II reaction center typical of the purple bacteria of today. When life first began on Earth, the atmosphere was mostly anaerobic, which made this type of reaction center perfect for use.

1.5 Evolution of Reaction Center

The reaction center complex is essential for the process of photosynthesis; consequently, plenty of research has been invested in understanding its development (Blankenship 2021). Biological pigment–protein structures comprising the RC accumulate sunlight, induce the separation of charges, and ultimately convert light energy into electrochemical potential. The photosystem I reaction center complex is one of them and confers a potent lowering capacity to diminish ferredoxin which ultimately transforms carbon dioxide into polysaccharides. One powerful oxidizing ability of splitting of water, for instance, by the photosystem II reaction center complex (Hall and Rao 1999; Nechushtai et al. 1996). The microfossil, dating back 3.5 billion years, that mimics current cyanobacteria represents one of the earliest known remnants of life.

Reaction centers of photosynthetic organisms are generally categorized into two broad categories based on the property of the final acceptor of electrons: type I (RC1), which uses a Fe-S cluster as the final recipient and is similar to PSI in addition to the RC-linked to Heliobacteria and green sulfur bacteria, whereas type II (RC2), which uses a dynamic quinone as the final recipient and is similar to PSII and the RC linked to green and purple-coloured filamentous bacteria (Heathcote et al. 2003). Phylogenetic investigations of PSI, as well as PSII reaction center core proteins throughout cyanobacteria and higher plants, showed certain notable similarities in structure, including the type II RC of purple bacteria and the type I RC associated with green sulfur bacteria (Nitschke and Rutherford 1991; Barber et al. 1999; Schubert et al. 1998). The shared characteristics reinforce the long-held theory that every single reaction center complex evolved from a common ancestor (Blankenship and Hartman 1998; Vermaas 1994; Schubert et al. 1998; Mulkidjanian and Junge 1997).

It is observed that the type II reaction center structure is almost similar to the ancestors of cyanobacterial PSII as it represents it. (Singhal et al. 2012). However, it was subsequently proposed that these centers additionally served as the ancestors of PSI (Baymann et al. 2001, 2003; Schubert et al. 1998). Nevertheless, it's likely that this RC2 also possessed certain type I traits. The primary protein of RC1 has both electron transport mechanisms and inherent antenna chlorophylls inside their 11 transmembrane helices, therefore distinguishing it from pre-existing RC2. The aforementioned functions are divided among multiple proteins, with the present RC2 (PSII) having fewer TM helices. Purple bacteria are currently considered the most primitive of all developing photosynthetic ancestral ancestries (Xiong et al. 2000).

BChl-g, the most closely related ancestral pigment predecessor to Chl-a, which is utilized by cyanobacteria as well as higher plants, is employed as a pigment by RC1-containing heliobacteria (Bryant 2006). Heliobacteria were therefore discovered to have a pigment precursor similar to that of Chl-a, but with the "inappropriate" reaction center type 1 to produce oxidation of water. The bulk of modern photosynthetic species have heterodimeric reaction centers, which are the consequence of gene duplication, according to sequence investigation of particularly type I as well as type II reaction centers (Heathcote et al. 2003; Blankenship 1992).

Comparable photosynthetic machinery, comprising both of the RC complexes, i.e., PS II and PS I, present in the membrane, is shared by cyanobacteria and plants. About 80 kDa polypeptides that constitute PS I RC's core protein have equivalent amino acid sequences of 80% (Kirsch et al. 1986). They have only approximately 15% identity compared with the RC polypeptides of green sulfur bacteria (Golbeck and Bryant 1991; Büttner et al. 1992), but almost little similarity to PS II's core polypeptides on the same membrane. However, PS II's primary protein shares 30% of its DNA with the purple bacterial gene (Golbeck and Bryant 1991; Snyder et al. 1993).

A wide range of records, encompassing structural, thermodynamic, spectroscopic, and molecular sequence research, unambiguously classify all reported RCs among two categories, called type I and type II complexes (Blankenship 2021). Through the development of photochemical reaction centers (RCs), photosynthesis progresses. Regarding the origin of RCs and photosynthesis, these are two distinct streams of thought (Olson and Blankenship 2004). There are two types of reaction center found in cyanobacteria that is type I and type II RC, whereas the remaining photosynthetic prokaryotes have only one RC. According to Blankenship et al. (2007), the many hypotheses for this occurrence may all be categorized as either selective loss scenarios or fusions or deviations thereof.

1.5.1 Role of Redox Switch in the Stability of Both Type I and Type II Reaction Centers

According to Allen (2005), in type I-containing photoauto-chemotrophic generations (e.g., *Heliobacillus* spp. and *Chlorobium*) as well as type II-containing photohetero-organotrophic ancestral organisms (*Rhodospirillum* spp., *Rhodobacter,* etc.), the type II (RC II) and type I (RC I) reaction centers separated, permitting particularization and inevitably the deactivation of the duplicated reaction center. Another theory holds that the DNA segment encoding the type I, as well as type II reaction centers, had been kept within a single anoxygenic ancestry but that their expression was controlled by a redox regulatory mechanism that prevented transcriptional interaction amongst the two chains of electron transport, each of which operated in a unique environment. The type I and type II reaction center-associated genes are still there in facultatively chemoautotrophic photosynthetic bacteria, despite their versatility (Fig. 1.1).

When hydrogen sulfide is present, dormant type II genes coincide with the transcription of type I reaction center genes in this putative progenitor of chloroplasts in addition to cyanobacteria. Once type I genes are suppressed in non-reducing conditions, type II genes are subsequently enhanced. Type I as well as type II reaction centers can ultimately coexist and serve complementary roles after regulatory power is removed. By releasing oxygen as well as generating electrons with the type I reaction center, functioning as photosystem I (PS I), the type II reaction center, operating exactly as photosystem II (PS II), degrades water rather than hydrogen sulfide.

The plan depicted in Fig. 1.1 further considers the potential for the reduction of the redox switch to become selectively favourable in the event that a water oxidation catalyst (Allen and Martin 2007) erroneously adheres to the type II reaction center's donor side. This configuration had the selective benefit of allowing electrons to move from water to ferredoxin and NADP+ for the first time.

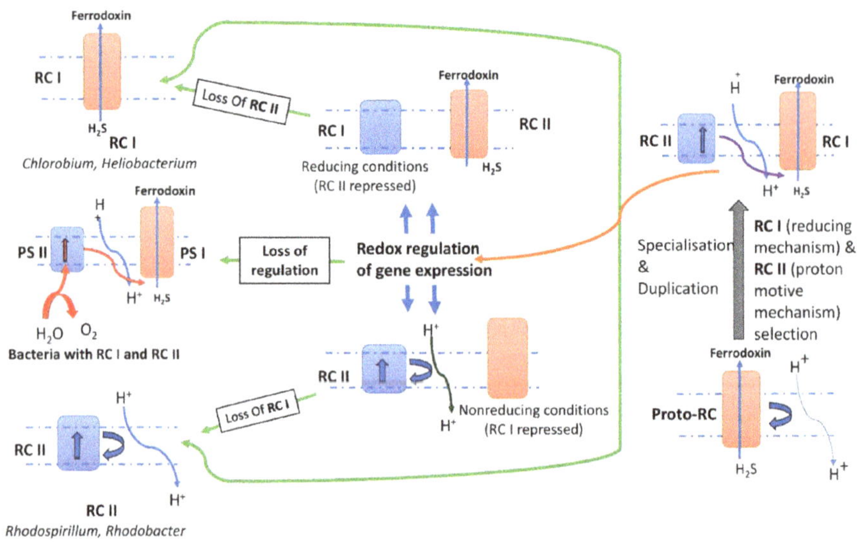

Fig. 1.1 Type I/type II centers stability as examined by a redox switch (Revised from Allen 2005)

1.5.2 Models for the Organization of Reaction Centers

1.5.2.1 The Selective Loss Model

The selective loss concept highlights a prehistoric photosynthetic living thing possessing two distinct kinds of reaction centers, analogous to oxygenic cyanobacteria (Olson and Pierson 1987a, b; Mauzerall 1992). In extant anoxygenic photosynthetic bacterial species, a single reaction center was found after any one of the two kinds of reaction centers vanished. The hypothesis proposes that inhabitants such as cyanobacteria existed during the primordial era, wherein life initially evolved. In 3.5 bya, microfossil investigation by Schopf and colleagues (Schopf and Packer 1987; Schopf 1993) provides preliminary evidence for this perspective.

Mulkidjanian et al. (2006) provided perhaps the most current evidence regarding the selective loss assertion, claiming that the broadened photosynthetic core genetic material set implies a cyanobacterial origin for the photosynthesis mechanism. The authors proposed a modified selective loss model in light of the fascinating geological scientific documentation that anoxygenic photosynthetic processes evolved earlier than oxygenic processes. They claimed that the earliest identifiable phototroph was a group known as "procyanobacteria," which resembled extant cyanobacteria in most cases but did not evolve oxygen. As a result, they transmitted genes for photosynthesis throughout additional anoxygenic photosynthetic communities of bacteria through both large-scale gene loss and lateral transfer of genes (Mulkidjanian et al. 2006).

1.5.2.2 The Fusion Model

The fusion model (Blankenship 1992; Mathis 1990) depicts, the type I as well as type II reaction centers possibly evolved individually within two distinct ancestral lineages (one into proteobacteria, or else Chloroflexi, while another into heliobacteria, or Chlorobi) prior to merging into a single lineage, forming the Duplex photosystem within cyanobacteria. The model depicts the apparatus of photosynthesis evolving from an elementary to a complex structure, which appears more plausible compared to the reverse conditions.

A group of workers (Olson and Pierson (1987a, b), Mathis (1990), Xiong et al. (1998), Blankenship (1992), Xiong and Bauer (2002a, b), and Meyer (1994) suggested that RC1 and RC2 emerged sequentially in various species before merging genetically to form the cyanobacterial photosynthetic line, which carries two photosystems sequentially. The evolutionary model proposed by Mathis (1990) involved the transmission of RC1 genes from the genome of a "heliobacterium" into the genetic sequence associated with a "purple bacterium" that carried RC2 genes. The cyanobacteria were the offspring of the resulting union.

Xiong and Bauer (2002a, b) established a significant modern manifestation regarding the fusion hypothesis. They proposed that RC2 (proteobacterial or purple bacterial) originated via a cytochrome b-like structural protein. Thus in three ways, this RC2 evolved. In the prior pathway, RC2 of purple bacteria gave rise to RC2 of the filamentous green bacteria. In the later pathway, RC2 of purple bacteria was established into RC2 of cyanobacteria and RC1 of heliobacteria in the consequence of a fusion between the genes encoding the original CP47/CP43 protein and the RC2 genes. Apparently, there is one evolutionary route from RC1 (heliobacterial) towards green sulfur bacterial RC1 and another to cyanobacterial RC1. Consequently, RC2 began as a proteobacterium and evolved into a cyanobacterium; RC1 originated from a heliobacterium. Thus findings conclude that RC proteins have a different evolutionary history than enzymes that synthesize pigments.

1.5.3 Organization of Reaction Center Motif

1.5.3.1 A Reaction Center Motif

Two categories of photosynthesis reaction centers are differentiated by the chemical makeup of the acceptors of electrons (Blankenship 1992). Pheophytin-quinone bacteria are represented by the filamentous green bacteria, purple bacteria, have photosystem II; iron-sulfur bacteria are represented by the heliobacteria, green sulfur bacteria, possess photosystem I. Anoxygenic bacteria possess just one form of photosystem, whereas cyanobacteria, along with plants, possess two types of photosystems. Within the reaction center particularly of purple bacteria, two symmetrically connected stems contain a carotenoid, two quinones, four bacterio-chlorophylls, two bacterio-pheophytins, and one non-heme iron (Feher et al. 1989; Deisenhofer and Michel 1989; Blankenship et al. 2006).

As described by Lu (2018), the assembly of photosystem I shows a high number of chlorophylls organized around three iron–sulfur clusters with a double symmetry about the axis. Three iron–sulfur complexes and six core tetrapyrroles appear to comprise the reaction center's photochemical core. However, the majority of those tetrapyrroles within the reaction center of the iron–sulfur type are engaged in light harvesting.

Generally, both kinds of reaction centers have a tetrapyrrole dimer compound joined with four monomeric tetrapyrrole compounds on one side and, additionally, an iron-binding site across the other. The amino acids from each of the core subunits act as ligands for the binding of the magnesium Mg^{+} atoms that make up the dimer, while the metal complex, with the dimer containing the iron atoms placed across the center in the two-fold axis in a symmetry. These investigations reveal the presence of a common reaction center motif that is enclosed through five helices made of two symmetric core subunits, two symmetric cofactor branches, a tetrapyrrole dimer, and a metal site.

1.5.3.2 Primary Electron Donor

Complexes comprising chlorophyll or bacteriochlorophyll are present in all photosynthetic systems and are responsible for gathering light across a wide surface area and concentrating it towards the reaction center. The main electron donor in these systems might theoretically be any of the tetrapyrroles, provided that they undergo oxidation. One distinct pair of chlorophylls, or bacteriochlorophylls, receives the light energy in every instance. Extensive experiments have demonstrated that the two contiguous bacteriochlorophylls depicted in the structure in three dimensions work as the prime electron donors within the reaction centers of purple bacteria. The dimeric structure of the principal electron donor gives rise to some distinctive properties.

For example, the donor among purple bacteria exhibits a particular spectral absorption band in the near-infrared region due to the excitation interaction produced by the bacterio-chlorophylls' restricted overlap. Differences in the tetrapyrroles' chemical makeup and their interactions against the neighbouring protein account for the observed variation in the donor's properties at various reaction sites. In purple bacteria, the wavelength corresponding to the optical band with the dimer spans from 860 nm to 900 nm, possessing reaction centers with broader wavelength transitions than those with bacteriochlorophyll a, when it comes to bacteriochlorophyll b.

1.5.3.3 Electron Transfer

The potential of reaction centers to accomplish the transfer of electrons with a quantum efficiency approaching unity constitutes a fundamental component of their photochemistry. Such a significant quantum yield is accomplished by the use of many intermediary acceptors of electrons. Each photosystem generates a steady charge-separated state around 30 ps following excitation. Although the mechanism of spectrum changes over time is complicated, the function of numerous variables influencing the primary electron transfer mechanism in purple bacteria has been established (Blankenship et al. 2006). Interpreting the optical changes in various types of reaction centers has been impeded by the growing amount of tetrapyrroles as well as the highly overlapped layout of the optical spectrum. Research is still ongoing to describe the electron transfer routes. The rate in purple bacteria appears susceptible to the free energy discrepancy among the activated and charge-separated phases yet insensitive toward the proportional distribution of electrons throughout the donor's two macrocycles (Van Brederode et al. 1997).

1.5.4 Ancestral Reaction Center

The key unanswered issues include how the ancestral reaction center would have appeared, where it might have come from, and what kind of selection pressure might have led to the development of two different kinds of reaction centers. The initial response to the center's role was most likely to simultaneously generate ATP along with its reducing substitutes. Green bacteria, heliobacteria, and purple non-sulfur bacteria all fit the description of possibly deriving from the photosynthetic progenitor from an evolutionary standpoint. The cyclic electron transfer pathway seen in purple bacteria, and perhaps in green bacteria and heliobacteria, allows electrons produced on the acceptor site to return from the source site via the cytochrome bc1 complex.

Consequently, a protons gradient is created that could be employed for the production of ATP, ion transportation, and various other purposes. According to a sequence analysis of enzymes associated with bacteriochlorophyll production in photosynthetic organisms, the oldest photosynthetic organisms with living offspring are purple bacteria (Xiong et al. 2000). On the other hand, the fundamental framework concerning the type II purple bacterial reaction center is significantly different from photosystems II and I, notably from that of the antenna system. Type II reaction centers and cytochrome b may have structural similarities (Xiong et al. 2000).

1.6 Origin of Antenna Pigments

The initial step in the process of photosynthesis is when molecules of antenna pigment absorb sun energy, which is capable of trapping and absorption of the direct source of energy of visible or near-infrared sunlight strongly. Numerous bacterial cells have undergone fusion and gene duplication events, and the combination of horizontal gene transfer is responsible for gains and losses of photochemical reactions and the process of photosynthesis across geological time. Only three types of pigments—phycobilins, carotenoids, and chlorophylls—are present in nature. The antenna complex, often referred to informally as the light-harvesting complex (LHC), is made up of pheophytin, carotenoids, xanthophylls, and chlorophyll molecules. It is affixed to thylakoid membrane proteins prevalent within the chloroplast's thylakoid membrane found in plants and cyanobacteria that convert absorbed surplus thermal energy into heat and transport effective energy from the sun to every chlorophyll component located in the reaction center of the photosystem (Jansson 1994).

Plant and green algae's antenna proteins are classified within the superfamily of proteins that bind to chlorophyll and carotenoid pigments. These proteins are peripheral to reaction centers, part of complexes that absorb light from membranes, and they catalyze the transfer of electrons from donating electrons to recipients (Boekema et al. 1995). Whereas peripheral antenna proteins are diverse, the core complex components have shown significant conservation across evolution (Alboresi et al. 2008).

The phycobilisomes found in cyanobacteria represented, around 3–3.5 bya, the first external antenna system to operate in a highly oxygenic environment (Gantt 1981). Lower-energy light is absorbed by PSI reaction centers as opposed to PSII. While Photosystem II is mostly found in the areas surrounding grana stacks, PSI is found in stroma-exposed thylakoids and grana end membranes. If the photosystems are not physically separated, it is possible that the quanta captured by the antenna will leave PSII and migrate to PSI. Therefore, in order to pack the increased amounts of chlorophyll inexpensively, the appressed layers are needed (Anderson 1999). Numerous other photoprotective internal processes, such as the movement of chloroplasts inside the cell, modifications to the size of the antenna, the build-up of anthocyanins, and modifications to the topology of the photosynthetic membrane, can also reduce the absorption of light. Photoreceptors, phytochrome, and cryptocrome are components of the light quality sensor (Neff et al. 2000; Cashmore et al. 1999). According to Stadnichuk and Tropin (2014), oxygenic photosynthesis is dependent on the coordinated activity of two membrane-bound photosystems, PS I and the water-oxidizing complex PS II, as well as their cooperation with a number of pigment antenna variations. Distinct wavelengths of light are absorbed by PSI and PSII. Antennae in plants are composed of proteins, including chlorophyll a/b and also chlorophyll a/c, and they work as a single reaction center for light energy absorption. The primary mechanism by which the chlorophyll a/b ratio varies in response to various irradiance levels is due to variations in LHCII content. Despite having the best characteristics for absorbing light, phycobilisomes were superseded by complexes for light gathering that had both chlorophyll a/b and a/c throughout the development of the green and chromophyte lineages within the chloroplasts.

PS I and the water-oxidizing complex PS II work together with various pigment antenna variations to dissipate heat thermally and are concerned with averting hazardous photo-oxidative reactions involving singlet oxygen along with quenching reactive oxygen species (Freer et al. 1996). The antenna pigment has the ability to boost the effectiveness of energy transmission to the response center by more than two orders of magnitude (Ruban 2015). According to Beddard and Porter (1976), the proximity of the excitation energies and spectrum similarities of the interacting pigments, as well as their mutual orientation, significantly affect the energy transfer rate.

Emerson and Arnold (1932), which employed microsecond light bursts to produce a single reaction center turnover (oxidation–reduction cycle) every burst, provided the first indication of the presence of the photosynthetic antenna. One of the primary requirements for the exchange of excitation energy and subsequent delivery of that energy entering the reaction center in the increasingly oxidizing environment is the near placement of light-collecting pigments in antenna proteins with respect to one another (Forster 1948).

The energetic electrons undergo a string of chemical processes before being accepted by the major co-factors, such as ferredoxin or plastoquinone. At least 10 light-harvesting complex (LHC) proteins are encoded in higher plants by a multigene family that includes nuclear genes. After that, these proteins get transported from the cytoplasm into the chloroplast and interact with PSI and PSII (Jansson 1999).

The vast family of complex proteins involved in light harvesting has a close relationship with the development of green organisms, including their photosynthetic apparatus. Several subclasses of this gene family have been identified for light-harvesting and photo-protection. These subclasses are high light-induced proteins (HLIPs), fucoxanthin chlorophyll a/c-binding proteins (FCP), promptly light-induced proteins (ELIPs), LHC-like proteins, for example, chlorophyll a/b-binding proteins, and LHCSR proteins.

Stress-related proteins (SRPs) called Light Harvesting complexes stress related (LHCSR) accumulate when illuminated with intense amounts of light and have a role in photoprotective processes (Dittami et al. 2010). According to recent transcriptome research on the stress response, oxidative stress, heat, light, and salt all cause changes in FCP levels (Hwang et al. 2008).

1.7 Evolution from Anoxygenic to Oxygenic Photosynthesis

Among the most important things to happen in Earth's history was oxygenic photosynthesis's emergence, which allowed for the emergence of aerobic metabolism as well as increasingly intricate life forms (Fig. 1.2) while also drastically altering the redox balance on the planet (Green 2011).

The hypothesis that is most often acknowledged as atmospheric oxygenation is the great oxidation event. In response to this, the atmosphere transitioned from nearly anaerobic conditions to the atmospheric conditions that exist today, maybe around 10^{-14} level (PAL) O_2 generated when water is photolyzed by UV light during a brief geological period prior to 2.32 bya (Buick 2008). Every known oxygen-evolving creature possesses chlorophyll a, pigment necessary for photosynthesis. In contrast, the reaction center pigments about the presence of more than one bacteriochlorophyll (a, b, or g) in anoxygenic photosynthetic bacteria, each of which absorbs light with a wavelength longer and lower energy than chlorophyll a. The majority of

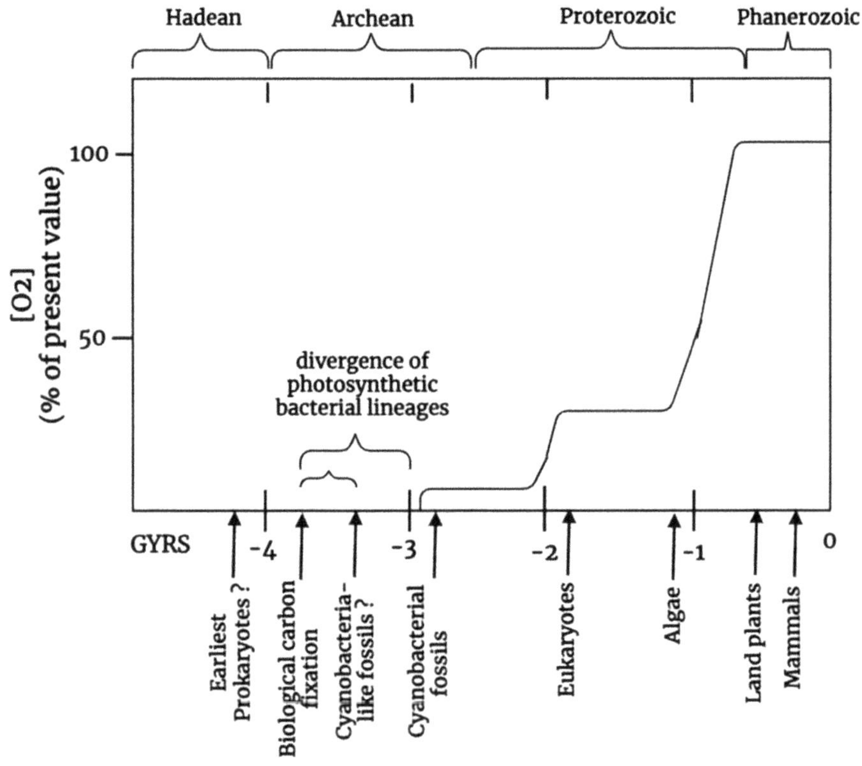

Fig. 1.2 Diagram showing the increase in oxygen levels on Earth in the early stages of life. Major geological periods are shown by brackets on the upper x-axis, and major evolutionary landmarks are shown by arrows on the lower x-axis, adapted from (Xiong and Bauer 2002a, b)

people proposed that oxygen has been present on Earth for 2.8–2.7 billion years when oxygenic photosynthesis first began (Davankov 2021).

The formation of an antioxidant with high enough oxidizing equivalents and the capacity to gather and store them, as well as the redox potential, which is produced through photochemical processes in the reaction center as well as the ability to transmit at least a single electron for each photon absorbed, are two crucial evolutionary developments that must take place before water oxidation can happen.

There are no known intermediate forms of anoxygenic photosynthetic reaction centers that possess either of these characteristics. The P680, photoactive chlorophyll-associated reaction center is associated with photosystem II, has a midpoint redox potential of more than 1 V, which is more than half a volt higher than the reaction centers present in anoxygenic bacteria that contain chlorophyll (Blankenship and Hartman 1998). Cardona (2019) made an interesting analogy and explained that the water-oxidizing Photosystem II enzyme behaves like a car, particularly in the Formula 1 racing category, while the anoxygenic Type II reaction center associated with phototrophic Proteobacteria as well as Chloroflexi is comparable to that of a Ford Model T automobile. This implies that the latter originated from the former.

As evident from research, oxygenic photosynthesis originated at last in the progress of photosynthesis evolution, and earlier, the Earth was captured by anoxygenic photosynthesis. The transition of metals in chemosynthetic organisms from greigite ($NiS_2[Fe_4S_4]S_2Fe$) and manganite ($CaMn_4O_8$) in anoxygenic organisms to Mn in oxygen-evolving organisms led to new potential gradients of oxygen in the environment (Russell et al. 2008). Restrictions arising from the photo-assembly of the water-oxidizing complex fuel containing the Mn before oxygenic photosynthesis evolved. Mn (II) had a significant function as a donor of electrons for anoxygenic photosynthesis (Fischer et al. 2015). Earlier, it was hydrogen peroxide (H_2O_2) and bicarbonate (HCO_{3^-}) that potentially participated as an electron donor to PSII (McKay and Hartman 1991), which was made up of the binuclear Mn-catalase enzyme and an anoxygenic reaction center (Banerjee et al. 2012). Blankenship and Hartman (1998) suggested that the original model for the modern OEC was a primitive form of Mn catalase. According to Borda et al. (2001), pyrite triggered H_2O_2 production, which was essential for the development of aerobic photosynthesis and the initial source of oxygen. The involvement of hydrogen peroxide as an electron donor may first come as a way to detoxify excess H_2O_2 from the environment. Another hypothesis proposed by Dismukes et al. (2001) stated that the thermodynamic analysis revealed that, bicarbonate would have been a more potent source of oxygen than water. Based on the theory of evolution, bacterial RC-II gave rise to PSII with a gradual increase in the redox potential of pigments (Williamson et al. 2011). Cardona (2016) and Sánchez-Baracaldo and Cardona (2020) speculated that photosystems I and II were both found in cyanobacteria, while anoxygenic organisms only contained either PSI or PSII and might be transferred by horizontally in a single organism so that both photosystems could oxidize the water for O_2 evolution.

However, there are no known native species that can oxidize water and have a single photosystem. More accurately, it is said that the cyanobacterium endosymbiont horizontally transferred genes into the host's nuclear genome of a non-photosynthetic eukaryotic unicellular organism.

Further, Cardona (2019) discussed the origin of the oxygenic photosystem which is more advanced than that of the anoxygenic photosystem just because it can oxidize water. This false presumption that Photosystem II is more advanced than anoxygenic reaction centers. Their hypothesis made it reasonable to understand that the theory of the evolution of photosynthetic processes are predicated on unverified hypotheses layered upon verifiable hypotheses. For example, the theory that anoxygenic reaction centers appear to be more primitive implies the presumption that primordial reaction centers failed to possess enough oxidizing power to split water, while this declaration has not been confirmed correct in any respect, his thoughts are straightforward in concluding that water oxidation is unlikely to be an ancestral trait (Cardona 2019).

1.8 Endosymbiosis (Origin of Chloroplast)

After cyanobacterium endosymbiosis, chloroplasts evolved in the eukaryotes, chloroplasts spread by diversifying their eukaryotic hosts and subsequently with different eukaryotic algae and ultimately form the photosynthetic eukaryotes that were previously nonphotosynthetic. Over a bya, a eukaryotic host first acquired a cyanobacterial endosymbiont, which led to the emergence of the first photosynthetic eukaryotes. This included glaucophytes, which have organelles called "cyanelles," red algae, which have "rhodoplasts," and plants and green algae, which have "chloroplasts" (Jensen and Leister 2014). According to evolutionary studies, eukaryotes have undergone three different types of endosymbiosis during their evolutionary history: primary, secondary, and tertiary (Fig. 1.3). Engulfment of a prokaryotic cyanobacterium resulted in the formation of a chloroplast in initial endosymbiosis, surrounded by two membranes, which is the progenitor of Plantae, it includes rhodophytes, glaucophytes, and chlorophyta (e.g., land plants and green algae) (McFadden 2001). During secondary endosymbiosis, eukaryotic algae that had previously been non-photosynthetic eukaryotes were consumed and converted into secondary chloroplasts. Three or four envelope membranes were present in secondary chloroplasts. The host cell's phagosomal membrane and the ingested alga's plasma membrane are thought to be represented by the third and fourth membranes, which are arrayed from inner to outer (Tomitani 2006).

Green algae's secondary endosymbiosis provides the chloroplasts that are contained by chlorarachniophytes and euglenids (Ponce-Toledo et al. 2018). The engulfment of an alga that possesses a secondary plastid is known as tertiary endosymbiosis, and it has only occurred in dinoflagellates (Yoon et al. 2005). According to endosymbiotic theory, the first as well as second envelope membranes connected to the contents of secondary chloroplasts are thought to divide in the same way as primary chloroplasts. Some researchers suggested that the evolution of the chloroplast genome is chimeric in origin as compared to a single instance of endosymbiosis, as the characteristics of cyanobacteria make chloroplast formation more complicated. The enzymatic components of the production of peptidoglycan in both plants and algae are monophyletic. However, most of these enzymes came from bacteria but not from cyanobacteria (Sato 2021). In the Rhizarian amoeba *Paulinella*, the second plastid primary endosymbiosis took place relatively recently, about 120 million years ago. An α-cyanobacterial prey and a heterotrophic ancestor of *Paulinella* were involved in this event which the latter gave rise to the "chromatophore," a photosynthetic organelle. Stephens et al. (2021) proposed a model for endosymbiosys called the "chassis and engine" based on their findings and previous reports related to *Paulinella*. The endosymbiotic stage of this photosynthetic *Paulinella* appears to be intermediate. The chromatophore functions as a vehicle's "engine," requiring "fuel" and other resources (like sunlight, water, proteins, and different nutrients including inorganic carbon to run the engine. The radiator removes excess heat produced by the engine through nonphotochemical quenching, photochemical quenching, and ROS detoxification. Internal feedback controls and regulatory mechanisms linked to the "chassis" (host cell) regulate the speed through various means, such as the Calvin-Benson-Bassham cycle, reductant, ATP synthesis, and electron transport control (Fig. 1.4). Motion is produced for functions like respiration, is sustained by the engine's power (photosynthate) with the help of power biosynthesis of macromolecules and metabolite sources. While some energy is required to create new engines and keep the ones that already exist running (Fig. 1.4).

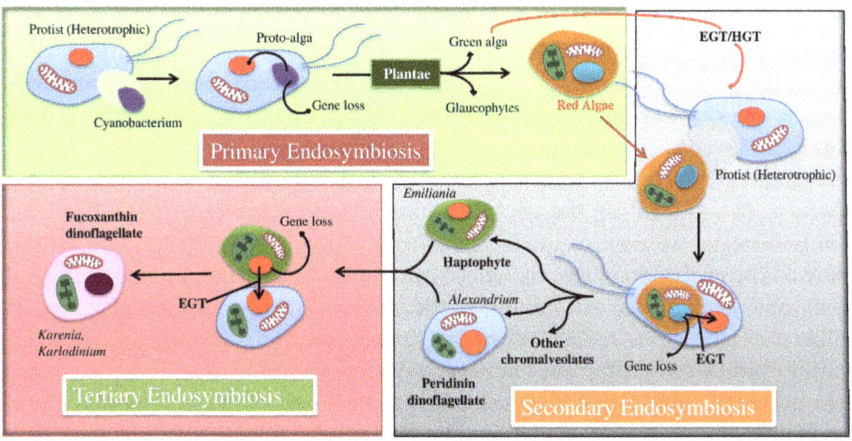

Fig. 1.3 Endosymbiotic mode of origin of chloroplast

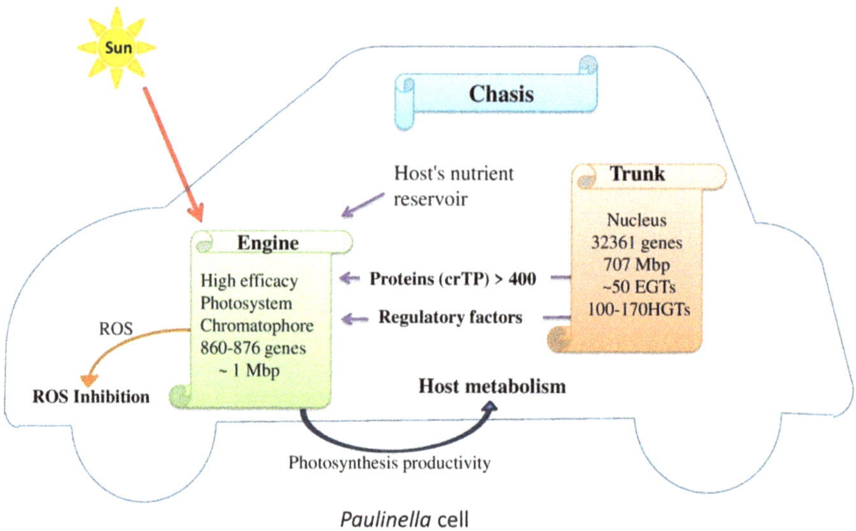

Fig. 1.4 The plastid endosymbiosis engine and chassis model for the evolution of chloroplast. The engine displays the chromatophore genome's approximate size as well as the total number of genes it encodes for all *Paulinella* species. *ROS* reactive oxygen species, *crTP* chromatophore-targeting peptide, *mt* mitochondrion, *EGT* endosymbiotic gene transfer

1.9 Conclusion

The emergence of photosynthesis is a complicated process with multiple origins and routes that necessitate a specific branching pathway. Different metabolic pathways may have produced components that were changed or recruited throughout different periods of evolutionary development. **Through structural gene modifications and a comparative study of the several tetrapyrrole's biosynthesis pathways discovered in reaction centers, evolutionary processes in Earth's environment, driven by photosynthesis from 3500 million years ago, may be tracked, demonstrating the complications of the evolution process.** Chl a and one Fe-S center were present in the ancestor of modern RCs. **Phase 1 and Phase 2 showed the evolution of the chromophore biosynthetic route as well as the quinone and cytochrome-containing cyclic electron transport systems.** This RC might use H_2 or H_2S as reductants in photo-assimilation or simply CO_2 fixation. **The major evolutionary difference between type I and type II reaction center's protein molecules, anoxygenic as well as oxygenic forms, and phototrophic bacteria affects the evolution of photosynthesis phases.** Antenna systems and carbon fixation routes were developed several times; photosynthetic RCs were all descended from the same source; and the earliest photosynthetic creatures were anoxygenic. Light-driven separation of charges across PSII's RC drives the reaction, and it is evident that this RC descended from the same progenitor as PSI, purple photosynthetic bacteria, along with green sulfur bacteria.

Acknowledgements KNT is thankful to Principal, Mahila Mahavidyalaya, Banaras Hindu University, Varanasi necessary facilities and infrastructural support.

Conflict of Interest Authors declare no conflict of interest.

References

Alboresi A, Caffarri S, Nogue F, Bassi R, Morosinotto T (2008) In silico and biochemical analysis of Physcomitrella patens photosynthetic antenna: identification of subunits which evolved upon land adaptation. PLoS One 3(4):e2033
Allen JF (2005) A redox switch hypothesis for the origin of two light reactions in photosynthesis. FEBS Lett 579(5):963–968
Allen JF, Martin W (2007) Out of thin air. Nature 445(7128):610–612
Anderson JM (1999) Insights into the consequences of grana stacking of thylakoid membranes in vascular plants: a personal perspective. Funct Plant Biol 26(7):625–639
Appel PWU (1980) On the early Archaean isua iron-formation, West Greenland. Precambrian Res 11(1):73–87
Awramik SM, Schopf JW, Walter MR (1983) Filamentous fossil bacteria from the Archean of Western Australia. In: Developments in precambrian geology, vol 7. Elsevier, pp 249–266
Banerjee M, Ballal A, Apte SK (2012) Mn-catalase (Alr0998) protects the photosynthetic, nitrogen-fixing cyanobacterium anabaena PCC7120 from oxidative stress. Environ Microbiol 14(11):2891–2900
Barber J, Nield J, Morris EP, Hankamer B (1999) Subunit positioning in photosystem II revisited. Trends Biochem Sci 24(2):43–45
Baymann F, Brugna M, Mühlenhoff U, Nitschke W (2001) Daddy, where did (PS) I come from? Biochimica et Biophysica Acta BBA)-Bioenergetics 1507(1–3):291–310
Baymann F, Lebrun E, Brugna M, Schoepp Cothenet B, Giudici Orticoni MT, Nitschke W (2003) The redox protein construction kit: pre-last universal common ancestor evolution of energy-conserving enzymes. Philos Trans R Soc Lond B Biol Sci 358(1429):267–274
Beddard GS, Porter G (1976) Concentration quenching in chlorophyll. Nature 260(5549):366–367
Ben-Shem A, Frolow F, Nelson N (2004) Evolution of photosystem I–from symmetry through pseudosymmetry to asymmetry. FEBS Lett 564(3):274–280
Blank CE (2004) Evolutionary timing of the origins of mesophilic sulphate reduction and oxygenic photosynthesis: a phylogenomic dating approach. Geobiology 2(1):1–20
Blankenship RE (1992) Origin and early evolution of photosynthesis. Photosynth Res 33:91–111
Blankenship RE (2001) Molecular evidence for the evolution of photosynthesis. Trends Plant Sci 6(1):4–6
Blankenship RE, Hartman H (1998) The origin and evolution of oxygenic photosynthesis. Trends Biochem Sci 23(3):94–97
Blankenship RE, Madigan MT, Bauer CE (eds) (2006) Anoxygenic photosynthetic bacteria, vol 2. Springer Science & Business Media
Blankenship RE, Sadekar S, Raymond J (2007) The evolutionary transition from anoxygenic to oxygenic photosynthesis. In: Evolution of primary producers in the sea, pp 21–35
Blankenship RE (2021) Molecular mechanisms of photosynthesis. John Wiley & Sons
Boekema EJ, Hankamer B, Bald D, Kruip J, Nield J, Boonstra AF, Rögner M (1995) Supramolecular structure of the photosystem II complex from green plants and cyanobacteria. Proc Natl Acad Sci 92(1):175–179
Borda MJ, Elsetinow AR, Schoonen MA, Strongin DR (2001) Pyrite-induced hydrogen peroxide formation as a driving force in the evolution of photosynthetic organisms on an early earth. Astrobiology 1(3):283–288
Brocks JJ, Love GD, Summons RE, Knoll AH, Logan GA, Bowden SA (2005) Biomarker evidence for green and purple Sulphur bacteria in a stratified Palaeoproterozoic Sea. Nature 437(7060):866–870

Bryant DA (ed) (2006) The molecular biology of cyanobacteria, vol 1. Springer Science & Business Media
Buick R (1984) Carbonaceous filaments from North pole, Western Australia: are they fossil bacteria in Archaean stromatolites? Precambrian Res 24(2):157–172
Buick R (1990) Microfossil recognition in Archean rocks: an appraisal of spheroids and filaments from a 3500 my old chert-barite unit at North Pole. Western Australia. Palaios 441–459
Buick R (2008) When did oxygenic photosynthesis evolve? Philos Trans R Soc B Biol Sci 363(1504):2731–2743
Büttner M, Xie DL, Nelson H, Pinther W, Hauska G, Nelson N (1992) Photosynthetic reaction center genes in green sulfur bacteria and in photosystem 1 are related. Proc Natl Acad Sci 89(17):8135–8139
Cardona T (2016) Reconstructing the origin of oxygenic photosynthesis: do assembly and photoactivation recapitulate evolution? Front Plant Sci 7:257
Cardona T (2019) Thinking twice about the evolution of photosynthesis. Open Biol 9(3):180246
Cashmore AR, Jarillo JA, Wu YJ, Liu D (1999) Cryptochromes: blue light receptors for plants and animals. Science 284(5415):760–765
Cloud P (1973) Paleoecological significance of the banded iron-formation. Econ Geol 68(7):1135–1143
Cloud P (1976) Beginnings of biospheric evolution and their biogeochemical consequences. Paleobiology 2(4):351–387
Davankov VA (2021) The riddle of atmospheric oxygen: photosynthesis or photolysis? Russ J Phys Chem A 95:1963–1970
Deisenhofer J, Michel H (1989) Nobel lecture. The photosynthetic reaction centre from the purple bacterium Rhodopseudomonas viridis. EMBO J 8(8):2149–2170
Des Marais DJ (2000) When did photosynthesis emerge on earth? Science 289(5485):1703–1705
Dismukes GC, Klimov VV, Baranov SV, Kozlov YN, DasGupta J, Tyryshkin A (2001) The origin of atmospheric oxygen on earth: the innovation of oxygenic photosynthesis. Proc Natl Acad Sci 98(5):2170–2175
Dittami SM, Michel G, Collén J, Boyen C, Tonon T (2010) Chlorophyll-binding proteins revisited-a multigenic family of light-harvesting and stress proteins from a brown algal perspective. BMC Evol Biol 10(1):1–14
Emerson R, Arnold W (1932) The photochemical reaction in photosynthesis. J Gen Physiol 16(2):191–205
Feher GAJP, Allen JP, Okamura MY, Rees DC (1989) Structure and function of bacterial photosynthetic reaction centres. Nature 339(6220):111–116
Fischer WW, Hemp J, Johnson JE (2015) Manganese and the evolution of photosynthesis. Orig Life Evol Biosph 45:351–357
Fischer WW, Hemp J, Johnson JE (2016) Evolution of oxygenic photosynthesis. Annu Rev Earth Planet Sci 44:647–683
Forster T (1948) Zwischenmolekulare energiewanderung und fluoreszenz. Ann Phys 437(1–2):55–75
Freer A, Prince S, Sauer K, Papiz M, Lawless AH, McDermott G, Isaacs NW (1996) Pigment–pigment interactions and energy transfer in the antenna complex of the photosynthetic bacterium Rhodopseudomonas acidophila. Structure 4(4):449–462
Gantt E (1981) Phycobilisomes. Annu Rev Plant Physiol 32(1):327–347
Giovagnetti V, Ruban AV (2018) The evolution of the photoprotective antenna proteins in oxygenic photosynthetic eukaryotes. Biochem Soc Trans 46(5):1263–1277
Golbeck JH, Bryant DA (1991) Photosystem I. Curr Top Bioenerg 16:83–177
Granick S (1965) Evolution of heme and chlorophyll. In: Evolving genes and proteins. Academic Press, pp 67–88
Green BR (2011) After the primary endosymbiosis: an update on the chromalveolate hypothesis and the origins of algae with Chl c. Photosynth Res 107:103–115
Hall DO, Rao K (1999) Photosynthesis. Cambridge University Press
He H, Wu X, Xian H, Zhu J, Yang Y, Lv Y, Konhauser KO (2021) An abiotic source of Archean hydrogen peroxide and oxygen that pre-dates oxygenic photosynthesis. Nat Commun 12(1):6611

Heathcote P, Jones MR, Fyfe PK (2003) Type I photosynthetic reaction centres: structure and function. Philos Trans R Soc Lond B Biol Sci 358(1429):231–243
Howe CJ, Barbrook AC, Nisbet RER, Lockhart PJ, Larkum AWD (2008) The origin of plastids. Philos Trans Royal Soc B Biol Sci 363(1504):2675–2685
Hwang YS, Jung G, Jin E (2008) Transcriptome analysis of acclimatory responses to thermal stress in Antarctic algae. Biochem Biophys Res Commun 367(3):635–641
Jansson S (1994) The light-harvesting chlorophyll ab-binding proteins. Biochim Biophys Acta 1184(1):1–19
Jansson S (1999) A guide to the Lhc genes and their relatives in Arabidopsis. Trends Plant Sci 4(6):236–240
Jensen PE, Leister D (2014) Chloroplast evolution, structure and functions. F1000prime Rep 6:1
Kirsch W, Seyer P, Herrmann RG (1986) Nucleotide sequence of the clustered genes for two P700 chlorophyll a apoproteins of the photosystem I reaction center and the ribosomal protein S14 of the spinach plastid chromosome. Curr Genet 10(11):843–855
Lazcano A, Miller SL (1999) On the origin of metabolic pathways. J Mol Evol 49:424–431
Lu Y (2018) Assembly and transfer of iron–sulfur clusters in the plastid. Front Plant Sci 9:336
Mathis P (1990) Compared structure of plant and bacterial photosynthetic reaction centers. Evolutionary implications. Biochimic Biophys Acta 1018(2–3):163–167
Mauzerall D (1992) Light, iron, Sam Granick and the origin of life. Photosynth Res 33:163–170
McFadden GI (2001) Primary and secondary endosymbiosis and the origin of plastids. J Phycol 37(6):951–959
McKay CP, Hartman H (1991) Hydrogen peroxide and the evolution of oxygenic photosynthesis. Orig Life Evol Biosph 21:157–163
Meyer TE (1994) Evolution of photosynthetic reaction centers and light harvesting chlorophyll proteins. Biosystems 33(3):167–175
Mulkidjanian AY, Junge W (1997) On the origin of photosynthesis as inferred from sequence analysis. Photosynth Res 51:27–42
Mulkidjanian AY, Koonin EV, Makarova KS, Mekhedov SL, Sorokin A, Wolf YI, Galperin MY (2006) The cyanobacterial genome core and the origin of photosynthesis. Proc Natl Acad Sci 103(35):13126–13131
Nechushtai R, Eden A, Cohen Y, Klein J (1996) Introduction to photosystem I: reaction center function, composition and structure. In: Oxygenic photosynthesis: the light reactions, pp 289–311
Neff MM, Fankhauser C, Chory J (2000) Light: an indicator of time and place. Genes Dev 14(3):257–271
Nelson N, Ben-Shem A (2005) The structure of photosystem I and evolution of photosynthesis. BioEssays 27(9):914–922
Nitschke W, Rutherford AW (1991) Photosynthetic reaction centres: variations on a common structural theme? Trends Biochem Sci 16:241–245
Olson JM, Blankenship RE (2004) Thinking about the evolution of photosynthesis. Photosynth Res 80:373–386
Olson JM, Pierson BK (1986) Photosynthesis 3.5 thousand million years ago. Photosynth Res 9:251–259
Olson JM, Pierson BK (1987a) Evolution of reaction centers in photosynthetic prokaryotes. Int Rev Cytol 108:209–248
Olson JM, Pierson BK (1987b) Origin and evolution of photosynthetic reaction centers. Orig Life Evol Biosph 17:419–430
Pflug HD (1967) Structured organic remains from the fig tree series (Precambrian) of the Barberton mountain land (South Africa). Rev Palaeobot Palynol 5(1–4):9–29
Ponce-Toledo RI, Moreira D, López-García P, Deschamps P (2018) Secondary plastids of Euglenids and Chlorarachniophytes function with a mix of genes of red and green algal ancestry. Mol Biol Evol 35(9):2198–2204
Rashby SE, Sessions AL, Summons RE, Newman DK (2007) Biosynthesis of 2-methylbacteriohopanepolyols by an anoxygenic phototroph. Proc Natl Acad Sci 104(38):15099–15104

Read JF (1976) Calcretes and their distinction from stromatolites. In: Developments in sedimentology, vol 20. Elsevier, pp 55–71

Ruban AV (2015) Evolution under the sun: optimizing light harvesting in photosynthesis. J Exp Bot 66(1):7–23

Russell MJ, Allen JF, Milner-White EJ (2008) Inorganic complexes enabled the onset of life and oxygenic photosynthesis. In: Photosynthesis. Energy from the sun: 14th international congress on photosynthesis. Springer, pp 1187–1192

Sánchez-Baracaldo P, Cardona T (2020) On the origin of oxygenic photosynthesis and cyanobacteria. New Phytol 225(4):1440–1446

Sato N (2021) Are cyanobacteria an ancestor of chloroplasts or just one of the gene donors for plants and algae? Genes 12(6):823

Schad M, Konhauser KO, Sánchez-Baracaldo P, Kappler A, Bryce C (2019) How did the evolution of oxygenic photosynthesis influence the temporal and spatial development of the microbial iron cycle on ancient earth? Free Radic Biol Med 140:154–166

Schidlowski M (1984) Biological modulation of the terrestrial carbon cycle: isotope clues to early organic evolution. Adv Space Res 4(12):183–193

Schidlowski M (1988) A 3,800-million-year isotopic record of life from carbon in sedimentary rocks. Nature 333(6171):313–318

Schopf JW (1983) Earth's earliest biosphere: its origin and evolution

Schopf JW (1993) Microfossils of the early Archean apex chert: new evidence of the antiquity of life. Science 260(5108):640–646

Schopf JW, Packer BM (1987) Early Archean (3.3-billion to 3.5-billion-year-old) microfossils from Warrawoona group, Australia. Science 237(4810):70–73

Schopf JW, Tripathi AB, Kudryavtsev AB (2006) Three-dimensional confocal optical imagery of Precambrian microscopic organisms. Astrobiology 6(1):1–16

Schubert WD, Klukas O, Saenger W, Witt HT, Fromme P, Krauß N (1998) A common ancestor for oxygenic and anoxygenic photosynthetic systems: a comparison based on the structural model of photosystem I. J Mol Biol 280(2):297–314

Singhal GS, Renger G, Sopory SK, Irrgang KD (2012) Concepts in photobiology: photosynthesis and photomorphogenesis. Springer Science & Business Media

Snyder SW, Thurnauer MC, Deisenhofer J, Norris JR (1993) The photosynthetic reaction center, vol 2. Academic Press, New York, p 285

Stadnichuk IN, Tropin IV (2014) Antenna replacement in the evolutionary origin of chloroplasts. Microbiology 83:299–314

Stephens TG, Gabr A, Calatrava V, Grossman AR, Bhattacharya D (2021) Why is primary endosymbiosis so rare? New Phytol 231(5):1693–1699

Summons RE, Jahnke LL, Hope JM, Logan GA (1999) 2-Methylhopanoids as biomarkers for cyanobacterial oxygenic photosynthesis. Nature 400(6744):554–557

Thrailkill J (1976) Speleothems. In: Developments in sedimentology, vol 20. Elsevier, pp 73–86

Tomitani A (2006) Origin and early evolution of chloroplasts. Paleontol Res 10(4):283–297

Van Brederode ME, Jones MR, Van Mourik F, Van Stokkum IH, Van Grondelle R (1997) A new pathway for transmembrane electron transfer in photosynthetic reaction centers of Rhodobacter sphaeroides not involving the excited special pair. Biochemistry 36(23):6855–6861

Vermaas W (1989) The structure and function of photosystem II. In: Techniques and new developments in photosynthesis research. Springer, Boston, pp 35–59

Vermaas WF (1994) Evolution of heliobacteria: implications for photosynthetic reaction center complexes. Photosynth Res 41:285–294

Walsh MW (1992) Microfossils and possible microfossils from the early archean Onverwacht group, Barberton mountain land, South Africa. Precambrian Res 54(2–4):271–293

Walter MR (1976) Geyserites of Yellowstone National Park: an example of abiogenic "stromatolites". In: Developments in sedimentology, vol 20. Elsevier, pp 87–112

Ward LM, Rasmussen B, Fischer WW (2019) Primary productivity was limited by electron donors prior to the advent of oxygenic photosynthesis. J Geophys Res Biogeo 124(2):211–226

Welander PV, Coleman ML, Sessions AL, Summons RE, Newman DK (2010) Identification of a methylase required for 2-methylhopanoid production and implications for the interpretation of sedimentary hopanes. Proc Natl Acad Sci 107(19):8537–8542
William Schopf J (2011) The paleobiological record of photosynthesis. Photosynth Res 107:87–101
Williamson A, Conlan B, Hillier W, Wydrzynski T (2011) The evolution of photosystem II: insights into the past and future. Photosynth Res 107:71–86
Xiong J, Bauer CE (2002a) A cytochrome b origin of photosynthetic reaction centers: an evolutionary link between respiration and photosynthesis. J Mol Biol 322(5):1025–1037
Xiong J, Bauer CE (2002b) Complex evolution of photosynthesis. Annu Rev Plant Biol 53(1):503–521
Xiong J, Inoue K, Bauer CE (1998) Tracking molecular evolution of photosynthesis by characterization of a major photosynthesis gene cluster from Heliobacillus mobilis. Proc Natl Acad Sci 95(25):14851–14856
Xiong J, Fischer WM, Inoue K, Nakahara M, Bauer CE (2000) Molecular evidence for the early evolution of photosynthesis. Science 289(5485):1724–1730
Yoon HS, Hackett JD, Van Dolah FM, Nosenko T, Lidie KL, Bhattacharya D (2005) Tertiary endosymbiosis driven genome evolution in dinoflagellate algae. Mol Biol Evol 22(5):1299–1308

Plant Responses to Hypoxia Stress: Signalling Cascades and Regulatory Mechanisms

2

Prashasti Pandey and Deepak Kumar

Abstract

Floods leading to waterlogged soil and submerged plants are one of the major abiotic stresses affecting plants across the globe. Extended time of waterlogged soil leads to hypoxic conditions and, in some cases, to anoxia as well. The untimely inundation of the soil causes a reduction in plant growth and yields of the crop. Major aspects affected are nutrient uptake efficiency, plant metabolism, reactive oxygen species (ROS) production, and water uptake efficiency. To comprehend a plant's response to this stress, it is important to understand the plant's perception of the stress, signalling of the perception, and its regulatory mechanism. In this chapter, the plant's signalling cascade under waterlogging stress via ROS, Ca2+, and NO-mediated will be reviewed. Regulation by hormones, transcription factors, regulatory RNAs, and metabolites will further be discussed to understand the plant's regulatory mechanism.

Keywords

Hypoxia · Ethylene responsive factors · ROS (reactive oxygen species) · Ca2+ · NO · Hormone

P. Pandey · D. Kumar (✉)
Department of Botany, Institute of Science, Banaras Hindu University, Varanasi, Uttar Pradesh, India
e-mail: deepakkumar@bhu.ac.in

A. K. Mishra (ed.), *Stress Biology in Photosynthetic Organisms*,
https://doi.org/10.1007/978-981-97-1883-2_2

2.1 Introduction

Global food security is at risk, and the delicate balance that supports our agricultural systems is upset by climate change's deep and complex effects on agriculture. A major obstacle to agricultural production is the increased frequency and severity of floods brought on by changes in precipitation patterns due to rising global temperatures. As a result of flooding conditions, plant root systems frequently face temporary or persistent low oxygen levels in the soil, causing hypoxia and/or anoxia. The conditions of oxygen limitation will significantly impact the growth, development, and survival of plants.

The effects of oxygen deprivation in plants have been studied for a long time now. These effects can vary depending on various factors. Plant growth under flooding can be impacted by the timing of the year, duration of flood, and developmental stages of the plant and the external temperatures (Dat et al. 2006). Hypoxic soil conditions confine plant growth and development while also damaging the cells, and these damages further worsen if the submergence of the plant is for a longer time or the environmental temperature is high as it will lead to increased utilization of oxygen due to plant respiration (Deutsch et al. 2015).

Hyperhydracity, a phenomenon of cell apoplast flooding, causes multiple morphological changes, leading to impaired gas exchange efficiency (van den Dries et al. 2013). Plants utilize escape pathways and quiescence pathways to dodge the hypoxia/anoxia stress and reach a normoxic condition. Escape pathways can involve the developing of certain organs like shoot elongation, aerenchyma formation, root development, and leaf acclimation (Mignolli et al. 2020; Parlanti et al. 2011). Quiescence pathways rely on ceasing the plant growth and saving their carbohydrate reserves until normoxia is restored (Fukao and Bailey-Serres 2008a, b).

Ethylene responsive factors (ERFs) belonging to group VII of proteins are known to play important roles in determining the molecular mechanisms of oxygen perception thereby making the plant stable under hypoxia (Licausi et al. 2011; Gibbs et al. 2011). Jasmonic acid and hypoxia crosstalk studies show that the jasmonate synthesis increases in initial stage of the hypoxia stress but decreases later due to reduced oxygen (Shukla et al. 2020). Ethylene signals have been shown to enhance the survival rate in *Arabidopsis*. H_2O_2 is produced as a secondary messenger in response to hypoxia, which can regulate seed germination and growth of the seedlings, leaf chlorophyll amount, and various hypoxia-inducible genes (Sasidharan and Voesenek 2015). Micro RNAs (miRNAs) which are known to have role in various cellular processes are identified in maize and teosinte under submergence stress condition (Sepúlveda-García et al. 2020).

In this chapter, we have discussed about key signalling pathways and various regulatory mechanisms that the plant employs to overcome the hypoxia stress.

2.2 Hypoxia Stress Signalling in Plants

Under hypoxic stress, plant signalling is a complex and dynamic process that is essential to the plant's ability to survive without oxygen. Plants use complex signalling cascades to initiate adaptive responses in response to reduced oxygen levels. Reactive oxygen species (ROC), calcium ions and nitric oxide contribute to the signalling network. Deciphering these signalling pathways might lead to improved crop resilience against environmental stresses and, eventually, sustainable agriculture in the face of shifting environmental conditions.

2.2.1 ROS-Dependent Plant Signalling

Hypoxic stress disrupts the plant cells' ability to produce reactive oxygen species (ROS) and the mitochondrial respiration pathway (Yamauchi et al. 2014; Bailey-Serres and Mittler 2006). Under oxygen shortage, ROS buildup has been observed in mitochondria and apoplasts through NADPH oxidases (Baxter-Burrell et al. 2002) (Fig. 2.1). In the apoplast, superoxide anions are produced by NADPH oxidases, which facilitate the transfer of electrons from NADPH to O_2 and are then quickly transformed into hydrogen peroxide (H_2O_2) (Sagi and Fluhr 2006). *RBOHD* (respiratory burst oxidase homologs) may have a function in anoxic stress as evidenced by the finding that knocking out *RBOHD* decreases superoxide generation and H_2O_2 buildup (Torres et al. 2002) causing poor adaptation under anoxia (Wang et al. 2016; Chen et al. 2015; Pucciariello et al. 2012).

Under hypoxic conditions, H_2O_2 functions as a signalling molecule in the *Arabidopsis* ROP (RHO-like small G-protein of plants) signal transduction pathway through an NADPH oxidase (Baxter-Burrell et al. 2002). Hypoxic condition sensing and ROS signalling can be connected together as found a recent study where hypoxia-responsive universal stress protein 1 (HRU1) which is induced by RAP2.12, interacts with RBOHD (Gonzali et al. 2015).

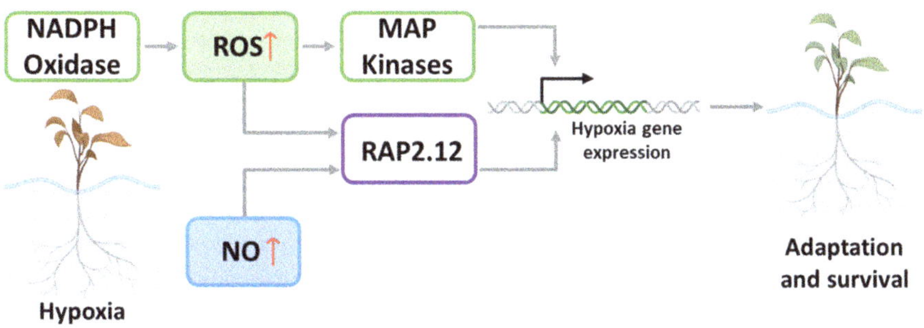

Fig. 2.1 ROS- and NO-mediated signalling lead to activation of MAP kinase pathways and RAP2.12, respectively, which initiates the gene expression of various hypoxia-responsive genes helping plants in survival

According to Chang et al. (2012), low oxygen levels in mitochondria trigger the ROS burst and MPK3 and MPK6 in *Arabidopsis*. Additionally, H_2O_2 produced in mitochondria has the potential to pass through the mitochondrial membrane and enter the cytosol (Hamanaka and Chandel 2009). Moreover, ROS inactivates Kin+ channels via Ca^{2+}, pH, and NO signalling pathways in several plant cells, as well as activate Ca^{2+} channels in plasma membrane and slow anion channels (Chen et al. 2017; Hedrich et al. 2010).

2.2.2 NO-Dependent Plant Signalling

In plants, nitric oxide (NO) generated by nitrate reductase and a plant-specific NOS (Chandok et al. 2003; Rockel et al. 2002) has been identified as a signalling molecule that responds to hypoxia as a secondary messenger causing guanylate cyclase activation and the synthesis of cyclic GMP and cyclic ADP ribose (Durner and Klessig 1999). Cytochrome oxidase is inhibited by NO, which results in protein nitrosylation. Moreover, this molecule combines with O^{2-} to generate $ONOO^-$ (peroxynitrite), a potent oxidant. H^2O^2-mediated signalling is known to be either agonistic or antagonistic to NO; the observed paradox most likely stems from variations in the concentration of these two signalling molecules (Wendehenne et al. 2001).

Plant class-I non-symbiotic hemoglobins (Hb), which increase in abundance in response to hypoxia, appear to be modulating the increase in NO levels in plants under this condition (Dordas et al. 2003a, b; Kundu et al. 2003). Compared to antisense Hb or non-transformed cell lines, maize and lucerne cell lines that overexpress a barley Hb retained greater amounts of ATP during hypoxia (Dordas et al. 2003a, b; Sowa et al. 1998). While overexpression of a Hb anticipated to have a weaker affinity for oxygen resulted in poorer stress tolerance, *Arabidopsis* seedlings that expressed a CaMV 35S:Hb transgene also demonstrated enhanced tolerance of oxygen shortage (Hunt et al. 2002). Hb may have a NO scavenger role by catalyzing the NAD(P)H-dependent process that converts NO to nitrate, according to recent in vitro research (Igamberdiev et al. 2004). Aerenchyma formation is also caused due to high ethylene content under lower expression of Hb gene (Zahra et al. 2021).

2.2.3 Ca^{+2}-Dependent Plant Signalling

It is well acknowledged that the primary regulatory signal governing plants' adaptive reactions to both biotic and abiotic stimuli is calcium ions (Ca^{2+}) (Zhu et al. 2015; Steinhorst and Kudla 2014; Reddy et al. 2011). Hypoxic stress quickly elevates Ca^{2+} in cells and this increase is essential for gene activation and adaptation responses at the cellular, tissue, and plant levels (Wang et al. 2016; Shabala et al. 2014; Subbaiah and Sachs 2003). The electron transport chain (ETC) is dampened by hypoxia, which causes the mitochondria to discharge Ca^{2+} into the cytosol, depolarizing the mitochondrial membrane. Under hypoxic circumstances, plasma membrane's depolarization inhibits K^+ channels, causing Ca^{2+} to enter the cytosol

(Fahling 2009; Lahiri et al. 2006). Many Ca^{2+}-sensing proteins, including CBL (calcineurin B-like), CIPKs (Ca^{2+}-independent protein kinases), CaM (calmodulin), CMLs (calmodulin-like proteins), and CDPKs (Ca2 + −dependent protein kinases), then detect a rise in cytosolic calcium (Reddy et al. 2011; Dodd et al. 2010) for reactions that are adaptive in plants.

2.3 Plant Regulatory Response Under Hypoxia Stress

Plants' hypoxic stress regulation mechanisms are intricately regulated systems that allow for environmental adaptation to low oxygen levels. Plants use complex transcriptional regulation to coordinate responses that are essential for their survival. Hormonal interplay in particular fine-tunes the plant's response to hypoxic stress. Regulatory RNAs and plant metabolome along with others work together and ensure a coordinated and efficient response from the plant making it adaptable to low oxygen environments.

2.3.1 Hormone-Regulated Response

Besides regulating developmental processes, plant hormones also control signalling responses under various stress conditions. Plants under abiotic stress produce endogenous hormones like ethylene, abscisic acid (ABA), indole-3-acetic acid (IAA), gibberellins (GA), and cytokinins (CTKs), which have a role in signal transduction and gene expression regulation (Xiong et al. 2002; Weyers and Paterson 2001) (Fig. 2.2). Of all the plant hormones, ethylene and abscisic acid (ABA) are primarily involved in water-related stress. Due to physical entrapment and active production of ethylene during the stress, this gaseous phytohormone quickly accumulates in the tissues of submerged plants causing a variety of acclimation responses, such as shoot elongation, adventitious root development, and glucose metabolism

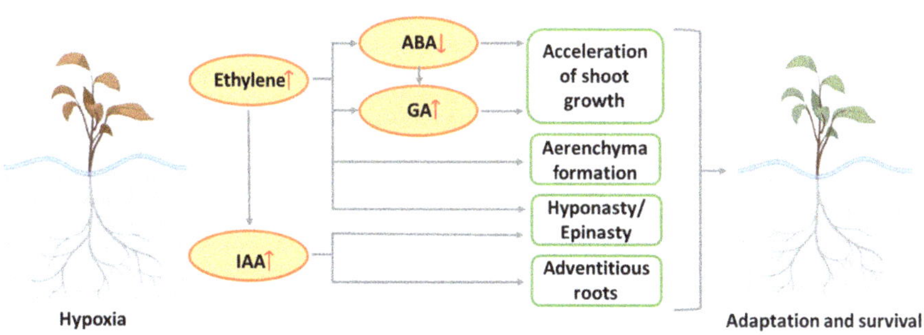

Fig. 2.2 Phytohormones like ethylene, IAA, ABA, and GA regulate various morphological adaptations including shoot growth, aerenchyma, and adventitious roots formation and hyponasty or epinasty helping the plant to escape the hypoxic condition as well as transport oxygen from shoot to root

(Fukao and Bailey-Serres 2008a, b). Exogenous application of ABA increases resilience against anoxia in *Arabidopsis* (Dat et al. 2004). Jasmonic acid levels are shown to increase in roots and shoots, whereas IAA in some findings is found to be lowered due to oxygen deficiency (Arbona and Gómez-Cadenas 2008; Reid and Wample 1985).

2.3.2 Transcription Factor Regulated Response

Hypoxia and anoxia under waterlogging and the metabolic components involved in waterlogging have also been studied using molecular biology techniques. The main areas of current study include metabolic processes, the identification of anaerobic proteins, or a combination of the two. Numerous anaerobic response elements, enzymes associated with the fermentation process (such as pyruvate decarboxylase (PDC) and alcohol dehydrogenase), and transcription factors have been linked to waterlogging (Dolferus et al. 2003; Sachs et al. 1996).

Genetic alterations of transcription factors in plants are currently being employed to generate plants which are more resilient to stresses (Table 2.1). During the state of hypoxia, ETHYLENE_RESPONSIVE FACTOR (ERFVII) is one the most important transcriptional factors that initiates gene expression (Gasch et al. 2016). Hypoxia responsive ERFs HRE1 and HRE2 are responsible to increase transcript levels as well as transcriptional processes (Licausi et al. 2011). Hypoxic conditions also trigger the MYB transcription factors expression where they also act like a signalling molecule along with calcineurin (CBL)-interacting protein kinase (CIPKs). CIPKs upregulate alcohol dehydrogenase gene under hypoxia stress and MYB activates α-amylase isozyme, which helps in enhancing plant's tolerance against stress (Lee et al. 2009). Increased MYB expression is also associated with increased expression of proline synthetase leading to increased accumulation of proline (Yang et al. 2012).

Under hypoxia, a number of other transcriptional factors are also controlled. For example, 180 transcriptional factor genes, MYB, bZIP (basic leucine zipper), bHLH (basic helix-loop-helix), and EFR are either elevated or downregulated (Licausi et al. 2010). Hypoxia tolerance is enhanced by transcriptional factors ZAT (zinc finger transcriptional factors) and HSF (heat shock factor). WRKY transcription factors are also known to confer immunity against submergence stress (Table 2.1) (Hsu et al. 2013).

2.3.3 Regulatory RNA-Mediated Response

RNA-related molecular processes and RNAs of various sizes and sources like trans-acting siRNAs, natural antisense siRNA (natsiRNA), and hypoxia-responsive miRNAs play important regulatory roles throughout the onset of hypoxia-induced reactions (Moldovan et al. 2010a, b). Long non-coding RNAs (lncRNAs) also regulate the resilience against hypoxia by their differential expression under

Table 2.1 Transcription factors and their function in plants

Sr. No.	Transcription factor	Plant	Function	Reference
1	MaRAP2–4	*Mentha arvensis*	Positive regulator, interacts with JARE *cis*-element, regulate oxidative stress response by modulating ROS scavenging	Phukan et al. (2018)
2	AdRAP2.3	*Actinidia deliciosa*	Improving expression levels of PDC and ADH genes	Pan et al. (2019)
3	HvERF2.11	Barley	Positive regulator, regulate the expression of the antioxidant enzymes	Luan et al. (2020)
4	HaWRKY76	*Helianthus annuus*		Raineri et al. (2015)
5	HaHB11	*Helianthus annuus*	Enhanced chlorophyll content and vascular bundle	Raineri et al. (2022)
6	PhERF2	Petunia	Aerenchyma development in the roots elevation in transcript abundances of PDC and ADH, LDH was down-regulated	Yin et al. (2019)
7	ZmEREB180	Maize	Formation of adventitious root. ZmEREB180 also enhances the mRNA level of genes encoding antioxidants to regulate ROS status via changing antioxidant levels to sustain cell homeostasis. *Arabidopsis thaliana*	Yu et al. (2019)
8	RAP2.2	*Arabidopsis thaliana*	Induces ADH1 gene expression	Hinz et al. (2010)
9	WRKY33 WRKY12	*Arabidopsis thaliana*	Can synergistically enhance ethylene response factor VII gene RAP2.2 and provide enhanced resistance to hypoxia	Tang et al. (2021)
10	WRKY53	*Arabidopsis thaliana*	Contributes to plant defense and root elongation along with long non-coding RNA	Li et al. (2020)
11	HRE1 and HRE2	*Arabidopsis thaliana*	Role on low oxygen level signalling and enhancing ethanolic fermentation and anaerobic gene expression to increase the plant's resistance	Licausi et al. (2010)
12	WRKY22	*Arabidopsis thaliana*	WRKY22 may play a role upstream of submergence signalling. It also mediates plant basal innate immunity	Hsu and Shih (2013)
13	SUB1A	*Oryza sativa*	Elevated ethylene and increased length of shoot in the seedling. Positively regulate ADH1 gene expression	Nishiuchi et al. (2012)
14	OsARD1	*Oryza sativa*	OsARD1 encodes an active ARD and acts as a key role in the synthesis of ethylene. Elevated ethylene and increased length of shoot in the seedling	Liang et al. (2019)
15	TaMyb1	*Triticum aestivum*	Role is response of varying oxygen concentrations around roots and various other abiotic stresses	Lee et al. (2007)

waterlogging stress (Yu et al. 2020). In low salicylic acid levels, WRKY transcription factors along with long non-coding RNA help in plant defense as well as root elongation (Li et al. 2020). Non-canonical mRNA generation and retention of poly(A) RNA in the nucleus are adaptive approaches that help the plants as hypoxia stress response (de Lorenzo et al. 2017; Niedojadło et al. 2016). Under hypoxic conditions, macromolecular RNA–protein complexes aid in the selective translation of stress-responsive gene transcripts (Lee and Bailey-Serres 2021). Post-transcriptional alternative splicing in rice germination under hypoxia is an important process (Chen et al. 2019). Through the confluence of AGONAUTE1 (AGO1) signalling with the AGO4-dependent RNA-directed DNA methylation pathway, RNA signalling regulates hypoxia-induced gene regulation in *Arabidopsis* (Loreti et al. 2020).

2.3.4 Metabolic Responses

Roots and shoots of a plant show different responses to oxygen deficiency due to hypoxia under waterlogging conditions. A genetic response to hypoxia stress can cause the primary metabolism of roots to be reorganized in various ways like increased sugar metabolism by raising the amount of sucrose synthase, increasing ethanolic fermentation, and by accumulating metabolites from alternate pathways, such as γ-aminobutyrate (GABA) and alanine (Alam et al. 2010; Drew 1997). Shoots usually show fewer symptoms of hypoxia as oxygen is generated via photosynthesis, whereas roots on the other hand show serious symptoms and produce alanine, GABA, and lactate (Mustroph et al. 2014).

Metabolomic studies reveal that hypoxia due to waterlogging can cause alterations in the amount of various organic acids, polyols, as well as amino acids, which are due to inhibited TCA pathway or catabolism (Irfan et al. 2010). Reduction in sugar export and nitrogen assimilation causes a significant rise in sugar content like hexose and polyol derivatives and a reduction in amino acids like alanine in plant leaves (Cui et al. 2019). Nitrate transport in the xylem is inhibited, causing reduced nitrate content in plant leaves (Cui et al. 2020), and ureides and glutamine transport via the xylem are reduced in the shoot (Amarante and Sodek 2006; Puiatti and Sodek 1999). Phloem (and xylem) sap disruption of metabolite exchange is likely the source of some of the significant effects of waterlogging on leaf metabolism, particularly nitrogen absorption.

2.4 Conclusion

In summary, the complex web of plant signalling cascades and their reactions to hypoxic stress highlight how remarkably adaptive plants are to environmental shocks. The intricacy of plants' molecular machinery is demonstrated by their capacity to detect low oxygen levels and transduce signals in response. The hypoxic stress response is mediated by intricate and diverse signalling pathways that involve several molecular players including as transcription factors, kinases, and hormones.

The activation of anaerobic metabolism, which permits energy synthesis in the absence of oxygen, is a crucial component of plant adaptation to hypoxia. A key signalling molecule, ethylene plays a pivotal role in coordinating a number of responses to hypoxia, including changes in root architecture, increased resistance to waterlogging, and regulation of gene expression. Reactive oxygen species and nitric oxide have a role in hypoxia signalling, which emphasises how interrelated these cascades are.

Furthermore, the complexity of the plant's response to hypoxic stress is revealed by the cross-talk across many signalling channels. Plants are more able to adapt to low oxygen levels when hormonal signals, such as ethylene and abscisic acid, are combined with other signalling elements. Comprehending these signalling cascades contributes to our understanding of plant biology and may help devise methods to improve crop resistance to environmental stressors like hypoxia and waterlogging.

Understanding of plant signalling cascades and response to hypoxic stress reveal the remarkably complex and adaptable processes of plants. By delving further into the molecular details of these processes, we open the door to new and creative ways to lessen the negative effects of environmental stress on plant development and agricultural output.

Acknowledgements DK is thankful to the Science and Engineering Research Board (Grant No: EEQ/2021/000593) and Institute of Eminence Seed Grant (Dev. scheme No. 6031 (B)), Banaras Hindu University, Varanasi for providing financial support to the laboratory. DK also acknowledges Centre of Advanced study in Botany, Institute of Science, Banaras Hindu University, for providing necessary facilities and infrastructural support. PP acknowledges the Prime Minister Research Fellowship (PMRF) for providing fellowship.

Conflict of Interest The authors declare no conflict of interest.

References

Alam I, Lee DG, Kim KH, Park CH, Sharmin SA, Lee H, Lee BH (2010) Proteome analysis of soybean roots under waterlogging stress at an early vegetative stage. J Biosci 35:49–62

Amarante LD, Sodek L (2006) Waterlogging effect on Xylem sap glutamine of nodulated soybean. Biol Plant 50:405–410

Arbona V, Gómez-Cadenas A (2008) Hormonal modulation of citrus responses to flooding. J Plant Growth Regul 27:241–250

Bailey-Serres J, Mittler R (2006) The roles of reactive oxygen species in plant cells. Plant Physiol 141:311–311

Baxter-Burrell A, Yang ZB, Springer PS, Bailey-Serres J (2002) RopGAP4-dependent Rop GTPase rheostat control of Arabidopsis oxygen deprivation tolerance. Science 296:2026–2028

Chandok MR, Ytterberg AJ, van Wijk KJ, Klessig DF (2003) The pathogen-inducible nitric oxide synthase (iNOS) in plants is a variant of the P protein of the glycine decarboxylase complex. Cell 113(4):469–482

Chang R, Jang CJH, Branco-Price C, Nghiem P, Bailey-Serres J (2012) Transient MPK6 activation in response to oxygen deprivation and reoxygenation is mediated by mitochondria and aids seedling survival in Arabidopsis. Plant Mol Biol 78:109–122

Chen L, Liao B, Qi H, Xie LJ, Huang L, Tan WJ et al (2015) Autophagy contributes to regulation of the hypoxia response during submergence in Arabidopsis thaliana. Autophagy 11:2233–2246

Chen ZH, Chen G, Dai F, Wang Y, Hills A, Ruan YL et al (2017) Molecular evolution of grass stomata. Trends Plant Sci 22:124–139

Chen MX, Zhu FY, Wang FZ, Ye NH, Gao B, Chen X, Liu YG (2019) Alternative splicing and translation play important roles in hypoxic germination in rice. J Exp Bot 70(3):817–833

Cui J, Davanture M, Zivy M, Lamade E, Tcherkez G (2019) Metabolic responses to potassium availability and waterlogging reshape respiration and carbon use efficiency in oil palm. New Phytol 223(1):310–322

Cui J, Lamade E, Fourel F, Tcherkez G (2020) δ15N values in plants are determined by both nitrate assimilation and circulation. New Phytol 226(6):1696–1707

Dat JF, Capelli N, Folzer H, Bourgeade P, Badot PM (2004) Sensing and signalling during plant flooding. Plant Physiol Biochem 42(4):273–282

Dat J, Folzer H, Parent C, Badot PM, Capelli N (2006) Hypoxia stress. Current understanding and perspectives. In: Floriculture, ornamental and plant biotechnology: advances and topical issues, vol 3, pp 664–674

de Lorenzo L, Sorenson R, Bailey-Serres J, Hunt AG (2017) Noncanonical alternative polyadenylation contributes to gene regulation in response to hypoxia. Plant Cell 29(6):1262–1277

Deutsch C, Ferrel A, Seibel B, Pörtner HO, Huey RB (2015) Climate change tightens a metabolic constraint on marine habitats. Science 348(6239):1132–1135

Dodd AN, Kudla J, Sanders D (2010) The language of calcium signaling. Annu Rev Plant Biol 61:593–620

Dolferus R, Klok EJ, Delessert C, Wilson S, Ismond KP, Good AG, Dennis ES (2003) Enhancing the anaerobic response. Ann Bot 91(2):111–117

Dordas C, Hasinoff BB, Igamberdiev AU, Manac'h N, Rivoal J, Hill RD (2003a) Expression of a stress-induced hemoglobin affects NO levels produced by alfalfa root cultures under hypoxic stress. Plant J 35(6):763–770

Dordas C, Rivoal J, Hill RD (2003b) Plant haemoglobins, nitric oxide and hypoxic stress. Ann Bot 91(2):173–178

Drew MC (1997) Oxygen deficiency and root metabolism: injury and acclimation under hypoxia and anoxia. Annu Rev Plant Biol 48(1):223–250

Durner J, Klessig DF (1999) Nitric oxide as a signal in plants. Curr Opin Plant Biol 2(5):369–374

Fahling M (2009) Cellular oxygen sensing, signalling and how to survive translational arrest in hypoxia. Acta Physiol 195:205–230

Fukao T, Bailey-Serres J (2008a) Ethylene—a key regulator of submergence responses in rice. Plant Sci 175(1–2):43–51

Fukao T, Bailey-Serres J (2008b) Submergence tolerance conferred by Sub1A is mediated by SLR1 and SLRL1 restriction of gibberellin responses in rice. Proc Natl Acad Sci 105(43):16814–16819

Gasch P, Fundinger M, Müller JT, Lee T, Bailey-Serres J, Mustroph A (2016) Redundant ERF-VII transcription factors bind to an evolutionarily conserved cis-motif to regulate hypoxia-responsive gene expression in Arabidopsis. Plant Cell 28(1):160–180

Gibbs DJ, Lee SC, Md Isa N, Gramuglia S, Fukao T, Bassel GW, Holdsworth MJ (2011) Homeostatic response to hypoxia is regulated by the N-end rule pathway in plants. Nature 479(7373):415–418

Gonzali S, Loreti E, Cardarelli F, Novi G, Parlanti S, Pucciariello C et al (2015) Universal stress protein HRU1 mediates ROS homeostasis under anoxia. Nat Plants 1:15151

Hamanaka RB, Chandel NS (2009) Mitochondrial reactive oxygen species regulate hypoxic signaling. Curr Opin Cell Biol 21:894–899. Els in plants. Physiol Rev 92: 1777–1811

Hedrich R, Kim TH, Bohmer M, Hu HH, Nishimura N, Schroeder JI (2010) Guard cell signal transduction network: advances in understanding abscisic acid, CO2, and Ca2+ signaling. Annu Rev Plant Biol 61:561–591. (2012) ion chann

Hinz M, Wilson IW, Yang J, Buerstenbinder K, Llewellyn D, Dennis ES, Dolferus R (2010) Arabidopsis RAP2. 2: an ethylene response transcription factor that is important for hypoxia survival. Plant Physiol 153(2):757–772

Hsu FC, Shih MC (2013) Plant defense after flooding. Plant Signal Behav 8(11):2699–2713
Hsu FC, Chou MY, Chou SJ, Li YR, Peng HP, Shih MC (2013) Submergence confers immunity mediated by the WRKY22 transcription factor in Arabidopsis. Plant Cell 25(7):2699–2713
Hunt PW, Klok EJ, Trevaskis B, Watts RA, Ellis MH, Peacock WJ, Dennis ES (2002) Increased level of hemoglobin 1 enhances survival of hypoxic stress and promotes early growth in Arabidopsis thaliana. Proc Natl Acad Sci 99(26):17197–17202
Igamberdiev AU, Seregelyes C, Manach N, Hill RD (2004) NADH-dependent metabolism of nitric oxide in alfalfa root cultures expressing barley hemoglobin. Planta 219:95–102
Irfan M, Hayat S, Hayat Q, Afroz S, Ahmad A (2010) Physiological and biochemical changes in plants under waterlogging. Protoplasma 241:3–17
Kundu S, Trent JT, Hargrove MS (2003) Plants, humans and hemoglobins. Trends Plant Sci 8(8):387–393
Lahiri S, Roy A, Baby SM, Hoshi T, Semenza GL, Prabhakar NR (2006) Oxygen sensing in the body. Prog Biophys Mol Biol 91:249–286
Lee TA, Bailey-Serres J (2021) Conserved and nuanced hierarchy of gene regulatory response to hypoxia. New Phytol 229(1):71–78
Lee TG, Jang CS, Kim JY, Kim DS, Park JH, Kim DY, Seo YW (2007) A Myb transcription factor (TaMyb1) from wheat roots is expressed during hypoxia: roles in response to the oxygen concentration in root environment and abiotic stresses. Physiol Plant 129(2):375–385
Lee KW, Chen PW, Lu CA, Chen S, Ho THD, Yu SM (2009) Coordinated responses to oxygen and sugar deficiency allow rice seedlings to tolerate flooding. Sci Signal 2(91):ra61
Li S, Nayar S, Jia H, Kapoor S, Wu J, Yukawa Y (2020) The Arabidopsis hypoxia inducible AtR8 long non-coding RNA also contributes to plant defense and root elongation coordinating with WRKY genes under low levels of salicylic acid. Noncoding RNA 6(1):8
Liang S, Xiong W, Yin C, Xie X, Jin YJ, Zhang S, Luan WJ (2019) Overexpression of OsARD1 improves submergence, drought, and salt tolerances of seedling through the enhancement of ethylene synthesis in rice. Front Plant Sci 10:1088
Licausi F, Van Dongen JT, Giuntoli B, Novi G, Santaniello A, Geigenberger P, Perata P (2010) HRE1 and HRE2, two hypoxia-inducible ethylene response factors, affect anaerobic responses in Arabidopsis thaliana. Plant J 62(2):302–315
Licausi F, Kosmacz M, Weits DA, Giuntoli B, Giorgi FM, Voesenek LA, Van Dongen JT (2011) Oxygen sensing in plants is mediated by an N-end rule pathway for protein destabilization. Nature 479(7373):419–422
Loreti E, Betti F, Ladera-Carmona MJ, Fontana F, Novi G, Valeri MC, Perata P (2020) ARGONAUTE1 and ARGONAUTE4 regulate gene expression and hypoxia tolerance. Plant Physiol 182(1):287–300
Luan H, Guo B, Shen H, Pan Y, Hong Y, Lv C, Xu R (2020) Overexpression of barley transcription factor HvERF2. 11 in Arabidopsis enhances plant waterlogging tolerance. Int J Mol Sci 21(6):1982
Mignolli F, Todaro JS, Vidoz ML (2020) Internal aeration and respiration of submerged tomato hypocotyls are enhanced by ethylene-mediated aerenchyma formation and hypertrophy. Physiol Plant 169(1):49–63
Moldovan D, Spriggs A, Dennis ES, Wilson IW (2010a) The hunt for hypoxia responsive natural antisense short interfering RNAs. Plant Signal Behav 5(3):247–251
Moldovan D, Spriggs A, Yang J, Pogson BJ, Dennis ES, Wilson IW (2010b) Hypoxia-responsive microRNAs and trans-acting small interfering RNAs in Arabidopsis. J Exp Bot 61(1):165–177
Mustroph A, Barding GA Jr, Kaiser KA, Larive CK, Bailey-Serres JULIA (2014) Characterization of distinct root and shoot responses to low-oxygen stress in Arabidopsis with a focus on primary C-and N-metabolism. Plant Cell Environ 37(10):2366–2380
Niedojadło J, Dełeńko K, Niedojadło K (2016) Regulation of poly (A) RNA retention in the nucleus as a survival strategy of plants during hypoxia. RNA Biol 13(5):531–543
Nishiuchi S, Yamauchi T, Takahashi H, Kotula L, Nakazono M (2012) Mechanisms for coping with submergence and waterlogging in rice. Rice 5:1–14

Pan DL, Wang G, Wang T, Jia ZH, Guo ZR, Zhang JY (2019) AdRAP2. 3, a novel ethylene response factor VII from Actinidia deliciosa, enhances waterlogging resistance in transgenic tobacco through improving expression levels of PDC and ADH genes. Int J Mol Sci 20(5):1189
Parlanti S, Kudahettige NP, Lombardi L, Mensuali-Sodi A, Alpi A, Perata P, Pucciariello C (2011) Distinct mechanisms for aerenchyma formation in leaf sheaths of rice genotypes displaying a quiescence or escape strategy for flooding tolerance. Ann Bot 107(8):1335–1343
Phukan UJ, Jeena GS, Tripathi V, Shukla RK (2018) Ma RAP 2-4, a waterlogging-responsive ERF from Mentha, regulates bidirectional sugar transporter at SWEET 10 to modulate stress response in Arabidopsis. Plant Biotechnol J 16(1):221–233
Pucciariello C, Parlanti S, Banti V, Novi G, Perata P (2012) Reactive oxygen species-driven transcription in Arabidopsis under oxygen deprivation. Plant Physiol 159:184–196
Puiatti M, Sodek L (1999) Waterlogging affects nitrogen transport in the xylem of soybean. Plant Physiol Biochem 37(10):767–773
Raineri J, Ribichich KF, Chan RL (2015) The sunflower transcription factor HaWRKY76 confers drought and flood tolerance to Arabidopsis thaliana plants without yield penalty. Plant Cell Rep 34:2065–2080
Raineri J, Caraballo L, Rigalli N, Portapila M, Otegui ME, Chan RL (2022) Expressing the sunflower transcription factor HaHB11 in maize improves waterlogging and defoliation tolerance. Plant Physiol 189(1):230–247
Reddy ASN, Ali GS, Celesnik H, Day IS (2011) Coping with stresses: roles of calcium- and calcium/calmodulin-regulated gene expression. Plant Cell 23:2010–2032
Reid DM, Wample RL (1985) Water relations and plant hormones. In: Hormonal regulation of development III: role of environmental factors. Springer, Berlin, Heidelberg, pp 513–578
Rockel P, Strube F, Rockel A, Wildt J, Kaiser WM (2002) Regulation of nitric oxide (NO) production by plant nitrate reductase in vivo and in vitro. J Exp Bot 53(366):103–110
Sachs MM, Subbaiah CC, Saab IN (1996) Anaerobic gene expression and flooding tolerance in maize. J Exp Bot 47(1):1–15
Sagi M, Fluhr R (2006) Production of reactive oxygen species by plant NADPH oxidases. Plant physiology 141(2):336–340
Sasidharan R, Voesenek LA (2015) Ethylene-mediated acclimations to flooding stress. Plant Physiol 169(1):3–12
Sepúlveda-García EB, Pulido-Barajas JF, Huerta-Heredia AA, Peña-Castro JM, Liu R, Barrera-Figueroa BE (2020) Differential expression of maize and teosinte microRNAs under submergence, drought, and alternated stress. Plan Theory 9(10):1367
Shabala S, Shabala L, Barcelo J, Poschenrieder C (2014) Membrane transporters mediating root signalling and adaptive responses to oxygen deprivation and soil flooding. Plant Cell Environ 37:2216–2233
Shukla V, Lombardi L, Pencik A, Novak O, Weits DA, Loreti E, Licausi F (2020) Jasmonate signalling contributes to primary root inhibition upon oxygen deficiency in Arabidopsis thaliana. Plan Theory 9(8):1046
Sowa AW, Duff SM, Guy PA, Hill RD (1998) Altering hemoglobin levels changes energy status in maize cells under hypoxia. Proc Natl Acad Sci 95(17):10317–10321
Steinhorst L, Kudla J (2014) Signaling in cells and organisms—calcium holds the line. Curr Opin Plant Biol 22:14–21
Subbaiah CC, Sachs MM (2003) Molecular and cellular adaptations of maize to flooding stress. Ann Bot 91:119–127
Tang H, Bi H, Liu B, Lou S, Song Y, Tong S, Liu H (2021) WRKY33 interacts with WRKY12 protein to up-regulate RAP2. 2 during submergence induced hypoxia response in Arabidopsis thaliana. New Phytol 229(1):106–125
Torres MA, Dangl JL, Jones JD (2002) Arabidopsis gp91phox homologues AtrbohD and AtrbohF are required for accumulation of reactive oxygen intermediates in the plant defense response. Proc Natl Acad Sci USA 99:517–522
van den Dries N, Giannì S, Czerednik A, Krens FA, de Klerk GJM (2013) Flooding of the apoplast is a key factor in the development of hyperhydricity. J Exp Bot 64(16):5221–5230

Wang F, Chen Z-H, Liu X, Colmer TD, Zhou M, Shabala S (2016) Tissue-specific root ion profiling reveals essential roles for the CAX and ACA calcium transport systems for hypoxia response in Arabidopsis. J Exp Bot 67:3747–3762

Wendehenne D, Pugin A, Klessig DF, Durner J (2001) Nitric oxide: comparative synthesis and signaling in animal and plant cells. Trends Plant Sci 6(4):177–183

Weyers JD, Paterson NW (2001) Plant hormones and the control of physiological processes. New Phytol 152(3):375–407

Xiong L, Schumaker KS, Zhu JK (2002) Cell signaling during cold, drought, and salt stress. Plant Cell 14(suppl_1):S165–S183

Yamauchi T, Watanabe K, Fukazawa A, Mori H, Abe F, Kawaguchi K et al (2014) Ethylene and reactive oxygen species are involved in root aerenchyma formation and adaptation of wheat seedlings to oxygen-deficient conditions. J Exp Bot 65:261–273

Yang A, Dai X, Zhang WH (2012) A R2R3-type MYB gene, OsMYB2, is involved in salt, cold, and dehydration tolerance in rice. J Exp Bot 63(7):2541–2556

Yin D, Sun D, Han Z, Ni D, Norris A, Jiang CZ (2019) PhERF2, an ethylene-responsive element binding factor, plays an essential role in waterlogging tolerance of petunia. Hortic Res 6:83

Yu F, Liang K, Fang T, Zhao H, Han X, Cai M, Qiu F (2019) A group VII ethylene response factor gene, ZmEREB180, coordinates waterlogging tolerance in maize seedlings. Plant Biotechnol J 17(12):2286–2298

Yu F, Tan Z, Fang T, Tang K, Liang K, Qiu F (2020) A comprehensive transcriptomics analysis reveals long non-coding RNA to be involved in the key metabolic pathway in response to waterlogging stress in maize. Genes 11(3):267

Zahra N, Hafeez MB, Shaukat K, Wahid A, Hussain S, Naseer R, Farooq M (2021) Hypoxia and anoxia stress: plant responses and tolerance mechanisms. J Agron Crop Sci 207(2):249–284

Zhu XY, Dunand C, Snedden W, Galaud JP (2015) CaM and CML emergence in the green lineage. Trends Plant Sci 20:483–489

Reactive Oxygen Species (ROS) and ROS Scavengers in Plant Abiotic Stress Response

3

Aakanksha Singh and P. K. Satheeshkumar

Abstract

The surroundings in which plants live experience fluctuations in every moment. Highly dynamic processes in the cells in response to these changes in the outside environment restore the equilibrium necessary for normal metabolism. Being sessile living beings' plants are affected by the changes in their surroundings much more than animals, who can move themselves to a more comfortable environment. Metabolic processes produce certain bi-products that play critical roles in living cells. They present in the cells under normal conditions within a permissible level and may increase their concentration under extreme conditions like stress. Reactive oxygen species or "ROS" is one of those by-products that have multiple roles in living cells. Nearly 1–2% of the oxygen that enters the cells is converted to ROS of different types such as singlet oxygen, superoxide radical, hydroxide radical, and hydrogen peroxide, which are inevitable components in activating many molecular pathways in cells. The sensitive balance in terms of the amount of ROS present in a cell at a time is taken care of by a group of molecules known as ROS scavengers, which are part of the antioxidant machinery in the cells. While the redox signalling pathways mediated by the ROS under normal conditions are beneficial for plant growth, and development, the pathways activated by the ROS under stress conditions have a negative effect on plants. Even though the ROS induces the ROS scavenger's production and activity, depending on the intensity and duration of the stress it may be beneficial or ineffective. This chapter discusses about different types of ROS and ROS scavengers, the conditions leading to their origin, the role played by them under normal and stressed conditions, and the major organelles involved in the ROS signalling pathways.

A. Singh · P. K. Satheeshkumar (✉)
Department of Botany, Banaras Hindu University, Varanasi, Uttar Pradesh, India
e-mail: satheesh.bot@bhu.ac.in

A. K. Mishra (ed.), *Stress Biology in Photosynthetic Organisms*,
https://doi.org/10.1007/978-981-97-1883-2_3

Keywords

Reactive oxygen species · Abiotic stress · Oxidative stress · Antioxidant system · ROS scavengers

3.1 Introduction

Plants, being the major players in maintaining a liveable environment on earth, are affected severely in terms of growth and productivity due to the changes happening to the climate and their surroundings, and the environmental services they provide are critically hampered (Bulgari et al. 2019). These changes negatively affect their ability to survive, thrive, and produce wealth. The stresses, such as heat shock, water scarcity, heavy metals, and xenobiotic compounds, contribute to a 50% loss in productivity (Khan et al. 2015). These environmental factors—ranging from extreme temperatures to irregular rainfall patterns and accumulation of pollutants—interact in natural settings, compounding their effects on plants' cellular, molecular, physiological, and metabolic functions (He et al. 2018).

Reactive oxygen species (ROS) are naturally synthesized in plant cells within controlled levels. However, the harsh environmental circumstances cause an excessive amount of ROS to be produced, causing damage to various cellular components through biomolecule oxidation, and contributing to plant mortality (Apel and Hirt 2004). Excessive accumulation of these molecules, stemming from the disruption of ROS redox homeostasis, can detrimentally impact the plant genome and its components, including DNA, RNA, and proteins. There are many effects of this ROS accumulation including chromosomal aberrations, mutations, and even cell death (Gechev and Petrov 2020). In aerobic organisms, many ROS such as hydroxyl radical (•OH), singlet oxygen ($^1O_2$), superoxide radical (O_2•−), and hydrogen peroxide (H_2O_2) were produced continuously from normal cellular metabolism (Mittler et al. 2011). To mitigate this, plants employ both enzymatic (e.g., superoxide dismutase (SOD), catalase (CAT), ascorbate peroxidase (APX), glutathione reductase (GR), and peroxidase (POX)) (Mittler et al. 2011) and non-enzymatic antioxidants (e.g., vitamins, flavonoids, and carotenoids) to regulate ROS levels and prevent oxidative harm (Jaleel et al. 2007).

The action of abiotic stressors on plants can be single or in parallel with more than one factor leads to the overproduction of ROS in cells, which makes it difficult for plants to grow and produce at their best. To reduce stress, or to increase tolerance against present and expected climatic conditions in the future, and the way plant signalling influences the economic productivity in commercial plants, it is of considerable significance to understand the underlying molecular processes of ROS signalling pathways. The current chapter addresses antioxidant-based defence mechanisms for controlling the oxidative stress created through ROS generation and mitigating oxidative stress.

3.2 Impact of Climate Change on ROS Generation

Abiotic stress, exacerbated by climate change, has notably affected plants on a global scale. The Food and Agriculture Organization (FAO) reported in 2019 that approximately 96.5% of cultivable land worldwide experiences some form of stress (Raza et al. 2019). The multifaceted impact of abiotic stressors on plants, compounded by climate change, poses challenges that plants struggle to mitigate (Fig. 3.1) (Table 3.1). Addressing the growing demand for food and the nutritional requirements needs a thorough understanding of the persistent effects of different abiotic stressors (both singly and in combination) on plants. Integrating such data into simulation models can aid in devising sustainable crop protection strategies that account for anticipated environmental changes.

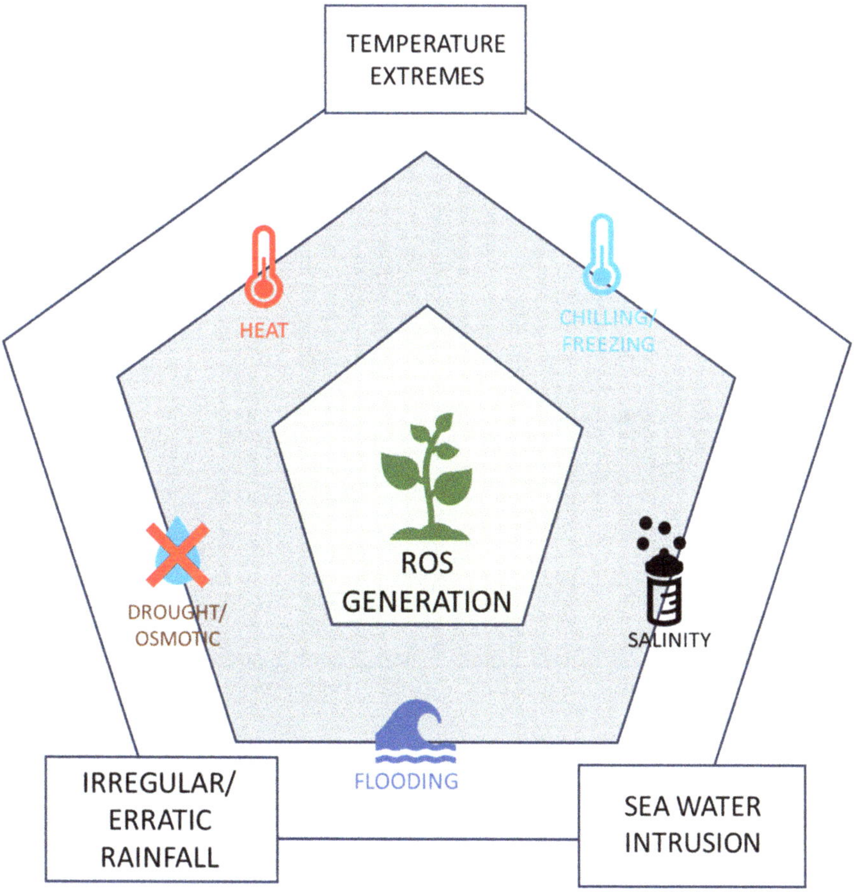

Fig. 3.1 Climate change gives rise to abiotic stresses which lead to ROS generation

Table 3.1 Abiotic stresses induce secondary stresses and their damaging effects on plant growth and activity

Abiotic stress	Induced secondary stresses	Effects on plants	Reference
Chilling/ freezing stress	Osmotic and oxidative stress	• Buildup of reactive oxygen species leading to oxidative damage; restriction of enzyme functions and disruption in metabolic equilibrium • Buildup of reactive oxygen species resulting in oxidative harm • Hindered enzyme functionality and disruption in metabolic equilibrium • Elevated cell desiccation and nutrient deprivation, aging, postponed maturation, impairment of photosystem II, and diminished photosynthetic function • Diminished growth and reduced overall productivity	Esim and Atici (2014), Megha et al. (2018)
Drought	Oxidative and osmotic stress	• Enhanced generation of reactive oxygen species (ROS) and heightened ion leakage were observed, leading to dehydration and a decline in turgor • Diminished uptake and movement of mineral nutrients occurred • Protein denaturation, enzyme activity loss, and decreased photosynthetic efficiency resulted from reduced chlorophyll levels and carbon dioxide assimilation • Elevated leaf temperature, premature abscission, necrosis, and inhibited plant growth were evident	Shao et al. (2008), Agurla et al. (2018)
Flooding/ waterlogging	Water deficiency stress, nutrient stress, oxidative stress	• Increased ROS and ethylene production and decreased antioxidant levels • Reduced stomatal conductance; abridged water and nutrient uptake • Reduced gaseous exchange, anoxia/hypoxia, increased anaerobic metabolism and inhibited root respiration; reduced photosynthetic activity due to the decreased chlorophyll content and damage of PS II • Stunted growth and senescence of leaf and inflorescence	Bailey-Serres and Voesenek (2008)
Heat stress	Water deficiency, oxidative and osmotic stress	• Elevated production of reactive oxygen species (ROS) causing oxidative damage, protein misfolding, and denaturation • Growth inhibition, foliar senescence, and abscission observed, accompanied by leaf and fruit discoloration • Reduction in carbon dioxide fixation • Disruption of both photosystem I (PS I) and photosystem II (PS II) • Disturbance in ion transport	Nievola et al. (2017), Sarkar et al. (2018), Demirel et al. (2020)

Heavy metals/ xenobiotic compounds	Nutrient and oxidative stress	• Enhanced production of reactive oxygen species (ROS) leading to oxidative damage • Disruption of the function and structure of enzymes, resulting in reduced stomatal conductance, CO_2 assimilation, and net photosynthesis rate • Decreased biomass accumulation, inhibition of seed germination, and impaired nutrient uptake	Shahzad et al. (2018)
Light/ radiation stress	Oxidative stress	• Enhanced production of reactive oxygen species (ROS) leading to oxidative damage • Disruption of the photosynthetic electron transport chain (ETC) and/or heightened activity of membrane-bound NADPH oxidase • Chlorophyll degradation resulting in reduced photosynthetic activity • Inhibition of epidermal cell expansion • Manifestations of leaf senescence, reduced rosette diameter, and a condensed inflorescence stem with an increased number of flowering stems	Noshi et al. (2016)
Salinity	Ionic imbalance, water scarcity, nutrient, oxidative, and osmotic stress	• Reactive oxygen species (ROS) production leads to oxidative damage, limiting the uptake and translocation of water and mineral nutrients. This causes Na + toxicity and decreases K+, Ca2+, and Mg2+ content, along with a reduction in soil water potential • Decreased stomatal opening, disorganized thylakoid ultrastructure, and reduced photosynthesis • Impaired seed germination, premature leaf senescence, and diminished overall growth and productivity	Ben Abdallah et al. (2016), Petridis et al. (2012), Golldack et al. (2014)

3.2.1 Temperature Stress

Extreme temperature fluctuations, either above or below the optimal range by 10–15 °C, induce heat or chilling/freezing stress, causing detrimental effects on various aspects of plant physiology, reproduction, and productivity. Research work on the effects of heat stress on morphological, molecular, and metabolic changes shows that it impairs major biological functions including photosynthesis, respiration, etc. The other effects can be a decrease in enzymatic activity, enhanced levels of calcium (Ca^{2+}) influx, and heat shock protein (HSP) production, thus intensifying reactive oxygen species (ROS) production (Bita and Gerats 2013). Plants use a defence mechanism known as the heat stress response to protect themselves from heat-induced damage when they sense a rise in temperature. Heat shock factors (HSFs) play an important role in controlling the expression of heat-responsive proteins as transcription factors. Heat shock protein (HSP) genes belong to different families in multiple isoforms and other stress-responsive genes are among the chaperone genes whose expression is stimulated by these HSFs. The plant's capacity to tolerate high temperatures is enhanced by this coordinated activation, which also gives improved resistance to high-temperature stress (Haider et al. 2021). Heat stress evokes the production of ROS and gathers it in different cellular locations within plant cells. According to Babbar et al. 2021, these locations include the nucleus, plasma membrane, mitochondria, cytosol, and chloroplasts. On the other hand, low-temperature chilling events cause changes in solubility, oxygen levels, and electron leakage within the electron transport chains, resulting in increased ROS production, affecting membrane fluidity, and enzyme function (Yu et al. 2020).

3.2.2 Water Stress

The evolving climatic conditions over recent decades have markedly influenced rainfall patterns, leading to erratic precipitation characterized by modified intensity and seasonal fluctuations (Feng et al. 2013). These alterations have spurred instances of both severe droughts and intense flooding in various regions across the world.

3.2.2.1 Drought

Drought stress can be experienced by plants when the soil and atmospheric humidity are low and the ambient air temperature is high. Droughts are the result of prolonged periods of inappropriate rainfall and can be severe or intermittent based on the continuity period. Drought leads to water scarcity causing stunted plant growth and decreased agricultural yield (Shao et al. 2008). Severe drought can be lethal as the plants will be completely devoid of water for their normal functioning. Water deficiency stress results in lowered water potential and turgor within plants, disrupting normal cellular functions. The physiological impact of water scarcity depends on its severity and duration. Under such stress, CO_2 fixation slows, leading to increased photorespiration, accelerated O_2 photoreduction in chloroplasts, and reduced stomatal opening (Agurla et al. 2018), leading to a decrease in photosystem activation, limited ATP synthesis, ROS accumulation, and oxidative damage in plants.

3.2.2.2 Waterlogging and Flooding

This stress results from excessive water accumulation in the soil due to prolonged heavy precipitation or inadequate drainage. The key characteristics of waterlogging include low light levels, hindered gaseous exchange, hypoxia, and anoxia around plant roots (Fukao et al. 2019). Anoxic conditions impede chloroplast and mitochondrial electron transport chains, resulting in ROS production (Sasidharan et al. 2018). Exposure to waterlogging in sesame exhibited an increased concentration of ROS and methylglyoxal and an elevated level of lipid peroxidation, indicating oxidative stress (Anee et al. 2019). When plants are flooded with clear water, the light exposure leads to photorespiration and generates peroxisomal H_2O_2 (Sasidharan et al. 2018). Furthermore, flooding alters soil pH, depletes essential nutrients, accumulates salts, and increases heavy metal availability, leading to nutritional deficiencies and additional stressors like salinity and heavy metal toxicity in plants (Steffens 2014).

3.2.3 Salinity

Salinity stress or salt stress is a condition in which the concentrations of sodium (Na^+) and chloride (Cl^-) ions in soil lead to an electrical conductivity value of 4 dS/ m^{-1} or more that severely hampers plant growth and yield. A high level of salt concentration causes both hypo-ionic and hyper-osmotic stress means the high concentration of Na^+ and Cl^- decreases the osmotic potential and prevents the uptake of essential ions like K^+ and Ca^{2+}. The generation of oxidative stress due to ROS accumulation under salinity has been reported extensively. To counteract this stress, plants have developed intricate mechanisms aimed at mitigating salt-induced damage. These mechanisms involve increasing Na^+ efflux, restricting Na^+ uptake, and sequestering Na^+ in vacuoles (Fu and Yang 2023). In response, plants employ multiple strategies to counteract this oxidative stress. The activation of the antioxidant system was employed to fight against oxidative stress.

3.3 Types of ROS

Oxygen in its natural state as triplet oxygen ($^3O_2$) possesses two unpaired electrons with parallel spins, which typically reduces its reactivity. However, various processes such as ionizing irradiation, ultraviolet-B light, electron transport chains (ETCs), and biochemical reactions, supply the necessary energy for oxygen to overcome its spin restriction, transforming it into ROS. Each type of ROS exhibits distinct chemical reactivity traits that are influenced by their lifespans and the distance they migrate from their cellular production sites (Waszczak et al. 2018).

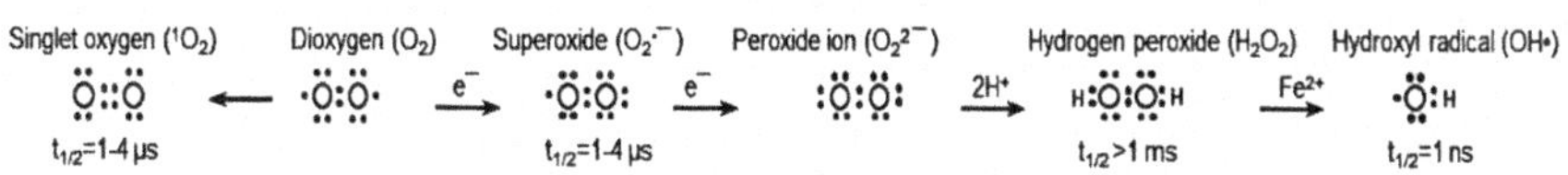

3.3.1 Singlet Oxygen

Singlet oxygen ($^1O_2$) arises as the first excited electronic state of molecular oxygen, resulting from the inversion of the spin of valence electrons in $^3O_2$. Unlike free radicals, $^1O_2$ doesn't carry unpaired electrons or a high-energy electron. Within photosynthetic membranes, interactions between triplet-state chlorophyll (chl) molecules and molecular oxygen form $^1O_2$ (Mittler 2002). The estimated amount of $^1O_2$ under ideal conditions is roughly ~10^{-13} M, and even minimal levels (~10^{-8} M) can cause damage to cell membranes (Kaur and Anastasio 2017). The concentrations ranging from ~10^{-5} to 10^{-4} M can result in cell death. Within living cells, the diffusion of $^1O_2$ is highly restricted to a limited distance (approximately 10–250 nm, from its production site) as they are short-living (about 4 μm) (Szechyńska-Hebda et al. 2022). Conditions such as light stress and lipoxygenase activity can enhance $^1O_2$ production (Dogra et al. 2018). Quenching of $^1O_2$ can occur through detoxification processes involving the non-enzymatic class of antioxidants such as β-carotene. Additionally, potential scavengers include plastoquinone and the D1 protein of photosystem II (PSII) and involved in the scavenging of $^1O_2$.

3.3.2 Superoxide Radical

Reduction of one electron from the molecular oxygen forms $O_2{\bullet}^-$, which often serves as a predecessor for another class of ROS, namely, H_2O_2. While the superoxide radical has almost the same life span as that of singlet oxygen (approximately 4 μm), it exhibits a greater reactivity than $^1O_2$ and disperses to a distance of a few micrometres from its production site (Janků and Petřivalský. 2019). An increase in $O_2{\bullet}^-$ often results in an elevation in the H_2O_2 levels, since two $O_2{\bullet}^-$ molecules along with two protons form H_2O_2. The concentration of $O_2{\bullet}^-$ within the cell under optimal conditions will always be much lower than the concentrations of H_2O_2. The iron–sulphur (Fe–S) complexes, part of proteins are one of the primary targets of $O_2{\bullet}^-$, causes the protein dysfunction, and release of Fe^{2+}, which can subsequently react with H_2O_2. The chemical reaction known as the Fenton reaction generates a highly hazardous ROS known as hydroxyl radicle when H_2O_2 reacts with Fe^{2+}. In addition to that, $O_2{\bullet}^-$ can also react with protons and produce $HO_2{\bullet}^-$. It exhibits greater reactivity, stability, and permeability through biological membranes compared to $O_2{\bullet}^-$ itself.

3.3.3 Hydrogen Peroxide

The non-radical ROS, H_2O_2 is a class of molecules whose concentration is tightly regulated intercellularly (10^{-9}–10^{-5} M) (Szechyńska-Hebda et al. 2022). Under certain conditions such as many abiotic stresses and external application of abscisic acid (maize seedling leaves) (Jiang and Zhang 2001) lead to the accumulation of H_2O_2. In plants, H_2O_2 is the mediator of many signalling pathways and relies on localized concentrations reaching approximately ~10^{-4} M. According to its chemical nature, H_2O_2 functions as a weak acid, exhibiting high stability (with a lifetime exceeding 1 ms) and diffusibility (spreading over distances beyond 1 μm, a typical range between organelles serving as relay stations). The major transporters involved in the movement of H_2O_2 are the tonoplast intrinsic protein (TIP) family and certain classes of aquaporins on the plasma membrane. Notably, while certain aquaporins facilitate H_2O_2 transport, the specific mechanisms governing this selectivity remain unclear (Nordzieke and Medraño-Fernandez 2018).

3.3.4 Hydroxyl Radical

The single unpaired electron present in the hydroxyl radical (•OH) is making them highly reactive compared to other ROS. This high reactivity leads to structural changes in proteins, lipid peroxidation, and membrane damage. Typically, •OH is present at incredibly low levels, estimated around ~10^{-15}–10^{-16} M, ensuring that the biological system is not being affected by these molecules. The deleterious effects of •OH signalling need concentration to reach 10^{-4} M (Kaur and Anastasio 2017; Szechyńska-Hebda et al. 2022). With an incredibly brief lifespan (approximately 1 ns) and limited diffusion span of around ~10 nm, •OH primarily exerts its effects at the site of its generation, lacking the capacity to diffuse beyond the cell or engage in cell-to-cell signalling (Hasanuzzaman et al. 2020). As per the present information, despite its destructive nature, •OH might possess roles beyond being solely detrimental. Its oxidative properties have been associated with many important processes in plant life including seed germination, reproduction, and adaptation to both biotic and abiotic stress conditions (Richards et al. 2015).

3.4 Cellular Defence Mechanism Against ROS: The Antioxidant System

In order to protect themselves against the toxic effects of ROS and to maintain its level to an optimum, plants have evolved two major antioxidant systems (enzymatic and non-enzymatic), which maintain an equilibrium between the generation ROS and removal of ROS from the system. The enzymatic antioxidant system consists of enzyme catalysed quenching of ROS in various steps which include the conversion of ROS into H_2O_2 and then to H_2O requiring presence of metallic co-factors. Superoxide dismutase (SOD), dehydroascorbate reductase (DHAR), catalases (CAT), ascorbate peroxidases (APX), monodehydroascorbate reductase (MDHAR), glutathione peroxidases (GPX), glutathione reductase (GR), etc. are some of the common enzymatic antioxidants recruited for ROS quenching. On the other hand, disrupting a free radical chain reaction is done by the antioxidants of the non-enzymatic class. During stress, many non-enzymatic antioxidants are involved in ROS scavenging, and some of them also works as osmoprotectant maintaining ionic balance in stresses which disrupts intracellular ionic balance like drought and salinity. Proline (Pro), glutathione (GSH), ascorbic acid (AsA), alpha tocopherol, carotenoids, flavonoids, phenolics, etc. are some of the common non-enzymatic antioxidants counteracting ROS (Table 3.2) (Sachdev et al. 2021).

Table 3.2 Antioxidant activity in response to abiotic stress in various plants

Abiotic stress(es)	Plants exposed	Antioxidant(s) activity	Reference
Chilling stress	*Cucumis sativus*(cucumber)	• The activity of superoxide dismutase (SOD), ascorbate peroxidase (APX), glutathione reductase (GR), and guaiacol peroxidase (GP) increased in leaves • Conversely, catalase (CAT) activity decreased in leaves	Lee and Lee (2000)
Drought	*Triticum aestivum* (wheat)	• Increased expression of ascorbate peroxidase (APX) and a well-maintained redox pool of ascorbic acid (AsA) and glutathione (GSH) strengthened the photosynthetic apparatus and mitochondria in acclimatized plants	Selote and Khanna-Chopra (2006)
Heavy metal (cu) stress	*Oryza sativa* (Rice)	• The activity of superoxide dismutase (SOD), guaiacol peroxidase (GP), ascorbate peroxidase (APX), glutathione reductase (GR), ascorbic acid (AsA), and glutathione (GSH) along with proline increased • Catalase (CAT) activity remained unchanged • Hydrogen peroxide (H_2O_2) levels and lipid peroxidation decreased	Thounaojam et al. (2012)
High-temperature stress	*Triticum aestivum* (wheat)	• The activity of superoxide dismutase (SOD), ascorbate peroxidase (APX), catalase (CAT), peroxidase (POX), and glutathione reductase (GR) increased in the tolerant genotype C306	Almeselmani et al. (2009)
High-temperature stress	*Spinacia oleracea* (spinach) seedling	• The overexpression of the gene encoding cytosolic heat shock 70 protein (SoHSC70) led to increased activity of superoxide dismutase (SOD), peroxidase (POX), catalase (CAT), and ascorbate peroxidase (APX) enzymes • This resulted in a decrease in oxidative membrane damage and a reduction in reactive oxygen species (ROS) accumulation	Qi et al. (2019)
Metalloid (boron) stress	*Artemisia annua*	• The activity of superoxide dismutase (SOD), peroxidase (POX), and catalase (CAT) was elevated	Aftab et al. (2010)
Salinity stress	*Oryza sativa* (Rice) seedling	• The application of exogenous manganese to stress-exposed seedlings resulted in an increase in both non-enzymatic antioxidants (phenolic compounds, flavonoids, and ascorbic acid—AsA) and enzymatic antioxidants (MDHAR, DHAR, superoxide dismutase—SOD, and catalase—CAT) content. This led to a reduction in the level of reactive oxygen species (ROS)	Rahman et al. (2016)

UV-B radiation	*Helianthus annuus* (sunflower) cotyledons	• The activity of catalase (CAT), glutathione dehydrogenase, guaiacol peroxidase (GP), and the ratio of reduced glutathione (GSH) to oxidized glutathione (GSSG) increased • The ratio of ascorbic acid (AsA) to dehydroascorbic acid (DHA), ascorbate peroxidase (APX), and glutathione reductase (GR) activity remained unchanged • Lipid peroxidation and oxidative damage in cotyledons decreased	Costa et al. (2002)
Low temperature + herbicide (isoproturon) stress	*Triticum aestivum* (wheat) seedling	• The foliar application of ascorbic acid (AsA) resulted in an increased activity of antioxidants such as superoxide dismutase (SOD), catalase (CAT), and peroxidase (POX) • The content of malondialdehyde (MDA) and the rate of reactive oxygen species (ROS) production declined	Wang et al. (2018)
Salinity + herbicide (2,4 dichlorophenoxyacetic acid) stress	*Oryza sativa* (Rice)	• Levels of enzymatic antioxidants (superoxide dismutase—SOD, catalase—CAT, ascorbate peroxidase—APX, and peroxidase—POX) and non-enzymatic antioxidants (phenolic compounds, total soluble phenols, proline, and sugars) were modulated • Content of hydrogen peroxide (H_2O_2) and superoxide anion ($O_2{\bullet}^-$) decreased, alleviating oxidative stress and lipid peroxidation	Islam et al. (2016)

3.4.1 Non-enzymatic Antioxidants

3.4.1.1 Ascorbic Acid (AsA)

Ascorbic acid or vitamin C shows antioxidant properties due to its ability to spare electrons for a wide array of chemical reactions (both enzymatic and non-enzymatic). Plants maintain a constant level of AsA concentration but the environmental stress conditions increase their concentration as per the severity and extent of stress. AsA is distributed across various subcellular compartments such as vacuoles, peroxisomes, chloroplasts, cytosol, apoplast, mitochondria, endoplasmic reticulum (ER), and nucleus. In chloroplasts particularly, ascorbic acid exists as the most predominant antioxidant under stress conditions (Dumanović et al. 2021). There are four major enzymes (APX, GR, DHAR, and MDHAR) involved in the ROS detoxification via AsA-GSH cycle. It operates across distinct cellular components including cytosol, chloroplast, peroxisomes, and mitochondria (Singh et al. 2022). The sustenance of AsA-GSH cycle is dependent heavily on NADPH, which is a reducing agent reducing agent present in the cell. AsA plays some additional roles in the plant cell other than their routine ROS detoxification activities. They react with $O_2{\bullet}^-$and H_2O_2, directly, and safeguarded the membrane integrity. AsA helps in maintaining the activities of those enzymes containing transition metal ions as prosthetic groups (Noctor and Foyer 1998).

3.4.1.2 Glutathione

The tripeptide glutathione (GSH in its reduced form) holds the distinction of being the most prevalent low-molecular-weight thiol-containing ROS scavenger within cells. It is ubiquitously distributed across various cell compartments, including the vacuoles, cytosol, peroxisomes, mitochondria, nuclei, chloroplasts, and endoplasmic reticulum. GSH has the ability to directly scavenge H_2O_2 or react nonenzymatically with $^1O_2$, $O_2{\bullet}^-$, and OH•. Furthermore, GSH along with AsA involved in the regeneration of α-tocopherol, a non-enzymatic antioxidant (Hasanuzzaman et al. 2020). Glutathionylation is a process of adding glutathione on DNA, proteins, and lipids, thereby protecting their integrity and regulating their function. Through a thiol–disulfide linkage, the reduced form of GSH, with its Cys thiol group, donates a reducing equivalent to ROS and proteins. This donation leads to the production of a radical form of GSH upon proton removal. Subsequently, the oxidized form of GSH (GSSG) is generated following the reaction between two GSH radicals (Singh et al. 2022). However, the primary function of GSH as an ROS scavenger primarily lies in its role in recycling the hydrophilic antioxidant ascorbic acid through the AsA-GSH cycle. Through a reaction catalysed by DHAR, AsA (oxidised) will be converted back to AsA (reduced) and GSH is actively involved in this process.

3.4.1.3 Carotenoids

Chemically, carotenoids are those molecules that poses polyene backbone in their structural composition. These compounds (like β-carotene, xanthophyll, lycopene, and zeaxanthin) serve as light-harvesting pigments that play a significant role in alleviating oxidative stress triggered by intense light exposure. They achieve this by intercepting excess energy and dispersing in the form of heat. They also act as protective agents, preventing the over-excitation of Photosystem II (PSII) by scavenging 1Chl*, 3Chl*, and $^1O_2$ (Sachdev et al. 2021). Furthermore, they can also interact with lipid peroxyl radical (LOO•) and neutralize them, thereby mitigating oxidative damage. AsA and tocopherol in a chain of reactions regenerate the lipid hydroperoxide and carotenoid radicals through the reduction steps (Yachandra et al. 1996).

3.4.1.4 Flavonoids

Plants use a defence mechanism known as the heat stress response to protect themselves from the heat-induced ROS generation by employing low-molecular-weight substances called flavonoids having hydroxyl groups, which highlights their antioxidant potential (Liu et al. 2014). They protect the cell membranes and prevent the damage of plant cells by scavenging the ROS, which may cause potential harm to the cell membrane through lipid peroxidation (Løvdal et al. 2010). Both AsA and peroxidases work in hand with flavonoids to scavenge $^1O_2$ generated in chloroplasts and H_2O_2 in the vacuole along with AsA and peroxidases in high light (Singh et al. 2022). In times of high temperatures, droughts, nutritional shortages, and UV radiation, flavonoids also work as antioxidants.

3.4.2 Enzymatic Antioxidants

3.4.2.1 Superoxide Dismutase (SOD: EC 1.15.1.1)

SOD is the first line of defence among the enzymatic antioxidants as it scavenge $O_2{\bullet}^-$, in a chain reaction that starts when the level of ROS in the cell rises, SOD breaks down into molecular oxygen and hydrogen peroxide ($2\ O_2{\bullet}^- + 2H^+ \rightarrow O_2 + H_2O_2$). The initial molecule formed during the reduction of monovalent oxygen (superoxide radical) is scavenged by superoxide dismutase (Dumanović et al. 2021). SOD is a metalloenzyme (metalloprotein) found in the plant genome within a small gene family differed based on the protein's affinity for the metal (copper (Cu), zinc (Zn), manganese (Mn), iron (Fe), and nickel (Ni)). The cellular localization of SOD varies and show distinct pattern of distribution across various organelles present in the cell. For example, Cu/Zn-SODs exhibit a localization preference within the apoplast, chloroplasts, mitochondria, nucleus, peroxisomes, and cytosol. Conversely, Fe-SOD resides in chloroplasts, mitochondria and peroxisomes. Meanwhile, Mn-SOD is primarily localized within peroxisomes and mitochondria (Sachdev et al. 2021). Indeed, multiple studies have consistently highlighted a significant surge in superoxide dismutase activity in plants when they face diverse abiotic stresses, including extreme light exposure, water deficit conditions, and metal toxicity. The increased SOD activity in plants frequently aligns with enhanced tolerance to environmental stressors (Szechyńska-Hebda et al. 2022).

3.4.2.2 Catalase (CAT, EC 1.11.1.6)

One of the most productive enzymes made by all living things while under stress is **catalase**. Hydrogen peroxide is directly impacted by this enzyme, which lessens the toxic effects. Catalase works on H_2O_2 molecules and reduces them to H_2O and O_2. The fact that catalase does not need a reducer to catalyse the reaction sets it apart from other enzymes (Sharma et al. 2012). The cellular localization of catalase is confined to the peroxisome, chloroplast, cytosol, and mitochondria. When the level of ROS rises under stressful situations, catalase takes part in scavenging. Only at high H_2O_2 concentrations catalase scavenge the H_2O_2 molecule without the use of energy due to its higher K_m for H_2O_2; at lower H_2O_2 concentrations, other peroxidases work with strong reducing agents like glutathione and ascorbate to do so (Palma et al. 2020). Catalase is highly expressed, particularly in all varieties of plant cells, and as a result, it is a crucial member of the plant's antioxidant system (Sofo et al. 2015).

3.4.2.3 Peroxidase (POD, EC 1.11.1.X)

In response to biotic and abiotic stressors, plants create a wide set of antioxidant enzymes known as **peroxidases** (You and Chan 2015). Peroxidases (POXs) play a crucial role as a secondary defence mechanism by eliminating excessive H_2O_2, aiding plants in maintaining ROS homeostasis. As the peroxidase enzyme's activity rises, ROS damage to the plant is reduced because it stops the generation of ROS by reducing H_2O_2 in the cell. In other words, peroxidase can counteract the negative effects of H_2O_2 by turning it into water and oxygen (Das and Roychoudhury 2014). Plant peroxidase's regular activity cycles can catalyse the reduction of H_2O_2. This set of enzymes is present in the cytosol, vacuole, apoplast, and chloroplasts. There are 3 classes of peroxidases present in plant cells. The class I peroxidases encompass cytochrome POX (cPOX), APX, and CAT-POX. Mn-dependent POX and ligninase represent class II peroxidases (Hiraga et al. 2001). The last class of peroxidase (Class III peroxidases (POX, EC 1.11.1.7)) operates at the extracellular space to scavenge the H_2O_2 and is the sole type of POX present in extracellular region (Rajput et al. 2021).

(a) **Ascorbate peroxidase (APX, EC 1.11.1.11)** is an enzyme categorized as a class I peroxidase with specificity for ascorbic acid. Its catalytic activity involves the reaction: 2 L-ascorbate + H_2O_2 + 2H+ → 2 MDHA +2 H_2O. This enzyme plays a dual role as both a scavenger of H_2O_2 and a sensor of redox changes within plant cells. It holds a widespread presence in various cellular components, and be named as the sAPX (in stroma) and tAPX (in thylakoids), mAPX (in microbodies such as glyoxysomes and peroxisomes), cAPX (in cytosol), and mitAPX (in mitochondria) (Dumanović et al. 2021). Ascorbate serves as the electron donor for H_2O_2 reduction through APX, leading to the production of monodehydroascorbate (MDHA) within the Foyer–Halliwell–AsA (AsA-GSH cycle).

(b) **Glutathione peroxidases (GPXs, EC 1.11.1.9)** are class of peroxidase enzymes within the antioxidant system, leveraging a highly reactive thiol group. They rely on reduced glutathione (GSH) and thioredoxins (TRXs) to scavenge H_2O_2 (Bela et al. 2015). Thus, GPXs play a crucial role within the overall antioxidant defence mechanisms.

3.4.2.4 Glutathione Reductase and Glutathione-S-Transferase

Glutathione reductase (GR; EC 1.6.4.2) play pivotal roles as the primary component within the ascorbate-glutathione (AsA-GSH) pathway, an essential part of the antioxidant defence system in plants. By using NADPH as an electron donor, GR catalyses the reaction of oxidised glutathione to its reduced form (glutathione (GSH)). Maintaining the requisite GSH/GSSG ratio is critical for an effective neutralisation of H_2O_2. Although the majority of glutathione reductase activity has been investigated in chloroplasts, plant cells' cytosolic isoenzymes also describe the mitochondrial forms of this enzyme. Glutathione reductase glyoxalase is also known to convert oxidized glutathione into reduced glutathione and aid in maintaining the reduced glutathione: oxidized glutathione ratio in plant cells (Gill et al. 2013).

Another crucial enzyme, **glutathione-s-transferase**, scavenges potentially harmful substances and shields plant cells. It has been discovered that glutathione-s-transferase catalyses the binding of glutathione to several harmful substances in cells. Additionally, it has been demonstrated that glutathione-s-transferases are crucial for defending plants against a variety of biotic and abiotic challenges, such as xenobiotic toxins, UV light, and photo-oxidation stress. The glutathione-s-transferase enzyme also binds the products of oxidative damage to the reduced glutathione and scavenges them. In order to sustain glutathione self-preservation, glutathione-S-transferase is regarded as a crucial enzyme (Singh et al. 2022; Rajput et al. 2021).

3.4.2.5 Water-Water Cycle (Mehler Cycle)

It starts with water and ends with water, as its name implies. This cycle only occurs in chloroplasts. Instead of transferring an electron from ferredoxin to NADP+ during this cycle, photosystem I transfers the electron to oxygen, converting it to superoxide (Exposito-Rodriguez et al. 2017). The superoxide dismutase enzyme immediately changes the superoxide radical into hydrogen peroxide. The ascorbate peroxidase enzymes totally break down the resultant hydrogen peroxide, converting it to water. The spread of active oxygen species is stopped by the continuous nature of these reactions. As a result, the harm that the ROS causes will be lessened (Slesak et al. 2007).

3.5 Oxidative Stress in Cellular Compartments Caused by Abiotic Stress

Oxidative stress is a unique and complex event that causes an imbalance in cellular redox equilibrium. Antioxidants' activity decreases under stressful circumstances, assisting ROS generation at an uncompensated level that causes oxidative burst and damage (Apel and Hirt 2004). Depending on the active enzymatic and nonenzymatic pathways, a certain compartment controls and localizes the formation of a particular ROS in a cell (Janků and Petřivalský 2019; Podgórska et al. 2017). Three different pathways primarily involved in the formation of ROS in plant cells are the electron transport chain (ETC), apoplast POXs, and plasma membrane-localized

NADPH oxidases (Podgórska et al. 2017). The two main processes, ROS generation and oxidative burst or retrograde signalling which normally evoked in plant cells under extreme environmental conditions. ETC, chlorophyll, proteins, and enzymes are all impacted by ROS that are produced in various organelles. However, further harm can be avoided by activating a mechanism in cell organelles that reduces ROS at the moment of initial generation. Tolerance strategies against stress by a plant are initiated once the ROS are channelled through signalling pathways which prevent oxidative damage and give protection to the plant against particular stress.

3.5.1 Chloroplast

The chloroplast, an immensely significant organelle found in plant cells, produces the energy necessary to sustain life on earth. It is vulnerable to harsh environments and is a key location for ROS production. This lowers photosynthetic efficiency, which results in declining growth and productivity. Investigating mechanisms that control excess ROS production and photosynthetic activity may stop negative outcomes. Unfavourable environmental factors cause stomatal conductance to decrease, CO_2 absorption to drop, and/or excited triplet chlorophyll (3Chl*) to form, leading to the disruption of the ETC, causing an excess of ROS to be produced, leading to photo-oxidation. The high energy intermediates, reductants, and O_2 near the reaction centre of PS I and II are the main causes of ROS production there (Asada 2006). The singlet chlorophyll (1Chl*) produced from the light-harvesting complexes (LHC) upon illumination by a high beam of light is a long-lived molecule that takes part in the charge separation process and transforms excitation energy into electrochemical energy. When the LHC of PS II receives a high energy excitation which is much above its utilization threshold limit the 3Chl* will be produced. When 3Chl* and O_2 interact, $^1O_2$ is produced, which is quite oxidizing. In addition to the extra light, other pressures like dryness cause an imbalance between the light's capture and use, which leads to the formation of $^1O_2$ (Pospíšil 2016). Depending on the severity of the stress a reduction of the PS II components, plastoquinone QA and QB prevent the charge distribution between P680 and pheophytin limiting the availability of CO_2 to Calvin's cycle. At the reaction centre of PS II, this process causes the generation of triplet chlorophyll (3P680), which ultimately results in the generation of $^1O_2$ (89). In PS I, under abiotic stress conditions, the electron leakage created by CO_2 under low concentration from the PTC reduces oxygen (O_2) via ferredoxin into $O{\bullet}^-$ (Mehler's reaction). While in PS II, $O_2{\bullet}^-$ is produced by reduction of ETC leading as a result of electron leakage from plastoquinone QA and QB to O_2. $O_2{\bullet}^-$ is also created when highly charged chlorophyll pigments excite oxygen (Cruz De Carvalho 2008). The Cu/Zn-SOD, then convert $O_2{\bullet}^-$, either spontaneously or by dismutation. The H_2O_2 produced has the potential to be a photo-inhibitor (at 10 μM concentration, 50% reduction in CO_2 fixation) since it can oxidize cysteine or methionine residues and thiol-modified Calvin's cycle enzymes (Foyer and Shigeoka 2011). The thylakoid membrane is overexcited as a result of the cooling stress, which results in photo-inhibition leading to the suppression of some of the photosynthetic components (Jung and Steffen 1997). Extreme salinity conditions damage the ultrastructure of chloroplast including the breakdown of the chloroplast membrane, bulging and disorganisation of the thylakoid, due to ROS: H_2O_2 and $O_2{\bullet}^-$ (Shu et al. 2013; Yamane et al. 2012). Exposure of plants to metals like cadmium, nickel, zinc, copper, and chromium inhibit photosystems' Calvin cycle enzymes, disrupts ETC, and produce ROS (Dietz et al. 1999). Exposure of chloroplast of pea plants to chromium [Cr (VI)] produced more $O_2{\bullet}^-$, H_2O_2, and •OH (Pandey et al. 2009).

3.5.2 Peroxisomes

Another important location for intracellular H_2O_2 synthesis is the peroxisome (Nyathi and Baker 2006). In addition, they control a number of crucial cellular processes, including the high oxidative metabolic pathways found in the majority of eukaryotic cells (Del Río and López-Huertas 2016). Along with H_2O_2, peroxisomes' matrix and/or membranes also create $O_2{\bullet}^-$, which is then released into the cytosol (Ortega-Galisteo et al. 2012). High salt levels increase photorespiration and H_2O_2 levels in peroxisomes (Yamane et al. 2012). Ascorbic acid and glutathione activity were shown to be decreased and lipid peroxidation to be elevated in tomato plants under salt stress, according to research by (Mittova et al. 2003). The formation of ROS by the activity of peroxisomal GOX is likely the source of the oxidative damage brought on by salt stress. Photorespiration is thought to produce more than 70% of the H_2O_2 produced during drought conditions (Voss et al. 2013). Additionally, the generation of H_2O_2 in peroxisomes is triggered by fatty acid oxidation, an imbalance in $O_2{\bullet}^-$ concentration, and by the actions of enzymes including flavin oxidases, UO, and XOD (Corpas et al. 2019; Del Río and López-Huertas 2016; Nyathi and Baker 2006; Ortega-Galisteo et al. 2012).

Chloroplasts release fatty acids in response to an abiotic stress caused by extended darkness, which is then processed by peroxisomal -oxidation (Su et al. 2019). In pea leaves, Cd and 2,4-dichlorophenoxyacetic acid (2,4-D) increased the formation of H_2O_2 (Ortega-Galisteo et al. 2012). Cd raises the level of H_2O_2 (GOX activity found to be enhanced) and lowers the activity of CAT, whereas 2,4-D increases the activities of ACX (−oxidation of fatty acids) and XOD. The number of peroxisomes per cell in plants is also increased under such abiotic stresses as salt, xenobiotic substances, heavy metals, and intense light (Del Río and López-Huertas 2016).

3.5.3 Mitochondria

Mitochondria is one of the major locations of synthesis of ROS: •OH, H_2O_2, and $O_2{\bullet}^-$, in plants. Under abiotic stress, photorespiration and mitochondrial ETC (mtETC) promote the generation of ROS. Two pathways—the cytochrome oxidase (COX) pathway which supports the synthesis with ATP and the alternative oxidase (AOX) pathway without ATP synthesis—perform the respiratory ETC which functions in the inner mitochondrial membranes (Keunen et al. 2011). When breathing is restricted under stress, mtETC is excessively reduced, which promotes the generation of $O_2^{\bullet -}$ due to electron leakage. When the mtETC is overloaded with electrons (the number of electrons entering exceeds its capacity) leads to the overreduction of the UQ pool, driving the production of ROS. It promotes the leakage of an electron to O_2 by creating ubisemiquinone radicals, which have high reducing properties, and reduce O_2 to $O_2^{\bullet -}$ (Rhoads et al. 2006).

3.6 Conclusions

Changing environmental conditions mainly due to anthropogenic activities put forth a big challenge to plant productivity, which is highly needed to feed an increasing population. The plants face more severe climatic conditions at present compared to a century ago and will become more serious in the future. ROS are the net result of these extreme environments that can hamper normal plant life. Studies conducted for many decades in the past have revealed the major classes of ROS produced in plant cells under extreme conditions and the mechanisms employed by the plant cells to counteract the production of ROS. The highly intricate balance maintained in the synthesis/detoxification cycle of ROS is significant in terms of its application in developing more stress-resilient plants. The impact of these molecules on the biochemical functions of cellular components and the proteins involved in the detoxification processes can be used to device certain mechanisms/strategy to develop stress-resistant plants with increased productivity.

Acknowledgements SKPK is thankful to the Head, Department of Botany Banaras Hindu University, Varanasi for providing conducive environment and infrastructure for research and Institute of Eminence Seed Grant Banaras Hindu University, Varanasi for providing financial support to the laboratory.

Conflict of Interest The authors declare no conflict of interest.

References

Aftab T, Khan MMA, Idrees M, Naeem M, Ram M (2010) Boron induced oxidative stress, antioxidant defence response and changes in artemisinin content in *Artemisia annua* L. J Agron Crop Sci 196(6):423–430. https://doi.org/10.1111/j.1439-037X.2010.00427.x

Agurla S, Gahir S, Munemasa S, Murata Y, Raghavendra AS (2018) Mechanism of stomatal closure in plants exposed to drought and cold stress. In: Iwaya-Inoue M, Sakurai M, Uemura M (eds) Survival strategies in extreme cold and desiccation, vol 1081. Springer, Singapore, pp 215–232. https://doi.org/10.1007/978-981-13-1244-1_12

Almeselmani M, Deshmukh P, Sairam R (2009) High temperature stress tolerance in wheat genotypes: role of antioxidant defence enzymes. Acta Agronomica Hungarica 57(1):1–14. https://doi.org/10.1556/AAgr.57.2009.1.1

Anee N, Rahman M, Bhuiyan A, Fujita, Hasanuzzaman (2019) Oxidative damage and antioxidant defense in Sesamum indicum after different waterlogging durations. Plan Theory 8(7):196. https://doi.org/10.3390/plants8070196

Apel K, Hirt H (2004) Reactive oxygen species: metabolism, oxidative stress, and signal transduction. Annu Rev Plant Biol 55(1):373–399. https://doi.org/10.1146/annurev.arplant.55.031903.141701

Asada K (2006) Production and scavenging of reactive oxygen species in chloroplasts and their functions. Plant Physiol 141(2):391–396. https://doi.org/10.1104/pp.106.082040

Babbar R, Karpinska B, Grover A, Foyer CH (2021) Heat-induced oxidation of the nuclei and cytosol. Front Plant Sci 11:617779. https://doi.org/10.3389/fpls.2020.617779

Bailey-Serres J, Voesenek LACJ (2008) Flooding stress: acclimations and genetic diversity. Annu Rev Plant Biol 59(1):313–339. https://doi.org/10.1146/annurev.arplant.59.032607.092752

Bela K, Horváth E, Gallé Á, Szabados L, Tari I, Csiszár J (2015) Plant glutathione peroxidases: emerging role of the antioxidant enzymes in plant development and stress responses. J Plant Physiol 176:192–201. https://doi.org/10.1016/j.jplph.2014.12.014

Ben Abdallah S, Aung B, Amyot L, Lalin I, Lachâal M, Karray-Bouraoui N, Hannoufa A (2016) Salt stress (NaCl) affects plant growth and branch pathways of carotenoid and flavonoid biosyntheses in Solanum nigrum. Acta Physiol Plant 38(3):72. https://doi.org/10.1007/s11738-016-2096-8

Bita CE, Gerats T (2013) Plant tolerance to high temperature in a changing environment: scientific fundamentals and production of heat stress-tolerant crops. Front Plant Sci 4:273. https://doi.org/10.3389/fpls.2013.00273

Bulgari R, Franzoni G, Ferrante A (2019) Biostimulants application in horticultural crops under abiotic stress conditions. Agronomy 9(6):306. https://doi.org/10.3390/agronomy9060306

Corpas FJ, Del Río LA, Palma JM (2019) Plant peroxisomes at the crossroad of NO and H_2O_2 metabolism. J Integr Plant Biol 61(7):803–816. https://doi.org/10.1111/jipb.12772

Costa H, Gallego SM, Tomaro ML (2002) Effect of UV-B radiation on antioxidant defense system in sunflower cotyledons. Plant Sci 162(6):939–945. https://doi.org/10.1016/S0168-9452(02)00051-1

Cruz De Carvalho MH (2008) Drought stress and reactive oxygen species: production, scavenging and signaling. Plant Signal Behav 3(3):156–165. https://doi.org/10.4161/psb.3.3.5536

Das K, Roychoudhury A (2014) Reactive oxygen species (ROS) and response of antioxidants as ROS-scavengers during environmental stress in plants. Front Environ Sci 2:1. https://doi.org/10.3389/fenvs.2014.00053

Del Río LA, López-Huertas E (2016) ROS generation in peroxisomes and its role in cell signaling. Plant Cell Physiol 076:pcw076. https://doi.org/10.1093/pcp/pcw076

Demirel U, Morris WL, Ducreux LJM, Yavuz C, Asim A, Tindas I, Campbell R, Morris JA, Verrall SR, Hedley PE, Gokce ZNO, Caliskan S, Aksoy E, Caliskan ME, Taylor MA, Hancock RD (2020) Physiological, biochemical, and transcriptional responses to single and combined abiotic stress in stress-tolerant and stress-sensitive potato genotypes. Front Plant Sci 11:169. https://doi.org/10.3389/fpls.2020.00169

Dietz K-J, Baier M, Krämer U (1999) Free radicals and reactive oxygen species as mediators of heavy metal toxicity in plants. In: Prasad MNV, Hagemeyer J (eds) Heavy metal stress in plants. Springer, Berlin, Heidelberg, pp 73–97. https://doi.org/10.1007/978-3-662-07745-0_4

Dogra V, Rochaix J, Kim C (2018) Singlet oxygen-triggered chloroplast-to-nucleus retrograde signalling pathways: an emerging perspective. Plant Cell Environ 41(8):1727–1738. https://doi.org/10.1111/pce.13332

Dumanović J, Nepovimova E, Natić M, Kuča K, Jaćević V (2021) The significance of reactive oxygen species and antioxidant defense system in plants: a concise overview. Front Plant Sci 11:552969. https://doi.org/10.3389/fpls.2020.552969

Esim N, Atici O (2014) Nitric oxide improves chilling tolerance of maize by affecting apoplastic antioxidative enzymes in leaves. Plant Growth Regul 72(1):29–38. https://doi.org/10.1007/s10725-013-9833-4

Exposito-Rodriguez M, Laissue PP, Yvon-Durocher G, Smirnoff N, Mullineaux PM (2017) Photosynthesis-dependent H_2O_2 transfer from chloroplasts to nuclei provides a high-light signalling mechanism. Nat Commun 8(1):49. https://doi.org/10.1038/s41467-017-00074-w

Feng X, Porporato A, Rodriguez-Iturbe I (2013) Changes in rainfall seasonality in the tropics. Nat Clim Chang 3(9):811–815. https://doi.org/10.1038/nclimate1907

Foyer CH, Shigeoka S (2011) Understanding oxidative stress and antioxidant functions to enhance photosynthesis. Plant Physiol 155(1):93–100. https://doi.org/10.1104/pp.110.166181

Fu H, Yang Y (2023) How plants tolerate salt stress. Curr Issues Mol Biol 45(7):5914–5934. https://doi.org/10.3390/cimb45070374

Fukao T, Barrera-Figueroa BE, Juntawong P, Peña-Castro JM (2019) Submergence and waterlogging stress in plants: a review highlighting research opportunities and understudied aspects. Front Plant Sci 10:340. https://doi.org/10.3389/fpls.2019.00340

Gechev T, Petrov V (2020) Reactive oxygen species and abiotic stress in plants. Int J Mol Sci 21(20):7433. https://doi.org/10.3390/ijms21207433

Gill SS, Anjum NA, Hasanuzzaman M, Gill R, Trivedi DK, Ahmad I, Pereira E, Tuteja N (2013) Glutathione and glutathione reductase: a boon in disguise for plant abiotic stress defense operations. Plant Physiol Biochem 70:204–212. https://doi.org/10.1016/j.plaphy.2013.05.032

Golldack D, Li C, Mohan H, Probst N (2014) Tolerance to drought and salt stress in plants: unraveling the signaling networks. Front Plant Sci 5:1. https://doi.org/10.3389/fpls.2014.00151

Haider S, Iqbal J, Naseer S, Yaseen T, Shaukat M, Bibi H, Ahmad Y, Daud H, Abbasi NL, Mahmood T (2021) Molecular mechanisms of plant tolerance to heat stress: current landscape and future perspectives. Plant Cell Rep 40(12):2247–2271. https://doi.org/10.1007/s00299-021-02696-3

Hasanuzzaman M, Bhuyan MHM, Zulfiqar F, Raza A, Mohsin S, Mahmud J, Fujita M, Fotopoulos V (2020) Reactive oxygen species and antioxidant defense in plants under abiotic stress: revisiting the crucial role of a universal defense regulator. Antioxidants 9(8):681. https://doi.org/10.3390/antiox9080681

He M, He C-Q, Ding N-Z (2018) Abiotic stresses: general defenses of land plants and chances for engineering multistress tolerance. Front Plant Sci 9:1771. https://doi.org/10.3389/fpls.2018.01771

Hiraga S, Sasaki K, Ito H, Ohashi Y, Matsui H (2001) A large family of class III plant peroxidases. Plant Cell Physiol 42(5):462–468. https://doi.org/10.1093/pcp/pce061

Islam F, Ali B, Wang J, Farooq MA, Gill RA, Ali S, Wang D, Zhou W (2016) Combined herbicide and saline stress differentially modulates hormonal regulation and antioxidant defense system in Oryza sativa cultivars. Plant Physiol Biochem 107:82–95. https://doi.org/10.1016/j.plaphy.2016.05.027

Jaleel CA, Manivannan P, Kishorekumar A, Sankar B, Gopi R, Somasundaram R, Panneerselvam R (2007) Alterations in osmoregulation, antioxidant enzymes and indole alkaloid levels in Catharanthus roseus exposed to water deficit. Colloids Surf B: Biointerfaces 59(2):150–157. https://doi.org/10.1016/j.colsurfb.2007.05.001

Janků L, Petřivalský. (2019) On the origin and fate of reactive oxygen species in plant cell compartments. Antioxidants 8(4):105. https://doi.org/10.3390/antiox8040105

Jiang M, Zhang J (2001) Effect of abscisic acid on active oxygen species, antioxidative defence system and oxidative damage in leaves of maize seedlings. Plant Cell Physiol 42(11):1265–1273. https://doi.org/10.1093/pcp/pce162

Jung S, Steffen KL (1997) Influence of photosynthetic photon flux densities before and during long-term chilling on xanthophyll cycle and chlorophyll fluorescence quenching in leaves of tomato (*Lycopersicon hirsutum*). Physiol Plant 100(4):958–966. https://doi.org/10.1111/j.1399-3054.1997.tb00023.x

Kaur R, Anastasio C (2017) Light absorption and the photoformation of hydroxyl radical and singlet oxygen in fog waters. Atmos Environ 164:387–397. https://doi.org/10.1016/j.atmosenv.2017.06.006

Keunen E, Remans T, Bohler S, Vangronsveld J, Cuypers A (2011) Metal-induced oxidative stress and plant mitochondria. Int J Mol Sci 12(10):6894–6918. https://doi.org/10.3390/ijms12106894

Khan MIR, Fatma M, Per TS, Anjum NA, Khan NA (2015) Salicylic acid-induced abiotic stress tolerance and underlying mechanisms in plants. Front Plant Sci 6:1. https://doi.org/10.3389/fpls.2015.00462

Lee DH, Lee CB (2000) Chilling stress-induced changes of antioxidant enzymes in the leaves of cucumber: in gel enzyme activity assays. Plant Sci 159(1):75–85. https://doi.org/10.1016/S0168-9452(00)00326-5

Liu S, Ju J, Xia G (2014) Identification of the flavonoid 3′-hydroxylase and flavonoid 3′,5′-hydroxylase genes from antarctic moss and their regulation during abiotic stress. Gene 543(1):145–152. https://doi.org/10.1016/j.gene.2014.03.026

Løvdal T, Olsen KM, Slimestad R, Verheul M, Lillo C (2010) Synergetic effects of nitrogen depletion, temperature, and light on the content of phenolic compounds and gene expression in leaves of tomato. Phytochemistry 71(5–6):605–613. https://doi.org/10.1016/j.phytochem.2009.12.014

Megha S, Basu U, Kav NNV (2018) Regulation of low temperature stress in plants by micro RNAs. Plant Cell Environ 41(1):1–15. https://doi.org/10.1111/pce.12956

Mittler R (2002) Oxidative stress, antioxidants and stress tolerance. Trends Plant Sci 7(9):405–410. https://doi.org/10.1016/S1360-1385(02)02312-9

Mittler R, Vanderauwera S, Suzuki N, Miller G, Tognetti VB, Vandepoele K, Gollery M, Shulaev V, Van Breusegem F (2011) ROS signaling: the new wave? Trends Plant Sci 16(6):300–309. https://doi.org/10.1016/j.tplants.2011.03.007

Mittova V, Tal M, Volokita M, Guy M (2003) Up-regulation of the leaf mitochondrial and peroxisomal antioxidative systems in response to salt-induced oxidative stress in the wild salt-tolerant tomato species *Lycopersicon pennellii*. Plant Cell Environ 26(6):845–856. https://doi.org/10.1046/j.1365-3040.2003.01016.x

Nievola CC, Carvalho CP, Carvalho V, Rodrigues E (2017) Rapid responses of plants to temperature changes. Temperature 4(4):371–405. https://doi.org/10.1080/23328940.2017.1377812

Noctor G, Foyer CH (1998) Ascorbate and glutathione: keeping active oxygen under control. Annu Rev Plant Physiol Plant Mol Biol 49(1):249–279. https://doi.org/10.1146/annurev.arplant.49.1.249

Nordzieke D, Medraño-Fernandez I (2018) The plasma membrane: a platform for intra- and intercellular redox signaling. Antioxidants 7(11):168. https://doi.org/10.3390/antiox7110168

Noshi M, Hatanaka R, Tanabe N, Terai Y, Maruta T, Shigeoka S (2016) Redox regulation of ascorbate and glutathione by a chloroplastic dehydroascorbate reductase is required for high-light stress tolerance in Arabidopsis. Biosci Biotechnol Biochem 80(5):870–877. https://doi.org/10.1080/09168451.2015.1135042

Nyathi Y, Baker A (2006) Plant peroxisomes as a source of signalling molecules. Biochim Biophys Acta 1763(12):1478–1495. https://doi.org/10.1016/j.bbamcr.2006.08.031

Ortega-Galisteo AP, Rodríguez-Serrano M, Pazmiño DM, Gupta DK, Sandalio LM, Romero-Puertas MC (2012) S-Nitrosylated proteins in pea (Pisum sativum L.) leaf peroxisomes: changes under abiotic stress. J Exp Bot 63(5):2089–2103. https://doi.org/10.1093/jxb/err414

Palma JM, Mateos RM, López-Jaramillo J, Rodríguez-Ruiz M, González-Gordo S, Lechuga-Sancho AM, Corpas FJ (2020) Plant catalases as NO and H2S targets Redox Biology 34101525. https://doi.org/10.1016/j.redox.2020.101525

Pandey V, Dixit V, Shyam R (2009) Chromium effect on ROS generation and detoxification in pea (Pisum sativum) leaf chloroplasts. Protoplasma 236(1–4):85–95. https://doi.org/10.1007/s00709-009-0061-8

Petridis A, Therios I, Samouris G, Tananaki C (2012) Salinity-induced changes in phenolic compounds in leaves and roots of four olive cultivars (Olea europaea L.) and their relationship to antioxidant activity. Environ Exp Bot 79:37–43. https://doi.org/10.1016/j.envexpbot.2012.01.007

Podgórska A, Burian M, Szal B (2017) Extra-cellular but extra-ordinarily important for cells: Apoplastic reactive oxygen species metabolism. Front Plant Sci 8:1353. https://doi.org/10.3389/fpls.2017.01353

Pospíšil P (2016) Production of reactive oxygen species by photosystem II as a response to light and temperature stress. Front Plant Sci 7:1. https://doi.org/10.3389/fpls.2016.01950

Qi C, Lin X, Li S, Liu L, Wang Z, Li Y, Bai R, Xie Q, Zhang N, Ren S, Zhao B, Li X, Fan S, Guo Y-D (2019) SoHSC70 positively regulates thermotolerance by alleviating cell membrane damage, reducing ROS accumulation, and improving activities of antioxidant enzymes. Plant Sci 283:385–395. https://doi.org/10.1016/j.plantsci.2019.03.003

Rahman A, Hossain MS, Mahmud J-A, Nahar K, Hasanuzzaman M, Fujita M (2016) Manganese-induced salt stress tolerance in rice seedlings: regulation of ion homeostasis, antioxidant defense and glyoxalase systems. Physiol Mol Biol Plants 22(3):291–306. https://doi.org/10.1007/s12298-016-0371-1

Rajput VD, Harish Singh RK, Verma KK, Sharma L, Quiroz-Figueroa FR, Meena M, Gour VS, Minkina T, Sushkova S, Mandzhieva S (2021) Recent developments in enzymatic antioxidant defence mechanism in plants with special reference to abiotic stress. Biology 10(4):267. https://doi.org/10.3390/biology10040267

Raza A, Razzaq A, Mehmood S, Zou X, Zhang X, Lv Y, Xu J (2019) Impact of climate change on crops adaptation and strategies to tackle its outcome: a review. Plan Theory 8(2):34. https://doi.org/10.3390/plants8020034

Rhoads DM, Umbach AL, Subbaiah CC, Siedow JN (2006) Mitochondrial reactive oxygen species. Contribution to oxidative stress and interorganellar signaling. Plant Physiol 141(2):357–366. https://doi.org/10.1104/pp.106.079129

Richards SL, Wilkins KA, Swarbreck SM, Anderson AA, Habib N, Smith AG, McAinsh M, Davies JM (2015) The hydroxyl radical in plants: from seed to seed. J Exp Bot 66(1):37–46. https://doi.org/10.1093/jxb/eru398

Sachdev S, Ansari SA, Ansari MI, Fujita M, Hasanuzzaman M (2021) Abiotic stress and reactive oxygen species: generation, signaling, and defense mechanisms. Antioxidants 10(2):277. https://doi.org/10.3390/antiox10020277

Sarkar J, Chakraborty B, Chakraborty U (2018) Plant growth promoting rhizobacteria protect wheat plants against temperature stress through antioxidant signalling and reducing chloroplast and membrane injury. J Plant Growth Regul 37(4):1396–1412. https://doi.org/10.1007/s00344-018-9789-8

Sasidharan R, Hartman S, Liu Z, Martopawiro S, Sajeev N, van Veen H, Yeung E, Voesenek LACJ (2018) Signal dynamics and interactions during flooding stress. Plant Physiol 176(2):1106–1117. https://doi.org/10.1104/pp.17.01232

Selote DS, Khanna-Chopra R (2006) Drought acclimation confers oxidative stress tolerance by inducing co-ordinated antioxidant defense at cellular and subcellular level in leaves of wheat seedlings. Physiol Plant 127(3):494–506. https://doi.org/10.1111/j.1399-3054.2006.00678.x

Shahzad B, Tanveer M, Che Z, Rehman A, Cheema SA, Sharma A, Song H, Rehman SU, Zhaorong D (2018) Role of 24-epibrassinolide (EBL) in mediating heavy metal and pesticide induced oxidative stress in plants: a review. Ecotoxicol Environ Saf 147:935–944. https://doi.org/10.1016/j.ecoenv.2017.09.066

Shao H-B, Chu L-Y, Jaleel CA, Zhao C-X (2008) Water-deficit stress-induced anatomical changes in higher plants. C R Biol 331(3):215–225. https://doi.org/10.1016/j.crvi.2008.01.002

Sharma P, Jha AB, Dubey RS, Pessarakli M (2012) Reactive oxygen species, oxidative damage, and antioxidative defense mechanism in plants under stressful conditions. J Bot 2012:1–26. https://doi.org/10.1155/2012/217037

Shu S, Yuan L-Y, Guo S-R, Sun J, Yuan Y-H (2013) Effects of exogenous spermine on chlorophyll fluorescence, antioxidant system and ultrastructure of chloroplasts in Cucumis sativus L. under salt stress. Plant Physiol Biochem 63:209–216. https://doi.org/10.1016/j.plaphy.2012.11.028

Singh A, Mehta S, Yadav S, Nagar G, Ghosh R, Roy A, Chakraborty A, Singh IK (2022) How to cope with the challenges of environmental stresses in the era of global climate change: an update on ROS stave off in plants. Int J Mol Sci 23(4):1995. https://doi.org/10.3390/ijms23041995

Slesak I, Libik M, Karpinska B, Karpinski S, Miszalski Z (2007) The role of hydrogen peroxide in regulation of plant metabolism and cellular signalling in response to environmental stresses. Acta Biochim Pol 54(1):39–50

Sofo A, Scopa A, Nuzzaci M, Vitti A (2015) Ascorbate peroxidase and catalase activities and their genetic regulation in plants subjected to drought and salinity stresses. Int J Mol Sci 16(12):13561–13578. https://doi.org/10.3390/ijms161613561

Steffens B (2014) The role of ethylene and ROS in salinity, heavy metal, and flooding responses in rice. Front Plant Sci 5:685. https://doi.org/10.3389/fpls.2014.00685

Su T, Li W, Wang P, Ma C (2019) Dynamics of peroxisome homeostasis and its role in stress response and signaling in plants. Front Plant Sci 10:705. https://doi.org/10.3389/fpls.2019.00705
Szechyńska-Hebda M, Ghalami RZ, Kamran M, Van Breusegem F, Karpiński S (2022) To be or not to be? Are reactive oxygen species, antioxidants, and stress signalling universal determinants of life or death? Cells 11(24):4105. https://doi.org/10.3390/cells11244105
Thounaojam TC, Panda P, Mazumdar P, Kumar D, Sharma GD, Sahoo L, Sanjib P (2012) Excess copper induced oxidative stress and response of antioxidants in rice. Plant Physiol Biochem 53:33–39. https://doi.org/10.1016/j.plaphy.2012.01.006
Voss I, Sunil B, Scheibe R, Raghavendra AS (2013) Emerging concept for the role of photorespiration as an important part of abiotic stress response. Plant Biol 15(4):713–722. https://doi.org/10.1111/j.1438-8677.2012.00710.x
Wang X, Wu L, Xie J, Li T, Cai J, Zhou Q, Dai T, Jiang D (2018) Herbicide isoproturon aggravates the damage of low temperature stress and exogenous ascorbic acid alleviates the combined stress in wheat seedlings. Plant Growth Regul 84(2):293–301. https://doi.org/10.1007/s10725-017-0340-x
Waszczak C, Carmody M, Kangasjärvi J (2018) Reactive oxygen species in plant signaling. Annu Rev Plant Biol 69(1):209–236. https://doi.org/10.1146/annurev-arplant-042817-040322
Yachandra VK, Sauer K, Klein MP (1996) Manganese cluster in photosynthesis: where plants oxidize water to dioxygen. Chem Rev 96(7):2927–2950. https://doi.org/10.1021/cr950052k
Yamane K, Taniguchi M, Miyake H (2012) Salinity-induced subcellular accumulation of H_2O_2 in leaves of rice. Protoplasma 249(2):301–308. https://doi.org/10.1007/s00709-011-0280-7
You J, Chan Z (2015) ROS regulation during abiotic stress responses in crop plants. Front Plant Sci 6:1. https://doi.org/10.3389/fpls.2015.01092
Yu J, Cang J, Lu Q, Fan B, Xu Q, Li W, Wang X (2020) ABA enhanced cold tolerance of wheat 'dn 1' via increasing ROS scavenging system. Plant Signal Behav 15(8):1780403. https://doi.org/10.1080/15592324.2020.1780403

Abiotic Stress-Induced Epigenetic Modifications in Plants

4

Preeti Patel, Archana Prasad, and Satya Shila Singh

Abstract

Plants are constantly exposed to various unfavourable environmental conditions, which drive them to adopt several survival strategies including epigenetic modifications. These modifications correlate with the expression of genes via transcriptional regulation deprived of altering the DNA sequences. Stress tolerance in plants is significantly conferred by epigenetic changes including small RNA-mediated silencing, histone alterations, and DNA methylation. These modifications also facilitate long-term stress tolerance adaptation via persistent gene regulation or transgenerational inheritance. Various advantageous traits in crop improvement are governed by the heritable epigenetic changes as epialleles. Recently, many studies emphasized the mechanism and regulation of epigenetic modifications adopted by plants in response to different stress. These modifications play an important role in governing stress tolerance, acclimatization, adaptation, and evolution practices. This chapter emphasizes the abiotic stress-induced epigenetic modifications and their regulation in plants. Further, the exploration of epigenetic changes leading to the production of stress-tolerant crops is also discussed.

Preeti Patel and Archana Prasad contributed equally with all other contributors.

P. Patel (✉)
Department of Plant Pathology, Ohio State University, Columbus, OH, USA

A. Prasad
Department of Fruit Tree Sciences, Institute of Plant Sciences, Volcani Center, Agricultural Research Organization, Rishon LeZion, Israel

S. S. Singh
Laboratory of Cyanobacterial Systematics and Stress Biology, Department of Botany, Institute of Science, Banaras Hindu University, Varanasi, India

A. K. Mishra (ed.), *Stress Biology in Photosynthetic Organisms*,
https://doi.org/10.1007/978-981-97-1883-2_4

Keywords

Stress tolerance · Histone modifications · DNA methylation/demethylation · Epigenetic changes · Crop improvement

4.1 Introduction

4.1.1 Stress: Types and Their Effects on Plants

Environmental stresses are a factor that plants are constantly subjected to numerous abiotic stresses, as a result of unpredictable environmental fluctuations which might be fatal to them or have a detrimental effect on their growth as well as the development of plants. Abiotic factors including drought, temperature extremes, salinity, dehydration, soil mineral deficiency, ultraviolet radiation, low nutrition, and heavy metal toxicity are the main factors contributing to crop losses and significantly reduce agricultural productivity by up to 70% as a result affecting global food security (Mittler 2006; Mantri et al. 2012; Vriet et al. 2015; Miryeganeh et al. 2021). Furthermore, it is anticipated that climate change is predicted to increase extreme weather events' frequency and intensity, which could have an impact on yields (Tubiello et al. 2007; Challinor et al. 2014). Nonetheless, a variety of distinct abiotic stressors are frequently applied to crops and other plants in the field (Naika et al. 2013; Borkotoky et al. 2013; Hauser et al. 2011; Gentry and Hennig 2014). As an example, a lot of crops in drought-affected areas must cope with both the drought and other stresses like heat or salt (Borkotoky et al. 2013; Naika et al. 2013).

4.1.2 Mechanisms Adopted by Plants to Cope Up the Stresses

A plant must adapt to the specific climatic circumstances it encounters in order to get acclimated to a particular abiotic stress. Consequently, the molecular, biochemical, and physiological reactions that are triggered by a particular stressor may not be the same as those that are triggered by a slightly different combination of environmental factors (Mittler 2006). In this particular instance, for plants to survive under severe stress, they have to be able to recognize modifications in their surroundings and mount defence mechanisms, such as the rapid onset of stress defence genes. To effectively respond to continuously unfavourable environmental conditions, gene expression may need to be carefully reprogrammed to restore the balance between growth, development, and survival (Vriet et al. 2015). Investigations using transcriptome profiling on plants exposed to different abiotic stress situations corroborate this. With minimal transcript expression overlap amongst the plant responses to factors like heat, cold, drought, bright light, salt, or mechanical stress, the plants responded differentially to each abiotic stress condition that was examined (Fig. 4.1) (Cheong et al. 2002; Fowler and Thomashow 2002; Kreps et al. 2002; Rossel et al. 2002; Rizhsky et al. 2004). Reactive oxygen species, or ROS, have been associated with a variety of abiotic and biotic stress conditions; however, it has been shown

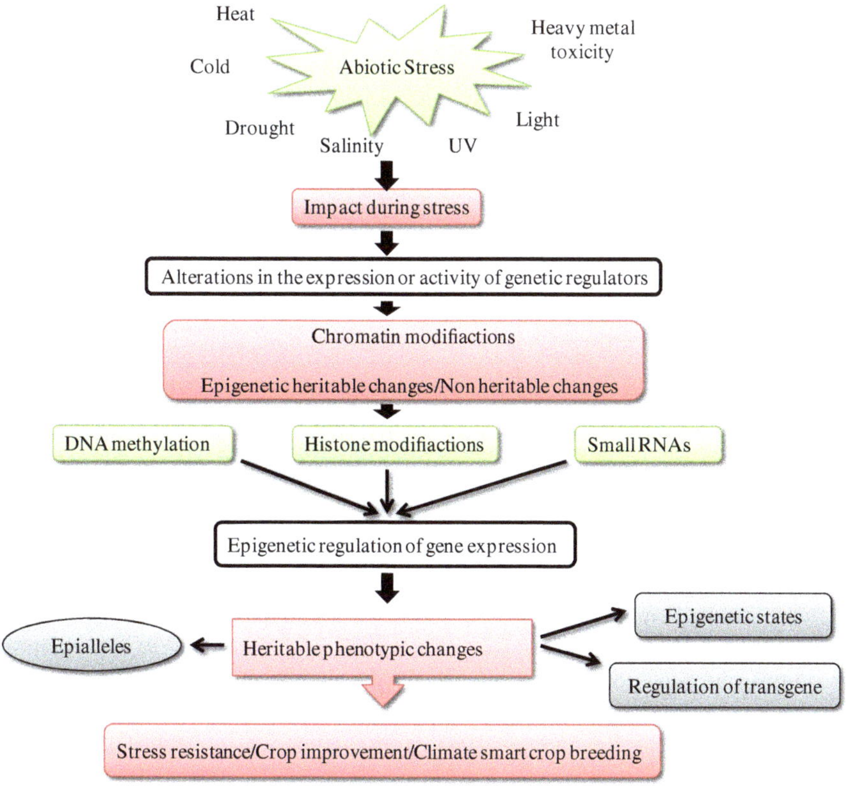

Fig. 4.1 Epigenetic modifications in plants against several abiotic stresses

that distinct genes within the *Arabidopsis* ROS gene network react differently to various stimuli (Mittler et al. 2004). According to these findings, each instance of abiotic stress calls for a distinctive acclimation response that is suited to the requirements of the plant. Furthermore, a reaction that differs from one response to another may be necessary when two or more independent stresses are combined. Apart from the fundamental variations between plants' acclimation reactions to various abiotic stress situations (Rizhsky et al. 2004; Kreps et al. 2002; Fowler and Thomashow 2002; Rossel et al. 2002; Cheong et al. 2002), different stresses may necessitate competing or antagonistic responses when combined. For instance, plants expand their stomata in response to heat stress to allow transpiration to cool their leaves. On the other hand, plants would not be able to open their stomata and their leaf temperature would increase if heat stress and drought were combined (Rizhsky et al. 2002). Plants may encounter comparable issues when heat stress is combined with salinity or heavy metal stress because higher transpiration may lead to increased salt or heavy metal uptake. The photosynthetic machinery produces more ROS when cold stress or drought is paired with high light levels due to the fact that under these circumstances, oxygen remains one of photosynthesis' principal reductive products and less CO_2 is accessible for the dark reaction (Mittler 2002). Drought and heavy metal stress are other instances of abiotic stresses that are antagonistic to one another and exacerbate each other's impacts (Barceló and Poschenrieder 1990). Therefore, a suitable response that is unique to every stressful circumstance that is occurring and intended to mitigate or rectify certain antagonistic elements of the stress combination would need to be put into place to aid plants in acclimating to an association of several abiotic stressors (Mittler 2006).

4.1.3 Epigenetic Modification Adopted by Plants to Regulate Stress

To respond and adapt to various environmental challenges, plants use a process referred to as transcriptional reprogramming, controlled by a set of transcription factors that are differentially activated or repressed (Feng et al. 2022). Stress signals are transmitted to target gene regulatory areas, altering specific levels of epigenetic changes and leading to differential transcript expression (Kouzarides 2007; Nakashima et al. 2014). Numerous essential biological activities, including DNA replication, transcription, and recombination as well as DNA damage repair, rely on chromatin as their functional template (Latrasse et al. 2008). Covalent shifts of the N-terminal tails of core histones, such as acetylation, methylation, phosphorylation, ubiquitination, sumoylation, and poly-ADP-ribosylation, impact nucleosome location and compression, which is why they are crucial for chromatin remodelling as well as gene regulation (Liu et al. 2014). Gene regulation in response to environmental stressors is largely dependent on the dynamics of epigenetic coding. For example, Histone modifications and DNA methylation are known to affect the chromatin state of stress-responsive genes, which in turn affects how those genes are expressed both transcriptionally and post-transcriptionally (Zhang et al. 2018; Luo and He 2020) (Fig. 4.2).

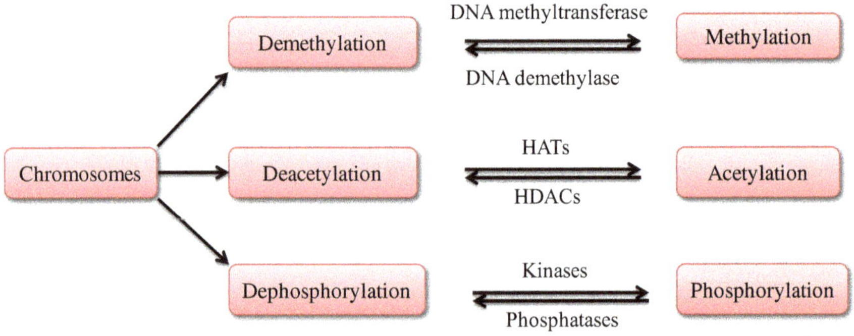

Fig. 4.2 Various epigenetic changes including (de)methylation, acetylation, and phosphorylation

4.1.4 Histone Modifications and DNA Methylation

The fundamental chromatin structural units known as nucleosomes are formed by the packaging of eukaryotic genomic DNA around the octamers of histones. A core histone octamer (two copies of each of the four core histone proteins: H2A, H2B, H3, and H4) plus 147 base pairs of DNA make up each nucleosome. These histones can be readily altered by a range of covalent changes at their N-terminal tails, notably phosphorylation, acetylation, ubiquitination, and methylation and this mechanism, known as histone post-translational modification (PTM) (Kouzarides 2007). Among these, histone methylation is a well-researched PTM. Histone methylation often takes place at arginine and lysine residues that have varying methyl numbers, such as mono-, di-, and tri-methylation. Histone acetylation is another extensively researched histone modification. Histone acetylation is widely thought to disrupt nucleosome binding, creating a looser chromatin state that facilitates transcriptional activation. (Shahbazian and Grunstein 2007). Acetylation and deacetylation of histones are crucial changes that regulate gene expression in plants. They act as critical switches that shift chromatin domains from permissive to restrictive states (Berger 2007). Apart from the changes in histone tails, the DNA strand can also undergo several modifications. The most notable one is the methylation of cytosine's carbon-5 (5-mC). DNA methylation can be found at promoters, introns, and transposable elements (TEs), and it can occur in a variety of sequence situations, such as asymmetrical CHH and symmetrical CG, as well as CHG (H stands for A, T, or C) (Henderson and Jacobsen 2007). These epigenetic marks are readily governed by several variables, such as the enzymes that can remove or catalyze (also known as "writers" or "erasers") the alteration to or from the histone, and the proteins that identify and link the modification with other molecules. In general, epigenetic changes have the ability to regulate such marked genes either transcriptionally or post-transcriptionally. Additionally, growing research indicates that the regulation of plant immunity depends on histone modification homeostasis (Xie and Duan 2023).

4.1.5 Stress Types Modifying Chromatin Parameters

Typically, plants can respond to environmental stimuli through the regulation of spatiotemporal gene expression by epigenetic changes. Even though reports of significant chromatin rearrangements have only included the aforementioned stressors, thus far, many other forms of stress can also cause molecular reactions. Chromatin changes are linked to drought, a severe disease that impacts plants and is brought on by the abscisic acid pathway (Han and Wagner 2014; Mehrotra et al. 2014; Kim et al. 2015). Osmotic stress or salinity is often linked with dehydration and also elicits chromatin-level responses (Deinlein et al. 2014). For higher plants, extreme temperatures cause heat stress and cold stress, two distinct reactions that alter chromatin structures (Kumar and Wigge 2010; Lang-Mladek et al. 2010; Pecinka et al. 2010; Tittel-Elmer et al. 2010). Chromatin alterations brought on by a lack of light are indicated by light perception factors (Tessadori et al. 2009; van Zanten et al. 2010, 2012). Chromatin modifications are caused by exposure to radiation with high energy or chemically induced DNA damage, and the effectiveness of DNA repair is supported by at least one chromatin remodelling factor (Rosa et al. 2013; Braszewska-Zalewska et al. 2014).

4.2 General Chromatin Responses

4.2.1 Stress-Induced DNA Methylation/Demethylation

DNA methylation and demethylation enzymes, histone variations, histone modification enzymes, chromatin remodelling complexes, and other chromatin-associated components are involved in plant stress responses at the chromatin level (Vriet et al. 2015). Stress can alter the methylation of DNA both globally and locally in addition to histone alterations. By maintaining a balance between methylating and demethylating processes, DNA methylation levels are regulated (Probst and Scheid 2015). Histone methylation and demethylation induce transcript expression in plants under abiotic stress (Pandey et al. 2016; Kong et al. 2020). This process is regulated by the HMT and HDM enzymes, which add or remove methyl groups from basic residues. Histone Lys methyl-transferases (HKMTs) and Arg methyl-transferases (PRMTs) are two examples of HMTs that methylate arginine and lysine histone residues (Cadavid et al. 2023).

DNA methylation is a very important covalent alteration that affects transcription and genomic stability (Table 4.1). The term "DNA methylation" describes the process by which members of the DNA methyltransferase (DNMT) families move a methyl group from S-adenosylmethionine to the carbon 5 of cytosine on DNA, creating m5C (5-methylcytosine) (Sahu et al. 2013). Plants have essential enzymes called m5C DNA glycosylases, also known as DNA demethylases, which remove m5C directly and start the process of replacing it during the active DNA demethylation pathway with an unmethylated cytosine (Parrilla-Doblas et al. 2019). In *Arabidopsis*, there are three chromosomal locations where DNA methylation takes place: CG, CHG, and CHH, where H stands for any nucleotide other than guanine. The primary function of the methyltransferase MET1 is to silence transposons, certain imprinted genes, and repetitive sequences. It is required to keep CG methylation stable. Together, the methyltransferases DRM2, CMT3, and CMT2 control non-CG methylation (Lindroth et al. 2001; Du et al. 2012; Zemach et al. 2013; Stroud et al. 2014). Additionally, small interfering RNAs (siRNAs) might contribute to the de novo synthesis of asymmetric CHH methylation and CG as well as CHG methylations through the RNA-directed DNA methylation (RdDM) process (Law and Jacobsen 2010). DRM2 and histone-modifying enzymes are enlisted by siRNAs generated by dicer-like (DCL) nucleases upon binding to argonaute (AGO) proteins in order to target DNA (Saze et al. 2012). Despite the widespread belief that cytosine DNA methylation is a permanent marker, active demethylation has been seen in

Table 4.1 DNA methylation/demethylation in response to different abiotic stresses

Epigenetic process	Abiotic stress	Plant	References
DNA methylation	Drought	Tomato	González et al. (2011), Huang et al. (2016)
Demethylation	Drought	*Arabidopsis*	Huang et al. (2019)
DNA methylation	Drought	Rice	Wang et al. (2011a, b)
Hypomethylation	Drought	Rice	Gayacharan and Joel (2013)
DNA methylation	Drought	Faba bean	Abid et al. (2017)
Demethylation and hypomethylation	Salinity	Soybean	Song et al. (2012)
Cytosine methylation	Salinity	*Arabidopsis*	Baek et al. (2011)
Hypomethylation	Salinity	*Arabidopsis*	Arikan et al. (2018)
DNA methylation	Salinity	Rice	Karan et al. (2012), Ferreira et al. (2019)
DNA methylation	Salinity	Rapeseed	Marconi et al. (2013)
DNA methylation	Salinity	Tomato	Huang et al. (2016)
DNA methylation	Cold	Tomato	Zhang et al. (2016)
Hypomethylation	Cold	Tobacco	Choi and Sano (2007), Palakolanu et al. (2022)
DNA methylation	Cold	*Chorispora bungeana*	Song et al. (2015)
DNA demethylation	Cold	Maize	Steward et al. (2000), Hu et al. (2012,)
DNA methylation	Heat	Rapeseed	Gao et al. (2014)
Hypermethylation	Copper	*Hydrilla verticillata*	Shi et al. (2017)

both plants and animals (Kohli and Zhang 2013). Plants use the DEMETER (DME) family of DNA glycosylases for DNA demethylation; this family includes the base excision repair-based REPRESSOR OF SILENCING (ROS1) in *Arabidopsis*. (Agius et al. 2006; Gehring et al. 2006; Penterman et al. 2007). DNA demethylation is also influenced by histone acetyltransferase IDM1, which creates a chromatin environment that promotes DNA glycosylase activity (Qian et al. 2012). The high degree of methylation found in silent heterochromatin DNA serves as an example of how DNA methylation has historically been linked to gene silencing. Stresses can alter gene expression by causing DNA to become hyper- or hypomethylated. In maize roots, a decline in methylation in the nucleosome core DNA was linked to the expression of ZmMI1 produced by cold stress (Steward et al. 2002). Increased DNA demethylation in the coding sequence of the glycerophosphodiesterase-like protein *NtGPDL* gene and the expression of the *NtGDPL* gene were found in tobacco exposed to stressors such aluminum, paraquat, salt, and cold (Choi and Sano 2007).

4.2.2 Histone Modification Related to Stress

Plants encounter various environmental factors during their development. Various biotic as well as abiotic factors regulate various physiological as well as developmental changes adopted by plants for their survival. This regulatory mechanism refers to the epigenetic changes adopted by plants to overcome various environmental stresses. The epigenetic changes include chromatin modifications for activation/inactivation of gene expression at transcriptional as well as post-transcriptional levels. These mechanisms regulate through the complex cascade via DNA methylation, non-coding small RNA, and histone modifications.

Histones are the nucleoproteins that are associated with DNA strands enabling the highly specific organization and complexation of chromatin. Nucleosomes are made up of approximately 147 DNA base pairs bounded around core histone proteins viz. H2A, H2B, H3, and H4. The histone proteins are rich in positively charged amino acids such as arginine and lysine which enables them to interact. They are modified with different chemical markers embedded at different places in the histone (Gates et al. 2017; Zhang et al. 2018; Demetriadou et al. 2020). The modifications such as methylation, phosphorylation, acetylation, ubiquitination etc. are exhibited by histone via lysine residue mainly present at its N-terminal regions. These modifications are known as post-translational modifications which are reversible processes regulated by specific enzymes that catalyze or remove specific marks. They play a significant role in altering the gene expression/activity associated with the core histones (Ruthenburg et al. 2007; Zhao et al. 2018). The addition and/or elimination of different chemical entities are catalysed by a specific class of enzymes depending on the type of chemical reactions. These enzymatic reactions include histone acetylation and deacetylation which are catalysed by histone acetyltransferases (HATs) and histone deacetylases (HDACs), respectively. Another modification include histone methylation and demethylation which are catalysed by histone methyl-transferases (HMTs) and histone demethylases (HDMs), respectively (Fig. 4.3) (Panara et al. 2022). These modifications are extensively studied in different plants under the influence of various environmental stimuli (Gates et al. 2017). The studies of histone modification in plants under the influence of different stresses are presented in Table 4.2. Recently, chromatin immunoprecipitation (ChIP) has been used as one of the most popular methods for quantifying and determining the specific histone markers linked with DNA regions. ChIP assay is an antibiotic-based method that recognizes specific histone modifications with DNA fragments (Bowler

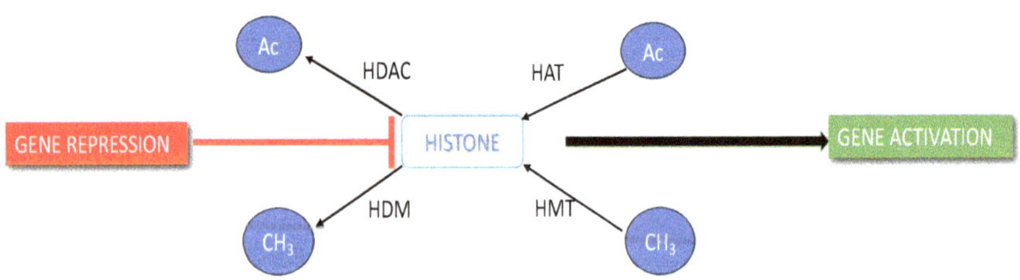

Fig. 4.3 Histone modifications associated with gene activation (right) and repression (left). Histone acetylation is mediated by histone acetyltransferase (HAT) and deacetylation involves histone deacetylases (HDAC). Histone methylation and demethylation are catalysed by histone methyl transferases (HMT) and demethylases (HDM), respectively

et al. 2004; Li et al. 2021). In addition to this plant biologists have worked on various technologies for studying chromatin changes that lead to genome modifications at multiple scales (Ouyang et al. 2020; Pei et al. 2021). These technologies include ChIP with high throughput chromosome conformation capture (HiChIP) topologically associated domains (TADs), chromosome conformation capture (3C) and chromatin loops with different imaging methods, allowing for the explicating chromosome territories (CTs), A/B compartments, (Mumbach et al. 2016; Grob and Grossniklaus 2017; Zheng et al. 2019; Golicz et al. 2020).

4.2.3 Histone Acetylation/Deacetylation Against Various Environmental Stresses

The supplementation of acetyl group to the histone is referred to as acetylation. It takes place when lysine residues present in the tail of histone proteins are attached to an acetyl group that is negatively charged (Boycheva et al. 2014). Acetylation of lysine is one of the important protein posttranslational modifications (PTMs) that is catalysed by different enzymes and important role in various cellular functions (Harachi et al. 2021). The neutralisation of the positive charge present on the histone tail leads to the weakening of the interaction between DNA and histone, which leads to the enhancement of the accessibility of transcription factors to the DNA (Kaiserli et al. 2018). Histone acetylation and deacetylation are the two major reversible epigenetic processes that are controlled by two sets of opposing enzymes histone acetyltransferases (HATs) and histone deacetylases (HDACs) (Shen et al. 2015; Dar et al. 2022). HATs catalyze the addition of acetyl group addition, while HDACs catalyze the removal of acetyl group. Different chromatin proteins TFs interrelate with HAT and/or HDAC to regulate specific gene expression that leads to regulating various signalling at cellular levels (Liu et al. 2014). In plants, four different types of HATs have been identified (1) CREB-binding protein (CBP)/p300, (2) GNAT (GCN5-related N-acetyltransferase), (3) and TATA-binding protein-associated factor 1 (TAF1), and (4) MOZ, Ybf2/Sas2 and Tip60 (MYST) (Shin et al. 2022). Different categories of HATs concerned with the acetylation of specific lysine residue such as HATs -HAG1 and HAG2 belong to the GNAT class involved in the catalysing the acetylation of H3K14 and H4K12, respectively. Similarly, HATs-HAM1 and HAM2 from the MYST class catalyse the acetylation of H4K5. Distinct HATs molecules lead to the acetylation of specific lysine catalysed by different groups of enzymatic entities (Hu et al. 2019). On the basis of subcellular distribution, HATs are further classified into two categories, viz., cytoplasmic-type B HATs that lead to acetylation of histone H4 at lys-5 and 12. It is described in maize and functions as a heterodimer. Another type is nucleus-based type A HATs that regulate assembly of chromatin and transcription of genes through acetylation of nuclear histones (Chen and Tian 2007). The role of HATs in response to salinity, heat, and drought stresses in *Arabidopsis*, rice, tomato, and poplar has been reviewed

Table 4.2 Histone modifications of plants in response to different stresses

Plant species	Stress condition	Type of modification	Histone markers/loci/regulator	References
Arabidopsis	Heat	Acetylayion	GCN5, AtHAC1	Hu et al. (2015), Roca Paixão et al. (2019)
	Salinity	Acetylation		Zheng et al. (2019)
Poplar	Drought	Acetylation		Li et al. (2019)
Arabidopsis	Drought and salinity	Deacetylation	HAD 9, HDA 15, HAD 19	Chen and Wu (2010), Zheng et al. (2016), Mehdi et al. (2016), Ueda et al. (2017, 2018), Lee and Seo (2019)
	Cold, drought, and salinity	Deacetylayion	HD2C, HD2D, AtSOS3, AtSOS1	Han et al. (2016), Sako et al. (2016), Park et al. (2019)
Rice	Salinity	Deacetylation	HDA705	Zhao et al. (2016)
Rice	Drought	Methylation Demethylation	H3K36me3, H3K4me3	Song et al. (2018), Chen et al. (2021)
Brachypodium	Drought	Deacetylation	BdHD1	Song et al. (2019)
Maize	Salinity	Acetylation	H3K9	Li et al. (2014)
Cassava	SAHA (histone deacetylase inhibitor)	Deacetylation	H3 and H4 hyperacetylation leads to upregulation of phytohormone genes	Patanun et al. (2017)
Tomato	Salinity, drought and ABA	Deacetylation	H3K27me3	Yu et al. (2018), Ding et al. (2022), Guo and Wang (2023)
Barley	Drought	Acetylation, methylation	H3K4me3, H3K9ac, H3 and H3K9me2	Temel et al. (2017)
Wheat	Salinity	Acetylation	TaGCN5	Hu et al. (2015), Zheng et al. (2019)
Wheat	Salinity	Deacetylation	TaHDA6	Liu et al. (2019)

by Ueda and Seki (2020). In *Arabidopsis*, a total of 12 HAT genes were recognised; five belong to the GNAT/MYST family, CBP family five genes, and two belong to the TAFII250 family (Pandey et al. 2002). In the soybean genome, 14 HATs were identified, 9 proteins from the GNAT/MYST, 3 proteins from CBP, and two from the TAFII250 group (Liew et al. 2013). Histone acetylation is also related to chilling stress (10 °C) in soybean roots. This study showed the occurrence of specific antibodies for H4K12acetyl and H3K9acetyl (Stępiński 2012). Under salt stress, the role of nuclear factor Y subunit GmNFYA leads to providing tolerance by regulation of histone acetylation in soybean (Lu et al. 2021). In one recent study, the role of SnRK1 kinase in the catalysation of histone acetylation has been identified in *Glycine soja* by Feng et al. (2022). They predict the protein SnRK1 kinase phosphorylate acetyltransferase GsMYST1 under salt stress conditions. They interact with GsNAC83 TF and form heterotrimeric complexes that lead to the activation of targeted genes. In addition to this, it also binds with the promoter of COR15B gene leading to its expression via acetylation of the Histone 4 (K5K8K12K16ace). However, more analyses are required to study the histone acetylation mechanism in soybeans as well as other plant species.

The HDACs in plants are categorised into three different families (1) Reduced Potassium Dependence3/Histone DeAcetylase1 (RPD3/HDA1) family, which is identical to yeast and widely found in eukaryotes (2) and Histone Deacetylase2 (HDT2) family, exclusively present in plants and first discovered in maize (3) Silent Information Regulator 2 (SIR2), it is (NAD-) nicotinamide adenine dinucleotide-dependent enzyme (Pandey et al. 2002). The role of HDACs in plants under stress related to ABA has been studied by Ueda and Seki (2020). In *Arabidopsis,* a total of sixteen HDACs were identified, out of which ten were from the RPD3/HDA1 family, 4 were from HDT and, 2 were from the SIR2 family referred to as SRT (Pandey et al. 2002). In *Glycine max* 28 HDAC genes were identified. Among these 28 HDACs, 18 are from the RPD3/HDA1 family referred to as GmHDA1 to GmHDA18, four belong to the SIR2 family and six HDACs are from the HD2 family showing highly conserved amino acid terminal (Yang et al. 2018). The RT-PCR expression analysis showed that GmHDAC genes play a significant role in various abiotic stress treatments. The RNA-Seq analysis of salt stress soybean showed the expression of three HDAC genes, viz. HDT 4, HDT2, and HDAC 17 get downregulated (Cadavid et al. 2020a). Another study conducted by the same group showed that the expression of the miRNA482bd-5p gene was directly or indirectly controlled by an HDAC under salt stress (Cadavid et al. 2020b).

Both HATs and HDACs act as coactivators and corepressors of transcription in the presence of specific protein complexes that leads to the regulation of various mechanism related to chromatin remodelling (Dar et al. 2022). Different studies have been conducted in *Arabidopsis*, tomato and rice for detailed classification and identification of HATs and HDACs superfamily. In general, the histone acetylation markers are accompanied by gene activation; however, deacetylation marks with the condensed chromatin structure and the repression of a gene (Fig. 4.3) (Kurdistani and Grunstein 2003).

4.2.4 Histone Methylation/Demethylation Under the Influence of Environmental Stimuli

Under different abiotic stresses, plants undergo alteration of gene expression due to histone methylation/demethylation (Pandey et al. 2016; Kong et al. 2020). Histone methylation and demethylation have been catalysed by histone methyltransferases (HMTs) and histone demethylases (HDM) for the addition and removal of methyl groups, respectively (Hu and Du 2022). The HMT methylates on different arginine and lysine entities of the histone (Liu et al. 2010). The enzymes that methylate arginine are referred to as Arg methyl-transferases (PRMTs) and the enzymes that methylate lysine are referred to as histone Lys methyl-transferases (HKMTs). The histone methylation results in the modification of chromatin structure that stimulates the accessibility of proteins to the targeted DNA for the regulation of the expression of a specific gene. In *Arabidopsis*, the role of HMTs under drought and salinity stresses has been studied (Ueda and Seki 2020). HMTs act as both repressor (symmetric H4R3me2, H3K27me3, and H3K9me2/3) as well as activation markers (asymmetric H4R3me2, H3K36me2/3, and H3K4me3) (Liu et al. 2010; Wang et al. 2016). The addition of methyl on specific lysine of the histone is mediated by HMTs with highly conserved SET domain group (SDG) protein (Ng et al. 2007; Chachar et al. 2022). On the other hand, the two main erasers that facilitate the removal of methyl from lysine and arginine residue by LSD1 (lysine-specific demethylases 1) and JMJ (jumonji-C domain-containing protein) family, respectively (Liu et al. 2017; Shin et al. 2022). In plants, the methylation mainly occurs at N-terminal of H3 and H4 histones moiety of lysine and arginine residue. Mostly, H3K4me3 and H3K36me3 are related to the transcriptional activators, while H3K9me2, H3K27me3, and H4R3 act as repressive markers (Berger 2007). In soybean, six LSD protein families were identified and characterised for their role as histone demethylase activity under abiotic stress (Liu et al. 2022). While 24 JmjC demethylases were identified for the floral initiation process by transcriptome analysis (Liew et al. 2013). The RNA-seq and ChIP-seq data analysis revealed that under salt stress, the histone methylation (H3K27me3) was highly correlated with the differential expression of genes (Sun et al. 2019). A similar kind of study performed by Song et al. (2012) showed that expression of MYB, AP2/DREB, and b-ZIP genes was correlated with enhanced levels of histone H3K4me3 and decline in H3K9me2 under salt stress. The role of histone modification H3K9me2 and H3K4me3 was also studied in the root tips of soybean under chilling stress (10 °C) using fluorescently tagged antibodies (Stępiński 2012). Although many workers elucidate the role of HMT and HDM genes influenced by different abiotic stress in plants but still the exploration of the regulatory mechanism for histone modification needs to be elucidated.

4.3 RNA Mechanism in Response to Epigenetic Stress

In plants, non-coding RNA (ncRNAs) are one the important components involved in the development, growth, reproduction, and various responses under different environmental stresses. They regulate different genes via both transcriptional as well as post-transcriptional mechanisms (Wang et al. 2017). Based on their length, the ncRNAs are categorised into types: short- and long-chain non-coding transcripts (Hou et al. 2019; Brant and Budak 2018). Further, short-chain ncRNAs include miRNAs (microRNAs), siRNAs (small interfering RNAs), tasiRNAs, and tRFs that vary an average length of 18–24 nucleotides (Aprile et al. 2019). The DICER-like proteins (DCLs) synthesize small RNA molecules by utilising 21–24 nucleotides of RNA molecules. However, the long non-coding RNAs (lncRNAs) are synthesized by more than 200 nucleotides (Mach 2017). In addition, circular RNA also regulates gene expression at post-transcriptional regulation (Bhogireddy et al. 2021; Li et al. 2021).

In plants, short-chain miRNAs control post-transcriptional mechanisms, while siRNA is involved in the silencing of genes at transcriptional and/or post-transcriptional levels (Wang et al. 2019). The synthesis of miRNA is mediated by its precursor that encodes the nuclear genome and is catalysed by the dsRNA-binding Hyponastic Leaves 1 (HYL1) proteins, Dicer-like 1 (DCL1), and the zinc-finger Serrate (SE) protein in the presence of nuclear-cap-binding complex (Miskiewicz et al. 2017). The RNAi (RNA interference) is involved in the post-transcriptional silencing of genes (Ravichandran et al. 2019). It is one of the most popular methods used by biologists to investigate the function of genes due to its high efficacy and specificity. However, siRNAs (small interfering RNAs) are transferred to the adjacent cell with the help of different carrier proteins using plasmodesmata and phloem to regulate various biological processes including epigenetic modification by RNA-dependent DNA methylation (RdDM) (Hyun et al. 2011; Xie and Yu 2015). Based on their origin and function siRNAs are further classified into ta-siRNAs (trans-acting siRNAs), hc-siRNAs (heterochromatin siRNAs), and nat-siRNAs (natural antisense siRNAs) (Guleria et al. 2011). In addition, IncRNAs (Intronic RNAs) also mediate Inc-RNA mediated chromatin modification that involves various developmental, reproduction, and epigenetic modification against various stresses including salinity, heat, salinity, pathogen, drought etc. (Eom et al. 2018). Although different workers showed the involvement of ncRNA in the regulation of genes against epigenetic stress. But still, its mechanism of synthesis and further regulatory system needs to be explored by plant biologists.

4.4 Conclusion and Future Prospectus

The world is facing an alarming situation due to the rise in population and global climate change. This challenge led to various tailing threats in the agriculture and food production sectors. Because of this, the investigation in the field of epigenetics became the leading of various plant biologists. In plants, various alterations at cellular levels such as DNA methylation, histone modifications, and small RNA/ncRNA mediated regulation are major stress-responsive epigenetic modifications. The in-depth knowledge and understanding of epigenetic regulation lead us to pave a better platform to overcome various environmental challenges procured by stress conditions. However, the study of the mechanism and regulators for epigenetic modifications is conducted in very limited plant species. It is essential to explore this promising field at a more elaborate level to generate more evidence and make it acceptable at the global level. The development of markers, stability of priming, and better understanding of regulators against different stresses will result in improved crop protection programs for the upcoming future.

Acknowledgements PP and AP would like to thank to Ohio State University, USA, and Agricultural Research Organization, Israel, respectively. SSS is grateful to Head Department of Botany BHU, India, for providing infrastructure facility for conducive research.

Conflict of Interest The authors declare no conflict of interest.

References

Abid G, Mingeot D, Muhovski Y, Mergeai G, Aouida M, Abdelkarim S, Aroua I, El Ayed M, M'hamdi M, Sassi K, Jebara M (2017) Analysis of DNA methylation patterns associated with drought stress response in faba bean (Vicia faba L.) using methylation-sensitive amplification polymorphism (MSAP). Environ Exp Bot 142:34–44

Agius F, Kapoor A, Zhu JK (2006) Role of the Arabidopsis DNA glycosylase/lyase ROS1 in active DNA demethylation. Proc Natl Acad Sci 103(31):11796–11801

Aprile A, Sabella E, Francia E, Milc J, Ronga D, Pecchioni N, Ferrari E, Luvisi A, Vergine M, De Bellis L (2019) Combined effect of cadmium and lead on durum wheat. Int J Mol Sci 20(23):5891

Arikan B, Özden S, Turgut-Kara N (2018) DNA methylation related gene expression and morpho-physiological response to abiotic stresses in Arabidopsis thaliana. Environ Exp Bot 149:17–26

Barceló JUAN, Poschenrieder C (1990) Plant water relations as affected by heavy metal stress: a review. J Plant Nutr 13(1):1–37

Baek D, Jiang J, Chung JS, Wang B, Chen J, Xin Z, Shi H (2011) Regulated AtHKT1 gene expression by a distal enhancer element and DNA methylation in the promoter plays an important role in salt tolerance. Plant Cell Physiol 52(1):149–161

Berger SL (2007) The complex language of chromatin regulation during transcription. Nature 447(7143):407–412

Bhogireddy S, Mangrauthia SK, Kumar R, Pandey AK, Singh S, Jain A, Budak H, Varshney RK, Kudapa H (2021) Regulatory non-coding RNAs: a new frontier in regulation of plant biology. Funct Integr Genomics 21:313–330

Borkotoky S, Saravanan V, Jaiswal A, Das B, Selvaraj S, Murali A, Lakshmi PTV (2013) The Arabidopsis stress responsive gene database. Int J Plant Genom 2013:949564

Bowler C, Benvenuto G, Laflamme P, Molino D, Probst AV, Tariq M, Paszkowski J (2004) Chromatin techniques for plant cells. Plant J 39(5):776–789

Boycheva I, Vassileva V, Iantcheva A (2014) Histone acetyltransferases in plant development and plasticity. Curr Genomics 15(1):28

Brant EJ, Budak H (2018) Plant small non-coding RNAs and their roles in biotic stresses. Front Plant Sci 9:1038

Braszewska-Zalewska A, Tylikowska M, Kwasniewska J, Szymanowska-Pulka J (2014) Epigenetic chromatin modifications in barley after mutagenic treatment. J Appl Genet 55:449–456

Cadavid IC, Guzman F, de Oliveira-Busatto L, de Almeida RMC, Margis R (2020a) Transcriptional analyses of two soybean cultivars under salt stress. Mol Biol Rep 47:2871–2888

Cadavid IC et al (2020b) HDAC inhibitor affects soybean miRNA482bd expression under salt and osmotic stress. J Plant Physiol 253:153261

Cadavid IC, Balbinott N, Margis R (2023) Beyond transcription factors: more regulatory layers affecting soybean gene expression under abiotic stress. Genet Mol Biol 46:e20220166

Chachar S, Chachar M, Riaz A, Shaikh AA, Li X, Li X, Guan C, Zhang P (2022) Epigenetic modification for horticultural plant improvement comes of age. Sci Hortic 292:110633

Challinor AJ, Watson J, Lobell DB, Howden SM, Smith DR, Chhetri N (2014) A meta-analysis of crop yield under climate change and adaptation. Nat Clim Chang 4(4):287–291

Chen ZJ, Tian L (2007) Roles of dynamic and reversible histone acetylation in plant development and polyploidy. Biochimic Biophys Acta 1769(5–6):295–307

Chen LT, Wu K (2010) Role of histone deacetylases HDA6 and HDA19 in ABA and abiotic stress response. Plant Signal Behav 5(10):1318–1320

Chen K, Du K, Shi Y, Yin L, Shen WH, Yu Y, Liu B, Dong A (2021) H3K36 methyltransferase SDG708 enhances drought tolerance by promoting abscisic acid biosynthesis in rice. New Phytol 230(5):1967–1984

Cheong YH, Chang HS, Gupta R, Wang X, Zhu T, Luan S (2002) Transcriptional profiling reveals novel interactions between wounding, pathogen, abiotic stress, and hormonal responses in Arabidopsis. Plant Physiol 129(2):661–677

Choi CS, Sano H (2007) Abiotic-stress induces demethylation and transcriptional activation of a gene encoding a glycerophosphodiesterase-like protein in tobacco plants. Mol Gen Genomics 277:589–600

Dar FA, Mushtaq NU, Saleem S, Rehman RU, Dar TUH, Hakeem KR (2022) Role of epigenetics in modulating phenotypic plasticity against abiotic stresses in plants. Int J Genom 2022:1

Deinlein U, Stephan AB, Horie T, Luo W, Xu G, Schroeder JI (2014) Plant salt-tolerance mechanisms. Trends Plant Sci 19(6):371–379

Demetriadou C, Koufaris C, Kirmizis A (2020) Histone N-alpha terminal modifications: genome regulation at the tip of the tail. Epigenetics Chromatin 13:1–13

Ding X, Zhang D, Gu D, Li Z, Liang H, Zhu H, Jiang Y, Duan X (2022) The histone H3K27 demethylase SlJMJ4 promotes dark-and ABA-induced leaf senescence in tomato. Hortic Res 9:uhab077

Du J, Zhong X, Bernatavichute YV, Stroud H, Feng S, Caro E, Vashisht AA, Terragni J, Chin HG, Tu A, Hetzel J (2012) Dual binding of chromomethylase domains to H3K9me2-containing nucleosomes directs DNA methylation in plants. Cell 151(1):167–180

Eom SH, Baek SA, Kim JK, Hyun TK (2018) Transcriptome analysis in Chinese cabbage (Brassica rapa ssp. pekinensis) provides the role of glucosinolate metabolism in response to drought stress. Molecules 23(5):1186

Feng P, Sun X, Liu X, Li Y, Sun Q, Lu H, Li M, Ding X, Dong Y (2022) Epigenetic regulation of plant tolerance to salt stress by histone acetyltransferase GsMYST1 from wild soybean. Front Plant Sci 13:860056

Ferreira LJ, Donoghue MT, Barros P, Saibo NJ, Santos AP, Oliveira MM (2019) Uncovering differentially methylated regions (DMRs) in a salt-tolerant rice variety under stress: one step towards new regulatory regions for enhanced salt tolerance. Epigenomes 3(1):4

Fowler S, Thomashow MF (2002) Arabidopsis transcriptome profiling indicates that multiple regulatory pathways are activated during cold acclimation in addition to the CBF cold response pathway. Plant Cell 14(8):1675–1690

Gao G, Li J, Li H, Li F, Xu K, Yan G, Chen B, Qiao J, Wu X (2014) Comparison of the heat stress induced variations in DNA methylation between heat-tolerant and heat-sensitive rapeseed seedlings. Breed Sci 64(2):125–133

Gates LA, Foulds CE, O'Malley BW (2017) Histone marks in the 'driver's seat': functional roles in steering the transcription cycle. Trends Biochem Sci 42(12):977–989

Gayacharan, Joel AJ (2013) Epigenetic responses to drought stress in rice (Oryza sativa L.). Physiol Mol Biol Plants 19:379–387
Gehring M, Huh JH, Hsieh TF, Penterman J, Choi Y, Harada JJ, Goldberg RB, Fischer RL (2006) DEMETER DNA glycosylase establishes MEDEA polycomb gene self-imprinting by allele-specific demethylation. Cell 124(3):495–506
Gentry M, Hennig L (2014) Remodelling chromatin to shape development of plants. Exp Cell Res 321(1):40–46
Golicz AA, Bhalla PL, Edwards D, Singh MB (2020) Rice 3D chromatin structure correlates with sequence variation and meiotic recombination rate. Commun Biol 3(1):235
González RM, Ricardi MM, Iusem ND (2011) Atypical epigenetic mark in an atypical location: cytosine methylation at asymmetric (CNN) sites within the body of a non-repetitive tomato gene. BMC Plant Biol 11(1):1–11
Grob S, Grossniklaus U (2017) Chromosome conformation capture-based studies reveal novel features of plant nuclear architecture. Curr Opin Plant Biol 36:149–157
Guleria P, Mahajan M, Bhardwaj J, Yadav SK (2011) Plant small RNAs: biogenesis, mode of action and their roles in abiotic stresses. Genom Proteom Bioinform 9(6):183–199
Guo JE, Wang H (2023) Histone deacetylase gene SlHDA1 regulates drought and salt tolerance in tomato (Solanum lycopersicum). Sci Hortic 313:111899
Han SK, Wagner D (2014) Role of chromatin in water stress responses in plants. J Exp Bot 65(10):2785–2799
Han Z, Yu H, Zhao Z, Hunter D, Luo X, Duan J, Tian L (2016) AtHD2D gene plays a role in plant growth, development, and response to abiotic stresses in Arabidopsis thaliana. Front Plant Sci 7:310
Harachi M, Masui K, Cavenee WK, Mischel PS, Shibata N (2021) Protein acetylation at the interface of genetics, epigenetics and environment in cancer. Meta 11(4):216
Hauser MT, Aufsatz W, Jonak C, Luschnig C (2011) Transgenerational epigenetic inheritance in plants. Biochimic Biophys Acta 1809(8):459–468
Henderson IR, Jacobsen SE (2007) Epigenetic inheritance in plants. Nature 447(7143):418–424
Hou J, Lu D, Mason AS, Li B, Xiao M, An S, Fu D (2019) Non-coding RNAs and transposable elements in plant genomes: emergence, regulatory mechanisms and roles in plant development and stress responses. Planta 250:23–40
Hu H, Du J (2022) Structure and mechanism of histone methylation dynamics in Arabidopsis. Curr Opin Plant Biol 67:102211
Hu Y, Zhang LU, He S, Huang MIN, Tan J, Zhao LIN, Yan S, Li HUI, Zhou KUN, Liang Y, Li L (2012) Cold stress selectively unsilences tandem repeats in heterochromatin associated with accumulation of H3K9ac. Plant Cell Environ 35(12):2130–2142
Hu Z, Song N, Zheng M, Liu X, Liu Z, Xing J, Ma J, Guo W, Yao Y, Peng H, Xin M (2015) Histone acetyltransferase GCN 5 is essential for heat stress-responsive gene activation and thermotolerance in Arabidopsis. Plant J 84(6):1178–1191
Hu Y, Lu Y, Zhao Y, Zhou DX (2019) Histone acetylation dynamics integrates metabolic activity to regulate plant response to stress. Front Plant Sci 10:1236
Huang W, Xian Z, Hu G, Li Z (2016) SlAGO4A, a core factor of RNA-directed DNA methylation (RdDM) pathway, plays an important role under salt and drought stress in tomato. Mol Breed 36:1–13
Huang S, Zhang A, Jin JB, Zhao B, Wang TJ, Wu Y, Wang S, Liu Y, Wang J, Guo P, Ahmad R (2019) Arabidopsis histone H3K4 demethylase JMJ 17 functions in dehydration stress response. New Phytol 223(3):1372–1387
Hyun TK, Uddin MN, Rim Y, Kim JY (2011) Cell-to-cell trafficking of RNA and RNA silencing through plasmodesmata. Protoplasma 248:101–116
Kaiserli E, Perrella G, Davidson ML (2018) Light and temperature shape nuclear architecture and gene expression. Curr Opin Plant Biol 45:103–111
Karan R, DeLeon T, Biradar H, Subudhi PK (2012) Salt stress induced variation in DNA methylation pattern and its influence on gene expression in contrasting rice genotypes. PLoS One 7(6):e40203

Kim JM, Sasaki T, Ueda M, Sako K, Seki M (2015) Chromatin changes in response to drought, salinity, heat, and cold stresses in plants. Front Plant Sci 6:114

Kohli RM, Zhang Y (2013) TET enzymes, TDG and the dynamics of DNA demethylation. Nature 502(7472):472–479

Kong L, Liu Y, Wang X, Chang C (2020) Insight into the role of epigenetic processes in abiotic and biotic stress response in wheat and barley. Int J Mol Sci 21(4):1480

Kouzarides T (2007) Chromatin modifications and their function. Cell 128(4):693–705

Kreps JA, Wu Y, Chang HS, Zhu T, Wang X, Harper JF (2002) Transcriptome changes for Arabidopsis in response to salt, osmotic, and cold stress. Plant Physiol 130(4):2129–2141

Kumar SV, Wigge PA (2010) H2A. Z-containing nucleosomes mediate the thermosensory response in Arabidopsis. Cell 140(1):136–147

Kurdistani SK, Grunstein M (2003) Histone acetylation and deacetylation in yeast. Nat Rev Mol Cell Biol 4(4):276–284

Lang-Mladek C, Popova O, Kiok K, Berlinger M, Rakic B, Aufsatz W, Jonak C, Hauser MT, Luschnig C (2010) Transgenerational inheritance and resetting of stress-induced loss of epigenetic gene silencing in Arabidopsis. Mol Plant 3(3):594–602

Latrasse D, Benhamed M, Henry Y, Domenichini S, Kim W, Zhou DX, Delarue M (2008) The MYST histone acetyltransferases are essential for gametophyte development in Arabidopsis. BMC Plant Biol 8:1–16

Law JA, Jacobsen SE (2010) Establishing, maintaining and modifying DNA methylation patterns in plants and animals. Nat Rev Genet 11(3):204–220

Lee HG, Seo PJ (2019) MYB96 recruits the HDA15 protein to suppress negative regulators of ABA signaling in Arabidopsis. Nat Commun 10(1):1713

Li H, Yan S, Zhao L, Tan J, Zhang Q, Gao F, Wang P, Hou H, Li L (2014) Histone acetylation associated up-regulation of the cell wall related genes is involved in salt stress induced maize root swelling. BMC Plant Biol 14:1–14

Li S, Lin YCJ, Wang P, Zhang B, Li M, Chen S, Shi R, Tunlaya-Anukit S, Liu X, Wang Z, Dai X (2019) The AREB1 transcription factor influences histone acetylation to regulate drought responses and tolerance in Populus trichocarpa. Plant Cell 31(3):663–686

Li Z, Tian P, Huang T, Huang J (2021) Noncoding-RNA-mediated regulation in response to macronutrient stress in plants. Int J Mol Sci 22(20):11205

Liew LC, Singh MB, Bhalla PL (2013) An RNA-seq transcriptome analysis of histone modifiers and RNA silencing genes in soybean during floral initiation process. PLoS One 8(10):e77502

Lindroth AM, Cao X, Jackson JP, Zilberman D, McCallum CM, Henikoff S, Jacobsen SE (2001) Requirement of CHROMOMETHYLASE3 for maintenance of CpXpG methylation. Science 292(5524):2077–2080

Liu C, Lu F, Cui X, Cao X (2010) Histone methylation in higher plants. Annu Rev Plant Biol 61:395–420

Liu X, Yang S, Zhao M, Luo M, Yu CW, Chen CY, Tai R, Wu K (2014) Transcriptional repression by histone deacetylases in plants. Mol Plant 7(5):764–772

Liu H, Wang C, Lee S, Deng Y, Wither M, Oh S, Ning F, Dege C, Zhang Q, Liu X, Johnson AM (2017) Clipping of arginine-methylated histone tails by JMJD5 and JMJD7. Proc Natl Acad Sci 114(37):E7717–E7726

Liu J, Zhi P, Wang X, Fan Q, Chang C (2019) Wheat WD40-repeat protein TaHOS15 functions in a histone deacetylase complex to fine-tune defense responses to Blumeria graminis f. sp. tritici. J Exp Bot 70(1):255–268

Liu M, Jiang J, Han Y, Shi M, Li X, Wang Y, Dong Z, Yang C (2022) Functional characterization of the lysine-specific histone demethylases family in soybean. Plan Theory 11(11):1398

Lu L, Wei W, Tao JJ, Lu X, Bian XH, Hu Y, Cheng T, Yin CC, Zhang WK, Chen SY, Zhang JS (2021) Nuclear factor Y subunit GmNFYA competes with GmHDA13 for interaction with GmFVE to positively regulate salt tolerance in soybean. Plant Biotechnol J 19(11):2362–2379

Luo X, He Y (2020) Experiencing winter for spring flowering: a molecular epigenetic perspective on vernalization. J Integr Plant Biol 62(1):104–117

Mach J (2017) The long-noncoding RNA ELENA1 functions in plant immunity, vol 29, p 916

Mantri N, Patade V, Penna S, Ford R, Pang E (2012) Abiotic stress responses in plants: present and future. In: Abiotic stress responses in plants: metabolism, productivity and sustainability, pp 1–19
Marconi G, Pace R, Traini A, Raggi L, Lutts S, Chiusano M, Guiducci M, Falcinelli M, Benincasa P, Albertini E (2013) Use of MSAP markers to analyse the effects of salt stress on DNA methylation in rapeseed (Brassica napus var. oleifera). PLoS One 8(9):e75597
Mehdi S, Derkacheva M, Ramström M, Kralemann L, Bergquist J, Hennig L (2016) The WD40 domain protein MSI1 functions in a histone deacetylase complex to fine-tune abscisic acid signaling. Plant Cell 28(1):42–54
Mehrotra R, Bhalothia P, Bansal P, Basantani MK, Bharti V, Mehrotra S (2014) Abscisic acid and abiotic stress tolerance–different tiers of regulation. J Plant Physiol 171(7):486–496
Miryeganeh M (2021) Plants' epigenetic mechanisms and abiotic stress. Genes 12(8):1106
Miskiewicz J, Tomczyk K, Mickiewicz A, Sarzynska J, Szachniuk M (2017) Bioinformatics study of structural patterns in plant microRNA precursors. Biomed Res Int 2017:6783010
Mittler R (2002) Oxidative stress, antioxidants and stress tolerance. Trends Plant Sci 7(9):405–410
Mittler R (2006) Abiotic stress, the field environment and stress combination. Trends Plant Sci 11(1):15–19
Mittler R, Vanderauwera S, Gollery M, Van Breusegem F (2004) Reactive oxygen gene network of plants. Trends Plant Sci 9(10):490–498
Mumbach MR, Rubin AJ, Flynn RA, Dai C, Khavari PA, Greenleaf WJ, Chang HY (2016) HiChIP: efficient and sensitive analysis of protein-directed genome architecture. Nat Methods 13(11):919–922
Naika M, Shameer K, Mathew OK, Gowda R, Sowdhamini R (2013) STIFDB2: an updated version of plant stress-responsive transcription factor database with additional stress signals, stress-responsive transcription factor binding sites and stress-responsive genes in Arabidopsis and rice. Plant Cell Physiol 54(2):e8–e8
Nakashima K, Yamaguchi-Shinozaki K, Shinozaki K (2014) The transcriptional regulatory network in the drought response and its crosstalk in abiotic stress responses including drought, cold, and heat. Front Plant Sci 5:170
Ng DW, Wang T, Chandrasekharan MB, Aramayo R, Kertbundit S, Hall TC (2007) Plant SET domain-containing proteins: structure, function and regulation. Biochim Biophys Acta-Gene Struct Expr 1769(5–6):316–329
Ouyang W, Xiong D, Li G, Li X (2020) Unraveling the 3D genome architecture in plants: present and future. Mol Plant 13(12):1676–1693
Palakolanu SR, Gupta S, Yeshvekar RK, Chakravartty N, Kaliamoorthy S, Shankhapal AR, Vempati AS, Kuriakose B, Lekkala SP, Philip M, Perumal RC (2022) Genome-wide miRNAs profiles of pearl millet under contrasting high vapor pressure deficit reveal their functional roles in drought stress adaptations. Physiol Plant 174(1):e13521
Panara V, Monteiro R, Koltowska K (2022) Epigenetic regulation of endothelial cell lineages during zebrafish development—new insights from technical advances. Front Cell Dev Biol 10:891538
Pandey R, MuÈller A, Napoli CA, Selinger DA, Pikaard CS, Richards EJ, Bender J, Mount DW, Jorgensen RA (2002) Analysis of histone acetyltransferase and histone deacetylase families of Arabidopsis thaliana suggests functional diversification of chromatin modification among multicellular eukaryotes. Nucleic Acids Res 30(23):5036–5055
Pandey G, Sharma N, Pankaj Sahu P, Prasad M (2016) Chromatin-based epigenetic regulation of plant abiotic stress response. Curr Genomics 17(6):490–498
Park HJ, Baek D, Cha JY, Liao X, Kang SH, McClung CR, Lee SY, Yun DJ, Kim WY (2019) HOS15 interacts with the histone deacetylase HDA9 and the evening complex to epigenetically regulate the floral activator GIGANTEA. Plant Cell 31(1):37–51
Parrilla-Doblas JT, Roldán-Arjona T, Ariza RR, Córdoba-Cañero D (2019) Active DNA demethylation in plants. Int J Mol Sci 20(19):4683
Patanun O, Ueda M, Matsui A, Tanaka M, Sakakibara H, Narangajavana J, Seki M (2017) The histone deacetylase inhibitor suberoylanilide hydroxamic acid alleviates salinity stress in cassava. Front Plant Sci 7:226406

Pecinka A, Dinh HQ, Baubec T, Rosa M, Lettner N, Scheid OM (2010) Epigenetic regulation of repetitive elements is attenuated by prolonged heat stress in Arabidopsis. Plant Cell 22(9):3118–3129

Pei L, Li G, Lindsey K, Zhang X, Wang M (2021) Plant 3D genomics: the exploration and application of chromatin organization. New Phytol 230(5):1772–1786

Penterman J, Zilberman D, Huh JH, Ballinger T, Henikoff S, Fischer RL (2007) DNA demethylation in the Arabidopsis genome. Proc Natl Acad Sci 104(16):6752–6757

Probst AV, Scheid OM (2015) Stress-induced structural changes in plant chromatin. Curr Opin Plant Biol 27:8–16

Qian W, Miki D, Zhang H, Liu Y, Zhang X, Tang K, Kan Y, La H, Li X, Li S, Zhu X (2012) A histone acetyltransferase regulates active DNA demethylation in Arabidopsis. Science 336(6087):1445–1448

Ravichandran S, Ragupathy R, Edwards T, Domaratzki M, Cloutier S (2019) MicroRNA-guided regulation of heat stress response in wheat. BMC Genomics 20(1):1–16

Rizhsky L, Liang H, Mittler R (2002) The combined effect of drought stress and heat shock on gene expression in tobacco. Plant Physiol 130(3):1143–1151

Rizhsky L, Liang H, Shuman J, Shulaev V, Davletova S, Mittler R (2004) When defense pathways collide. The response of Arabidopsis to a combination of drought and heat stress. Plant Physiol 134(4):1683–1696

Roca Paixão JF, Gillet FX, Ribeiro TP, Bournaud C, Lourenço-Tessutti IT, Noriega DD, Melo BPD, de Almeida-Engler J, Grossi-de-Sa MF (2019) Improved drought stress tolerance in Arabidopsis by CRISPR/dCas9 fusion with a histone AcetylTransferase. Sci Rep 9(1):8080

Rosa M, Von Harder M, Aiese Cigliano R, Schlögelhofer P, Mittelsten Scheid O (2013) The Arabidopsis SWR1 chromatin-remodeling complex is important for DNA repair, somatic recombination, and meiosis. Plant Cell 25(6):1990–2001

Rossel JB, Wilson IW, Pogson BJ (2002) Global changes in gene expression in response to high light in Arabidopsis. Plant Physiol 130(3):1109–1120

Ruthenburg AJ, Li H, Patel DJ, David Allis C (2007) Multivalent engagement of chromatin modifications by linked binding modules. Nat Rev Mol Cell Biol 8(12):983–994

Sahu PP, Pandey G, Sharma N, Puranik S, Muthamilarasan M, Prasad M (2013) Epigenetic mechanisms of plant stress responses and adaptation. Plant Cell Rep 32:1151–1159

Sako K, Kim JM, Matsui A, Nakamura K, Tanaka M, Kobayashi M, Saito K, Nishino N, Kusano M, Taji T, Yoshida M (2016) Ky-2, a histone deacetylase inhibitor, enhances high-salinity stress tolerance in Arabidopsis thaliana. Plant Cell Physiol 57(4):776–783

Saze H, Tsugane K, Kanno T, Nishimura T (2012) DNA methylation in plants: relationship to small RNAs and histone modifications, and functions in transposon inactivation. Plant Cell Physiol 53(5):766–784

Shahbazian MD, Grunstein M (2007) Functions of site-specific histone acetylation and deacetylation. Annu Rev Biochem 76:75–100

Shen Y, Wei W, Zhou DX (2015) Histone acetylation enzymes coordinate metabolism and gene expression. Trends Plant Sci 20(10):614–621

Shi D, Zhuang K, Xia Y, Zhu C, Chen C, Hu Z, Shen Z (2017) Hydrilla verticillata employs two different ways to affect DNA methylation under excess copper stress. Aquat Toxicol 193:97–104

Shin H, Choi WL, Lim JY, Huh JH (2022) Epigenome editing: targeted manipulation of epigenetic modifications in plants. Genes Genom 44:1–9

Song Y, Ji D, Li S, Wang P, Li Q, Xiang F (2012) The dynamic changes of DNA methylation and histone modifications of salt responsive transcription factor genes in soybean. PLoS One 7(7):e41274

Song Y, Liu L, Feng Y, Wei Y, Yue X, He W, Zhang H, An L (2015) Chilling-and freezing-induced alterations in cytosine methylation and its association with the cold tolerance of an alpine subnival plant, Chorispora bungeana. PLoS One 10(8):e0135485

Song T, Zhang Q, Wang H, Han J, Xu Z, Yan S, Zhu Z (2018) OsJMJ703, a rice histone demethylase gene, plays key roles in plant development and responds to drought stress. Plant Physiol Biochem 132:183–188

Song J, Henry HA, Tian L (2019) Brachypodium histone deacetylase BdHD1 positively regulates ABA and drought stress responses. Plant Sci 283:355–365
Stępiński D (2012) Levels of DNA methylation and histone methylation and acetylation change in root tip cells of soybean seedlings grown at different temperatures. Plant Physiol Biochem 61:9–17
Steward N, Kusano T, Sano H (2000) Expression of ZmMET1, a gene encoding a DNA methyltransferase from maize, is associated not only with DNA replication in actively proliferating cells, but also with altered DNA methylation status in cold-stressed quiescent cells. Nucleic Acids Res 28(17):3250–3259
Steward N, Ito M, Yamaguchi Y, Koizumi N, Sano H (2002) Periodic DNA methylation in maize nucleosomes and demethylation by environmental stress. J Biol Chem 277(40):37741–37746
Stroud H, Do T, Du J, Zhong X, Feng S, Johnson L, Patel DJ, Jacobsen SE (2014) Non-CG methylation patterns shape the epigenetic landscape in Arabidopsis. Nat Struct Mol Biol 21(1):64–72
Sun L, Song G, Guo W, Wang W, Zhao H, Gao T, Lv Q, Yang X, Xu F, Dong Y, Pu L (2019) Dynamic changes in genome-wide histone3 lysine27 trimethylation and gene expression of soybean roots in response to salt stress. Front Plant Sci 10:1031
Temel A, Janack B, Humbeck K (2017) Drought stress-related physiological changes and histone modifications in barley primary leaves at HSP17 gene. Agronomy 7(2):43
Tessadori F, van Zanten M, Pavlova P, Clifton R, Pontvianne F, Snoek LB, Millenaar FF, Schulkes RK, van Driel R, Voesenek LA, Spillane C (2009) Phytochrome B and histone deacetylase 6 control light-induced chromatin compaction in Arabidopsis thaliana. PLoS Genet 5(9):e1000638
Tittel-Elmer M, Bucher E, Broger L, Mathieu O, Paszkowski J, Vaillant I (2010) Stress-induced activation of heterochromatic transcription. PLoS Genet 6(10):e1001175
Tubiello FN, Soussana JF, Howden SM (2007) Crop and pasture response to climate change. Proc Natl Acad Sci 104(50):19686–19690
Ueda M, Seki M (2020) Histone modifications form epigenetic regulatory networks to regulate abiotic stress response. Plant Physiol 182(1):15–26
Ueda M, Matsui A, Tanaka M, Nakamura T, Abe T, Sako K, Sasaki T, Kim JM, Ito A, Nishino N, Shimada H (2017) The distinct roles of class I and II RPD3-like histone deacetylases in salinity stress response. Plant Physiol 175(4):1760–1773
Ueda M, Matsui A, Nakamura T, Abe T, Sunaoshi Y, Shimada H, Seki M (2018) Versatility of HDA19-deficiency in increasing the tolerance of Arabidopsis to different environmental stresses. Plant Signal Behav 13(8):e1475808
van Zanten M, Tessadori F, McLoughlin F, Smith R, Millenaar FF, van Driel R, Voesenek LA, Peeters AJ, Fransz P (2010) Photoreceptors CRYTOCHROME2 and phytochrome B control chromatin compaction in Arabidopsis. Plant Physiol 154(4):1686–1696
van Zanten M, Tessadori F, Peeters AJ, Fransz P (2012) Shedding light on large-scale chromatin reorganization in Arabidopsis thaliana. Mol Plant 5(3):583–590
Vriet C, Hennig L, Laloi C (2015) Stress-induced chromatin changes in plants: of memories, metabolites and crop improvement. Cell Mol Life Sci 72:1261–1273
Wang W, Zhao X, Pan Y, Zhu L, Fu B, Li Z (2011a) DNA methylation changes detected by methylation-sensitive amplified polymorphism in two contrasting rice genotypes under salt stress. J Genet Genomics 38(9):419–424
Wang WS, Pan YJ, Zhao XQ, Dwivedi D, Zhu LH, Ali J, Fu BY, Li ZK (2011b) Drought-induced site-specific DNA methylation and its association with drought tolerance in rice (Oryza sativa L.). J Exp Bot 62(6):1951–1960
Wang J, Meng X, Yuan C, Harrison AP, Chen M (2016) The roles of cross-talk epigenetic patterns in Arabidopsis thaliana. Brief Funct Genomics 15(4):278–287
Wang J, Meng X, Dobrovolskaya OB, Orlov YL, Chen M (2017) Non-coding RNAs and their roles in stress response in plants. Genomics Proteomics Bioinformatics 15(5):301–312
Wang J, Mei J, Ren G (2019) Plant microRNAs: biogenesis, homeostasis, and degradation. Front Plant Sci 10:360
Xie SS, Duan CG (2023) Epigenetic regulation of plant immunity: from chromatin codes to plant disease resistance. aBIOTECH 4:1–16

Xie M, Yu B (2015) siRNA-directed DNA methylation in plants. Curr Genomics 16(1):23–31
Yang C, Shen W, Chen H, Chu L, Xu Y, Zhou X, Liu C, Chen C, Zeng J, Liu J, Li Q (2018) Characterization and subcellular localization of histone deacetylases and their roles in response to abiotic stresses in soybean. BMC Plant Biol 18(1):1–13
Yu X, Gao Q, Chen G, Guo JE, Guo X, Tang B, Hu Z (2018) SlHDA5, a tomato histone deacetylase gene, is involved in responding to salt, drought, and ABA. Plant Mol Biol Report 36:36–44
Zemach A, Kim MY, Hsieh PH, Coleman-Derr D, Eshed-Williams L, Thao K, Harmer SL, Zilberman D (2013) The Arabidopsis nucleosome remodeler DDM1 allows DNA methyltransferases to access H1-containing heterochromatin. Cell 153(1):193–205
Zhang B, Tieman DM, Jiao C, Xu Y, Chen K, Fei Z, Giovannoni JJ, Klee HJ (2016) Chilling-induced tomato flavor loss is associated with altered volatile synthesis and transient changes in DNA methylation. Proc Natl Acad Sci 113(44):12580–12585
Zhang H, Lang Z, Zhu JK (2018) Dynamics and function of DNA methylation in plants. Nat Rev Mol Cell Biol 19(8):489–506
Zhao J, Li M, Gu D, Liu X, Zhang J, Wu K, Zhang X, da Silva JAT, Duan J (2016) Involvement of rice histone deacetylase HDA705 in seed germination and in response to ABA and abiotic stresses. Biochem Biophys Res Commun 470(2):439–444
Zhao S, Zhang B, Yang M, Zhu J, Li H (2018) Systematic profiling of histone readers in Arabidopsis thaliana. Cell Rep 22(4):1090–1102
Zheng Y, Ding Y, Sun X, Xie S, Wang D, Liu X, Su L, Wei W, Pan L, Zhou DX (2016) Histone deacetylase HDA9 negatively regulates salt and drought stress responsiveness in Arabidopsis. J Exp Bot 67(6):1703–1713
Zheng M, Liu X, Lin J, Liu X, Wang Z, Xin M, Yao Y, Peng H, Zhou DX, Ni Z, Sun Q (2019) Histone acetyltransferase GCN 5 contributes to cell wall integrity and salt stress tolerance by altering the expression of cellulose synthesis genes. Plant J 97(3):587–602

The Role of microRNA in Stress Signaling and Adaptive Response in Plants

5

Madhu Tiwari

Abstract

In the sphere of agriculture, the profound challenge of ensuring sustainable food production, is vital for our very survival, remains front and center. Crop productivity, essential for feeding the world's population, is consistently threatened by a range of environmental stressors, both abiotic and biotic. However, the study of sustainable agriculture, particularly through the lens of microRNA research, offers a promising avenue for enhancing agricultural practices. MicroRNAs (miRNAs), small noncoding RNA molecules, serve as pivotal regulators in modulating gene expression, providing valuable insights into the intricate molecular mechanisms underlying plant responses to abiotic stressors. Plants have evolved sophisticated adaptive mechanisms to cope with adverse environmental conditions such as drought, high temperatures, cold, salinity, and nutrient scarcity. In this context, miRNAs emerge as essential contributors, playing a central role in alleviating the deleterious effects of these environmental stresses on plant physiology and molecular processes. Furthermore, miRNAs play a significant role in enhancing plant defenses against biotic stress, thereby fortifying resistance to pathogens. Comprehending these mechanisms enables researchers to harness the potential of microRNAs (miRNAs) for enhancing crop resilience and sustainability in agriculture. This chapter provides a thorough examination of miRNA biogenesis, regulation, and their involvement in shaping plant responses to environmental stress. Various miRNAs have been recognized as pivotal regulators influencing genes in plants and crops, as well as impacting transcription factors and stress response pathways. By manipulating these miRNAs, researchers aim to strengthen crops against challenging conditions and ensure stable yields in adverse circumstances. With molecular insights into miRNAs, we are better pre-

M. Tiwari (✉)
Aix Marseille University, CEA, CNRS UMR 7265, Bioscience and Biotechnology Institute of Aix Marseille, CEA/Cadarache, Saint-Paul-lez-Durance, France

A. K. Mishra (ed.), *Stress Biology in Photosynthetic Organisms*,
https://doi.org/10.1007/978-981-97-1883-2_5

pared to navigate a time marked by unpredictable climate patterns and increasing food demands. This chapter directs readers' attention to the promising potential of microRNAs (miRNAs) in enhancing plants' resilience to environmental stresses, thereby revolutionizing agricultural practices amid uncertainties in climate conditions.

Keyword

Environmental stress · miRNA · Gene expression · Plant adaptation · Food security

5.1 Introduction

5.1.1 Understanding Plant Stress Responses

The escalating global population and heightened demand for livestock necessitate a substantial surge in food and fodder production. As per the projections from the United Nations Population Division, the world's population is anticipated to reach 8.3 billion by 2030 (Datta et al. 2019; Begum 2022). Ensuring that we provide higher quality food to sustain this growing population is a crucial mission for scientists in the twenty-first century. Additionally, changing climate conditions are harming agricultural productivity worldwide. Extreme weather events are a major factor in causing both nonliving and living stressors, leading to an annual loss of more than 50% of crop yields globally (Chinweze 2023; Al-Tawaha et al. 2022).

It has been noted that plants manifest a diverse array of adaptive responses to environmental stressors, showcasing their ability to thrive in adverse conditions (Sun et al. 2019; Zhang et al. 2022a, d; Begum 2022; Liu et al. 2023). When faced with drought, they often close their stomata to reduce water loss and may alter their root systems to seek moisture deeper in the soil.

Under heat stress conditions, the induction of heat shock proteins and antioxidant enzymes occurs, serving to safeguard cellular structures and prevent damage (Wani et al. 2020; Zhang et al. 2022d). Conversely, in cold conditions, plants undergo cold acclimation, adjusting the composition of their cell membranes and producing antifreeze proteins to prevent ice formation. The imposition of salinity stress on plants induces ion exclusion mechanisms and prompts the accumulation of compatible solutes as a means to uphold osmotic equilibrium. In a parallel manner, oxidative stress triggers the synthesis of antioxidant enzymes and the accumulation of nonenzymatic antioxidants. In flooding stress conditions, plants necessitate aerenchyma formation and adaptations to low oxygen levels. Light stress triggers photoprotective mechanisms and shade avoidance responses (Barczak et al. 2022). In a comparable context, biotic stresses, encompassing viral, fungal, bacterial, insect, vector, and nematode factors, have the potential to significantly impair the productivity of horticultural crops and the overall physiological mechanisms of the plant (Kumar et al. 2021). In the presence of pathogens, plants deploy phytochemicals and pathogenesis-related proteins to deter attackers (Šečić et al. 2021). Additionally, nutrient stress leads to the regulation of nutrient uptake and storage strategies. The nuanced stress responses manifesting distinctiveness among plant species are of paramount significance in orchestrating plant survival strategies and facilitating adaptation to dynamically evolving environmental conditions. This adaptative capability is integral for preserving ecosystem resilience and sustaining agricultural productivity within the context of contemporary climate variability.

Abiotic responses do vary from plant to plant but are vital for their survival and how they adapt to changes in their surroundings. These factors pose significant challenges to agriculture and can greatly reduce crop yields, compromise food security, and contribute to economic losses for farmers and nations (Kerchev et al. 2020; Chinweze 2023). Comprehending these responses is imperative for formulating strategies aimed at enhancing crop resilience amidst evolving environmental conditions and enhancing overall agricultural productivity (Sharma et al. 2022). The deleterious effects of these stresses on plant growth and development manifest through induced physiological and molecular alterations.

Researchers have discovered that during times of environmental stress, plants produce specific microRNAs (miRNAs) which interact with stress-related genes to enhance their survival (Praveen et al. 2023; Zhang et al. 2022a, d; Begum 2022; Liu et al. 2023). These miRNAs are now the focus of scientific investigation as a means to gain deeper insights into how plants respond to environmental pressures. MicroRNAs (miRNAs) play an important role in the regulation of gene expression, and our objective is to elucidate their mechanisms in plants with the aim of enhancing crop resilience to various stressors. Although this research is still in its early phases, it holds the potential to bolster crops' ability to withstand challenging environmental conditions.

5.1.2 The Emergence of microRNAs (miRNAs)

Over recent years, microRNAs (miRNAs) have surfaced as a potent genetic tool with the capacity to unveil molecular insights into both abiotic and biotic stress tolerance. Their potential extends to the regulation of stress responses in plants at the molecular level (Verma and Sarkar 2021). Plant microRNAs (miRNAs) frequently target diverse families of transcription factors (TFs), which oversee the expression of genes associated with specific traits. Recognized as integral components within gene regulatory networks, a thorough comprehension of miRNAs is acknowledged to substantially propel our grasp of plant responses to environmental stresses. The miRNA-mediated stress regulatory networks are currently being acknowledged as innovative tools for enhancing both abiotic and biotic stress tolerance in crops (Raza et al. 2023).

MicroRNAs (miRNAs) constitute a distinct class of single-stranded, noncoding RNA molecules characterized by high conservation and typically measuring 20–24 nucleotides in length. Their primary function involves post-transcriptional gene regulation, exerting their influence across various organisms, including plants (Begum 2022; Pandita 2022; Singh et al. 2023; Mittal et al. 2023; Shang et al.

2023). In the realm of gene expression control, miRNAs engage in binding interactions with complementary messenger RNAs (mRNAs), thereby initiating either target mRNA degradation or the inhibition of its translation into a functional protein. This regulatory paradigm enables miRNAs to finely modulate a diverse array of cellular processes, encompassing responses to both biotic and abiotic stresses in plants.

The initial identification of miRNA genes in *Caenorhabditis elegans* by Lee et al. in 1993 marked the inception, and since then, there has been a notable surge in the discovery of an expanding array of plant miRNA families, rendering them a prominent research focus. The initial exploration of miRNAs in plants centered on their involvement in responding to environmental stresses. Among the earliest stress-related miRNAs uncovered in plants is miR395. This milestone discovery dates back to 2005 when scientists were examining how *Arabidopsis thaliana*, a model plant, copes with sulfate deficiency, a condition that induces stress. Sulfur is an essential nutrient for plants, and when they experience sulfate deficiency, it can lead to stress. During this study, scientists found that miR395 is upregulated in response to sulfate starvation. This specific miRNA is recognized for its regulatory role in modulating the expression of genes associated with sulfate uptake and assimilation, thereby facilitating plants in adapting to conditions characterized by sulfur deficiency (Jones-Rhoades and Bartel 2004).

5.1.3 Biogenesis and Regulation of miRNAs

In the realm of miRNAs, "biogenesis" denotes the intricate process by which these regulatory molecules are generated within a cell. It is established knowledge that miRNAs, short regulatory RNAs, are encoded by endogenous miRNA genes (MIR-genes) (Wang et al. 2019b). In plants, the genome contains more than 100 *MIR* genes, and they are found in diverse genomic locations, including within genes (intronic), between genes (intergenic), or in clusters (Yu et al. 2017; Axtell et al. 2011). The regulation of MIR gene expression is meticulously governed by a variety of transcription factors (TFs), as mentioned previously. The initiation of plant miRNA biogenesis commences with the transcription of primary miRNA transcripts (pri-miRNAs), which are generally several hundred nucleotides in length and transcribed by RNA polymerase II enzymes (Bajczyk et al. 2023). Within the nucleus, the primary miRNA transcript (pri-miRNA) undergoes processing orchestrated by the Drosha-DGCR8 (DiGeorge syndrome critical region 8) complex. Drosha enzymatically cleaves the pri-miRNA, resulting in the formation of a shorter hairpin structure recognized as the precursor miRNA (pre-miRNA), typically spanning around 70 nucleotides. Subsequent to cleavage, the pre-miRNA translocates from the nucleus to the cytoplasm, facilitated by the exportin-5 protein (Ergin and Çetinkaya 2022). Upon reaching the cytoplasm, the pre-miRNA undergoes additional processing facilitated by an enzyme known as Dicer. Dicer, an enzyme of the ribonuclease III class, cleaves the pre-miRNA, generating a double-stranded RNA duplex comprising the mature miRNA (guide strand) and its complementary strand.

Subsequently, the guide strand is incorporated into the RNA-induced silencing complex (RISC), housing Argonaute proteins. The RISC complex utilizes the guide strand to selectively target and regulate specific messenger RNAs (mRNAs) (Li and Yu 2021; Bajczyk et al. 2023).

The regulation of miRNA gene expression can occur at the transcriptional level. Transcription factors and other regulatory proteins have the capability to bind to the promoter regions of MIR genes, thereby modulating and orchestrating their transcriptional activity in response to diverse cellular signals. Once miRNAs are processed and functional, their activity can be regulated by other molecules. Proteins such as RNA-binding proteins and competing endogenous RNAs (ceRNAs) can interact with miRNAs to modulate their availability and effectiveness. miRNAs can be degraded by exonucleases, which control the stability of mature miRNAs. This turnover process helps regulate miRNA abundance and activity (Bajczyk et al. 2023).

5.2 Environmental Stressors and miRNA Regulatory Mechanisms

5.2.1 Abiotic Stress

Within the domain of plant biology, which centers on growth and development, it is firmly established that gene expression patterns in the plant system are markedly influenced by both biotic and abiotic stress factors. The regulation of this gene expression is intricately governed by regulatory elements, predominantly constituted by transcription factors and miRNAs (Ebrahimi-Khaksefidi et al. 2015; Raza et al. 2023). An accumulating body of compelling empirical evidence substantiates the pivotal role ascribed to the regulation of miRNA levels in orchestrating the reprogramming of plant responses to the dynamic challenges imposed by their environment, encompassing a spectrum of both abiotic and biotic stressors (Brant and Budak 2018; Nogoy et al. 2018). Let's delve into the specifics of various stressors that plants face and explore how miRNAs orchestrate their responses, ensuring their resilience and survival in a changing environment.

5.2.1.1 Salinity and Drought Stress

Drought and salinity stresses pose significant challenges to plant growth and yield, prompting intensive investigation by botanists seeking to comprehend their adverse effects and the adaptive mechanisms employed by plants. The exploration of microRNAs (miRNAs) has emerged as a key focus for gaining a deeper understanding of the molecular mechanisms underlying drought stress. There is a growing body of evidence indicating the study of drought-responsive miRNAs in various dicotyledonous and monocotyledonous species. In response to drought stress, the phytohormone abscisic acid (ABA) plays a central role in initiating adaptive responses to cope with water depletion, primarily through the regulation of stomatal conductance. ABA also engages in collaborative interactions with other plant hormones, including salicylic acid (SA), auxin (AUX), gibberellins (GA), ethylene

(ET), brassinosteroid (BR), jasmonic acid (JA), among others, collectively conferring resistance against drought stress.

Significantly, small RNA sequencing studies have identified numerous drought-responsive miRNAs in *Arabidopsis*, tomato, maize, barley, and rice, highlighting the conservation and diversity of these regulatory elements across different plant species (Liu et al. 2008; Aravind et al. 2017; Ferdous et al. 2017; Zhou et al. 2010). A plethora of drought-inducible miRNAs has been discovered and characterized in plants. These miRNAs primarily target regulatory genes that exert influence over various physiological processes, including growth, development, osmotic stress tolerance, antioxidant defense mechanisms, phytohormone-mediated signaling pathways, and the regulation of senescence, collectively contributing to the plant's adaptive responses under conditions of drought stress (Singh et al. 2023).

In *Arabidopsis*, the research conducted by Liu et al. (2008) identified 14 stress-inducible miRNAs, including miR168, miR171, and miR396, responsive to various stresses such as salt, drought, and cold. In a separate investigation by Yu et al. (2019), the overexpression of gma-miR169c in *Arabidopsis* heightened drought stress sensitivity, resulting in leaf water deficit and reduced plant survival rate. Similarly, in rice exposed to drought conditions, miR156, miR159, miR160, miR164, miR167, miR169, miR319, miR396, and miR397 were reported to mitigate negative effects by regulating hormones, development, physiology, and stress responses (Nadarajah and Kumar 2019).

In *Triticum turgidum*, under drought conditions, a study by Islam et al. (2022) revealed the differential expression of 438 miRNAs, among which 13 exhibited significant regulation in response to drought, with several targeting transcription factors.

In maize, drought stress prompted the fine-tuning of target gene expression by several reported miRNAs, including miR159, miR160, miR167, miR389a, miR393, miR397b, miR402, and miR474. Notably, miR398a-MYB101/MYB33 and miR164-NAC were downregulated, whereas miR393-MYB101/MYB33 and miR402-MYB101/MYB33 were upregulated, as reported by Das and Mondal (2021).

In *Medicago truncatula*, Zhang et al. (2022c) observed the upregulation of miR398a/b and miR408 under drought stress, with subsequent downregulation of their target transcripts.

Regarding salt stress, Melino and Tester (2023) highlighted its complex impact on plants through hyperosmotic stress, ion toxicities, and imbalances, as well as oxidative stress. Studies in *Arabidopsis* and *Setaria viridis* have focused on miR397. Overexpression of SvMIR397 or AtMIR397 in *Arabidopsis* under 150 mM sodium chloride (NaCl) treatment led to the repression of the target gene LACCASE (LAC), resulting in decreased lignin content and increased salt stress sensitivity (Nguyen et al. 2020).

5.2.1.2 Nutrient Stress

For the optimal growth and development of plants, it is crucial to maintain a sustained and sufficient supply of soil nutrients. Several miRNAs have been identified for their specific roles in facilitating nutrient absorption and transport in plants during nutrient scarcity (Shahzad et al. 2021). Potassium (K+) levels are critical determinants of plant growth, development, and stress responses, and they are typically maintained at stable concentrations within plant cells. MiR171 has been implicated in plant responses to low-K+ stress, as evidenced by the significant upregulation of miR171 expression in wheat (*Triticum aestivum* L.) subjected to low-K+ treatment for 4 days (Zhang et al. 2022b). In sugarcane, the expression patterns of miRNAs under different species, stress types, tissues (seedlings, leaves, spikelets, and roots), and growth conditions (field, greenhouse, hydroponic culture system) have shown differential expression of MiR396 and MiR171 in most cases (Shriram et al. 2016).

Nitrogen (N) is crucial for plant growth and survival, being a key component of nucleic acids, proteins, and chlorophyll. Its deficiency results in severe growth retardation (Bhat et al. 2023). Recent findings have unveiled miRNA-mediated regulation of plant growth and metabolism under N deficiency. Liang et al. (2012) reported differential expression of N-deficiency-responsive miRNAs in *Arabidopsis*. N-starvation increased the expression of miR160, miR780, miR842, and miR846, while miR169, miR398, miR399, miR408, miR827, and miR857 were downregulated in *Arabidopsis* (Liang et al. 2012). In peanut, N deficiency significantly reduced lateral root and primary root length, associated with upregulation of miR396, miR160, miR393, and miR167, and downregulation of GRF and AFB3 (Li et al. 2021). Fontana et al. (2020) studied the effects of phosphorus (P) deficiency on the expression of miRNAs and their associated targets in cotton. Interestingly, miR390, miR160, and miR164 were found to play roles in regulating lateral root development, showing increased expression during early stages of phosphorus (P) deficiency treatment, leading to improved root growth and differentiation in plants.

5.2.1.3 UV Radiation Stress

The environmental challenge of ultraviolet (UV) radiation stress is a significant concern for plants. UV radiation encompasses UV-A (long-wave), UV-B (medium-wave), and UV-C (short-wave) radiation within the electromagnetic spectrum. While UV-C is predominantly filtered out by the Earth's atmosphere, UV-A and UV-B can directly impact plants. Among these radiations, UV-B (280–320 nm) has been identified as particularly detrimental to plant growth and development, primarily attributed to membrane damage induced by the generation of reactive oxygen species (ROS) (Podolec et al. 2021). Plants have evolved diverse strategies to alleviate the damage induced by UV radiation and adapt to this environmental stressor. In *Arabidopsis*, several UV-B stress-induced miRNAs have been identified, including miR319, miR157, miR170, miR169, miR401, miR171, miR393, miR172, and miR1398. These miRNAs are implicated in the regulatory network that helps plants respond to and cope with the challenges posed by UV-B radiation (Raza et al. 2023). In tomato leaves subjected to various light qualities, Dong et al. (2020) identified a total of 108 known miRNAs and 142 novel miRNAs. Similarly, Fang et al. (2019) observed improved antioxidant activity and alleviation of UVB-induced damage through the overexpression of Lsi1 (low silicon rice 1 gene). In their study, two types of genetically modified rice, one with reduced Lsi1 (Lsi1-RNAi) and the other with elevated Lsi1 (Lsi1-OX), exhibited distinct miRNA patterns compared to regular rice when exposed to UV-B radiation.

5.2.1.4 Extreme Temperature Stress

Extreme temperatures, whether high or low, exert substantial effects on miRNA expressions in plants. MiR398 is a recognized miRNA responsive to ambient temperature changes (Lee et al. 2010). Under high-temperature conditions, certain miRNAs such as miR398 and miR319 are upregulated in plants. For example, in wheat, miR398 is induced in response to heat stress, regulating genes associated with antioxidant defense (Zhang et al. 2015). Additionally, Ragupathy et al. (2016) observed downregulation of miR156, Tae-miR818, miR169, miR528, and miR398 in wheat under heat stress conditions. The heat-responsive miR528 and miR9662 were found to play a role in regulating antioxidant activity and stress-responsive mitochondrial proteins in wheat (Ravichandran et al. 2019). Lin et al. (2018) demonstrated that overexpression of miR160 significantly enhanced thermotolerance in transgenic *Arabidopsis* by affecting target genes such as ARF10, ARF16, and ARF17, which control the production of heat shock proteins essential for plant survival in hot conditions.

Low temperatures and frost also pose challenges to plant survival, resulting in growth retardation and yield loss. In *Chrysanthemum dichrum*, freezing treatment reduced the expression of miR398, while its target genes, CSD1 and CSD2, showed increased expression (Li et al. 2022). Conversely, in canola (*Brassica napus*), another dicotyledonous plant, miR398 was upregulated in leaves under cold stress conditions (Megha et al. 2018). Recent research on *Arabidopsis* roots revealed a link between the Aux/IAA14 gene and the plant's response to cold stress. Advanced sequencing techniques identified 180 known and 71 new cold-responsive miRNAs in *Arabidopsis*, highlighting the downregulation of miR169, particularly its variants miR169a, miR169d, and miR169h, in slr1 plants compared to the wild type after a 12-h cold treatment at 4 °C (Aslam et al. 2020).

5.2.1.5 Heavy Metal Stress

Heavy metals (HMs) can lead to soil contamination, adversely affecting plant growth. These metals are categorized as essential (e.g., Cu, Fe, Zn, and Mn) and nonessential (e.g., Cd, Al, and Hg), with both posing risks to plants. Essential metals play vital roles in enzymatic and biochemical reactions at low concentrations but become toxic at higher levels, while nonessential metals are toxic even at low concentrations. The entry of HMs into the human food chain through contaminated crops poses a serious health threat. Recent investigations have revealed miRNA-mediated transcriptional and post-transcriptional regulations of gene expression in plants subjected to HM stress.

miRNAs play crucial roles in mediating plant responses to heavy metal stress by regulating genes responsible for detoxification and chelation. For example, miR398 is associated with the regulation of metallothioneins and other metal-binding proteins, aiding in the sequestration and detoxification of heavy metals within plant cells. Exposure to heavy metals often leads to the generation of reactive oxygen species (ROS), causing oxidative damage to cellular components. miRNAs like miR398 can regulate the expression of antioxidant enzymes, such as superoxide dismutases (SOD), to help plants combat heavy metal-induced oxidative stress.

Heavy metals like Cu2+, Fe3+, and Ni2+ can generate hydroxyl radicals, leading to the downregulation of miR398 in various plant species when exposed to heavy metals. miR398 is highly conserved in terrestrial plants, and its family members (miR398a, miR398b, miR398c) in *Arabidopsis* are observed to be regulated under Cu2+ toxicity. Additionally, miRNAs, including miR398, miR395, and miR408, have been reported to target transporters involved in Cd uptake and sequestration, regulating genes associated with Cd transport and binding. Cd exposure can induce ROS generation, and miRNAs like miR398, targeting Cu/Zn SODs, can modulate antioxidant enzyme activities to mitigate Cd-induced oxidative stress.

Differential miRNA expression responses to Hg exposure have been observed in multiple plant species. Deep sequencing studies identified altered expression in conserved and non-conserved miRNAs in *Medicago truncatula* exposed to Hg stress. Furthermore, in *Indica* rice roots, miRNA responses to arsenic (As) stress were identified, with increased expression of miR408, miR528, and miR397b, and repression of miR1316 and miR390. The toxicity induced by arsenite in plant tissues manifests in diminished rates of photosynthesis, perturbed carbohydrate metabolism, the generation of reactive oxygen species (ROS), and lipid peroxidation (Yang et al. 2022; Liu and Zhang 2012).

5.2.2 Biotic Stress

Biotic stress in plants, caused by living organisms such as pathogens and herbivores, prompts sophisticated defense mechanisms, and miRNAs are pivotal players in these responses. For instance, miR156/miR172 and miR393 are involved in plant development and flowering, as well as auxin signaling, respectively. During pathogen attacks, elevated levels of miR156, miR172, and decreased miR393 may redirect resources towards defense mechanisms in *Arabidopsis*.

Additionally, miR162 targets Dicer-like 1 (DCL1) in miRNA biogenesis. Herbivore attacks can decrease miR162 levels, resulting in increased DCL1 and production of more defense-related miRNAs in rice. Similarly, miR398 levels decrease during herbivore attacks, leading to increased superoxide dismutase (SOD) expression, mitigating oxidative stress caused by feeding.

In rice, Osa-miR528 plays a role in defenses against rice stripe virus (RSV), while miR482 and miR2118 negatively regulate bacterial and oomycete infections in tomato. Bacterial infections induce miRNA167, miRNA160, miRNA390, miRNA393, and miRNA846 in tomato plants. In *Cassava* infected with *Xanthomonas axonopodis* pv. *manihotis* (Xam), miRNA397, miRNA398, miRNA408, miRNA535, miRNA395, and miRNA482 show decreased expression.

Climate change intensifies biotic stresses, particularly fungal diseases. *Fusarium* wilt in cucumber plants shows elevated expression of miR319a and miR6300, regulating immunity against *Fusarium oxysporum* f.sp. *cucumerinum* (Foc). Maize infected with *Fusarium verticillioides* exhibits increased miR408 expression. Cotton plants exposed to *Verticillium dahliae* infection show increased ghr-miR393 levels. Sugarcane response to smut disease involves differentially expressed miRNAs, including miRNA5671, miRNA5054, miRNA5783, miRNA5221, and miRNA6478.

Nematodes, such as root-knot nematodes (RKNs) and cyst nematodes (CNs), also induce miRNA responses. Cotton plants infected with RKNs show increased miR319 expression, regulating the immune response against RKN. *Arabidopsis* infected with cyst nematode *Heterodera schachtii* exhibits induced expression of miR827, enhancing susceptibility to nematode infection. A comprehensive list of stress-responsive miRNA families in plants is provided in Table 5.1.

Table 5.1 miRNA families in plants responding to abiotic and biotic stresses

Plant	miRNAs families	Stress		Targeted gene	References
Apple (*Malus domestica*)	miR156/SPL	Abiotic	Salt	*MdWRKY100*	Ma et al. (2021)
Arabidopsis thaliana	miR160 miR156 miR398 gma-miR169c miR319, miR157, miR160, miR159, miR165, miR166, miR170	Abiotic	Salt Heat Temperature Sulphur dioxide Drought UVB	*ARF6 and ARF8* *ARF10, ARF16 and ARF17* SPL9 *CSD1 and CSD2*	Jagadhesan et al. (2022), Lin et al. (2018), Sang et al. (2023), Li et al. (2017), Yu et al. (2019), Raza et al. (2023)
	miRNA773	Biotic	*Plectosphaerrella cucumerina*	MET2	Salvador-Guirao et al. (2018)
	miR827		(*Heterodera schachtii*) cyst nematode	–	Hewezi et al. (2016)
Rice (*Oryza sativa L.)*	miR168 miR319 miR393 miR156, miR159, miR160, miR164, miR167, miR169, miR319, miR396	Abiotic	Salt Cold Cadmium Drought	OsAGO1a OsTCP1, OsTCP8, OsTCP11, OsTCP13, and OsTCP21 TIR1	Wan et al. (2022), Gull et al. (2023), Jiang et al. (2022), Nadarajah and Kumar (2019)

(continued)

Table 5.1 (continued)

Plant	miRNAs families	Stress		Targeted gene	References
	Osa-miR528	Biotic	Rice stripe virus (RSV)	–	Kumar et al. (2022)
	miR482		*Bradyrhizobium japonicum (Rhizobia bacteria)*		Kumar et al. (2022)
Rice (Indica ssp)	miR408, miR528, miR397b	Abiotic	*Arsenic toxicity*	–	Liu and Zhang (2012)
Tomato (*Solanum lycopersicum L*)	miR482 miR2118b	Biotic	*Pseudomonas syringae Phytophthora infestans*	–	Zhang et al. (2022a, d)
Cotton (*Gossypium hirsutum*)	ghr-miR 414c	Abiotic	Salt	GhFSD1	Wang et al. (2019a)
	miR319	Biotic	Root-knot nematodes (RKN)	TCP4 gene	Raza et al. (2023)
	ghr-miR393		*Verticillium dahliae*	–	Shi et al. (2022)
Setaria viridis	SvMIR397	Abiotic	Salt	LACCASE (LAC):LAC2, LAC4, LAC17	Nguyen et al. (2020)
Tobacco (*Nicotiana tabacum L. cv. Xanthi*)	miR167	Abiotic	Salt	TcARF6	Ye et al. (2020)
Soybean (*Glycine max L.*)	gma-miR394a	Abiotic	Drought	*GmFBX176*	Yu et al. (2023)
Maize (*Zea mays*)	Zma-miR156b, Zma-miR156k, Zma-miR166d, Zma-miR167f, Zma-miR171b, Zma-miR393b	Abiotic	Cadmium	*B73 and Mo17*	Gao et al. (2021)
	miR408	Biotic	*Fusarium verticillioides*	–	Zhou et al. (2020), Beccaccioli et al. (2021)

(continued)

Table 5.1 (continued)

Plant	miRNAs families	Stress		Targeted gene	References
Sugarcane *(Saccharum officinarum L.)*	miRNA5671, miRNA5054, miRNA5783, miRNA5221,	Biotic	*Sporisorium scitamineum*	–	Su et al. (2017)
Peanut (Arachis hypogaea)	miR396, miR160, miR393, miR167	Abiotic	Nitrogen deficiency	GRF and AFB3	Li et al. (2021)
Chrysanthemum *(Chrysanthemum dichrum)*	miR398	Abiotic	Low temperature	CSD1 and CSD2	Li et al. (2022)
Canola *(Brassica napus)*	miR398	Abiotic	Low temperature	–	Megha et al. (2018)
Barrel medic (Medicago truncatula)	miR319	*Abiotic*	Mercury (Hg) stress	–	Li et al. (2023)

This table offers a comprehensive compilation of miRNA families in plants that demonstrate responses to a range of abiotic and biotic stress factors. It summarizes miRNA families, their associated stressors, and includes reference citations for further exploration

5.3 Applications of miRNA in Sustainable Crop Agriculture

Sustainable agriculture is a critical focus in addressing the challenges of feeding a growing population amidst climate change and environmental degradation. MicroRNAs (miRNAs), diminutive yet potent regulators within the realm of plant biology, are increasingly gaining recognition for their pivotal involvement in diverse aspects, including plant growth, fruit development, agronomic traits, and adaptive responses to stress conditions. Harnessing miRNA biotechnology, particularly through genetically modified (GM) crops and genome editing techniques like CRISPR/Cas9, presents promising avenues for crop improvement and enhanced resilience to environmental stresses.

Recent research has demonstrated the potential of manipulating miRNAs for developing stress-tolerant crops. Knocking out specific miRNAs, such as miR166 in rice, has shown improved drought tolerance. CRISPR/Cas9 has been applied to various crops, including rice, wheat, soybean, maize, and tomato, to enhance resilience to adverse environmental conditions. For example, salinity-resistant rice was developed by knocking out the OsRR22 gene using CRISPR/Cas9.

Minimizing mineral toxicity has also been addressed through CRISPR/Cas9-mediated genome editing. In rice, the targeted knockout of OsARM1 and OsNramp5 genes demonstrated enhanced arsenic tolerance and reduced cadmium accumulation, respectively. In tomatoes, CRISPR/Cas9 mutants lacking SlymiR482e-3p displayed increased resistance against *Fusarium oxysporum*.

The CRISPR/Cas9 system has been effectively used in banana production, addressing bacterial and fungal diseases. The application of CRISPR/Cas9 with the *Phytoene desaturase* (PDS) gene in banana demonstrated successful genome editing, providing resistance against diseases and streamlining editing procedures.

In the realm of synthetic biology, CRISPR technology is transformative, offering innovative solutions for plant enhancement. Researchers explore strategies like bypassing photorespiration or reengineering the core enzyme *Rubisco* to enhance photosynthetic efficiency and increase plant biomass. Genome editing techniques contribute to a deeper understanding of miRNA roles in plant development and stress resilience, paving the way for targeted modifications to improve crop yield and sustainability.

5.4 Emerging Technologies and Tools for miRNA Research

In the ever-evolving landscape of plant and molecular biology, a remarkable array of cutting-edge tools and technologies has emerged, revolutionizing our capacity to explore the intricate world of miRNA research and its profound implications in plant health, development, and stress responses (Ergin and Çetinkaya 2022). These tools and technologies are pivotal for advancing our knowledge of miRNA-mediated gene regulation in plants (Raza et al. 2023; Zhang et al. 2017; Motameny et al. 2010). They facilitate the discovery of novel miRNAs, their targets, and their roles in plant biology. This knowledge has implications for improving crop yield, enhancing stress tolerance, and addressing various challenges in agriculture and plant science. Below mentioned are few of these techniques followed in today's plant research:

- **Next-Generation Sequencing (NGS):** NGS technologies, including small RNA sequencing, have transformed miRNA research by providing high-throughput and comprehensive profiling of miRNAs. This approach enables the discovery of novel miRNAs and the quantification of miRNA expression levels (Liu et al. 2020; Hu et al. 2017; Motameny et al. 2010).
- **CRISPR-Cas9 Gene Editing:** The CRISPR-Cas9 system has been harnessed to manipulate miRNA expression, facilitating the functional characterization of specific miRNAs in plants and other organisms (Zhong et al. 2023; Mujtaba et al. 2021; Gao 2021).
- **Bioinformatics Tools:** Advanced computational tools and databases are invaluable for miRNA research. Tools like miRBase and TargetScan assist in the prediction of miRNA targets, functional annotations, and the identification of regulatory networks (Kozomara et al. 2019).
- **Fluorescent Reporters:** The use of fluorescent proteins and markers allows researchers to visualize miRNA activity in real time, providing insights into miRNA function in living plants cells and tissues (Mujtaba et al. 2021).

The integration of these advanced tools and technologies into miRNA research not only unlocks the secrets of plant gene regulation but also holds the promise of enhancing crop resilience, yield, and sustainability in a rapidly changing world, securing a brighter future for agriculture and plant science.

5.4.1 miRNA Database

miRNA databases and target databases are essential resources in the field of molecular biology and genetics. These databases compile information on miRNAs, their sequences, expression patterns, and functions, as well as their interactions with target messenger RNAs (mRNAs). Researchers and scientists use these databases to access valuable information on miRNAs and their roles in gene regulation, development, diseases, and other biological processes. These databases facilitate the exploration and understanding of the complex world of miRNAs and their impact on various aspects of biology and medicine. Here are some of the commonly used miRNA databases used worldwide:

- **miRBase:** miRBase is one of the most comprehensive and widely used miRNA databases. It provides information on miRNA sequences, families, and annotations across various species (Griffiths-Jones 2006). miRBase is available at https://mirbase.org/,
- **miRTarBase:** This database focuses on miRNA–target interactions. It provides experimentally validated miRNA–target relationships, which can be crucial for understanding miRNA function (Kariuki et al. 2023). miRTarBase is available at http://mirtarbase.cuhk.edu.cn/,
- **miRDB:** miRDB is a database that offers predictions for miRNA targets based on machine learning models. It provides target gene predictions for miRNAs in multiple species (Wong and Wang 2015).
- **TarBase:** TarBase represents a database centering on experimentally substantiated miRNA–target interactions, offering comprehensive details on the intricacies of these molecular associations.
- **miRGate:** miRGate is a database that offers insights into the regulatory roles of miRNAs in various diseases. It provides data on miRNA–disease associations and interactions (Andrés-León et al. 2017).
- **miRNEST:** This database focuses on miRNA expression data. It offers information about miRNA expression profiles in different tissues and under various conditions. miRGate is available at http://mirgate.bioinfo.cnio.es/.
- **Plant MPSS Database:** For plant-specific miRNA research, the Plant MPSS (Massively Parallel Signature Sequencing) database provides information on small RNA expression profiles in various plant species (Meyers et al. 2004).

5.5 Conclusion and Future Prospects

Extensive research has elucidated the intricate mechanisms by which miRNAs regulate gene expression in response to various abiotic stressors, including drought, heat, cold, salinity, and nutrient deficiency. The manipulation of specific miRNAs has shown promise in enhancing crop resilience and improving tolerance to challenging environmental conditions. This breakthrough provides a potential avenue for mitigating the adverse impacts of climate change on agricultural productivity.

In the realm of increasing disease resistance in plants, microRNAs (miRNAs) play a pivotal role by selectively targeting and regulating the expression of genes associated with defense responses. This modulation contributes significantly to the development of plant immunity, highlighting the integral function of miRNAs in enhancing the resistance mechanisms against various diseases. This innovative approach not only reduces reliance on chemical pesticides but also minimizes the associated environmental risks. miRNAs have demonstrated their potential in optimizing nutrient uptake and utilization in crops, contributing to nutrient management in sustainable agriculture. By fine-tuning the expression of genes related to nutrient transport and metabolism, miRNAs can improve nutrient use efficiency, reduce runoff, and minimize environmental pollution.

Photosynthesis, a fundamental process in crop productivity, can be enhanced through miRNA-mediated regulation of photosynthesis-related genes. By modulating the expression of key genes involved in light capture, carbon fixation, and energy conversion, miRNAs offer a novel approach to increasing photosynthetic efficiency and, consequently, boosting crop yields.

MiRNA-based strategies have enabled the development of crops with traits such as drought resistance and enhanced nutrient content. Manipulating miRNA expression patterns has led to the successful engineering of crops with improved water-use efficiency, addressing critical challenges in global agriculture.

While the potential of miRNAs in agriculture is vast, it is essential to address safety concerns and navigate regulatory frameworks to ensure responsible and ethical implementation. Engaging various stakeholders, including farmers, consumers, and policymakers, is crucial for fostering acceptance and understanding of this revolutionary approach. In conclusion, miRNAs represent a revolutionary approach to addressing pressing challenges in agriculture. Their versatile applications offer immense potential for sustainable agriculture, and collaborative efforts can pave the way for a prosperous and sustainable future in agriculture.

Acknowledgements I extend my sincere gratitude to Dr. Debasis Chakrabarty at CSIR-NBRI, India, for providing valuable suggestions for the successful completion of this chapter.

References

Al-Tawaha AR, Mrabet R, Bayanati M, Santhosh B, Benkeblia N, Imran A, Khalid S, Al-Tawaha AR, Alatrash H, Jabbour Y (2022) Role of biotechnology in climate-resilient agriculture. In: Climate change and agriculture: perspectives, sustainability and resilience, pp 78–94

Andrés-León E, Gómez-López G, Pisano DG (2017) Prediction of miRNA–mRNA interactions using miRGate. In: MicroRNA detection and target identification: methods and protocols, pp 225–237

Aravind J, Rinku S, Pooja B, Shikha M, Kaliyugam S, Mallikarjuna MG, Kumar A, Rao AR, Nepolean T (2017) Identification, characterization, and functional validation of drought-responsive microRNAs in subtropical maize inbreds. Frontiers in plant science 8:261876

Aslam M, Sugita K, Qin Y, Rahman A (2020) Aux/IAA14 regulates microRNA-mediated cold stress response in *Arabidopsis* roots. Int J Mol Sci 21(22):8441

Axtell MJ, Westholm JO, Lai EC (2011) Vive la différence: biogenesis and evolution of microRNAs in plants and animals. Genome Biol 12(4):1–3

Bajczyk M, Jarmolowski A, Jozwiak M, Pacak A, Pietrykowska H, Sierocka I, Swida-Barteczka A, Szewc L, Szweykowska-Kulinska Z (2023) Recent insights into plant miRNA biogenesis: multiple layers of miRNA level regulation. Plan Theory 12(2):342

Barczak-Brzyżek A, Brzyżek G, Koter M, Siedlecka E, Gawroński P, Filipecki M (2022) Plastid retrograde regulation of miRNA expression in response to light stress. BMC Plant Biol 22(1):150

Beccaccioli M, Salustri M, Scala V, Ludovici M, Cacciotti A, D'Angeli S, Brown DW, Reverberi M (2021) The effect of *Fusarium verticillioides* fumonisins on fatty acids, sphingolipids, and oxylipins in maize germlings. Int J Mol Sci 22(5):2435

Begum Y (2022) Regulatory role of microRNAs (miRNAs) in the recent development of abiotic stress tolerance of plants. Gene 821:146283

Bhat MA, Mishra AK, Jan S, Bhat MA, Kamal MA, Rahman S, Shah AA, Jan AT (2023) Plant growth promoting rhizobacteria in plant health: a perspective study of the underground interaction. Plan Theory 12(3):629

Brant EJ, Budak H (2018) Plant small non-coding RNAs and their roles in biotic stresses. Front Plant Sci 9:392912

Chinweze C (2023) Desertification, land degradation and drought resilience. Cuvillier Verlag

Das R, Mondal SK (2021) Plant miRNAs: Biogenesis and its functional validation to combat drought stress with special focus on maize. Plant Gene 27:100294

Datta NR, Rogers S, Bodis S (2019) Challenges and opportunities to realize "the 2030 agenda for sustainable development" by the United Nations: implications for radiation therapy infrastructure in low-and middle-income countries. Int J Radiat Oncol Biol Phys 105(5):918–933

Dong F, Wang C, Dong Y, Hao S, Wang L, Sun X, Liu S (2020) Differential expression of microRNAs in tomato leaves treated with different light qualities. BMC Genomics 21(1):1–1

Ebrahimi Khaksefidi R, Mirlohi S, Shiran B, Fallahi H, Budak H (2015) Differential expression of seven conserved microRNAs in response to abiotic stress and their regulatory network in Helianthus annuus. Front Plant Sci 6:149903

Ergin K, Çetinkaya R (2022) Regulation of microRNAs. miRNomics: microRNA biology and computational analysis, pp 1–32

Fang C, Li L, Zhang P, Wang D, Yang L, Reza BM, Lin W (2019) *Lsi1* modulates the antioxidant capacity of rice and protects against ultraviolet-B radiation. Plant Sci 278:96–106

Ferdous J, Sanchez-Ferrero JC, Langridge P, Milne L, Chowdhury J, Brien C, Tricker PJ (2017) Differential expression of microRNAs and potential targets under drought stress in barley. Plant Cell Environ 40(1):11–24

Fontana JE, Wang G, Sun R, Xue H, Li Q, Liu J, Davis KE, Thornburg TE, Zhang B, Zhang Z, Pan X (2020) Impact of potassium deficiency on cotton growth, development and potential microRNA-mediated mechanism. Plant Physiol Biochem 153:72–80

Gao C (2021) Genome engineering for crop improvement and future agriculture. Cell 184(6):1621 1635

Gao Y, Li SJ, Zhang SW, Feng T, Zhang ZY, Luo SJ, Mao HY, Borkovich KA, Ouyang SQ (2021) SlymiR482e-3p mediates tomato wilt disease by modulating ethylene response pathway. Plant Biotechnol J 19(1):17

Griffiths-Jones S (2006) miRBase: the microRNA sequence database. Methods Mol Biol 342:129–138

Gull S, Uddin S, Hussain HA, Wang S, Bayar J, Liu J (2023) Genome-wide analysis reveals the TCP-miR159-miR319 module is crucial for rice (*Oryza sativa* L.) growth and response to drought and salinity. Plant Stress 10:100215

Hewezi T, Piya S, Qi M, Balasubramaniam M, Rice JH, Baum TJ (2016) *Arabidopsis* miR827 mediates post-transcriptional gene silencing of its ubiquitin E3 ligase target gene in the syncytium of the cyst nematode *Heterodera schachtii* to enhance susceptibility. Plant J 88(2):179–192
Hu Y, Lan W, Miller D (2017) Next-generation sequencing for microRNA expression profile. In: Bioinformatics in microRNA research, pp 169–177
Islam W, Idrees A, Waheed A, Zeng F (2022) Plant responses to drought stress: microRNAs in action. Environ Res 215:114282
Jagadhesan B, Das S, Singh D, Jha SK, Durgesh K, Sathee L (2022) Micro RNA mediated regulation of nutrient response in plants: the case of nitrogen. Plant Physiol Rep 27(3):345–357
Jiang J, Zhu H, Li N, Batley J, Wang Y (2022) The miR393-target module regulates plant development and responses to biotic and abiotic stresses. Int J Mol Sci 23(16):9477
Jones-Rhoades MW, Bartel DP (2004) Computational identification of plant microRNAs and their targets, including a stress-induced miRNA. Mol Cell 14(6):787–799
Kariuki D, Asam K, Aouizerat BE, Lewis KA, Florez JC, Flowers E (2023) Review of databases for experimentally validated human microRNA–mRNA interactions. Database 2023:baad014
Kerchev P, Van Der Meer T, Sujeeth N, Verlee A, Stevens CV, Van Breusegem F et al (2020) Molecular priming as an approach to induce tolerance against abiotic and oxidative stresses in crop plants. Biotechnol Adv 40:107503. https://doi.org/10.1016/j.biotechadv.2019.107503
Kozomara A, Birgaoanu M, Griffiths-Jones S (2019) miRBase: from microRNA sequences to function. Nucleic Acids Res 47(D1):D155–D162
Kumar R, Tiwari RK, Jeevalatha A, Siddappa S, Shah MA, Sharma S et al (2021) Potato apical leaf curl disease: current status and perspectives on a disease caused by tomato leaf curl New Delhi virus. J Plant Dis Prot 128:897–911. https://doi.org/10.1007/S41348-021-00463-W/METRICS
Kumar K, Mandal SN, Neelam K, de Los Reyes BG (2022) MicroRNA-mediated host defense mechanisms against pathogens and herbivores in rice: balancing gains from genetic resistance with trade-offs to productivity potential. BMC Plant Biol 22(1):351
Lee RC, Feinbaum RL, Ambros V (1993) The C elegans heterochronic gene lin-4 encodes small RNAs with antisense complementarity to lin-14. Cell 75(5):843–854
Lee H, Yoo SJ, Lee JH, Kim W, Yoo SK, Fitzgerald H, Carrington JC, Ahn JH (2010) Genetic framework for flowering-time regulation by ambient temperature-responsive miRNAs in *Arabidopsis*. Nucleic Acids Res 38(9):3081–3093
Li M, Yu B (2021) Recent advances in the regulation of plant miRNA biogenesis. RNA Biol 18(12):2087–2096
Li L, Yi H, Xue M, Yi M (2017) miR398 and miR395 are involved in response to SO2 stress in *Arabidopsis thaliana*. Ecotoxicology 26:1181–1187
Li L, Li Q, Davis KE, Patterson C, Oo S, Liu W, Liu J, Wang G, Fontana JE, Thornburg TE, Pratt IS (2021) Response of root growth and development to nitrogen and potassium deficiency as well as microRNA-mediated mechanism in peanut (*Arachis hypogaea* L.). Front Plant Sci 12:695234
Li J, Song Q, Zuo ZF, Liu L (2022) MicroRNA398: a master regulator of plant development and stress responses. Int J Mol Sci 23(18):10803
Li H, Qiu T, Zhou Z, Kang L, Chen R, Zeng L, Yu H, Wang Y, Song J (2023) Genome-wide analysis of the growth-regulating factor family in *Medicago truncatula*. J Plant Growth Regul 42(4):2305–2316
Liang G, He H, Yu D (2012) Identification of nitrogen starvation-responsive microRNAs in *Arabidopsis thaliana*. PLoS One 7(11):e48951
Lin JS, Kuo CC, Yang IC, Tsai WA, Shen YH, Lin CC, Liang YC, Li YC, Kuo YW, King YC, Lai HM (2018) MicroRNA160 modulates plant development and heat shock protein gene expression to mediate heat tolerance in *Arabidopsis*. Front Plant Sci 9:68
Liu HH, Tian X, Li YJ, Wu CA, Zheng CC (2008) Microarray-based analysis of stress-regulated microRNAs in Arabidopsis thaliana. RNA 14(5):836–843
Liu Q, Zhang H (2012) Molecular identification and analysis of arsenite stress-responsive miRNAs in rice. J Agric Food Chem 60(26):6524–6536

Liu K, Tong H, Li T, Wang X, Chen Y (2020) Research progress in molecular biology related quantitative methods of microRNA. Am J Transl Res 12(7):3198

Liu J, Wei L, Feng S (2023) Research progress of non-coding RNAs in vegetable responses to abiotic stresses. Gene 8:147537

Ma Y, Xue H, Zhang F, Jiang Q, Yang S, Yue P, Wang F, Zhang Y, Li L, He P, Zhang Z (2021) The miR156/SPL module regulates apple salt stress tolerance by activating MdWRKY100 expression. Plant Biotechnol J 19(2):311–323

Megha S, Basu U, Joshi RK, Kav NN (2018) Physiological studies and genome-wide microRNA profiling of cold-stressed *Brassica napus*. Plant Physiol Biochem 132:1–7

Melino V, Tester M (2023) Salt-tolerant crops: time to deliver. Annu Rev Plant Biol 74:671–696

Meyers BC, Vu TH, Tej SS, Ghazal H, Matvienko M, Agrawal V, Ning J, Haudenschild CD (2004) Analysis of the transcriptional complexity of *Arabidopsis thaliana* by massively parallel signature sequencing. Nat Biotechnol 22(8):1006–1011

Mittal U, Kumar V, Kukreja S, Singh B, Goutam U (2023) Scope of small RNA technology to develop biotic stress tolerant food crops. In: Plant small RNA in food crops. Academic Press, pp 545–569

Motameny S, Wolters S, Nürnberg P, Schumacher B (2010) Next generation sequencing of miRNAs–strategies, resources and methods. Genes 1(1):70–84

Mujtaba M, Wang D, Carvalho LB, Oliveira JL, Espirito Santo Pereira AD, Sharif R, Jogaiah S, Paidi MK, Wang L, Ali Q, Fraceto LF (2021) Nanocarrier-mediated delivery of miRNA, RNAi, and CRISPR-Cas for plant protection: current trends and future directions. ACS Agric Sci Technol 1(5):417–435

Nadarajah K, Kumar IS (2019) Drought response in rice: the miRNA story. Int J Mol Sci 20(15):3766

Nguyen DQ, Brown CW, Pegler JL, Eamens AL, Grof CP (2020) Molecular manipulation of microRNA397 abundance influences the development and salt stress response of *Arabidopsis thaliana*. Int J Mol Sci 21(21):7879

Nogoy FM, Niño MC, Song JY, Jung YJ, Kang KK, Nou I, Cho YG (2018) Plant microRNAs in molecular breeding. Plant biotechnology reports 12:15–25

Pandita D (2022) MicroRNAs shape the tolerance mechanisms against abiotic stress in maize. In: Omics approach to manage abiotic stress in cereals. Springer Nature, Singapore, pp 479–493

Podolec R, Demarsy E, Ulm R (2021) Perception and signaling of ultraviolet-B radiation in plants. Annu Rev Plant Biol 72:793–822

Praveen A, Dubey S, Singh S, Sharma VK (2023) Abiotic stress tolerance in plants: a fascinating action of defense mechanisms. 3 Biotech 13(3):102

Ragupathy R, Ravichandran S, Mahdi MS, Huang D, Reimer E, Domaratzki M, Cloutier S (2016) Deep sequencing of wheat sRNA transcriptome reveals distinct temporal expression pattern of miRNAs in response to heat, light and UV. Sci Rep 6(1):39373

Ravichandran S, Ragupathy R, Edwards T, Domaratzki M, Cloutier S (2019) MicroRNA-guided regulation of heat stress response in wheat. BMC Genomics 20(1):1–6

Raza A, Charagh S, Karikari B, Sharif R, Yadav V, Mubarik MS, Habib M, Zhuang Y, Zhang C, Chen H, Varshney RK (2023) miRNAs for crop improvement. Plant Physiol Biochem 24:107857

Salvador-Guirao R, Baldrich P, Weigel D, Rubio-Somoza I, San Segundo B (2018) The microRNA miR773 is involved in the Arabidopsis immune response to fungal pathogens. Mol Plant-Microbe Interact 31(2):249–259

Sang Q, Fan L, Liu T, Qiu Y, Du J, Mo B, Chen M, Chen X (2023) MicroRNA156 conditions auxin sensitivity to enable growth plasticity in response to environmental changes in *Arabidopsis*. Nat Commun 14(1):1449

Šečić E, Kogel KH, Ladera-Carmona MJ (2021) Biotic stress-associated microRNA families in plants. J Plant Physiol 263:153451

Shahzad A, Ullah S, Dar AA, Sardar MF, Mehmood T, Tufail MA, Shakoor A, Haris M (2021) Nexus on climate change: agriculture and possible solution to cope future climate change stresses. Environ Sci Pollut Res 28:14211–14232

Shang R, Lee S, Senavirathne G, Lai EC (2023) MicroRNAs in action: biogenesis, function and regulation. Nat Rev Genet 28:1–8
Sharma M, Kumar P, Verma V, Sharma R, Bhargava B, Irfan M (2022) Understanding plant stress memory response for abiotic stress resilience: molecular insights and prospects. Plant Physiol Biochem 179:10–24
Shi G, Wang S, Wang P, Zhan J, Tang Y, Zhao G, Li F, Ge X, Wu J (2022) Cotton miR393-TIR1 module regulates plant defense against *Verticillium dahliae* via auxin perception and signaling. Front Plant Sci 13:888703
Shriram V, Kumar V, Devarumath RM, Khare TS, Wani SH (2016) MicroRNAs as potential targets for abiotic stress tolerance in plants. Front Plant Sci 7:817
Singh A, Jain D, Pandey J, Yadav M, Bansal KC, Singh IK (2023) Deciphering the role of miRNA in reprogramming plant responses to drought stress. Crit Rev Biotechnol 43(4):613–627
Su Y, Zhang Y, Huang N, Liu F, Su W, Xu L, Ahmad W, Wu Q, Guo J, Que Y (2017) Small RNA sequencing reveals a role for sugarcane miRNAs and their targets in response to Sporisorium scitamineum infection. BMC Genomics 18:1–19
Sun X, Lin L, Sui N (2019) Regulation mechanism of microRNA in plant response to abiotic stress and breeding. Mol Biol Rep 46:1447–1457
Verma S, Sarkar AK (2021) miRNA-mediated regulation of biotic and abiotic stress responses in plants. In: Agricultural biotechnology: latest research and trends, pp 463–492
Wan J, Meng S, Wang Q, Zhao J, Qiu X, Wang L, Li J, Lin Y, Mu L, Dang K, Xie Q (2022) Suppression of microRNA168 enhances salt tolerance in rice (*Oryza sativa* L.). BMC Plant Biol 22(1):1–1
Wang W, Liu D, Chen D, Cheng Y, Zhang X, Song L, Hu M, Dong J, Shen F (2019a) MicroRNA414c affects salt tolerance of cotton by regulating reactive oxygen species metabolism under salinity stress. RNA Biol 16(3):362–375
Wang J, Mei J, Ren G (2019b) Plant microRNAs: biogenesis, homeostasis, and degradation. Front Plant Sci 10:360
Wani SH, Kumar V, Khare T, Tripathi P, Shah T, Ramakrishna C, Aglawe S, Mangrauthia SK (2020) miRNA applications for engineering abiotic stress tolerance in plants. Biologia 75:1063–1081
Wong N, Wang X (2015) miRDB: an online resource for microRNA target prediction and functional annotations. Nucleic Acids Res 43(D1):D146–D152
Yang Y, Huang J, Sun Q, Wang J, Huang L, Fu S, Qin S, Xie X, Ge S, Li X, Cheng Z (2022) MicroRNAs: key players in plant response to metal toxicity. Int J Mol Sci 23(15):8642
Ye Y, Wang J, Wang W, Xu LA (2020) ARF family identification in *Tamarix chinensis* reveals the salt responsive expression of TcARF6 targeted by miR167. PeerJ 8:e8829
Yu Y, Jia T, Chen X (2017) The 'how'and 'where'of plant microRNAs. New Phytol 216(4):1002–1017
Yu Y, Ni Z, Wang Y, Wan H, Hu Z, Jiang Q, Sun X, Zhang H (2019) Overexpression of soybean miR169c confers increased drought stress sensitivity in transgenic *Arabidopsis thaliana*. Plant Sci 285:68–78
Yu Y, Wang P, Wan H, Wang Y, Hu H, Ni Z (2023) The Gma-miR394a/GmFBX176 module is involved in regulating the soybean (*Glycine max* L.) response to drought stress. Plant Sci 29:111879
Zhang B (2015) MicroRNA: a new target for improving plant tolerance to abiotic stress. J Exp Bot 66(7):1749–1761
Zhang Y, Huang H, Zhang D, Qiu J, Yang J, Wang K, Zhu L, Fan J, Yang J (2017) A review on recent computational methods for predicting noncoding RNAs. Biomed Res Int 2017:9139504
Zhang Y, Waseem M, Zeng Z, Xu J, Chen C, Liu Y, Zhai J, Xia R (2022a) MicroRNA482/2118, a miRNA superfamily essential for both disease resistance and plant development. New Phytol 233(5):2047–2057
Zhang N, Feng X, Zeng Q, Lin H, Wu Z, Gao X, Huang Y, Wu J, Qi Y (2022b) Integrated analysis of miRNAs associated with sugarcane responses to low-potassium stress. Front Plant Sci 12:750805

Zhang F, Yang J, Zhang N, Wu J, Si H (2022c) Roles of microRNAs in abiotic stress response and characteristics regulation of plant. Front Plant Sci 13:919243

Zhang Y, Zhou Y, Zhu W, Liu J, Cheng F (2022d) Non-coding RNAs fine-tune the balance between plant growth and abiotic stress tolerance. Front Plant Sci 13:965745

Zhong Z, Liu G, Tang Z, Xiang S, Yang L, Huang L, He Y, Fan T, Liu S, Zheng X, Zhang T (2023) Efficient plant genome engineering using a probiotic sourced CRISPR-Cas9 system. Nat Commun 14(1):6102

Zhou L, Liu Y, Liu Z, Kong D, Duan M, Luo L (2010) Genome-wide identification and analysis of drought-responsive microRNAs in Oryza sativa. J Exp Bot 61(15):4157–4168

Zhou Z, Cao Y, Li T, Wang X, Chen J, He H, Yao W, Wu J, Zhang H (2020) MicroRNAs are involved in maize immunity against *Fusarium verticillioide*s ear rot. Genomics Proteomics Bioinformatics 18(3):241–255

Cellular Responses Against Abiotic Stress-Induced Reactive Oxygen Species

6

Deepankar Mondal, Dhananjay Shukla,
Naveen K. Vishvakarma, Rajani Prabha, and Vineeta Dixit

Abstract

This chapter explores molecular mechanisms underlying behind molecular pathways triggered by excessive reactive oxygen species (ROS) production due to stress signalling. Abiotic stressors that disrupt redox equilibrium and jeopardise cellular integrity usually includes dehydration, salinity, heavy metals, high temperatures, and UV radiation. These conditions also cause ROS buildup. The chapter examines the various pathways and regulatory mechanisms that underlie the generation of reactive oxygen species (ROS) in plants exposed to abiotic stress, focusing on the critical roles played by peroxisomes, mitochondria, NADPH oxidases, and chloroplasts. It draws attention to the diverse networks of signalling and stress-responsive pathways that are involved in ROS-mediated signalling, including calcium signalling, transcription factors like AP2/ERF, WRKY, and NAC, and mitogen-activated protein kinases, that coordinates with gene expression and stress responses. The chapter also covers the complex systems of antioxidant defense, which include both enzymatic and non-enzymatic antioxidants, and how they work together to reduce oxidative damage and scavenges ROS buildup. Understanding the underlying effects of ROS-induced cellular damage on membranes, proteins, DNA, and organelles is also discussed, along with an explanation of the consequences for plant growth, development,

D. Mondal · D. Shukla · N. K. Vishvakarma
Department of Biotechnology, Guru Ghasidas Vishwavidyalaya, Bilaspur, Chhattisgarh, India

R. Prabha
Department of Botany, Balrampur Girls Degree College, Balrampur, Uttar Pradesh, India

V. Dixit (✉)
Department of Botany, Nilamber-Pitamber University, Garhwa, Jharkhand, India

A. K. Mishra (ed.), *Stress Biology in Photosynthetic Organisms*,
https://doi.org/10.1007/978-981-97-1883-2_6

and yield. This thorough investigation of the generation of reactive oxygen species (ROS) and molecular reactions triggered by abiotic stress lays the groundwork for comprehending the mechanisms underlying plant resilience and presents opportunities for developing stress-tolerant crops to maintain global food security in the face of shifting environmental conditions.

Keywords
Abiotic stress · Reactive oxygen species · ROS · Cellular response · Oxidative stress

6.1 Introduction

Climate change is proving as a greatest disruptor of environmental balance and a major source of abiotic stress factors in plants that has significantly decreased their productivity making them susceptible to extinction. Alterations in temperature, precipitation patterns, waterlogging or drought, salinity, presence of nutrition, xenobiotics stressors, heavy metals, and extreme weather events brought on by climate change often create unsupportive conditions for plant growth (Chaudhry and Sidhu 2022). These stressors trigger production of many molecules that exhibits harmful effect on plant cell, one of them is the reactive oxygen species (ROS). The plant cells are evolving through these stressors to develop molecular pathways that are designed to respond to ROS production and thus can avoid ROS-mediated harmful effects (Singh et al. 2019).

Reactive oxygen species (ROS) is a group of oxygen-derived chemical species that are highly reactive, such as superoxide radicals, hydrogen peroxide, and hydroxyl radicals. They get formed as natural by-products of various kinds of metabolic processes ongoing in cells, particularly during respiration and photosynthesis (Gechev and Petrov 2020). During ambient conditions, their concentration remains low within a threshold limit, not enough to mount any stress response (Sharma et al. 2021). Reactive oxygen species has many functions in a cell including as a signalling molecule (D'Autréaux and Toledano 2007) in defence mechanism (Patlevič et al. 2016) and stress responses that include pathways involved in the development and growth (Tsukagoshi 2016). However, extreme conditions that trigger excessive production of ROS mounts an oxidative damage in the cells too, such as oxidizing their biomolecules (Qamer et al. 2021) including lipids, proteins, and carbohydrates, which could lead to eventual death of the plant cell. To avoid this damage, plant cells have many strategies, one of them is the production of anti-oxidants, which can control ROS production and neutralise its oxidative damage on the cells (Singh et al. 2019). Therefore, cellular ROS levels in optimum concentration can harmonise plant growth and development, whereas abiotic stress-induced excessive ROS production is dangerous and detrimental to plant growth, triggering many stress and anti-oxidant responses (Sharma et al. 2021).

6.2 Abiotic Stress and ROS Generation

Stress is anything that destabilises cellular homeostasis and stimulates certain molecular pathways as a result. Abiotic stress in particular is occurrence of extreme physical conditions such as high or low temperature, salinity, waterlogging, or drought (Fig. 6.1) (He et al. 2018). These extreme physical conditions cause exponential increase in ROS production that in turn causes disbalance in cellular redox homeostasis (Nadarajah 2020). In stress conditions, pathways generating ROS gain momentum over pathways that scavenge ROS, causing an oxidative burst leading to oxidative damage. ROS production pathways in cells are highly localised, even at

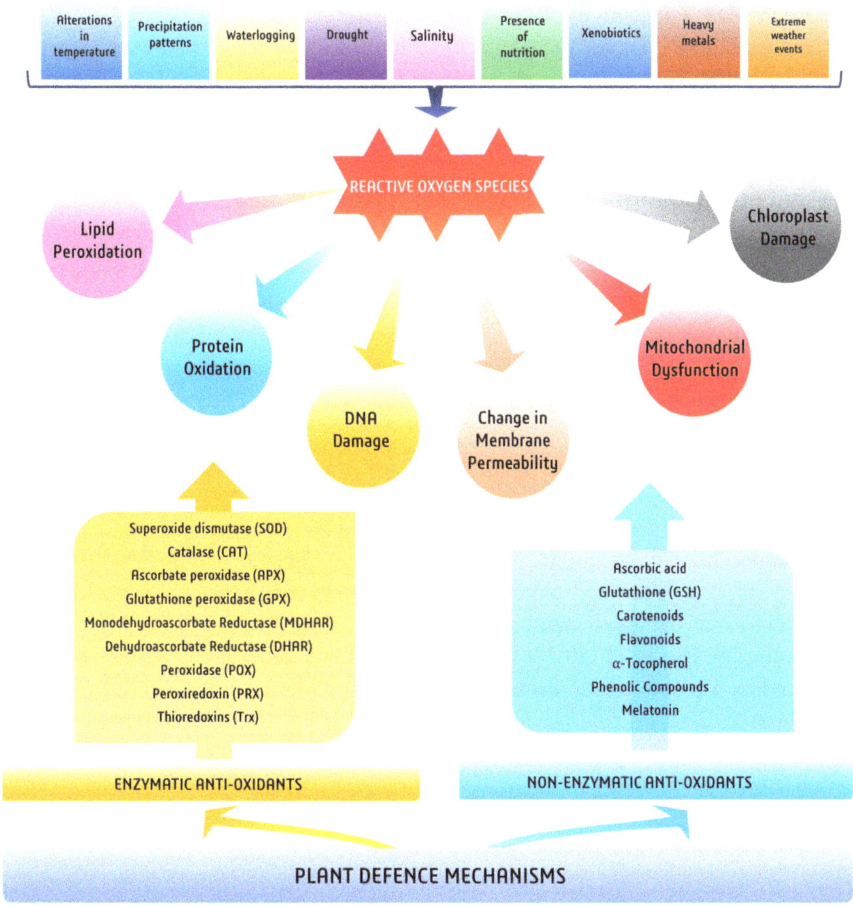

Fig. 6.1 A brief overview of the relationship between abiotic stress, ROS, and how it damages biomolecular structures including various plant defence mechanisms against it

the compartment level where ROS status may be depending upon currently stimulated pathways (Das and Roychoudhury 2014). For instance, in chloroplast, an excited chlorophyll (Chl*) can produce singlet oxygen, a superoxide radical, a hydroxyl radical, or an hydrogen peroxide species by PSI and PSII of its electron transport chain (ETC) called Mehler's reaction. Here, reduced iron and copper can also neutralise the superoxide radical as a response, which is called Fenton reaction. Similarly in peroxisomes, superoxide radicals and hydrogen peroxide are produced by ETCs using photorespiration-specific enzyme and NADH. Mitochondrial ETC also produces superoxide radical, which may get dismutated by SOD (superoxide dismutase) producing hydrogen peroxide. In plasma membrane, the enzyme NADPH oxidase can convert oxygen to superoxide radical that later moves to apoplast and may dis-mutate into hydrogen peroxide by the action of superoxide dismutase (SOD) (Zheng et al. 2023). Therefore, a low or moderate generation of ROS in various cellular compartments may modulate various cellular pathways; however, a high ROS production will result in an oxidative stress and which later causes oxidative damage (Schieber and Chandel 2014).

6.3 Biomolecular Damage by Reactive Oxygen Species

Under adverse environmental conditions, ROS gets overproduced that leads to oxidative burst and later causes oxidative damage. This may cause severe damage to various biomolecules such as damage to membrane integrity, nucleic acids, chlorophyll degradation, or even stimulation of apoptotic and PCD pathways and also causes peroxidation of lipids and protein oxidation (Fig. 6.2) (Sharma et al. 2012).

6.3.1 Lipid Peroxidation

Unsaturated fatty acids found in cell membranes can be attacked by ROS, for instance singlet oxygen and species such as hydroxyl radicals. Lipid peroxidation is the process that damages membrane lipids, impairing the fluidity and integrity of the membrane (Su et al. 2019). Lipid peroxides and other reactive aldehydes are produced as a result, which eventually causes membrane rupture and malfunction (Endale et al. 2023). Lipid peroxidation is also a result of a photo-oxidative stress due to ROS buildup in the chloroplast.

6.3.2 Protein Oxidation

Amino acid residues of any protein structure can be oxidised by ROS, which may change the structure and functionality of the protein. Protein aggregation and breakdown can also result from oxidative alteration of proteins, which can also degrade

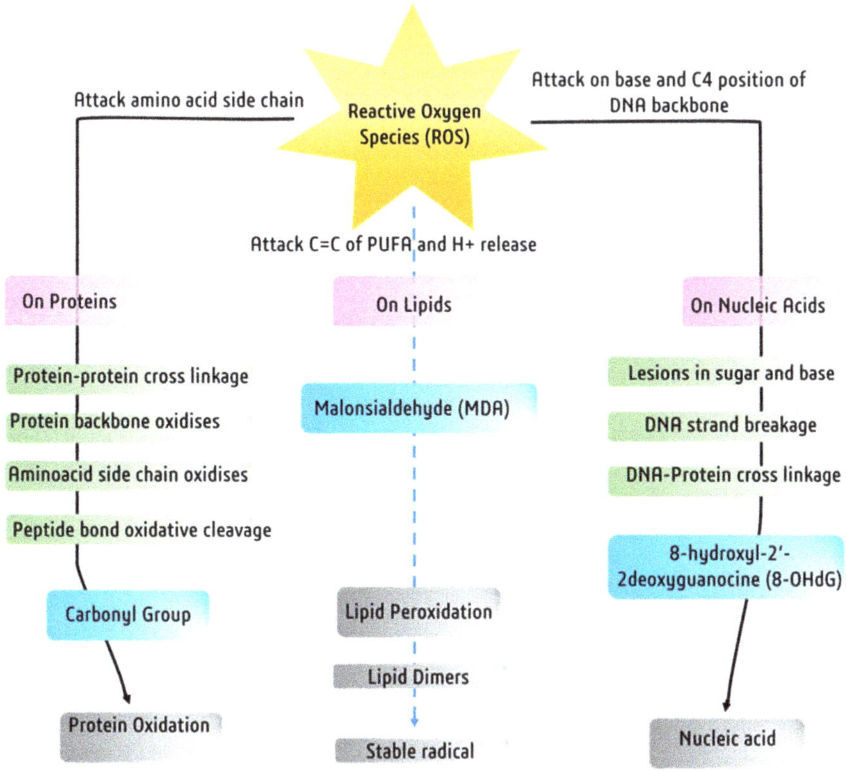

Fig. 6.2 Illustration of ROS mediated biomolecular damage, including that on proteins, lipids, and nucleic acids

enzyme performance, interfere with signalling pathways, and alter protein folding (Cecarini et al. 2007). Protein oxidative damage can impair many cellular activities involved in plant's growth and stress adaptation (Davies 2000).

6.3.3 DNA Damage

It has been a known fact that damage to DNA integrity can trigger ROS production (Rowe et al. 2008), but the opposite is also possible at certain circumstances. For instance, excessive ROS in the cytoplasm can lead to oxidative lesions in the DNA of plant cells, which can result in strand breakage, base alterations, and DNA cross-linking. Plant growth and development may be impacted by this ROS-mediated DNA damage, which may eventually lead to mutations, genomic instability, and impaired DNA replication or transcription (Renaudin 2021).

6.3.4 Change in Membrane Permeability

ROS-induced lipid peroxidation can damage membranes' lipid bilayer structure (Jin et al. 2021), impairing their ability to pass ions and molecules selectively through them. This change in membrane permeability has the potential to interfere with cellular homeostasis and impacts a number of physiological functions (Cordeiro 2014).

6.3.5 Mitochondrial Dysfunction

ROS has the potential to harm mitochondria by oxidatively destroying its biomolecules such as proteins, mitochondrial DNA, and lipids localised within the mitochondria (Zorov et al. 2014). Due to impaired mitochondrial function, ATP production is compromised, which in turn may cause more ROS creation. This vicious cycle can result in disrupted energy status of cells and the metabolism of plants as a whole (Zhao et al. 2019).

6.3.6 Chloroplast Damage

The buildup of ROS, which mostly got generated during photosynthesis, can bring about oxidative damage to the constituent parts of chloroplasts (Li and Kim 2022). Plant productivity and development are impacted by this damage, which also compromises electron transport chains, damages photosynthetic equipment, and lowers photosynthetic efficiency (Kim 2020).

6.4 Oxidative Stress Defence Mechanisms

Within a cell, reactive oxygen species (ROS) play plenty of roles, including signalling, regulatory, and potentially harmful ones. Because of their regulatory function, ROS play a critical role as mediators in cellular signalling pathways, influencing a number of physiologically vital pathways and processes including that of cell division, proliferation, and apoptosis (Kim 2020). It also takes part in the redox signalling cascades that regulate gene expression and cellular reactions to stress, where they function as signalling molecules. On the other hand, ROS can damage biomolecules like lipids, proteins, and DNA when their synthesis outpaces the antioxidant defence mechanisms of the cell (Schieber and Chandel 2014). This can result in oxidative stress and cellular malfunction. Therefore, plant cells have evolved numerous pathways that are effectively anti-oxidative mechanisms as shown in Fig. 6.3 to preserve cellular homeostasis and avoid oxidative damage by carefully balancing beneficial and harmful effects of ROS (Gechev and Petrov 2020). These anti-oxidative mechanisms can be of two types (a) enzymatic anti-oxidants and (b) non-enzymatic antioxidants.

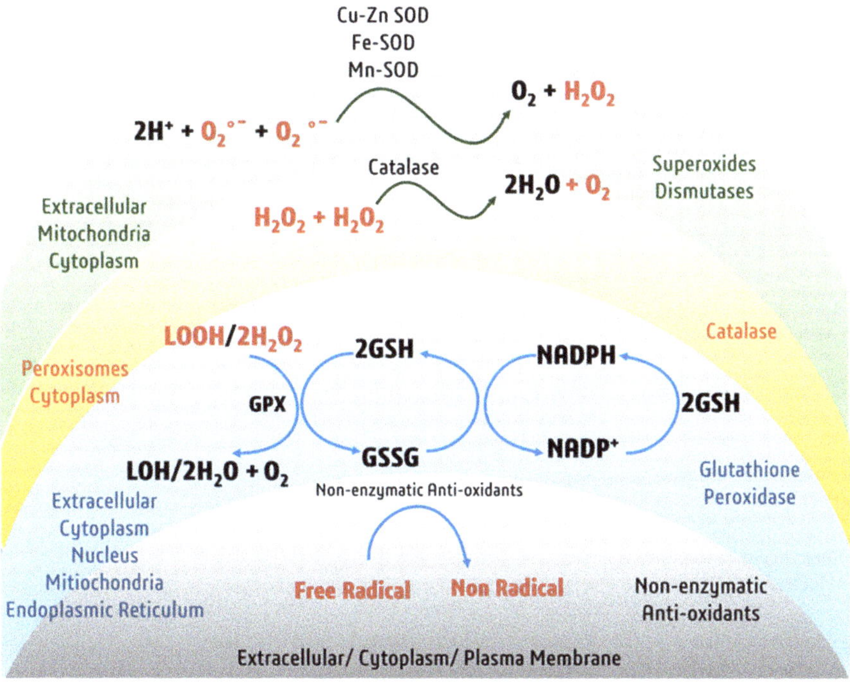

Fig. 6.3 A brief description of enzymatic and non-enzymatic antioxidants, their function and operative cell compartment

6.4.1 Enzymatic Antioxidants

Many enzymatic anti-oxidants are synthesised during stress conditions, which actively participate to reduce immediate effects of ROS and avoid oxidative damage (Fig. 6.3). Most enzymatic antioxidants first convert the ROS species to H_2O_2 and then H_2O_2 is later converted to water (H_2O) (Table 6.1). Following enzymatic anti-oxidants play most important roles.

6.4.1.1 Superoxide Dismutase (SOD)

Under stress exposure, stress-responsive transcription factors function to activate genes as feedback to several stress signals. For instance, transcription factors like AP2/ERF, MYB, WRKY, and NAC families are known to control the expression of SOD genes in response to various abiotic stressors. After exposure to stress, plants often respond by increasing the generation of SOD, which is considered the first barrier against toxic effects of ROS (Miao and St Clair 2009). SOD, which is a metalloenzyme, helps to lessen the damaging effects of superoxide accumulation by catalyzing the disproportionate conversion of one superoxide radical into hydrogen peroxide (H_2O_2) and other into molecular oxygen (O_2). SOD is present in three

Table 6.1 List of enzymatic anti-oxidants with their generalised function and example, if applicable

Name	Function	Example
Enzymatic antioxidants		
Superoxide dismutase (SOD)	Catalyses dismutation of superoxide radicals into oxygen and hydrogen peroxide, reducing superoxide levels	Cu/Zn-SOD Mn-SOD Fe-SOD
Catalase (CAT)	Catalyses decomposition of hydrogen-peroxide into water and oxygen, preventing its accumulation	Catalase
Ascorbate peroxidase (APX)	Reduces hydrogen peroxide to water, utilizing ascorbate as an electron donor, protecting cells against the increasing oxidative damage	Ascorbate peroxidase (APX)
Glutathione peroxidase (GPX)	Reduces hydrogen peroxide and organic hydroperoxides, using glutathione as a reducing agent	Glutathione peroxidase (GPX)
Glutathione reductase (GR)	Regeneration of reduced glutathione from oxidised glutathione, thus maintaining the pool of GSH for antioxidant activity	Glutathione reductase (GR)
Monodehydroascorbate reductase (MDHAR)	Reduces monodehydroascorbate to ascorbate, recycling, and maintaining the antioxidant activity of ascorbate	–
Dehydroascorbate reductase (DHAR)	Reduces dehydroascorbate back to ascorbate, restoring its antioxidant capacity	–
Peroxidase (POX)	Catalyses reduction of hydrogen peroxide and organic hydroperoxides, contributing to ROS detoxification	–
Peroxiredoxin (PRX)	Reduces hydrogen peroxide and organic hydroperoxides, protecting cells from oxidative damage	–
Thioredoxin	Participates in disulfide bond reduction in proteins, maintaining proper structure and function under oxidative conditions	Trx-f, Trx-m
Non-enzymatic anti-oxidants		
Ascorbic acid (vitamin C)	Acts as a free radical scavenger, donating electrons to neutralise ROS	Ascorbic acid (vitamin C)
Glutathione (GSH)	Participates in redox reactions, directly scavenging free radicals, and maintaining redox homeostasis	–
Carotenoids	Act as antioxidants by quenching singlet oxygen and neutralizing free radicals	Beta-carotene, lutein, zeaxanthin
Flavonoids	Scavenge free radicals, chelate metals, and modulate pathways engaged in stress responses	Quercetin, kaempferol, catechins
α-Tocopherol (vitamin E)	Acts as a lipid-soluble antioxidant, reducing lipid peroxidation	α-Tocopherol (vitamin E)
Phenolic compounds	Scavenge free radicals, chelate metals, and modulate cellular signalling pathways	Resveratrol, lignans, phenolic acids
Melatonin	Functions as a potent free radical scavenger which activates antioxidant enzymes that are protective against oxidative stress	Melatonin

isoforms, namely (a) Cu-Zn-SOD localised in cytosol, nuclei, mitochondria, chloroplast, peroxisomes, and apoplast; (b) Fe-SOD present in mitochondria, chloroplast, and peroxisomes; and (c) Mn-SOD present in mitochondria, peroxisomes, and vascular tissues (Rojo et al. 2004).

6.4.1.2 Catalase (CAT)

Hydrogen peroxide is produced in most ROS production pathways can be broken down by catalase (CAT) into water and oxygen, which keeps this potentially harmful molecule from accumulating (Nantapong et al. 2019).

6.4.1.3 Ascorbate Peroxidase (APX) and Glutathione Peroxidase (GPX)

By using ascorbate as an electron donor, APX is vital for detoxifying hydrogen peroxide and preventing oxidative damage to cells. And GPX is essential for employing glutathione as a reducing agent to reduce hydrogen peroxide and organic hydroperoxides (He et al. 2017).

6.4.1.4 Monodehydroascorbate Reductase (MDHAR) and Dehydroascorbate Reductase (DHAR)

MDHAR catalyses the transformation of monodehydroascorbate (MDHA) to ascorbate, or vitamin C, and is a necessary constituent of the ascorbate-glutathione cycle. By recycling ascorbate, a powerful antioxidant, this enzymatic activity contributes in conservation of a strong antioxidant defence system in plants against oxidative stress (Hasanuzzaman et al. 2019). **Dehydroascorbate Reductase (DHAR):** DHAR restores and preserves the antioxidant activity of ascorbate in the cell by converting dehydroascorbate (DHA) back to ascorbate. DHAR is a complementary enzyme to MDHAR. Maintaining the antioxidant properties of ascorbate and refilling its pool depend on this enzymatic activity (Filiz et al. 2019).

6.4.1.5 Peroxidase (POX) and Peroxiredoxin (PRX)

These two enzymes are involved in the scavenging of ROS such as hydrogen peroxide (H_2O_2). By catalyzing the transformation of H_2O_2 to water, POX considerably reduces the consequences of oxidative stress. In addition, PRX enzymes effectively lowers organic hydroperoxides and H_2O_2, strengthening the cell's defence against oxidative damage (Choudhury et al. 2017).

6.4.1.6 Thioredoxins (Trx)

An essential part of antioxidant defence mechanisms and redox signalling pathways. They take part in the reduction of disulfide bonds in proteins, preserving their appropriate structure and functionality in the face of oxidation. Furthermore, Trx modulates many cellular responses involved in oxidative stress by controlling a number of enzymes involved in redox processes (Kritsiligkou et al. 2018).

6.4.2 Nonenzymatic Anti-Oxidants

In plants, under stress non-enzymatic antioxidants are essential as they provide an additional line of defence in addition to enzymatic antioxidants. Due to their varied structures and functions, these molecules work together to lessen oxidative damage brought on by a range of stressors. Ascorbic acid (vitamin C), glutathione, carotenoids (beta-carotene and xanthophylls), flavonoids (quercetin and kaempferol), α-tocopherol (vitamin E), and phenolic compounds (resveratrol, lignans, and melatonin) are examples of these antioxidants, which vary in structure and function (Ahmad et al. 2010) (Table 6.1). They prevent cellular damage as they scavenge reactive oxygen species (ROS) and are involved in capturing free radicals and reactive chemicals (Waśkiewicz et al. 2014). For example, carotenoids scavenge singlet oxygen and absorb surplus light energy, whereas flavonoids bind metals and alter stress-related signalling pathways. Cell membranes are shielded from lipid peroxidation by α-tocopherol, whereas phenolic substances control cellular activities and scavenge free radicals (Nantapong et al. 2019). By preserving redox equilibrium, these non-enzymatic antioxidants greatly increase plants' resilience and improve their capacity to withstand abiotic challenges such as heavy metals, drought, and salinity (Choudhury et al. 2017).

6.4.2.1 Vitamin C

Ascorbic acid, also called vitamin C, is the main water-soluble antioxidant produced in plants. It directly scavenges ROS, neutralizing free radicals like hydroxyl and superoxide radicals (Mellidou et al. 2021). It conserves the redox status of the cell and aids in the regeneration of other antioxidants like vitamin E under stress. But in some cases, it is also observed that high vitamin C content has reduced plant stress tolerance levels (Xiao et al. 2021). Foliar application of vitamin C is also reported to produce enhanced tolerance to salinity and osmoprotection in barley (Hassan et al. 2021).

6.4.2.2 Glutathione (GSH)

An essential low-molecular-weight thiol, glutathione is involved in detoxifying reactive nitrogen species or RNS and free radicals that effectively preserves the redox balance in any cell (Hasanuzzaman et al. 2017). In addition, GSH takes part in signalling pathways particularly in AsA (ascorbic acid)-GSH pathway, which is a major pathway to scavenge cellular H_2O_2. Glutathione participates in the detoxification process of methylglyoxal (MG), which in particular is produced at low levels by aerobic metabolism under normal circumstances, just like reactive oxygen species (ROS) (Dorion et al. 2021).

6.4.2.3 Carotenoids

Beta-carotene, lutein, and zeaxanthin are examples of carotenoids, which are lipophilic pigments with antioxidant properties. By absorbing extra light energy and scavenging singlet oxygen and free radicals produced during photosynthesis, they shield plants from photooxidative damage (Giani and Martínez-Espinosa 2020; Zuluaga et al. 2017).

Flavonoids are a broad class of polyphenolic chemicals distinguished by their anti-oxidant characteristics. They bind metal ions that stimulate the generation of ROS, scavenge free radicals, and alter signalling pathways linked to stress reactions (Tay et al. 2013). They also shield cellular components from oxidative stress and oxidative stress particularly from reactive oxygen species (ROS) (Shomali et al. 2022).

6.4.2.4 α-Tocopherol or Vitamin E

As a lipophilic antioxidant, vitamin E mainly shields membranes from ROS-induced lipid peroxidation. It functions as a popular antioxidant that breaks the chain, stops the spread of free radical presents, and prevails within cell membranes (Hasanuzzaman et al. 2014).

6.4.2.5 Phenolic Compounds

Resveratrol, lignans, and phenolic acids are examples of phenolic complex that have antioxidant qualities. They enhance overall stress tolerance by chelating metals, scavenging free radicals, and regulating different cellular processes involved in stress responses (Naikoo et al. 2019).

6.4.2.6 Melatonin

Known for its function in circadian cycles, melatonin is a potent antioxidant in plants. It shields plants from oxidative stress by scavenging free radicals, activating antioxidant enzymes, and assisting in the maintaining a cellular redox equilibrium (Morvaridzadeh et al. 2020).

6.5 Conclusion

In order to increase plant tolerance against variety of abiotic stressors, non-enzymatic antioxidants work in tandem with enzymatic antioxidants to scavenge-free radicals, regulate redox balance, protect cellular structures, and modify signalling pathways. Their diverse functions play a major part in defining defence mechanisms of plants against oxidative stress triggered by unfavourable environmental conditions. The chapter emphasised the role that enzymatic and non-enzymatic antioxidant defence systems play in preventing ROS-induced damage and preserving redox equilibrium. The significant effects of oxidative stress on plant growth, development, and overall fitness have become clear because of new discoveries on ROS-induced damage to cellular components. Comprehending these chemical reactions not only elucidates the complexities of plant stress adaptation but also presents encouraging pathways for cultivating crops that are resistant to stress. By utilizing this information, research projects in the future might concentrate on specific interventions to increase plant resistance to environmental stressors, guaranteeing agricultural productivity that is sustainable in the face of changing climatic conditions.

Conflict of Interest The authors declare no conflict of interest.

References

Ahmad P, Jaleel CA, Salem MA, Nabi G, Sharma S (2010) Roles of enzymatic and nonenzymatic antioxidants in plants during abiotic stress. Crit Rev Biotechnol 30(3):161–175. https://doi.org/10.3109/07388550903524243

Cecarini V, Gee J, Fioretti E, Amici M, Angeletti M, Eleuteri AM, Keller JN (2007) Protein oxidation and cellular homeostasis: emphasis on metabolism. Biochimic Biophys Acta 1773(2):93–104. https://doi.org/10.1016/j.bbamcr.2006.08.039

Chaudhry S, Sidhu GPS (2022) Climate change regulated abiotic stress mechanisms in plants: a comprehensive review. Plant Cell Rep 41(1):1–31. https://doi.org/10.1007/s00299-021-02759-5

Choudhury FK, Rivero RM, Blumwald E, Mittler R (2017) Reactive oxygen species, abiotic stress and stress combination. Plant J 90(5):856–867. https://doi.org/10.1111/tpj.13299

Cordeiro RM (2014) Reactive oxygen species at phospholipid bilayers: distribution, mobility and permeation. Biochimic Biophys Acta 1838(1, Part B):438–444. https://doi.org/10.1016/j.bbamem.2013.09.016

D'Autréaux B, Toledano MB (2007) ROS as signalling molecules: mechanisms that generate specificity in ROS homeostasis. Nat Rev Mol Cell Biol 8(10):813–824. https://doi.org/10.1038/nrm2256

Das K, Roychoudhury A (2014) Reactive oxygen species (ROS) and response of antioxidants as ROS-scavengers during environmental stress in plants. Front Environ Sci 2:53. https://doi.org/10.3389/fenvs.2014.00053

Davies KJA (2000) Oxidative stress, antioxidant defenses, and damage removal, repair, and replacement systems. IUBMB Life 50(4–5):279–289

Dorion S, Ouellet JC, Rivoal J (2021) Glutathione metabolism in plants under stress: beyond reactive oxygen species detoxification. Meta 11(9):641. https://doi.org/10.3390/metabo11090641

Endale HT, Tesfaye W, Mengstie TA (2023) ROS induced lipid peroxidation and their role in ferroptosis. Front Cell Develop Biol 11:1. https://doi.org/10.3389/fcell.2023.1226044

Filiz E, Ozyigit II, Saracoglu IA, Uras ME, Sen U, Yalcin B (2019) Abiotic stress-induced regulation of antioxidant genes in different Arabidopsis ecotypes: microarray data evaluation. Biotechnol Biotechnol Equip 33(1):128–143. https://doi.org/10.1080/13102818.2018.1556120

Gechev T, Petrov V (2020) Reactive oxygen species and abiotic stress in plants. Int J Mol Sci 21(20):21. https://doi.org/10.3390/ijms21207433

Giani M, Martínez-Espinosa RM (2020) Carotenoids as a protection mechanism against oxidative stress in Haloferax mediterranei. Antioxidants 9(11):9. https://doi.org/10.3390/antiox9111060

Hasanuzzaman M, Nahar K, Fujita M (2014) Chapter 12—role of tocopherol (vitamin E) in plants: abiotic stress tolerance and beyond. In: Ahmad P et al (eds) Emerging technologies and management of crop stress tolerance. Academic Press, pp 267–289. https://doi.org/10.1016/B978-0-12-800875-1.00012-0

Hasanuzzaman M, Nahar K, Anee TI, Fujita M (2017) Glutathione in plants: biosynthesis and physiological role in environmental stress tolerance. Physiol Mol Biol Plants 23(2):249–268. https://doi.org/10.1007/s12298-017-0422-2

Hasanuzzaman M, Bhuyan MHMB, Anee TI, Parvin K, Nahar K, Mahmud JA, Fujita M (2019) Regulation of ascorbate-glutathione pathway in mitigating oxidative damage in plants under abiotic stress. Antioxidants 8(9):8. https://doi.org/10.3390/antiox8090384

Hassan A, Fasiha Amjad S, Hamzah Saleem M, Yasmin H, Imran M, Riaz M, Ali Q, Ahmad Joyia F, Mobeen A, Ali S, Abdullah Alsahli A, Nasser Alyemeni M (2021) Foliar application of ascorbic acid enhances salinity stress tolerance in barley (*Hordeum vulgare* L.) through modulation of morpho-physio-biochemical attributes, ions uptake, osmo-protectants and stress response genes expression. Saudi J Biol Sci 28(8):4276–4290. https://doi.org/10.1016/j.sjbs.2021.03.045

He L, He T, Farrar S, Ji L, Liu T, Ma X (2017) Antioxidants maintain cellular redox homeostasis by elimination of reactive oxygen species. Cell Physiol Biochem 44(2):532–553. https://doi.org/10.1159/000485089

He M, He C-Q, Ding N-Z (2018) Abiotic stresses: general defenses of land plants and chances for engineering multistress tolerance. Front Plant Sci 9:1. https://doi.org/10.3389/fpls.2018.01771

Jin T, Wang X, Deng Z, Liu X, Liang D (2021) ROS-induced dramatic lipid changes in *Arabidopsis*. Redox Rep 26(1):190–196. https://doi.org/10.1080/13510002.2021.2002001

Kim C (2020) ROS-driven oxidative modification: its impact on chloroplasts-nucleus communication. Front Plant Sci 10:1. https://doi.org/10.3389/fpls.2019.01729

Kritsiligkou P, Rand JD, Weids AJ, Wang X, Kershaw CJ, Grant CM (2018) Endoplasmic reticulum (ER) stress–induced reactive oxygen species (ROS) are detrimental for the fitness of a thioredoxin reductase mutant. J Biol Chem 293(31):11984–11995. https://doi.org/10.1074/jbc.RA118.001824

Li M, Kim C (2022) Chloroplast ROS and stress signaling. Plant Commun 3(1):100264. https://doi.org/10.1016/j.xplc.2021.100264

Mellidou I, Koukounaras A, Kostas S, Patelou E, Kanellis AK (2021) Regulation of vitamin C accumulation for improved tomato fruit quality and alleviation of abiotic stress. Genes 12(5):12. https://doi.org/10.3390/genes12050694

Miao L, St Clair DK (2009) Regulation of superoxide dismutase genes: implications in disease. Free Radic Biol Med 47(4):344–356. https://doi.org/10.1016/j.freeradbiomed.2009.05.018

Morvaridzadeh M, Sadeghi E, Agah S, Nachvak SM, Fazelian S, Moradi F, Persad E, Heshmati J (2020) Effect of melatonin supplementation on oxidative stress parameters: a systematic review and meta-analysis. Pharmacol Res 161:105210. https://doi.org/10.1016/j.phrs.2020.105210

Nadarajah KK (2020) ROS homeostasis in abiotic stress tolerance in plants. Int J Mol Sci 21(15):5208. https://doi.org/10.3390/ijms21155208

Naikoo MI, Dar MI, Raghib F, Jaleel H, Ahmad B, Raina A, Khan FA, Naushin F (2019) Chapter 9—role and regulation of plants phenolics in abiotic stress tolerance: an overview. In: Khan MIR, Reddy PS, Ferrante A, Khan NABT-PSM (eds) Plant signaling molecules. Woodhead Publishing, pp 157–168. https://doi.org/10.1016/B978-0-12-816451-8.00009-5

Nantapong N, Murata R, Trakulnaleamsai S, Kataoka N, Yakushi T, Matsushita K (2019) The effect of reactive oxygen species (ROS) and ROS-scavenging enzymes, superoxide dismutase and catalase, on the thermotolerant ability of Corynebacterium glutamicum. Appl Microbiol Biotechnol 103(13):5355–5366. https://doi.org/10.1007/s00253-019-09848-2

Patlevič P, Vašková J, Švorc PJ, Vaško L, Švorc P (2016) Reactive oxygen species and antioxidant defense in human gastrointestinal diseases. Integr Med Res 5(4):250–258. https://doi.org/10.1016/j.imr.2016.07.004

Qamer Z, Chaudhary MT, Du X, Hinze L, Azhar MT (2021) Review of oxidative stress and antioxidative defense mechanisms in *Gossypium hirsutum* L. in response to extreme abiotic conditions. J Cotton Res 4(1):9. https://doi.org/10.1186/s42397-021-00086-4

Renaudin X (2021) Chapter three—reactive oxygen species and DNA damage response in cancer. In: Weyemi U et al (eds) Chromatin and genomic instability in cancer, vol 364. Academic Press, pp 139–161. https://doi.org/10.1016/bs.ircmb.2021.04.001

Rojo AI, Salinas M, Martín D, Perona R, Cuadrado A (2004) Regulation of cu/Zn-superoxide dismutase expression via the phosphatidylinositol 3 kinase/Akt pathway and nuclear factor-κB. J Neurosci 24(33):7324–7334. https://doi.org/10.1523/JNEUROSCI.2111-04.2004

Rowe LA, Degtyareva N, Doetsch PW (2008) DNA damage-induced reactive oxygen species (ROS) stress response in Saccharomyces cerevisiae. Free Radic Biol Med 45(8):1167–1177. https://doi.org/10.1016/j.freeradbiomed.2008.07.018

Schieber M, Chandel NS (2014) ROS function in redox signaling and oxidative stress. Curr Biol 24(10):R453–R462. https://doi.org/10.1016/j.cub.2014.03.034

Sharma P, Jha AB, Dubey RS, Pessarakli M (2012) Reactive oxygen species, oxidative damage, and antioxidative defense mechanism in plants under stressful conditions. J Bot 2012:217037. https://doi.org/10.1155/2012/217037

Sharma A, Sharma P, Kumar R, Sharma V, Bhardwaj R, Sharma I (2021) In: Singh VP, Singh S, Tripathi DK, Prasad SM, Bhardwaj R et al (eds) 10—Role of reactive oxygen species in the

regulation of abiotic stress tolerance in legumes. Academic Press, pp 217–243. https://doi.org/10.1016/B978-0-12-815355-0.00012-6
Shomali A, Das S, Arif N, Sarraf M, Zahra N, Yadav V, Aliniaeifard S, Chauhan DK, Hasanuzzaman M (2022) Diverse physiological roles of flavonoids in plant environmental stress responses and tolerance. Plants (Basel, Switzerland) 11(22):11. https://doi.org/10.3390/plants11223158
Singh A, Kumar A, Yadav S, Singh IK (2019) Reactive oxygen species-mediated signaling during abiotic stress. Plant Gene 18:100173. https://doi.org/10.1016/j.plgene.2019.100173
Su L-J, Zhang J-H, Gomez H, Murugan R, Hong X, Xu D, Jiang F, Peng Z-Y (2019) Reactive oxygen species-induced lipid peroxidation in apoptosis, autophagy, and ferroptosis. Oxidative Med Cell Longev 2019:5080843. https://doi.org/10.1155/2019/5080843
Tay WM, da Silva GFZ, Ming L-J (2013) Metal binding of flavonoids and their distinct inhibition mechanisms toward the oxidation activity of Cu2+–β-amyloid: not just serving as suicide antioxidants! Inorg Chem 52(2):679–690. https://doi.org/10.1021/ic301832p
Tsukagoshi H (2016) Control of root growth and development by reactive oxygen species. Curr Opin Plant Biol 29:57–63. https://doi.org/10.1016/j.pbi.2015.10.012
Waśkiewicz A, Beszterda M, Goliński P (2014) Chapter 7—nonenzymatic antioxidants in plants. In: Oxidative damage to plants. Academic Press, pp 201–234. https://doi.org/10.1016/B978-0-12-799963-0.00007-1
Xiao M, Li Z, Zhu L, Wang J, Zhang B, Zheng F, Zhao B, Zhang H, Wang Y, Zhang Z (2021) The multiple roles of ascorbate in the abiotic stress response of plants: antioxidant, cofactor, and regulator. Front Plant Sci 12:1. https://doi.org/10.3389/fpls.2021.598173
Zhao R, Jiang S, Zhang L, Yu Z (2019) Mitochondrial electron transport chain, ROS generation and uncoupling (review). Int J Mol Med 44(1):3–15. https://doi.org/10.3892/ijmm.2019.4188
Zheng M, Liu Y, Zhang G, Yang Z, Xu W, Chen Q (2023) The applications and mechanisms of superoxide dismutase in medicine, food, and cosmetics. Antioxidants 12(9):12. https://doi.org/10.3390/antiox12091675
Zorov DB, Juhaszova M, Sollott SJ (2014) Mitochondrial reactive oxygen species (ROS) and ROS-induced ROS release. Physiol Rev 94(3):909–950. https://doi.org/10.1152/physrev.00026.2013
Zuluaga M, Gueguen V, Pavon-Djavid G, Letourneur D (2017) Carotenoids from microalgae to block oxidative stress. BioImpacts 7(1):1–3. https://doi.org/10.15171/bi.2017.01

Unveiling Stress Tolerance Mechanisms in Photosynthetic Cyanobacterial Symbionts: A Comprehensive Review

7

Veda Sai Kalavalapalli, Bandaru Aishwarya, and R. Gyana Prasuna

Abstract

Mangrove ecosystems, characterized by their unique and dynamic intertidal environments, are home to a diverse array of organisms, including cyanobacterial symbionts. These photosynthetic cyanobacteria play a pivotal role in these coastal habitats, contributing to their resilience and ecological functioning. In the face of environmental stressors, cyanobacterial symbionts employ a range of adaptive strategies, encompassing ecological, molecular, and biochemical aspects, to ensure their own survival while benefiting their host mangrove trees. There seems to be an intricate world of stress tolerance mechanisms in mangrove-associated cyanobacterial symbionts. Drawing from an extensive body of research reported by various authors, the present study explores the multifaceted strategies that these symbionts employ to thrive in the face of stress. This also sheds some light on their ecological significance. The paper begins by examining the ecological aspects of stress tolerance in these cyanobacterial symbionts. Moving further, it elucidates their roles in nutrient cycling, soil stabilization, and carbon sequestration. Their ability to fix atmospheric nitrogen, through a complex web of mutualistic relationships with host plants, is highlighted as a crucial contribution to the nitrogen-poor mangrove soils. Furthermore, their capacity to support food webs by serving as a fundamental component of the trophic chain underscores their role in the overall stability of mangrove ecosystems. Shifting to the molecular realm, the paper uncovers the genetic adaptations and molecular mechanisms that underpin stress tolerance in cyanobacterial symbionts. Specific pathways such as base excision repair (BER), nucleotide excision repair (NER), and mismatch repair (MMR) are explored through the lens of specific examples,

V. S. Kalavalapalli · B. Aishwarya · R. G. Prasuna (✉)
Department of Microbiology and FST, GITAM School of Science, GITAM Deemed to be University, Visakhapatnam, India
e-mail: gravi@gitam.edu

A. K. Mishra (ed.), *Stress Biology in Photosynthetic Organisms*,
https://doi.org/10.1007/978-981-97-1883-2_7

elucidating how these mechanisms preserve genomic integrity. These adaptations, including the production of heat shock proteins (HSPs) and osmoprotectants, contribute to the resilience of these symbionts under high-temperature conditions and osmotic stress. The bioactive compounds produced by cyanobacterial symbionts when facing stress, serve diverse functions, ranging from acting as antioxidants that mitigate oxidative stress to functioning as osmoprotectants that preserve cellular turgor in saline environments. Their role in deterring herbivory and protecting the host mangrove is discussed, emphasizing the dynamic interactions within the ecosystem. Chemical signaling, such as quorum sensing and mutualistic signaling, in the communication and coordination of cyanobacterial symbionts is also very essential for their survival. Through examples like *Nodularia spumigena*'s quorum sensing and the flavonoid-mediated nutrient exchange in symbiotic relationships, the paper showcases the complex web of chemical communication that enhances the ecological functioning of mangrove ecosystems. In the broader context of climate change and increasing environmental stress, understanding the stress tolerance mechanisms of cyanobacterial symbionts is paramount. As these remarkable organisms continue to shape the ecological dynamics of mangrove ecosystems, the insights gathered from these offer a valuable perspective on the intricate web of interactions and adaptations that enable their survival and ecological contributions. In conclusion, the dynamic interplay of ecological, molecular, and biochemical strategies employed by mangrove photosynthetic cyanobacterial symbionts underpins their resilience the extraordinary adaptations that make them a cornerstone of mangrove resilience.

Keywords

Mangrove ecosystems · Cyanobacterial symbionts · Stress tolerance mechanisms · Ecological adaptations · Molecular responses · Biochemical strategies · Nitrogen fixation · DNA repair genes · Quorum sensing · Mutualistic relationships · Bioactive compounds · Environmental stress · Resilience · Climate change · Coastal habitats · Ecological functioning

7.1 Introduction

Mangroves are the coastal guardians of biodiversity and stand as resilient testaments to the intricacies of life in some of the most challenging ecological environments on the planet (Spalding et al. 2014). They occupy the intertidal zones along tropical and subtropical coastlines, braving relentless assaults from the sea including surging tides, saline waters, and the harsh embrace of the sun (Mayers 1991). To thrive in such adversity, mangroves have evolved unique strategies and allies, with one of the most intriguing being their partnership with photosynthetic cyanobacterial symbionts (Boden 2022).

The importance of mangroves cannot be overstated (Veitayaki et al. 2017). These coastal ecosystems offer a multitude of ecological services (Barbier et al. 2011). Ranging from providing critical nursery grounds for countless marine species to sequestering significant amounts of carbon thus contributing to the global fight against climate change (Rossi and Rizzo 2020). However, it is their endurance in the face of adversities that truly sets them apart. This ability to thrive in challenging conditions is in no small part due to their remarkable relationships with microorganisms and notably the photosynthetic cyanobacterial symbionts (Mutalipassi et al. 2021).

Cyanobacteria among the oldest and most ancient photosynthetic organisms known to wisdom have contributed to the Earth's oxygen-rich atmosphere and the preservation of life (Knoll 2003). However, it is the symbiotic variants of these organisms that continue to allure scientists and experimenters across multiple disciplines. These associations are marked by intricate ecological connections. sustained by mutualistic or commensal associations, where cyanobionts give essential resources to their hosts in exchange for a stable terrain and access to nutrients (Decelle et al. 2015).

Photosynthetic cyanobacterial symbionts, often referred to simply as "cyanobionts," stand as exceptional examples of nature's versatility and the intricate dance of coexistence within the microbial world (Pawlowski 2009). These symbiotic microorganisms, a subset of the larger group of cyanobacteria, have forged remarkable partnerships with an array of life forms, from plants and fungi to corals and sponges (Blackall et al. 2015). These symbionts are typically found associated with the root systems of mangroves, establishing symbiotic associations (Farrer et al. 2022). These associations are considered mutualistic because they provide benefits to both parties involved (Mutalipassi et al. 2021). The cyanobacterial symbionts harness sunlight through photosynthesis, converting carbon dioxide and water into organic compounds, such as sugars, while releasing oxygen (Subashchandrabose et al. 2011). Their ability to photosynthesize, converting sunlight into energy and oxygen, forms the basis of their unique role in symbiotic relationships, making them fundamental players in the ecology and evolution of our planet (Fischer et al. 2016). They possess the unique ability to convert atmospheric nitrogen gas (N_2) into biologically available forms of nitrogen, such as ammonia (NH_3). This nitrogen-fixing capability is particularly significant in mangrove ecosystems, where nitrogen is often a limiting nutrient.

These symbionts thrive in a mutually salutary relationship with the mangrove trees (Yadav et al. 2021). In return for the stable and protected terrain handed by the tree root system (Preti 2013). Cyanobacterial symbionts contribute significantly to the tree's nutrition and overall health (Erwin and Thacker 2008). Likewise the specific types of cyanobacteria forming these associations may vary depending on geographic position and original environmental conditions (Dvořák et al. 2015).

The presence of cyanobacterial symbionts might contribute to the overall health and sustainability of mangrove ecosystems (El-Sayed et al. 2023). Mangroves survive where many other plant species would struggle due to the challenges posed by high salinity and other environmental stressors (Cabot et al. 2014). The symbionts probably enhance the resilience of mangroves, supporting them through nutrients, growth regulators, or antagonism towards other pathogenic microorganisms (Haydon et al. 2021).

For many years, ecologists, molecular biologists, and environmental scientists have been fascinated by the strength of this collaboration. However, despite a substantial amount of study some of the mysteries surrounding these symbionts stress tolerance processes are still unknown. This review study aims to close this information gap by providing an in-depth analysis of the ecological, genetic, and biochemical aspects that enable symbionts to overcome the obstacles posed by their coastal environment.

Stress tolerance mechanisms are pivotal to the survival of mangrove photosynthetic cyanobacterial symbionts, as they encounter an array of environmental stressors in their daily existence (Fitt et al. 2001). High salinity a hallmark of mangrove ecosystems necessitates mechanisms for osmotic regulation, antioxidant production, and nutrient uptake (Ben Hamed et al. 2018). Elevated temperatures especially in tropical and subtropical regions demand adaptations to heat stress alterations in metabolic pathways, and protective proteins (Ahmad et al. 2021). UV radiation, intensified by the tropical sun, calls for strategies to combat photo-oxidative damage, necessitating the synthesis of UV-absorbing compounds and adjustments in photosynthetic processes (Honeth 2018). The investigation into these stress tolerance mechanisms transcends the microcosm of mangrove ecosystems (Wang and Gu 2021). It has implications for understanding broader ecological principles the adaptation of microorganisms to extreme conditions, and the conservation of critical coastal habitats (Bang et al. 2018). Additionally, it offers insights into climate change resilience, carbon sequestration, and the preservation of biodiversity (Rodrigues et al. 2023).

Mangrove forests represent an ecosystem of immense ecological and economic importance, acting as a buffer between terrestrial and marine environments (Asari et al. 2021).

A thorough knowledge of mangroves has the potential to influence conservation efforts, foster innovation, and inspire new research avenues in the pursuit of ecological sustainability.

7.2 Ecological Aspects

The ecological significance of mangrove ecosystems is undeniable. These coastal habitats are renowned for their unique position at the intersection of terrestrial and marine environments, giving rise to ecosystems of unparalleled complexity and importance (Briggs and Bowen 2013).

The foundation of mangrove ecosystems is, of course, the mangrove trees themselves (Primavera et al. 2019). These halophytes, specialized to tolcrate high salinity, are known for their distinct aerial roots and adaptability to a range of environmental stressors (Rahman et al. 2021). However, the tale of mangrove resilience is incomplete without the symbiotic microorganisms nestled within the plant structures. These microorganisms which include cyanobacterial symbionts reside in cavities such as pneumatophores and leaf glands offering mutualistic benefits in return for shelter and nutrients (Saenger and Saenger 2002).

7.2.1 The Role of Mangroves and Their Ecosystems

Mangroves serve as a bridge between the terrestrial and marine worlds, creating a transition zone where a unique blend of freshwater and saltwater ecosystems coexist (Finkl and Makowski 2021). They act as nature's filters, trapping sediments and pollutants while providing critical habitat for a diverse array of wildlife (Levin et al. 2001). These coastal forests function as essential nursery grounds for countless marine species, supporting fisheries that are vital to the livelihoods of coastal communities (Ayyam et al. 2019).

The significance of mangroves does not end there. They are carbon sequestration powerhouses, storing substantial amounts of carbon in their biomass and sediments (Azman et al. 2023). As the world grapples with the effects of climate change, understanding the factors that enable mangroves to thrive under challenging conditions is of paramount importance (Dalby 2009). The cyanobacteria also contribute significantly to the carbon sequestration. The symbiotic association, together perhaps is capable of enhancing the carbon binding (Devaprakash et al. 2024).

7.2.2 Ecological Stresses in Mangrove Ecosystems

Mangrove forests are no strangers to adversity. They endure a suite of ecological stresses that would challenge even the most robust ecosystems. The most prominent of these stresses include:

7.2.2.1 Salinity Fluctuations

Mangroves are subjected to tidal fluctuations that can result in substantial variations in salinity (Wang et al. 2022). Symbiotic cyanobacteria in the root structures of mangroves must adapt to these saline swings, making them key players in salt tolerance (Wulff 2012).

Salinity fluctuations represent one of the most defining characteristics of mangrove ecosystems (Feller et al. 2010). These habitats endure the constant ebb and flow of tides, resulting in varying salinity levels (Friedrichs and Perry 2001). During high tide, mangroves can be inundated with saltwater, subjecting them to hypersaline conditions (Zaman et al. 2014). During low tide, the mangroves face desiccation and high salinity in the soil due to evaporation (Peters et al. 2021). This cyclic change in salinity presents a significant ecological stress (Yadav et al. 2011).

Cyanobacterial symbionts are exposed to these extreme salinity fluctuations, necessitating mechanisms for osmoregulation and adaptation (Rosic 2022). They must modulate their cellular osmotic pressure to withstand the saline onslaught during high tide while avoiding desiccation during low tide (Kumar et al. 2019a). Molecular mechanisms that enable osmoregulation are pivotal for their survival in this stressful environment (Choe 2013).

7.2.2.1.1 Hypersaline Conditions During High Tide

During high tide, mangrove areas become submerged in saltwater, often resulting in hypersaline conditions (Pratolongo 2022). Cyanobacterial symbionts must have effective osmoregulation mechanisms to maintain the right balance of ions and solutes within their cells (Hagemann 2016). This helps them adapt to high salinity conditions. Genetic adaptations that render cyanobacterial symbionts halotolerant are crucial (Bergi and Trivedi 2020). These adaptations include the expression of specific genes that allow them to thrive in saline environments by regulating salt concentrations within their cells (Polle and Chen 2015) or activation of genes involved in salt stress responses, which may lead to the synthesis of osmoprotectants and stress-related proteins (Mansour and Hassan 2022).

7.2.2.1.2 Desiccation and High Salinity During Low Tide

Conversely, during low tide, the receding water levels lead to desiccation and elevated soil salinity due to evaporation (McMahon 1988/1990). These mechanisms help mangrove trees and cyanobacterial symbionts conserve water by closing stomata or modifying cell structures to reduce water loss (Kathiresan and Bingham 2001). Understanding the molecular mechanisms that enable cyanobacterial symbionts to respond to high soil salinity is essential (Rosic 2022). These mechanisms may involve the regulation of ion transporters and the activation of stress-related genes (Shah et al. 2021).

The symbiotic relationship between cyanobacterial symbionts and mangrove trees may involve genetic adaptations that facilitate mutual tolerance to salt stress, which is vital for the continued success of this partnership (Devaprakash et al. 2024).

7.2.2.2 Extreme Temperatures

Coastal environments can experience dramatic temperature fluctuations, with high temperatures during the day and lower temperatures at night (Lima and Wethey 2012). These thermal extremes demand robust stress response mechanisms in the symbionts (Cziesielski et al. 1877).

Nutrient Limitations

The availability of essential nutrients is often limited in the mangrove environment (Reef et al. 2010). Cyanobacterial symbionts play a role in nutrient cycling and must adapt to nutrient-poor conditions (Díez and Ininbergs 2014).

7.2.2.3 Pathogen and Herbivore Pressure

Like all ecosystems, mangroves face biotic pressures in the form of pathogens and herbivores (Cannicci et al. 2008). The interactions between symbionts and these biotic factors further underscore the complexity of mangrove ecology (Krauss et al. 2008).

7.2.2.4 Nutrient Limitations: Thriving on Scarcity

Nutrient availability in mangrove ecosystems is often limited. The brackish water and waterlogged soils of these habitats make nutrient uptake a challenging endeavor

(Krauss et al. 2008). While cyanobacterial symbionts play a crucial role in enhancing nutrient availability for their host mangrove trees through nitrogen fixation, they, too, face nutrient limitations (Benavides et al. 2017).

Mangrove ecosystems, characterized by their brackish waters and waterlogged soils, present an ecological conundrum regarding nutrient availability (Bann 1998). Nutrient limitations are a common feature of these coastal habitats, which can make nutrient uptake a challenging endeavor for both mangrove trees and their associated cyanobacterial symbionts (Wulff 2012). The ability to efficiently utilize scarce resources is an essential ecological adaptation (Bernai 2013).

Genes related to nutrient acquisition and utilization are central to their ecological success (Richardson et al. 2018). The molecular mechanisms governing nutrient uptake, nitrogen fixation, and resource allocation are critical components of their response to nutrient limitations (Nacry et al. 2013).

Some of the key aspects of the molecular adaptations to nutrient limitations in cyanobacteria include:

Transporter Proteins: Cyanobacterial symbionts often possess specialized transporter proteins that can scavenge and transport essential nutrients efficiently (Ford et al. 2021). These transporters are finely regulated to ensure maximum resource utilization (Douglas and Raven 2003).

Nitrate Transporters (NRT): These transporters are responsible for the uptake of nitrate (NO_3^-), which is a significant source of nitrogen for many cyanobacteria (Swapnil et al. 2021; Watzer et al. 2019). Nitrate is an essential nutrient for protein synthesis and growth (Tischner and Kaiser 2007).

Phosphate Transporters (Pst and Pit Systems): Cyanobacterial symbionts utilize phosphate transporters to acquire phosphate (PO_4^{3-}) from the environment (Sanz-Luque et al. 2020). Phosphate is a vital nutrient for processes such as DNA replication and energy transfer in cells (Tiwari et al. 2015).

Iron Transporters (Fut, Feo, and others): Iron is a critical micronutrient, and various transporter proteins help cyanobacterial symbionts acquire iron from their surroundings (Chakraborty et al. 2019; Schoffman et al. 2016). Iron is essential for photosynthesis and other biochemical reactions (Roy et al. 2020).

Sulfate Transporters (SulT): Sulfate transporters facilitate the uptake of sulfate ions (SO_4^{2-}) from the environment (Takahashi 2019). Sulfate is a source of sulfur, which is essential for amino acid synthesis and other cellular processes (Takahashi 2010).

Amino Acid Transporters: Cyanobacterial symbionts often have specialized transporters for amino acids, which are fundamental building blocks for protein synthesis and other cellular functions.

Sugar Transporters: Transporters for sugars and other organic compounds play a role in acquiring carbon and energy sources from the environment.

Nutrient Uptake Regulation: The molecular regulation of nutrient uptake pathways is vital. Cyanobacterial symbionts can modulate the activity of nutrient transporters and channels in response to changing nutrient availability.

Efficient Resource Allocation: They have evolved mechanisms for efficient allocation of acquired nutrients within their cells, ensuring that essential processes receive priority when resources are limited.

7.2.2.4.1 Nitrogen Fixation and Symbiotic Exchange

One of the most remarkable features of cyanobacterial symbionts in mangrove ecosystems is their ability to fix atmospheric nitrogen and provide it to their host mangrove trees (Díez and Ininbergs 2014). This unique nitrogen-fixing capability has both ecological and molecular implications:

Nitrogenase Enzyme Complex: The molecular machinery for nitrogen fixation is centered around the nitrogenase enzyme complex (Hoffman et al. 2014).

Symbiotic Signaling: Cyanobacterial symbionts and mangrove trees engage in sophisticated symbiotic signaling to enable nutrient exchange (Biradar et al. 2023). Molecular insights into this signaling process shed light on the genetic basis of this crucial mutualistic interaction (Wernegreen 2002).

7.2.2.4.2 Resource Allocation and Competition

In nutrient-limited environments, efficient resource allocation is of paramount importance (Burnap 2015). Cyanobacterial symbionts have evolved strategies to compete with other microorganisms for the limited resources available (Usher et al. 2007). This competition, which also involves predators and grazers, influences the molecular adaptations of cyanobacterial symbionts (O'Neil 1999):

Genetic Adaptations for Resource Competition: Cyanobacterial symbionts may possess specific genetic adaptations that allow them to outcompete rivals for scarce resources (Guerrero and Berlanga 2016). Some cyanobacterial symbionts produce allelopathic compounds, which are chemicals that inhibit the growth of other microorganisms (Gantar et al. 2008). Genetic adaptations related to the synthesis of these compounds enable the symbionts to deter competitors (Blackstone 2020). These compounds can interfere with the growth or reproduction of rival microorganisms, creating a niche for the cyanobacterial symbionts (Robertson and Groffman 2024). Understanding these genetic strategies deepens our appreciation of their ecological success. Genetic adaptations related to regulatory genes enable cyanobacterial symbionts to sense changes in resource availability and environmental conditions (Badger et al. 2006). These genes allow the symbionts to adjust their metabolic activities and resource utilization strategies based on environmental cues (Cassier-Chauvat et al. 2021). When nutrients become scarce, these regulatory genes can trigger specific responses to enhance resource competition (Rai et al. 2000). Nostoc commune possesses a suite of stress-responsive genes that are activated when it encounters challenging conditions, such as desiccation (Yadav et al. 2022). These genes trigger the synthesis of protective compounds and stress-related proteins, allowing Nostoc to survive in harsh environments (Gupta et al. 2024). These genetic adaptations collectively provide cyanobacterial symbionts with a competitive advantage in nutrient-limited environments (Díez et al. 2023).

Defensive Mechanisms: To protect themselves and their nutrient resources, cyanobacterial symbionts may have developed molecular defenses (Lindquist et al. 2005). These defenses can include the production of chemical compounds that detect competitors and grazers (Tan 2023). *Lyngbya majuscula*, a filamentous marine cyanobacterium, strengthens its cell wall, making it resistant to physical stressors like wave action and grazing (Kazmi et al. 2022). This enhanced cell resilience offers protection against mechanical damage (Pietra 1990). Various *Synechococcus* species engage in niche differentiation by specializing in different light and nutrient conditions within the water column (Six et al. 2021). Some occupy surface waters, where light is abundant, while others thrive in deeper, nutrient-rich layers. This resource specialization reduces direct competition among them (Six et al. 2007). *Microcystis aeruginosa*, a freshwater cyanobacterium, produces microcystins (Oberholster et al. 2004). These potent toxins not only deter competitors but also serve as a defense mechanism against zooplankton grazers (Zurawell et al. 2005). By releasing microcystins, *Microcystis* defends its resources and occupies its ecological niche (Rastogi et al. 2014). *Anabaena flos-aquae* is known for its production of neurotoxic saxitoxins (Christensen and Khan 2020). These toxins are a defense mechanism against grazers like Daphnia (Pawlik-Skowrońska and Bownik 2021). When consumed, the toxins detect predators and protect the cyanobacterium (Wilken et al. 2014).

Chemical Signaling: Molecular communication through chemical signaling plays a role in resource competition (Keller and Surette 2006). Cyanobacterial symbionts may use chemical cues to influence the behavior of other microorganisms in the ecosystem (Zhou et al. 2016). *Nodularia spumigena,* a filamentous cyanobacterium commonly found in brackish water environments, employs quorum sensing as a form of chemical communication (Suikkanen et al. 2005). When the population reaches a certain density, individual cells release specific signaling molecules (e.g., *Nodularia spumigena*-derived factors) into the surrounding environment (Neilan et al. 2013).

7.2.2.5 Temperature Extremes: Heat and Resilience

Mangrove ecosystems are subjected to temperature extremes, with some regions experiencing scorching heat (Jennerjahn et al. 2017). Cyanobacterial symbionts must endure high temperatures and, in some cases, thermal stress, while maintaining their metabolic activities (Díez and Ininbergs 2014). Their ability to mount heat shock responses and protect their proteins from denaturation is vital for survival (Rosic 2022).

Genes related to heat shock responses, such as chaperone proteins and stress-responsive transcription factors, are essential for their resilience in the face of temperature extremes (Haider et al. 2022). Molecular insights into these genes shed light on the strategies that cyanobacterial symbionts employ to thrive in this challenging thermal environment (Lauritano et al. 2020).

7.2.2.6 Competition and Predation: Ecological Dynamics

Mangrove ecosystems are not isolated landscapes but are intertwined with broader ecological dynamics (Feller et al. 2010). Cyanobacterial symbionts must navigate competition with other microorganisms for resources (Mazard et al. 2016). This competitive landscape also includes predators and grazers that may prey on these microorganisms (Dimijian 2000).

Genes related to resource utilization, defense mechanisms, and chemical signaling play a pivotal role in shaping their ecological interactions (Zhou et al. 2016).

7.2.2.7 Anthropogenic Pressures: Human Impacts

Human activities, such as coastal development, pollution, and overexploitation of resources, pose significant ecological stresses on mangrove ecosystems (Abuodha and Kairo 2001). These pressures can disrupt the delicate balance of these habitats and threaten the survival of their inhabitants, including cyanobacterial symbionts (Stock et al. 2021).

Molecular research can provide insights into the genetic responses of cyanobacterial symbionts to anthropogenic stresses (Cassier-Chauvat et al. 2021). Understanding how these microorganisms adapt to and mitigate the impacts of human activities is essential for their long-term conservation (Dubey et al. 2019).

7.2.3 Contributions of Cyanobacterial Symbionts to Mangrove Resilience

Cyanobacterial symbionts play a pivotal role in nutrient cycling, enhancing nutrient availability in these otherwise nutrient-poor environments (Adams et al. 2012). The nitrogen-fixing abilities of these symbionts are particularly critical, as nitrogen is often a limiting nutrient in mangrove ecosystems (Thatoi et al. 2013). Their photosynthetic activities contribute to carbon inputs and, subsequently, carbon sequestration.

In times of stress, these symbionts may engage in protective mechanisms, such as the production of bioactive compounds, to support the host mangrove.

7.2.3.1 Ecosystem Stability

In the dynamic and often turbulent intertidal zone of mangrove ecosystems, stability is paramount. The presence of cyanobacterial symbionts enhances the overall stability of these habitats in several ways:

Soil Stabilization: The intertwining of mangrove roots and the contribution of organic matter from cyanobacterial symbionts facilitate soil stabilization (Rocha et al. 2020). This minimizes erosion, maintaining the integrity of the coastline (Alongi 2020).

Nutrient Cycling: Symbionts play a pivotal role in nutrient cycling, enhancing the availability of essential elements, such as nitrogen, to mangrove trees (Yu et al.

2021). This nutrient enrichment supports the growth and vitality of the host trees, further contributing to ecosystem stability (Thatoi et al. 2013).

7.2.3.2 Carbon Sequestration

Mangrove ecosystems are renowned for their remarkable capacity to sequester carbon (Alongi 2014). The combined efforts of mangrove trees and their symbionts result in the efficient capture and storage of atmospheric carbon dioxide, making them vital players in climate change mitigation:

- Photosynthetic Powerhouse: Cyanobacterial symbionts, with their photosynthetic prowess, fix carbon dioxide and produce organic matter (Khanna et al. 2016). This photosynthetic activity sequesters carbon and facilitates carbon storage in mangrove sediments (Zhu and Yan 2022).
- Blue Carbon Reservoir: The term "blue carbon" refers to the carbon stored in coastal ecosystems like mangroves (Macreadie et al. 2021). The symbiotic partnership is instrumental in the accumulation of blue carbon, making mangrove ecosystems key players in carbon sequestration efforts (Siikamäki et al. 2013).

7.2.4 Summary of Ecological Aspects

In this section, are outlined the ecological context of mangrove forests and the stresses they face.

The ecological significance of this relationship extends beyond the boundaries of mangrove ecosystems. It has implications for carbon sequestration, climate change mitigation, and the preservation of biodiversity in coastal environments. Understanding the ecological roles and contributions of cyanobacterial symbionts is fundamental to appreciating the broader significance of these unique and vital ecosystems.

7.3 Molecular Mechanisms of Stress Tolerance

Cyanobacterial symbionts survival in mangrove ecosystems relies on intricate interactions and signaling pathways (Devaprakash et al. 2024). In this section, we explore the molecular mechanisms, genes, and research insights that elucidate how these microorganisms respond and adapt to the dynamic coastal environment.

7.3.1 Symbiotic Interactions: Genes and Nutrient Exchange

Cyanobacterial symbionts orchestrate complex symbiotic interactions with both mangrove trees and heterotrophic microbes, driven by specific genes and molecular processes (Petersen and Yuen 2021).

7.3.1.1 Symbiotic Associations with Mangrove Trees: A Molecular Nutrient Exchange

The symbiotic relationships with mangrove trees are facilitated by a suite of genes and signaling pathways (Farnsworth 2004). Notably, nitrogen fixation genes, including nifH and nifD, enable the conversion of atmospheric nitrogen into ammonia (Li and Chen 2020). The process, regulated by the Nod factor pathway, fosters nutrient exchange, benefiting both symbiotic partners.

7.3.1.2 Interactions with Heterotrophic Microbes: Chemical Signaling and Genes

Heterotrophic microbes within the microbial community engage in intricate interactions with cyanobacterial symbionts, often governed by chemical signaling (Zhou et al. 2016). Research has unveiled genes such as those responsible for quorum sensing molecules and secondary metabolite production, shedding light on the molecular basis of these interactions. This molecular dialog influences nutrient cycling and biofilm formation (Mohan et al. 2018).

7.3.2 Environmental Sensing and Signaling: Insights into Adaptation

Cyanobacterial symbionts possess a molecular arsenal for environmental sensing and signaling, allowing them to adapt swiftly to changing conditions (Amin et al. 2012).

7.3.2.1 Light Perception: Photoreceptors and Phototaxis

Light is a fundamental environmental factor for cyanobacterial symbionts, crucial for photosynthesis and their overall survival (Ting et al. 2002). Recent research has provided valuable insights into the genes and photoreceptors that govern their light perception (Fankhauser and Staiger 2002). Photoreceptor families, including phytochromes and cryptochromes, have been extensively studied. These studies illuminate the molecular basis of phototaxis, allowing the microorganisms to optimize their positioning for photosynthesis in response to changing light conditions (Losi and Gärtner 2012).

7.3.2.2 Salt Sensing: Osmosensors and Osmoregulation Genes

High salinity is a characteristic feature of mangrove ecosystems. Cyanobacterial symbionts rely on osmosensors and osmoregulation genes to maintain cellular turgor pressure and ion homeostasis in the face of fluctuating salinity levels (Wijayasinghe et al. 2017). Recent research has unveiled the genetic underpinnings of salt sensing mechanisms, including the Ktr potassium transport system. These genetic insights provide a detailed understanding of how these microorganisms adapt to the saline conditions of their habitat (Kumar et al. 2020).

7.3.2.3 Temperature Sensitivity: Thermosensors and Heat Shock Responses

Elevated temperatures pose a significant challenge to cyanobacterial symbionts in mangrove ecosystems. Their ability to mount rapid heat shock responses is critical (Díez and Ininbergs 2014). Recent research has delved into the genes and molecular mechanisms responsible for temperature sensing and heat shock responses. Genes such as dnaK, groEL, and htpG have been studied in detail (Vaze 2014). These genetic findings enhance our understanding of how these microorganisms safeguard their proteins from thermal stress.

7.3.3 Signaling Pathways in Stress Responses: Unraveling Molecular Cascades

Complex signaling pathways transduce environmental cues into adaptive actions, shaping cyanobacterial symbionts stress responses (Calderon and Strand 2021).

7.3.3.1 MAPK Signaling: Coordinating Stress Responses

Mitogen-activated protein kinase (MAPK) signaling pathways serve as central coordinators of stress responses (Beere 2001). They relay extracellular stress signals to intracellular responses, orchestrating adaptations to high salinity, temperature fluctuations, and UV radiation. Current research highlights the intricate network of genes and molecular cascades involved in MAPK signaling (He et al. 2018).

7.3.3.2 Quorum Sensing: Influence on Community Dynamics

Quorum sensing is a cornerstone of intermicrobial communication within the microbial community of mangrove ecosystems (Carlson 2015). Recent research has unraveled the molecular basis of quorum sensing in cyanobacterial symbionts and their interactions with heterotrophic microbes. *Trichodesmium erythraeum* employs quorum sensing to coordinate its behavior. When faced with nutrient limitations, these cyanobacteria communicate with one another to form larger colonies (Frischkorn 2018). This collective behavior offers protection from grazers and competitors, enhancing their survival (Held 2020). Genes encoding quorum-sensing molecules, such as acyl-homoserine lactones (AHLs) and autoinducer-2 (AI-2), have been extensively studied (Roy et al. 2011). These genetic insights offer a comprehensive view of the chemical dialogues that govern microbial interactions and community dynamics (LaSarre and Federle 2013).

7.3.3.3 Secondary Metabolites: Chemical Arsenal for Communication

Cyanobacterial symbionts are known producers of secondary metabolites, some of which have ecological roles as allelopathic compounds or antimicrobial agents (Kumar et al. 2019b). These secondary metabolites influence the composition and dynamics of microbial communities in the rhizosphere and biofilms, potentially shaping the diversity and interactions within these ecosystems. Beyond quorum sensing, secondary metabolites serve as a chemical arsenal that underpins microbial interactions. Recent studies have elucidated the genes responsible for the production of secondary metabolites in cyanobacterial symbionts (Zhang et al. 2022). These genes, including those encoding polyketide synthases and non-ribosomal peptide synthetases, contribute to the synthesis of compounds with ecological significance. Understanding the genetic regulation of secondary metabolite production sheds light on their role in shaping microbial interactions (Reen et al. 2015).

7.3.3.4 Biofilm Formation: The Genetic Blueprint

Biofilm formation is a crucial outcome of microbial interactions in mangrove ecosystems (Yang et al. 2022). Recent research endeavors have delved into the genetic blueprint of biofilm formation. Genes responsible for adhesion factors, exopolysaccharide production, and extracellular matrix components have been identified and studied. The genetic insights provide a deeper understanding of how these microorganisms orchestrate the formation of biofilms and their ecological implications (Gupta et al. 2019).

7.3.3.5 Nutrient Cycling: Genetic Insights into Ecological Impact

Microbial interactions within the community have far-reaching implications for nutrient cycling (Philippot et al. 2023). Recent research has traced the genetic aspects of nutrient cycling, particularly the genes involved in the decomposition of organic matter, nutrient uptake, and metabolic pathways (Zehr and Ward 2002). Understanding the genetic regulation of nutrient cycling processes expands our knowledge of the ecological impact of these microbial interactions (Isobe and Ohte 2014).

7.3.4 Cross-Talk Between Molecular Mechanisms: Integration and Regulation

Interactions, environmental sensing, and signaling pathways are intricately intertwined. The integration of signals and cross-regulation of molecular processes fine-tunes the stress responses of cyanobacterial symbionts (Nair et al. 2019).

7.3.4.1 Integrating Signals: cAMP and Second Messengers

Cyclic adenosine monophosphate (cAMP) and second messenger systems are central to signal integration. These molecular processes prioritize responses to multiple stressors and optimize resource allocation. Recent studies have elucidated the molecular basis of signal integration (Hengge 2011).

7.3.4.2 Cross-Regulation of Stress Responses: Genetic Insights

Signaling pathways cross-regulate one another, enhancing adaptability. For example, salt sensing pathways can modulate the heat shock response. Research has unveiled the genetic and molecular basis of these cross-regulatory mechanisms, shedding light on the fine-tuning of stress responses (Kissoudis 2016).

7.3.5 Symbiotic Interactions with Mangrove Trees: Nutrient Exchange Mechanisms

Cyanobacterial symbionts engage in highly evolved symbiotic interactions with mangrove trees, facilitating nutrient exchange. This subsection delves into the molecular mechanisms, genes, and recent research findings that underpin this critical aspect of their ecological relationships (Bocanegra et al. 2021).

7.3.5.1 The Nod Factor Pathway: Molecular Dialogue for Nutrient Exchange

The Nod factor pathway is at the heart of the nutrient exchange between cyanobacterial symbionts and mangrove trees. Recent research has unraveled the intricate molecular dialogue facilitated by this pathway. Genes involved in nodulation (nod) and rhizobial synthesis (nol) have been identified, shedding light on the genetic orchestration of this exchange (Chakraborty et al. 2022). Additionally, the genetic basis of nifH and nifD, responsible for nitrogen fixation, has been explored in detail. Understanding these genes and their regulation provides insights into the mutualistic associations that drive nutrient exchange within this ecosystem (Geddes and Oresnik 2016).

7.3.5.2 Genetic Regulation of Nutrient Exchange: Insights from Transcription Factors

Transcription factors play a central role in regulating nutrient exchange processes. Recent research has uncovered the molecular players that govern the expression of genes involved in nutrient exchange. Genes such as the LysR family transcriptional regulators have been identified as key regulators in this context. A deeper understanding of the genetic regulation of nutrient exchange sheds light on the fine-tuning of this symbiotic relationship (Seitz and Blokesch 2013).

7.3.5.3 Chemical Signaling: Beyond the Nod Factor Pathway

While the Nod factor pathway plays a pivotal role, chemical signaling extends beyond this pathway. Recent studies have explored the production of secondary metabolites and other signaling molecules that facilitate the communication between cyanobacterial symbionts and mangrove trees (Petersen et al. 2020). These signaling molecules, including flavonoids and lipochitooligosaccharides, are integral to the establishment and maintenance of symbiotic interactions. Research findings elucidate the diverse chemical dialogues that underpin these relationships (Martin et al. 2017).

7.3.5.4 Role of Mycorrhizal Associations: A Tripartite Symbiosis

Mycorrhizal associations add a layer of complexity to the symbiotic interactions. Recent research has illuminated the genes and molecular mechanisms that underlie mycorrhizal associations within this tripartite symbiosis (Chiu and Paszkowski 2019). Genes encoding mycorrhizal symbiotic proteins and enzymes, such as mycorrhiza-induced small secreted proteins (MiSSPs), have been explored. These genetic insights expand our understanding of the intricate dynamics of symbiotic relationships within mangrove ecosystems (Kumar 2022).

7.3.5.5 Ecological Implications: Nutrient Cycling and Biodiversity

The nutrient exchange between cyanobacterial symbionts and mangrove trees has profound ecological implications. Recent research has delved into the ecological consequences of this exchange, particularly in the context of nutrient cycling and biodiversity (Wahab et al. 2023). By tracing the flow of fixed nitrogen and carbon compounds, researchers have uncovered how these interactions influence nutrient availability and microbial diversity within the ecosystem. This ecological perspective offers a holistic understanding of the broader significance of these nutrient exchange mechanisms (Jackson et al. 2008).

7.4 Conclusion and Broader Significance

Section 7.4: Conclusion and Discussion—Unveiling the Ecological and Molecular Significance of Cyanobacterial Symbionts in Mangrove Ecosystems (Alvarenga et al. 2015).

7.4.1 Synthesizing Ecological and Molecular Insights

In this concluding and discussion section, we undertake a more extensive exploration of the ecological and molecular significance of cyanobacterial symbionts in mangrove ecosystems. By merging ecological and molecular perspectives, we aim to provide a comprehensive understanding of the broader implications of their presence in these unique coastal habitats (Kelly et al. 2017).

7.4.2 Ecosystem Stability: Cyanobacterial Symbionts as Stewards of Mangroves

The ecological significance of cyanobacterial symbionts in mangrove ecosystems is profoundly rooted in their role as stewards of these delicate coastal habitats (Mukherji et al. 2019). Their ability to engage in complex nutrient exchange with mangrove trees, primarily mediated by the Nod factor pathway, has far-reaching implications for ecosystem stability. Through nitrogen fixation, these microorganisms actively enrich the nutrient pool within the mangrove soil, fostering an environment resilient to the challenges posed by fluctuating environmental stressors (Hall and Union 2005).

From a molecular perspective, the genes underpinning this nutrient exchange mechanism are crucial. Understanding the genetic regulation of nitrogen fixation not only unveils the intricacies of the symbiotic relationship between cyanobacterial symbionts and mangrove trees but also highlights their importance in maintaining the vitality of this ecosystem. Research on nifH, nifD, and nodulation genes is invaluable in comprehending the molecular basis of nutrient exchange (Sun et al. 2021).

The significance of this ecological stability extends to coastal protection. Mangroves act as natural barriers against storm surges and coastal erosion, safeguarding both human communities and the broader ecosystem (Asari et al. 2021). Cyanobacterial symbionts, by supporting the health of mangroves through nutrient exchange, play an indirect yet vital role in preserving coastal resilience (Farrer et al. 2022).

7.4.3 Carbon Sequestration: A Molecular Lens on Climate Mitigation

Cyanobacterial symbionts contribute to carbon sequestration in mangrove ecosystems, underscoring their molecular and ecological significance in mitigating climate change. These microorganisms optimize their positioning for photosynthesis through phototaxis, a molecular mechanism rooted in specific photoreceptor genes, such as phytochromes and cryptochromes (Bhatla and Lal 2023).

Molecular research into the genetic underpinnings of phototaxis offers a nuanced understanding of their ability to respond to variations in light conditions, thereby enhancing carbon capture. The optimization of photosynthesis, driven by phototaxis, is a vital component of the carbon sequestration capacity of mangrove ecosystems (Clegg et al. 2021).

In addition to phototaxis, their role in nutrient cycling, often facilitated by the Nod factor pathway, ensures the availability of essential nutrients like nitrogen for the overall ecosystem (Habig et al. 2015). The molecular insights into nutrient cycling mechanisms provide a deeper understanding of the role of cyanobacterial symbionts in enhancing carbon sequestration (Badger et al. 2006).

These molecular processes are not only integral to their survival but also have a direct impact on the carbon dynamics within mangroves. As global efforts to combat climate change intensify, understanding the molecular intricacies of carbon sequestration in mangrove ecosystems becomes increasingly relevant (Feller et al. 2010).

7.4.4 Biodiversity Conservation: Microbial Communities as Guardians

Microbial communities, shaped by the chemical signaling, genetic interactions, and nutrient cycling orchestrated by cyanobacterial symbionts, serve as guardians of biodiversity within mangrove ecosystems (Ansari et al. 2023). The molecular insights into quorum sensing, secondary metabolite production, and the genetic basis of nutrient cycling offer an enhanced view of their role in shaping microbial diversity and community dynamics (Urvoy et al. 2022).

Quorum sensing, a cornerstone of microbial communication, influences microbial interactions and biofilm formation, which, in turn, have ecological consequences for the broader ecosystem (Monds and O'Toole 2009). Understanding the molecular basis of quorum sensing, including the genetic regulation of quorum-sensing molecules like AHLs and AI-2, provides valuable insights into the chemical dialogues that underpin community dynamics (Wang et al. 2020).

Secondary metabolites, often produced by cyanobacterial symbionts, serve as a chemical arsenal for microbial interactions. The genetic insights into the production of secondary metabolites contribute to our understanding of their role in shaping the microbial community, nutrient cycling, and the overall diversity of the ecosystem (Raimundo et al. 2018).

As biodiversity conservation gains prominence in the face of global environmental challenges, the role of microbial communities within mangrove ecosystems becomes increasingly significant. By deciphering the genetic and molecular basis of these interactions, we gain valuable knowledge that can inform conservation strategies (Bardgett 2005).

7.4.5 Future Horizons: Avenues for Research and Preservation

The complexity of cyanobacterial symbionts interactions, their environmental adaptation, and the genetic regulation of essential processes provide numerous avenues for future research and preservation efforts. Investigating the intricate molecular processes, genes, and regulatory networks that govern their responses to light, salinity, and temperature is central to furthering our understanding of their resilience in mangrove ecosystems.

Future research can delve into the specific genetic regulators of environmental responses, offering a deeper understanding of how these microorganisms adapt to the dynamic coastal environment. Research into the cross-talk between different environmental sensing mechanisms is an exciting frontier that promises to provide insights into their integrated responses to multiple stressors.

Preservation efforts should not only focus on safeguarding mangroves but also the microbial communities that underpin their ecological dynamics. Conservation strategies informed by molecular and ecological insights will be essential in maintaining the health and vitality of these coastal ecosystems.

As we stand on the precipice of unprecedented environmental challenges, the ecological and molecular significance of cyanobacterial symbionts in mangrove ecosystems is a testament to the delicate interplay between these microorganisms, their environment, and the broader ecological context. The molecular insights enhance our understanding of their role in ecosystem stability, carbon sequestration, and biodiversity conservation, making them invaluable stewards of the coastal environment.

Conflict of Interest The authors declare no conflict of interest.

References

Abuodha P, Kairo J (2001) Human-induced stresses on mangrove swamps along the Kenyan coast. Hydrobiologia 458:255–265

Adams DG, Duggan PS, Jackson O (2012) Cyanobacterial symbioses. In: Ecology of cyanobacteria II: their diversity in space and time. Springer, Berlin, pp 593–647

Ahmad M, Waraich EA, Skalicky M, Hussain S, Zulfiqar U, Anjum MZ et al (2021) Adaptation strategies to improve the resistance of oilseed crops to heat stress under a changing climate: an overview. Front Plant Sci 12:767150

Alongi DM (2014) Carbon cycling and storage in mangrove forests. Annu Rev Mar Sci 6:195–219

Alongi DM (2020) Coastal ecosystem processes. CRC Press, Boca Raton

Alvarenga DO, Rigonato J, Branco LHZ, Fiore MF (2015) Cyanobacteria in mangrove ecosystems. Biodivers Conserv 24:799–817

Amin SA, Parker MS, Armbrust EV (2012) Interactions between diatoms and bacteria. Microbiol Mol Biol Rev 76(3):667–684

Ansari M, Devi BM, Sarkar A, Chattopadhyay A, Satnami L, Balu P et al (2023) Microbial exudates as biostimulants: role in plant growth promotion and stress mitigation. J Xenobiotics 13(4):572–603

Asari N, Suratman MN, Mohd Ayob NA, Abdul Hamid NH (2021) Mangrove as a natural barrier to environmental risks and coastal protection. In: Mangroves: ecology, biodiversity and management. Springer, Berlin, pp 305–322

Ayyam V, Palanivel S, Chandrakasan S, Ayyam V, Palanivel S, Chandrakasan S (2019) Coastal ecosystems and services. In: Coastal ecosystems of the tropics-adaptive management. Springer, Berlin, pp 21–47

Azman MS, Sharma S, Hamzah ML, Zakaria RM, Palaniveloo K, MacKenzie RA (2023) Total ecosystem blue carbon stocks and sequestration potential along a naturally regenerated mangrove forest chronosequence. For Ecol Manag 527:120611

Badger MR, Price GD, Long BM, Woodger FJ (2006) The environmental plasticity and ecological genomics of the cyanobacterial CO2 concentrating mechanism. J Exp Bot 57(2):249–265

Bang C, Dagan T, Deines P, Dubilier N, Duschl WJ, Fraune S et al (2018) Metaorganisms in extreme environments: do microbes play a role in organismal adaptation? Zoology 127:1–19

Bann C (1998) Economic valuation of mangroves: a manual for researchers. EEPSEA special paper/IDRC Regional Office for Southeast and East Asia, Economy and Environment Program for Southeast Asia

Barbier EB, Hacker SD, Kennedy C, Koch EW, Stier AC, Silliman BR (2011) The value of estuarine and coastal ecosystem services. Ecol Monogr 81(2):169–193

Bardgett R (2005) The biology of soil: a community and ecosystem approach. Oxford University Press, Oxford

Beere HM (2001) Stressed to death: regulation of apoptotic signaling pathways by the heat shock proteins. Sci STKE 2001(93):re1

Ben Hamed K, Dabbous A, El Shaer H, Abdely C (2018) Salinity responses and adaptive mechanisms in halophytes and their exploitation for producing salinity tolerant crops. In: Salinity responses and tolerance in plants, volume 2: exploring RNAi, genome editing and systems biology. Springer, Berlin, pp 1–19

Benavides M, Bednarz VN, Ferrier-Pagès C (2017) Diazotrophs: overlooked key players within the coral symbiosis and tropical reef ecosystems? Front Mar Sci 4:10

Bergi J, Trivedi R (2020) Bioremediation of saline soil by cyanobacteria. In: Microbial bioremediation & biodegradation. Springer, Berlin, pp 447–465

Bernai RR (2013) Managing the risks of extreme events and disasters to advance climate change adaptation. Econ Energy Environ Policy 2(1):101–113

Bhatla SC, Lal MA (2023) Light perception and transduction. In: Plant physiology, development and metabolism. Springer, Berlin, pp 363–390

Biradar S, Ramya R, Sankaranarayanan A (2023) Mycobionts interactions in lichen. In: Microbial symbionts. Elsevier, Amsterdam, pp 215–233

Blackall LL, Wilson B, Van Oppen MJ (2015) Coral—the world's most diverse symbiotic ecosystem. Mol Ecol 24(21):5330–5347

Blackstone NW (2020) Chemiosmosis, evolutionary conflict, and eukaryotic symbiosis. In: Symbiosis: cellular, molecular, medical and evolutionary aspects. Springer, Berlin, pp 237–251

Bocanegra A, Macho-González A, Garcimartín A, Benedí J, Sánchez-Muniz FJ (2021) Whole alga, algal extracts, and compounds as ingredients of functional foods: composition and action mechanism relationships in the prevention and treatment of type-2 diabetes mellitus. Int J Mol Sci 22(8):3816

Boden J (2022) Evolution of cyanobacteria through geological time. University of Bristol, Bristol

Briggs JC, Bowen BW (2013) Marine shelf habitat: biogeography and evolution. J Biogeogr 40(6):1023–1035

Burnap RL (2015) Systems and photosystems: cellular limits of autotrophic productivity in cyanobacteria. Front Bioeng Biotechnol 3:1

Cabot C, Sibole JV, Barceló J, Poschenrieder C (2014) Lessons from crop plants struggling with salinity. Plant Sci 226:2–13

Calderon RH, Strand Å (2021) How retrograde signaling is intertwined with the evolution of photosynthetic eukaryotes. Curr Opin Plant Biol 63:102093

Cannicci S, Burrows D, Fratini S, Smith TJ III, Offenberg J, Dahdouh-Guebas F (2008) Faunal impact on vegetation structure and ecosystem function in mangrove forests: a review. Aquat Bot 89(2):186–200

Carlson SN (2015) Secondary metabolite regulation and drug-lead discovery from aquatic actinomycetes. University of Illinois at Chicago, Chicago

Cassier-Chauvat C, Blanc-Garin V, Chauvat F (2021) Genetic, genomics, and responses to stresses in cyanobacteria: biotechnological implications. Genes 12(4):500

Chakraborty S, Verma E, Singh SS (2019) Cyanobacterial siderophores: ecological and biotechnological significance. In: Cyanobacteria. Elsevier, Amsterdam, pp 383–397

Chakraborty S, Valdés-López O, Stonoha-Arther C, Ané J-M (2022) Transcription factors controlling the rhizobium–legume symbiosis: integrating infection, organogenesis and the abiotic environment. Plant Cell Physiol 63(10):1326–1343

Chiu CH, Paszkowski U (2019) Mechanisms and impact of symbiotic phosphate acquisition. Cold Spring Harb Perspect Biol 11(6):a034603

Choe KP (2013) Physiological and molecular mechanisms of salt and water homeostasis in the nematode Caenorhabditis elegans. Am J Phys Regul Integr Comp Phys 305(3):R175–R186

Christensen VG, Khan E (2020) Freshwater neurotoxins and concerns for human, animal, and ecosystem health: a review of anatoxin-a and saxitoxin. Sci Total Environ 736:139515

Clegg MR, Wacker A, Spijkerman E (2021) Phenotypic diversity and plasticity of photoresponse across an environmentally contrasting family of phytoflagellates. Front Plant Sci 12:707541

Cziesielski MJ, Liew YJ, Cui G, Schmidt-Roach S, Campana S, Marondedze C et al (1877) Multiomics analysis of thermal stress response in a zooxanthellate cnidarian reveals the importance of associating with thermotolerant symbionts. Proc R Soc B Biol Sci 2018(285):20172654

Dalby S (2009) Security and environmental change. Polity, Cambridge

Decelle J, Colin S, Foster RA (2015) Photosymbiosis in marine planktonic protists. In: Marine protists: diversity and dynamics. Springer, Berlin, pp 465–500

Devaprakash M, Thirumalaivasan R, Sivakumar N, Shyamkumar R (2024) Cyanobacterial interactions and symbiosis. In: Cyanobacteria. Elsevier, Amsterdam, pp 425–489

Díez B, Ininbergs K (2014) Ecological importance of cyanobacteria. In: Cyanobacteria: an economic perspective. Wiley, New York, pp 41–63

Díez J, López-Lozano A, Domínguez-Martín M, Gómez-Baena G, Muñoz-Marín M, Melero-Rubio Y et al (2023) Regulatory and metabolic adaptations in the nitrogen assimilation of marine picocyanobacteria. FEMS Microbiol Rev 47(1):fuac043

Dimijian GG (2000) Evolving together: the biology of symbiosis, part 1. Proc (Baylor Univ Med Cent) 13(3):217

Douglas AE, Raven JA (2003) Genomes at the interface between bacteria and organelles. Philos Trans R Soc Lond B Biol Sci 358(1429):5–18
Dubey A, Malla MA, Khan F, Chowdhary K, Yadav S, Kumar A et al (2019) Soil microbiome: a key player for conservation of soil health under changing climate. Biodivers Conserv 28:2405–2429
Dvořák P, Poulíčková A, Hašler P, Belli M, Casamatta DA, Papini A (2015) Species concepts and speciation factors in cyanobacteria, with connection to the problems of diversity and classification. Biodivers Conserv 24:739–757
El-Sayed WM, Abdrabo MA, Hamed MM (2023) Mangrove ecosystem components and benefits. In: Marine ecosystems: a unique source of valuable bioactive compounds, vol 3. Bentham Science, Sharjah, pp 155–183
Erwin PM, Thacker RW (2008) Phototrophic nutrition and symbiont diversity of two Caribbean sponge–cyanobacteria symbioses. Mar Ecol Prog Ser 362:139–147
Fankhauser C, Staiger D (2002) Photoreceptors in Arabidopsis thaliana: light perception, signal transduction and entrainment of the endogenous clock. Planta 216:1–16
Farnsworth E (2004) Hormones and shifting ecology throughout plant development. Ecology 85(1):5–15
Farrer EC, Van Bael SA, Clay K, Smith MK (2022) Plant-microbial symbioses in coastal systems: their ecological importance and role in coastal restoration. Estuar Coasts 45(7):1805–1822
Feller IC, Lovelock CE, Berger U, McKee KL, Joye SB, Ball M (2010) Biocomplexity in mangrove ecosystems. Annu Rev Mar Sci 2:395–417
Finkl CW, Makowski C (2021) Coastal ecosystems of the world. J Coast Res:1–230
Fischer WW, Hemp J, Johnson JE (2016) Evolution of oxygenic photosynthesis. Annu Rev Earth Planet Sci 44:647–683
Fitt WK, Brown BE, Warner ME, Dunne RP (2001) Coral bleaching: interpretation of thermal tolerance limits and thermal thresholds in tropical corals. Coral Reefs 20:51–65
Ford BA, Sullivan GJ, Moore L, Varkey D, Zhu H, Ostrowski M et al (2021) Functional characterisation of substrate-binding proteins to address nutrient uptake in marine picocyanobacteria. Biochem Soc Trans 49(6):2465–2481
Friedrichs CT, Perry JE (2001) Tidal salt marsh morphodynamics: a synthesis. J Coast Res 27:7–37
Frischkorn KR (2018) Physiological ecology of Trichodesmium and its microbiome in the oligotrophic ocean. Columbia University, New York
Gantar M, Berry JP, Thomas S, Wang M, Perez R, Rein KS (2008) Allelopathic activity among cyanobacteria and microalgae isolated from Florida freshwater habitats. FEMS Microbiol Ecol 64(1):55–64
Geddes BA, Oresnik IJ (2016) The mechanism of symbiotic nitrogen fixation. In: The mechanistic benefits of microbial symbionts. Springer, Berlin, pp 69–97
Guerrero R, Berlanga M (2016) From the cell to the ecosystem: the physiological evolution of symbiosis. Evol Biol 43:543–552
Gupta P, Pruthi PA, Pruthi V (2019) Role of exopolysaccharides in biofilm formation. In: Introduction to biofilm engineering. ACS Publications, Washington, DC, pp 17–57
Gupta N, Bhattacharjee S, Mishra AK (2024) Stress biology and signal perceptions in cyanobacteria. In: Cyanobacteria. Elsevier, Amsterdam, pp 383–423
Habig J, Hassen A, Swart A (2015) Application of microbiology in conservation agriculture. In: Conservation agriculture. Springer, Cham, pp 525–557
Hagemann M (2016) Coping with high and variable salinity: molecular aspects of compatible solute accumulation. In: The physiology of microalgae. Springer, Berlin, pp 359–372
Haider S, Raza A, Iqbal J, Shaukat M, Mahmood T (2022) Analyzing the regulatory role of heat shock transcription factors in plant heat stress tolerance: a brief appraisal. Mol Biol Rep 49(6):5771–5785
Hall G, Union M (2005) Undergraduate symposium 2006
Haydon TD, Seymour JR, Raina J-B, Edmondson J, Siboni N, Matthews JL et al (2021) Rapid shifts in bacterial communities and homogeneity of Symbiodiniaceae in colonies of Pocillopora acuta transplanted between reef and mangrove environments. Front Microbiol 12:756091

He M, He C-Q, Ding N-Z (2018) Abiotic stresses: general defenses of land plants and chances for engineering multistress tolerance. Front Plant Sci 9:1771
Held NA (2020) Protein regulation in Trichodesmium and other marine bacteria: observational and interpretive biomarkers of biogeochemical processes. Massachusetts Institute of Technology, Cambridge
Hengge R (2011) Stationary-phase gene regulation in escherichia coli §. EcoSal Plus 4(2). https://doi.org/10.1128/ecosalplus.5.6.3
Hoffman BM, Lukoyanov D, Yang Z-Y, Dean DR, Seefeldt LC (2014) Mechanism of nitrogen fixation by nitrogenase: the next stage. Chem Rev 114(8):4041–4062
Honeth C (2018) The impact of UV-light on grapevine berry and wine metabolites. Stellenbosch University, Stellenbosch
Isobe K, Ohte N (2014) Ecological perspectives on microbes involved in N-cycling. Microbes Environ 29(1):4–16
Jackson LE, Burger M, Cavagnaro TR (2008) Roots, nitrogen transformations, and ecosystem services. Annu Rev Plant Biol 59:341–363
Jennerjahn TC, Gilman E, Krauss KW, Lacerda L, Nordhaus I, Wolanski E (2017) Mangrove ecosystems under climate change. In: Mangrove ecosystems: a global biogeographic perspective: structure, function, and services. Springer, Berlin, pp 211–244
Kathiresan K, Bingham BL (2001) Biology of mangroves and mangrove ecosystems. Adv Mar Biol 40:81–251
Kazmi SSUH, Yapa N, Karunarathna SC, Suwannarach N (2022) Perceived intensification in harmful algal blooms is a wave of cumulative threat to the aquatic ecosystems. Biology 11(6):852
Keller L, Surette MG (2006) Communication in bacteria: an ecological and evolutionary perspective. Nat Rev Microbiol 4(4):249–258
Kelly RP, Closek CJ, O'Donnell JL, Kralj JE, Shelton AO, Samhouri JF (2017) Genetic and manual survey methods yield different and complementary views of an ecosystem. Front Mar Sci 3:283
Khanna N, Raleiras P, Lindblad P (2016) Fundamentals and recent advances in hydrogen production and nitrogen fixation in cyanobacteria. In: The physiology of microalgae. Springer, Berlin, pp 101–127
Kissoudis C (2016) Genetics and regulation of combined abiotic and biotic stress tolerance in tomato. Wageningen University and Research, Wageningen
Knoll AH (2003) The geological consequences of evolution. Geobiology 1(1):3–14
Krauss KW, Lovelock CE, McKee KL, López-Hoffman L, Ewe SM, Sousa WP (2008) Environmental drivers in mangrove establishment and early development: a review. Aquat Bot 89(2):105–127
Kumar A (2022) Microbial biocontrol: sustainable agriculture and phytopathogen management: volume 1. Springer Nature, Berlin
Kumar A, Mann A, Lata C, Kumar N, Sharma P (2019a) Salinity-induced physiological and molecular responses of halophytes. In: Research developments in saline agriculture. Springer, Berlin, pp 331–356
Kumar G, Teli B, Mukherjee A, Bajpai R, Sarma B (2019b) Secondary metabolites from cyanobacteria: a potential source for plant growth promotion and disease management. In: Secondary metabolites of plant growth promoting rhizomicroorganisms: discovery and applications. Springer, Berlin, pp 239–252
Kumar S, Paul D, Bhushan B, Wakchaure G, Meena KK, Shouche Y (2020) Traversing the "Omic" landscape of microbial halotolerance for key molecular processes and new insights. Crit Rev Microbiol 46(6):631–653
LaSarre B, Federle MJ (2013) Exploiting quorum sensing to confuse bacterial pathogens. Microbiol Mol Biol Rev 77(1):73–111
Lauritano C, Rizzo C, Lo Giudice A, Saggiomo M (2020) Physiological and molecular responses to main environmental stressors of microalgae and bacteria in polar marine environments. Microorganisms 8(12):1957
Levin LA, Boesch DF, Covich A, Dahm C, Erséus C, Ewel KC et al (2001) The function of marine critical transition zones and the importance of sediment biodiversity. Ecosystems 4:430–451

Li Q, Chen S (2020) Transfer of nitrogen fixation (nif) genes to non-diazotrophic hosts. ChemBioChem 21(12):1717–1722
Lima FP, Wethey DS (2012) Three decades of high-resolution coastal sea surface temperatures reveal more than warming. Nat Commun 3(1):704
Lindquist N, Barber PH, Weisz JB (2005) Episymbiotic microbes as food and defence for marine isopods: unique symbioses in a hostile environment. Proc R Soc B Biol Sci 272(1569):1209–1216
Losi A, Gärtner W (2012) The evolution of flavin-binding photoreceptors: an ancient chromophore serving trendy blue-light sensors. Annu Rev Plant Biol 63:49–72
Macreadie PI, Costa MD, Atwood TB, Friess DA, Kelleway JJ, Kennedy H et al (2021) Blue carbon as a natural climate solution. Nat Rev Earth Environ 2(12):826–839
Mansour MMF, Hassan FA (2022) How salt stress-responsive proteins regulate plant adaptation to saline conditions? Plant Mol Biol 108(3):175–224
Martin FM, Uroz S, Barker DG (2017) Ancestral alliances: plant mutualistic symbioses with fungi and bacteria. Science 356(6340):eaad4501
Mayers TW (1991) Sarasota Bay mangroves 1991: past alterations and future possibilities. New College of the University of South Florida
Mazard S, Penesyan A, Ostrowski M, Paulsen IT, Egan S (2016) Tiny microbes with a big impact: the role of cyanobacteria and their metabolites in shaping our future. Mar Drugs 14(5):97
McMahon RF (1988/1990) Thermal tolerance, evaporative water loss, air-water oxygen consumption and zonation of intertidal prosobranchs: a new synthesis. In: Progress in littorinid and muricid biology: proceedings of the second European meeting on littorinid biology, Tjärnö Marine Biological Laboratory, Sweden, July 4–8. Springer, Berlin
Mohan R, Benton M, Dangelmaier E, Fu Z, Chandra Sekhar A (2018) Quorum sensing and biofilm formation in pathogenic and mutualistic plant-bacterial interactions. In: Implication of quorum sensing system in biofilm formation and virulence. Springer, Berlin, pp 133–160
Monds RD, O'Toole GA (2009) The developmental model of microbial biofilms: ten years of a paradigm up for review. Trends Microbiol 17(2):73–87
Mukherji S, Haldar S, Ghosh A (2019) Investigation of the structural and functional microbial diversity in Indian mangroves. In: Microorganisms in saline environments: strategies and functions. Springer, Berlin, pp 93–130
Mutalipassi M, Riccio G, Mazzella V, Galasso C, Somma E, Chiarore A et al (2021) Symbioses of cyanobacteria in marine environments: ecological insights and biotechnological perspectives. Mar Drugs 19(4):227
Nacry P, Bouguyon E, Gojon A (2013) Nitrogen acquisition by roots: physiological and developmental mechanisms ensuring plant adaptation to a fluctuating resource. Plant Soil 370:1–29
Nair A, Chauhan P, Saha B, Kubatzky KF (2019) Conceptual evolution of cell signaling. Int J Mol Sci 20(13):3292
Neilan BA, Pearson LA, Muenchhoff J, Moffitt MC, Dittmann E (2013) Environmental conditions that influence toxin biosynthesis in cyanobacteria. Environ Microbiol 15(5):1239–1253
O'Neil JM (1999) Grazer interactions with nitrogen-fixing marine cyanobacteria: adaptation for N-acquisition? Bull Inst Oceanogr Monaco Spec 19:293–318
Oberholster P, Botha A, Grobbelaar J (2004) Microcystis aeruginosa: source of toxic microcystins in drinking water. Afr J Biotechnol 3(3):159–168
Pawlik-Skowrońska B, Bownik A (2021) Cyanobacterial anabaenopeptin-B, microcystins and their mixture cause toxic effects on the behavior of the freshwater crustacean Daphnia magna (Cladocera). Toxicon 198:1–11
Pawlowski K (2009) Prokaryotic symbionts in plants. Springer Science & Business Media, Berlin
Peters R, Lovelock C, López-Portillo J, Bathmann J, Wimmler M-C, Jiang J et al (2021) Partial canopy loss of mangrove trees: mitigating water scarcity by physical adaptation and feedback on porewater salinity. Estuar Coast Shelf Sci 248:106797
Petersen JM, Yuen B (2021) The symbiotic "all-rounders": partnerships between marine animals and chemosynthetic nitrogen-fixing bacteria. Appl Environ Microbiol 87(5):e02129-20
Petersen L-E, Kellermann MY, Schupp PJ (2020) Secondary metabolites of marine microbes: from natural products chemistry to chemical ecology. In: YOUMARES, vol 9, pp 159–180

Philippot L, Chenu C, Kappler A, Rillig MC, Fierer N (2023) The interplay between microbial communities and soil properties. Nat Rev Microbiol:1–14

Pietra F (1990) A secret world. In: Natural products of marine life. Birkhäuser, Basel, pp 232–234

Polle A, Chen S (2015) On the salty side of life: molecular, physiological and anatomical adaptation and acclimation of trees to extreme habitats. Plant Cell Environ 38(9):1794–1816

Pratolongo PD (2022) Salt marshes and mangroves: tidal saline wetlands dominated by vascular plants. In: Marine biology: a functional approach to the oceans and their organisms, vol 211. Routledge, New York

Preti F (2013) Forest protection and protection forest: tree root degradation over hydrological shallow landslides triggering. Ecol Eng 61:633–645

Primavera JH, Friess DA, Van Lavieren H, Lee SY (2019) The mangrove ecosystem. In: World seas: an environmental evaluation. Elsevier, Amsterdam, pp 1–34

Rahman MM, Mostofa MG, Keya SS, Siddiqui MN, Ansary MMU, Das AK et al (2021) Adaptive mechanisms of halophytes and their potential in improving salinity tolerance in plants. Int J Mol Sci 22(19):10733

Rai A, Söderbäck E, Bergman B (2000) Tansley review no. 116 cyanobacterium–plant symbioses. New Phytol 147(3):449–481

Raimundo I, Silva SG, Costa R, Keller-Costa T (2018) Bioactive secondary metabolites from octocoral-associated microbes—new chances for blue growth. Mar Drugs 16(12):485

Rastogi RP, Sinha RP, Incharoensakdi A (2014) The cyanotoxin-microcystins: current overview. Rev Environ Sci Biotechnol 13:215–249

Reef R, Feller IC, Lovelock CE (2010) Nutrition of mangroves. Tree Physiol 30(9):1148–1160

Reen FJ, Romano S, Dobson AD, O'Gara F (2015) The sound of silence: activating silent biosynthetic gene clusters in marine microorganisms. Mar Drugs 13(8):4754–4783

Richardson EJ, Bacigalupe R, Harrison EM, Weinert LA, Lycett S, Vrieling M et al (2018) Gene exchange drives the ecological success of a multi-host bacterial pathogen. Nat Ecol Evol 2(9):1468–1478

Robertson GP, Groffman P (2024) Nitrogen transformations. In: Soil microbiology, ecology and biochemistry. Elsevier, Amsterdam, pp 407–438

Rocha F, Esteban Lucas-Borja M, Pereira P, Muñoz-Rojas M (2020) Cyanobacteria as a nature-based biotechnological tool for restoring salt-affected soils. Agronomy 10(9):1321

Rodrigues CID, Brito LM, Nunes LJ (2023) Soil carbon sequestration in the context of climate change mitigation: a review. Soil Syst 7(3):64

Rosic N (2022) Molecular mechanisms of stress tolerance in cyanobacteria. In: Ecophysiology and biochemistry of cyanobacteria. Springer, Berlin, pp 131–153

Rossi S, Rizzo L (2020) Marine animal forests as carbon immobilizers or why we should preserve these three-dimensional alive structures. In: Perspectives on the marine animal forests of the world. Springer, Berlin, pp 333–400

Roy V, Adams BL, Bentley WE (2011) Developing next generation antimicrobials by intercepting AI-2 mediated quorum sensing. Enzym Microb Technol 49(2):113–123

Roy R, Reinders A, Ward JM, McDonald TR (2020) Understanding transport processes in lichen, Azolla–cyanobacteria, ectomycorrhiza, endomycorrhiza, and rhizobia–legume symbiotic interactions. F1000Res 9:F1000 Faculty Rev-39

Saenger P, Saenger P (2002) Biotic interactions and mangrove performance. In: Mangrove ecology, silviculture and conservation. Springer, Berlin, pp 147–182

Sanz-Luque E, Bhaya D, Grossman AR (2020) Polyphosphate: a multifunctional metabolite in cyanobacteria and algae. Front Plant Sci 11:938

Schoffman H, Lis H, Shaked Y, Keren N (2016) Iron–nutrient interactions within phytoplankton. Front Plant Sci 7:1223

Seitz P, Blokesch M (2013) Cues and regulatory pathways involved in natural competence and transformation in pathogenic and environmental Gram-negative bacteria. FEMS Microbiol Rev 37(3):336–363

Shah WH, Rasool A, Saleem S, Mushtaq NU, Tahir I, Hakeem KR et al (2021) Understanding the integrated pathways and mechanisms of transporters, protein kinases, and transcription factors in plants under salt stress. Int J Genomics 2021:5578727

Siikamäki J, Sanchirico JN, Jardine S, McLaughlin D, Morris D (2013) Blue carbon: coastal ecosystems, their carbon storage, and potential for reducing emissions. Environ Sci Policy Sustain Dev 55(6):14–29

Six C, Finkel ZV, Irwin AJ, Campbell DA (2007) Light variability illuminates niche-partitioning among marine picocyanobacteria. PLoS One 2(12):e1341

Six C, Ratin M, Marie D, Corre E (2021) Marine Synechococcus picocyanobacteria: light utilization across latitudes. Proc Natl Acad Sci 118(38):e2111300118

Spalding M, Mcivor A, Tonneijck F, Tol S, van Eijk P (2014) Mangroves for coastal defence

Stock W, Callens M, Houwenhuyse S, Schols R, Goel N, Coone M et al (2021) Human impact on symbioses between aquatic organisms and microbes. Aquat Microb Ecol 87:113–138

Subashchandrabose SR, Ramakrishnan B, Megharaj M, Venkateswarlu K, Naidu R (2011) Consortia of cyanobacteria/microalgae and bacteria: biotechnological potential. Biotechnol Adv 29(6):896–907

Suikkanen S, Fistarol GO, Granéli E (2005) Effects of cyanobacterial allelochemicals on a natural plankton community. Mar Ecol Prog Ser 287:1–9

Sun W, Shahrajabian MH, Cheng Q (2021) Nitrogen fixation and diazotrophs—a review. Rom Biotechnol Lett 26(4):2834–2845

Swapnil P, Meena M, Rai AK (2021) Molecular interaction of nitrate transporter proteins with recombinant glycinebetaine results in efficient nitrate uptake in the cyanobacterium Anabaena PCC 7120. PLoS One 16(11):e0257870

Takahashi H (2010) Regulation of sulfate transport and assimilation in plants. Int Rev Cell Mol Biol 281:129–159

Takahashi H (2019) Sulfate transport systems in plants: functional diversity and molecular mechanisms underlying regulatory coordination. J Exp Bot 70(16):4075–4087

Tan LT (2023) Impact of marine chemical ecology research on the discovery and development of new pharmaceuticals. Mar Drugs 21(3):174

Thatoi H, Behera BC, Mishra RR, Dutta SK (2013) Biodiversity and biotechnological potential of microorganisms from mangrove ecosystems: a review. Ann Microbiol 63(1):1–19

Ting CS, Rocap G, King J, Chisholm SW (2002) Cyanobacterial photosynthesis in the oceans: the origins and significance of divergent light-harvesting strategies. Trends Microbiol 10(3):134–142

Tischner R, Kaiser W (2007) Nitrate assimilation in plants. In: Biology of the nitrogen cycle. Elsevier, Amsterdam, pp 283–301

Tiwari B, Singh S, Kaushik MS, Mishra AK (2015) Regulation of organophosphate metabolism in cyanobacteria. A review. Microbiology 84:291–302

Urvoy M, Labry C, L'helguen S, Lami R (2022) Quorum sensing regulates bacterial processes that play a major role in marine biogeochemical cycles. Front Mar Sci 9:834337

Usher KM, Bergman B, Raven JA (2007) Exploring cyanobacterial mutualisms. Annu Rev Ecol Evol Syst 38:255–273

Vaze ND (2014) Inactivation of airborne bacteria by direct interaction with non-thermal dielectric barrier discharge plasma: the involvement of reactive oxygen species. Drexel University, Philadelphia

Veitayaki J, Waqalevu V, Varea R, Rollings N (2017) Mangroves in small island development states in the Pacific: an overview of a highly important and seriously threatened resource. In: Participatory mangrove management in a changing climate: perspectives from the Asia-Pacific. Springer, Berlin, pp 303–327

Wahab A, Muhammad M, Munir A, Abdi G, Zaman W, Ayaz A et al (2023) Role of arbuscular mycorrhizal fungi in regulating growth, enhancing productivity, and potentially influencing ecosystems under abiotic and biotic stresses. Plants 12(17):3102

Wang Y-S, Gu J-D (2021) Ecological responses, adaptation and mechanisms of mangrove wetland ecosystem to global climate change and anthropogenic activities. Int Biodeterior Biodegradation 162:105248

Wang S, Payne GF, Bentley WE (2020) Quorum sensing communication: molecularly connecting cells, their neighbors, and even devices. Annu Rev Chem Biomol Eng 11:447–468

Wang C, Wong S, Liao T, Weng J, Chen M, Huang M et al (2022) Photosynthesis in response to salinity and submergence in two Rhizophoraceae mangroves adapted to different tidal elevations. Tree Physiol 42(5):1016–1028
Watzer B, Spät P, Neumann N, Koch M, Sobotka R, Macek B et al (2019) The signal transduction protein PII controls ammonium, nitrate and urea uptake in cyanobacteria. Front Microbiol 10:1428
Wernegreen JJ (2002) Genome evolution in bacterial endosymbionts of insects. Nat Rev Genet 3(11):850–861
Wijayasinghe YS, Tyagi A, Poddar NK (2017) Regulation of cell volume by osmolytes. In: Cellular osmolytes: from chaperoning protein folding to clinical perspectives. Springer, Berlin, pp 195–228
Wilken S, Verspagen JM, Naus-Wiezer S, Van Donk E, Huisman J (2014) Biological control of toxic cyanobacteria by mixotrophic predators: an experimental test of intraguild predation theory. Ecol Appl 24(5):1235–1249
Wulff J (2012) Ecological interactions and the distribution, abundance, and diversity of sponges. Adv Mar Biol 61:273–344
Yadav S, Irfan M, Ahmad A, Hayat S (2011) Causes of salinity and plant manifestations to salt stress: a review. J Environ Biol 32(5):667
Yadav G, Srivastva R, Gupta P (2021) Endophytes and their applications as biofertilizers. In: Microbial technology for sustainable environment. Springer, Berlin, pp 95–123
Yadav P, Singh RP, Rana S, Joshi D, Kumar D, Bhardwaj N et al (2022) Mechanisms of stress tolerance in cyanobacteria under extreme conditions. Stresses 2(4):531–549
Yang L, Yang Q, Lin L, Luan T, Yang Q, Li P et al (2022) Effects of wetland, tide, and season on benthic biofilms and related sediment properties in mangrove ecosystems. Front Mar Sci 9:1043826
Yu H, Liu X, Yang C, Peng Y, Yu X, Gu H et al (2021) Co-symbiosis of arbuscular mycorrhizal fungi (AMF) and diazotrophs promote biological nitrogen fixation in mangrove ecosystems. Soil Biol Biochem 161:108382
Zaman S, Bhattacharyya SB, Pramanick P, Raha AK, Chakraborty S, Mitra A (2014) Rising water salinity: a threat to mangroves of Indian Sundarbans. In: Water insecurity: a social dilemma, vol 13. Emerald Group Publishing Limited, Bradford, pp 167–183
Zehr JP, Ward BB (2002) Nitrogen cycling in the ocean: new perspectives on processes and paradigms. Appl Environ Microbiol 68(3):1015–1024
Zhang Y, Gallant É, Park J-D, Seyedsayamdost MR (2022) The small-molecule language of dynamic microbial interactions. Ann Rev Microbiol 76:641–660
Zhou J, Lyu Y, Richlen ML, Anderson DM, Cai Z (2016) Quorum sensing is a language of chemical signals and plays an ecological role in algal-bacterial interactions. Crit Rev Plant Sci 35(2):81–105
Zhu J-J, Yan B (2022) Blue carbon sink function and carbon neutrality potential of mangroves. Elsevier, Amsterdam, p 153438
Zurawell RW, Chen H, Burke JM, Prepas EE (2005) Hepatotoxic cyanobacteria: a review of the biological importance of microcystins in freshwater environments. J Toxic Environ Health, Part B 8(1):1–37

Photosynthetic Resilience Under Stress: Unraveling the Role of Phytohormones and Phytometabolites

8

Vandana Sindhu and Satya Shila Singh

Abstract

Photosynthetic organisms grow in diverse habitats and also continuously deal with the fluctuating environmental conditions. The changes in the environmental factors such as intensity and quality of light, salt and nutrient concentrations, temperature extremes, and heavy metal contaminations widely affect the growth and photosynthetic efficiency of the phototrophs. However, these organisms display outstanding ability to cope up with such adverse conditions present on the earth, which is the result of resilience in the photosynthesis. Phototrophs show resistance against ever changing environmental conditions up to a certain level by maintaining their photosynthetic yield. Some phytochemicals such as secondary metabolites and phytohormones play crucial role in the resilient behavior of photosynthesis. Among phytometabolites, alpha tocopherol plays a potent role in scavenging reactive oxygen species under adverse conditions. In phototrophs, during the synthesis of alpha-tocopherol, gamma tocopherol, a phytometabolite is converted into alpha tocopherol via gamma tocopherol methyl transferase (gamma TMT) enzyme in final rate-limiting step. This enzyme has been successfully characterized in plants, but annotation and characterization are still lacking in cyanobacteria. So, diversity and phylogenetic analyses of methyltransferase domain, responsible for gamma TMT activity, have been done in this chapter. After bioinformatic analysis, it was found that some amino acid residues are conserved among domain found in all three groups of phototrophs, and these residues may be crucial for the activity of gamma TMT. Additionally, molecular docking was performed between gamma TMT-containing domain protein (*Nostoc sphaeroides* CCNUC1) and gamma-tocopherol to validate the interaction between them. Moreover, there are some literature that individually define

V. Sindhu · S. S. Singh (✉)
Laboratory of Cyanobacterial Systematics and Stress Biology, Department of Botany, Institute of Science, Banaras Hindu University, Varanasi, India

A. K. Mishra (ed.), *Stress Biology in Photosynthetic Organisms*,
https://doi.org/10.1007/978-981-97-1883-2_8

the role of phytohormones and phytometabolites in photosynthetic resilience, but here in this chapter we have compiled all these reports and discussed their roles. Besides, bioinformatic study of gamma TMT in cyanobacteria has also been done.

Keywords

Phytometabolites · Phytohormones · Photosynthetic organisms · Alpha-tocopherol · Gamma tocopherol methyl transferase · Resilience

8.1 Introduction

The whole ecosystem of the earth is dependent on the photosynthesis for food (fixed carbon) and energy either directly or indirectly. Photosynthesis is the process through which photoautotrophs such as cyanobacteria, algae, and plants fixed carbon from the atmosphere using sunlight as the source of energy and water as the source of electron (Nelson and Ben-Shem 2004). This fixed carbon further transferred to the consumers and continuous transfer occurs in the food chain of the ecosystem. Photoautotrophs grow in defined habitats and their metabolism works according to their habitats. But on the earth, environmental conditions continuously fluctuate, for example, temperature, humidity, soil salinity, water deficit that pose a major threat to the survival of photosynthetic organisms including higher plants and cyanobacteria (Fig. 8.1). In natural conditions, fluctuation of light intensity occurs, even in normal day, and photosynthetic system adjusts itself according to prevailing light conditions (Vialet-Chabrand et al. 2017; Slattery et al. 2018; Gjindali et al. 2021). These varying conditions result in overall declining in growth, photosynthetic efficiency, yield, biomass, at a global scale (Fig. 8.1). Apart from the fluctuating environmental factors, other abiotic stresses such as contamination of heavy metals and toxic organic compounds and biotic stresses (virus, insect, bacteria) also affect the photosynthetic process (Kusk and Nyholm 1992; Papazoglou et al. 2005; Hattab et al. 2009; Bilgin et al. 2010). Heavy metals and salinity affect the root growth resulting in the reduction in water and nutrients uptake (Kul et al. 2020). Among all stresses, drought stress is considered as the most detrimental cause of plant growth inhibition as 80–85% plant biomass is made up of water (Seleiman et al. 2021). Under drought stress, basic structural organization of photosynthetic apparatus is affected that causes decline in the rate of photosynthesis (Zhang et al. 2020). CO_2 fixation via Calvin cycle is negatively influenced by environmental conditions such as extreme temperature and salt stresses (Sharma et al. 2020). Due to decrease in photosynthetic carbon fixation, light absorbed by plants is not fully utilized so the extra light leads to spike in reactive oxygen species ($O_2^{\bullet-}$ and H_2O_2) production, which in turn hinder the synthesis of D1 protein in photosystem II (Takahashi and Murata 2008). Moreover, D1 protein inhibition results in the down-regulation of photosynthetic electron transport chain. Since photosynthetic organisms are sessile, they must have to deal with unfavorable conditions like fluctuating

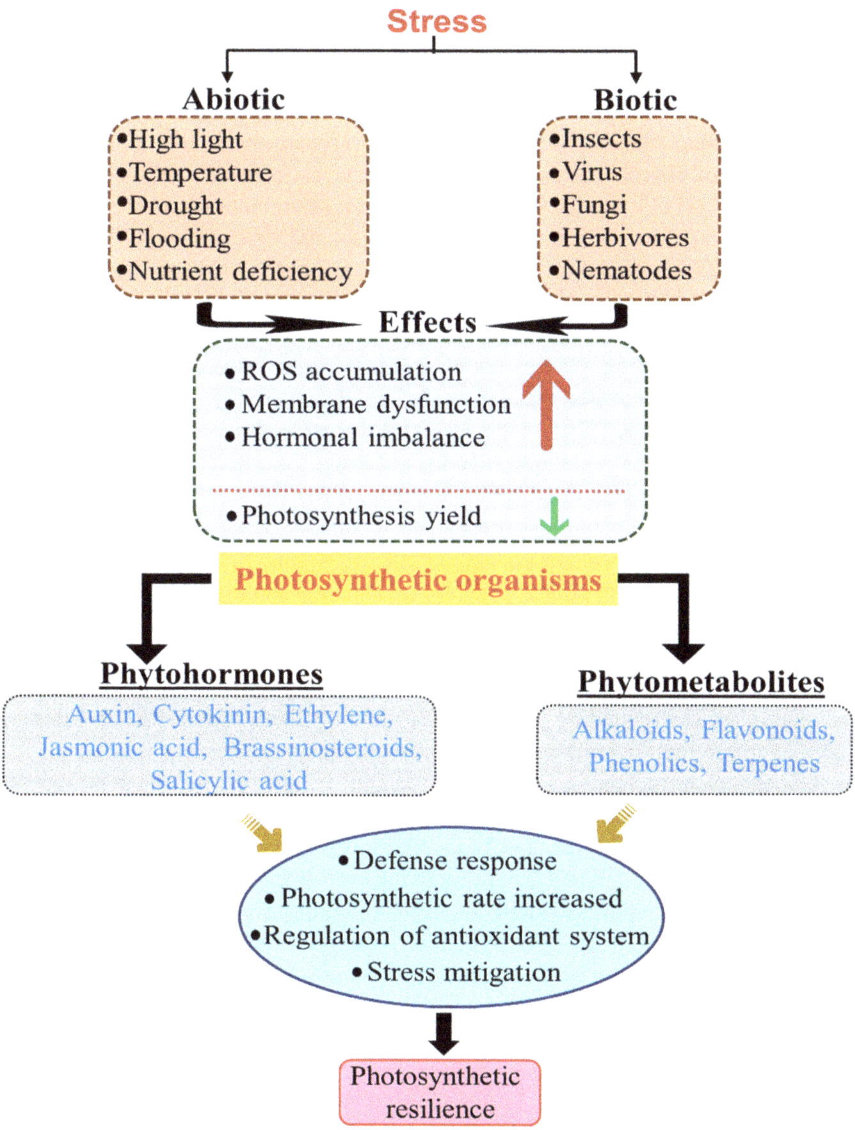

Fig. 8.1 Schematic diagram showing the effect of stress on the photosynthetic organisms (cyanobacteria, algae and plants) and consequent changes in the phytohormones and phytometabolites for resilience

light, water, and temperature that are detrimental to their growth and development (Johnová et al. 2016). To cope up with these extreme conditions, photosynthetic organisms have well-developed system for sensing, signaling, and production of phytohormones and phytometabolites (Khan et al. 2015; Zheng et al. 2023). The phytohormonal signaling pathways and induction of phytometabolites are linked in a complex network that regulates plant physiology in order to withstand with varying environmental conditions.

While in algae and cyanobacteria, several evidences are available that suggest the presence of phytohormone-like active substances from their extract (Sergeeva et al. 2002; Tsavkelova et al. 2006a, b). However, phytometabolites are crucial for protecting the photosynthetic apparatus in algae and cyanobacteria (Maeda and DellaPenna 2007). Phytohormones often known as PHs, with a low molecular weight, found in very low concentration (nano or micromolar concentration) and act locally (in cyanobacteria, thallophytes, and higher plants) and long distances (in higher plants) (Grzegorczyk-Karolak et al. 2023). So far, nine classes of phytohormones have been discovered that are, auxin, first discovered phytohormone, cytokinins (CKs), gibberellins (GAs), salicylates (SA), brassinosteroids (BRs), jasmonates (JA), abscisic acid (ABA), ethylene (ET), and strigolactones (SL), the last discovered PHs (Gomez-Roldan et al. 2008; Fahad et al. 2015).

Phytometabolites, including primary and secondary metabolites, are integral to the physiological processes of photosynthetic organisms. Primary metabolites like amino acids and carbohydrates serve as fundamental building blocks for cellular functions and growth. In contrast, secondary metabolites have specialized functions crucial for plant adaptation and resilience in changing environmental conditions, particularly during photosynthesis. Phenolic acids, flavonoids, stilbenes, and tocopherol are key secondary metabolites in photosynthetic organisms, contributing significantly to their overall metabolic and ecological strategies.

All these phytometabolites have immense antioxidative power that continuously mitigates the ROS generated during photosynthesis in both optimum and fluctuating conditions (Fig. 8.1) (Ozturk and Unal 2023). Their production in cells is either upregulated or downregulated according to the ever-changing environmental conditions. Among all metabolites, alpha tocopherol (vitamin E) is one of the major phytometabolites found in all photosynthetic organisms, but studies related to the characterization of enzymes involved in its synthesis have not yet been performed in cyanobacteria and algae. Till date, the gene encoding gamma-tocopherol methyltransferase (γ-TMT), involved in the conversion of gamma-tocopherol to alpha-tocopherol, has not been studied in cyanobacteria and algae, but they are well characterized in higher plants. In this chapter, the distribution of phytohormones and phytometabolites along with their crucial roles (both exogenously applied and endogenously produced) in photosynthetic resilience among photosynthetic organisms has been discussed. Additionally, bioinformatic analysis of the γ-TMT domain responsible for alpha-tocopherol synthesis has been conducted to assess its diversity and identify conserved residues in cyanobacteria, algae, and higher plants. Moreover, for the first time, molecular docking was conducted to check the binding of gamma TMT-containing domain protein (*Nostoc sphaeroides* CCNUC1) with gamma-tocopherol to prove the interaction of this protein (γ-TMT-containing domain protein) with gamma tocopherol.

8.2 Major Phytohormones: Their Distribution in Photosynthetic Organisms

8.2.1 Auxin

For over a century, auxin has remained the prominent center of investigations due to its fascinating roles. Charles Darwin discovered the auxin. In the book "The Power of movement in Plants" in 1881, Charles Darwin and Francis proposed the hypothesis that a "messenger" is produced in the tip of an oat coleoptile and transmitted in a downward direction leads the plant to bend towards directed light (Darwin and Darwin 1880). Darwin's experiments were based on previous studies related to roots bending toward gravity (Ciesielski 1871). The term auxin originated from the Greek word "auxien," which means "to grow" (Teale et al. 2006), and first-time auxin was examined from the human urine sample (Kögl and Haagen-Smit 1931). The indole acetic acid (IAA) was the first discovered form of auxin (Went and Thimann 1937; Zažímalová et al. 2014). The existence of auxin has been found from cyanobacteria to thallophytes and higher plants. Due to simpler body organization of cyanobacteria and algae than vascular plants, they lack the elaborate hormonal system unlike higher land plants. Therefore, the distribution and role of auxin is not well understood in lower forms. In twentieth century, auxins and their inactive analogs were extracted from green (*Chlorella*, *Enteromorpha,* and *Cladophora*), brown (*Laminaria* and *Macrocystis*), and red (*Botryocladia*) algae and cyanobacteria (*Oscillatoria*) (Schiewer 1967). Light affects the concentrations of auxin, abscisic acid, and salicylic acid in the *Anabaena* sp. (Wenz et al. 2019). Recent researches have revealed the presence of auxin-like substances from the extract of certain species of brown algae, e.g., *Fucus visiculosus* and *Fucus distichus*, green and characean algae. The concentration of auxin in algal thalli is relatively lower than higher plants, and its concentration is also varied upon seasonal fluctuations (Tarakhovskaya et al. 2007). These chemicals are not true auxins, such as indole-3-acetic acid (IAA), which is present in higher plants, but they can have similar growth-promoting properties under certain environmental conditions. The distribution and function of these compounds in algae and cyanobacteria are not well documented or understood unlike higher plants. Only few reports are available that showed the presence of auxin like substances in the extract of cyanobacteria and algae (Table 8.1). Hence, further research is needed to trace their particular roles and regulatory mechanisms of auxin-related chemicals in distinct algal and cyanobacterial species.

8.2.2 Cytokinin

Cytokinin was first time isolated from autoclaved DNA of herring sperm and calf thymus, named as kinetin (6-furfuryl aminopurine) (Miller et al. 1955). Later kinetin and related compounds named as cytokinins due to their ability to promote

Table 8.1 Distribution of phytohormones among photosynthetic organisms including cyanobacteria, algae, and higher plants

S. no.	Phytohormone	Cyanobacteria	Algae	Bryophytes and tracheophytes
1	Auxin	*Oscillatoria*, *Chlorogloea*, *Scenedesmus*, *Spirulina*, *Synechococcus* (Ahmad and Winter 1969; Provasoli and Carlucci 1974; Cruz et al. 2023) *Synechocystis* sp., *Chroococcidiopsis* sp., *Anabaena* sp., *Phormidium* sp., *Oscillatoria* sp., *Nostoc* sp. (Sergeeva et al. 2002; Hussain et al. 2010; Mazhar et al. 2013)	*Enteromorpha*, *Chlorella*, *Cladophora*, *Caulerpa*, *Macrocystis*, *Laminaria*, *Fucus*, *Ascophyllum*, *Botryocladia*, *Porphyra* (Schiewer 1967; Provasoli and Carlucci 1974; Jacobs 1993; Stirk and Van Staden 1997; Cooke et al. 2002; Sitnik et al. 2003; Polevoi et al. 2003)	In all angiosperms and most in most genera of bryophytes
2	Cytokinin	*Arthronema*, *Calothrix* (Stirk et al. 2002; Ördög et al. 2004). *Synechocystis* sp., *Chroococcidiopsis* sp., *Anabaena* sp., *Phormidium* sp., *Oscillatoria* sp., *Calothrix* sp., *Chlorogloeopsis* sp., *Rhodospirillum* sp. (Tsavkelova et al. 2006a, b; Hussain et al. 2010)	*Protococcus*, *Chlorella*, *Scenedesmus*, *Chlamydomonas*, *Fucus*, *Ascophyllum*, *Sargassum*, *Macrocystis*, *Porphyra*, *Chara*, *Euglena* (Provasoli and Carlucci 1974; Farooqi et al. 1990; Swaminathan et al. 1977; Stirk et al. 2002, 2003; Ördög et al. 2004)	In all angiosperms and in most genera of bryophytes
3	Gibberellin	*Anabaenopsis* sp., *Cylindrospermum* sp., *Phormidium foveolarum* (Gupta and Agarwal 1973; Tsavkelova et al. 2006a, b)	*Caulerpa*, *Fucus* (Radley 1961; Jacobs 1993). *Hypnea musciformis*, *Ecklonia radiata*, *Chlorella* sp., *Chlamydomonas reinhardtii* (Jennings 1967, 1968)	In angiosperms and gymnosperms Rare reports on bryophytes and pteridophytes
4	Ethylene	*Synechococcus* sp., *Anabaena* sp., *Nostoc* sp., *Calothrix* sp., *Scytonema*, *Cylindrospermum* sp. (Tsavkelova et al. 2006a, b)	*Porphyratenera*, *Padina arborescens*, *Ecklonia maxima*, *Chlorella pyrenoidosa* (Watanabe and Kondo 1976; Kreslavsky et al. 1997)	In angiosperms and gymnosperms *Pellia* (Thomas et al. 1983). *Funaria* (Rohwer and Bopp 1985)

5	Jasmonates	*Spirulina* (Stirk et al. 2003)	*Dunaliella, Chlorella, Gelidium* (Arnold et al. 2001; Sitnik et al. 2003)	In angiosperms and gymnosperms (Wasternack 2007) Rare reports on bryophytes and Pteridophytes
6	ABA	*Synechococcus Leopoliensis*, *Nostoc muscorum*, *Trichormus variabilis*, *Anabaena variabilis* (Zahradníčková et al. 1991; Maršálek et al. 1992; Hartung 2010)	*Chlorella, Dunaliella, Haematococcus, Ascophyllum, Laminaria* (Tominaga et al. 1993; Kobayashi et al. 1997; Nimura and Mizuta 2002)	In angiosperms, gymnopserms and pteridophytes (McAdam et al. 2016). In bryophytes (Takezawa et al. 2011)
7	Brasssinosteriods	–	*Hydrodictyon* (Yokota et al. 1987) Chlorophyceae, trebouxiophyceae, ulvophyceae, charophyceae, pheophyceae (Stirk et al. 2013)	In angiosperms and gymnosperms; in few pteridophytes and bryophytes (Park et al. 1999; Bajguz and Hayat 2009; Stirk et al. 2013)
8.	Lunularic acid	–	*Enteromorpha* (Pryce 1972; Niemann and Dörffling 1980)	In liverworts, *Hydrangea macrophylla* (Gorham 1977)
9.	Rhodomorphin	–	*Griffithsia* (Waaland and Watson 1980; Watson and Waaland 1983)	–

cytokinesis (Skoog and Armstrong 1970). After about 40 years of discovery, they have been isolated from plant material such as fresh coconut endosperm and also in root nodules of Australian pine (*Casuarina equisetifolia*) infected by the *Frankia* (Barciszewski et al. 1996; Raman and Elumalai 1996; Ge et al. 2005). Miller and Letham independently isolated the naturally occurring cytokinin, zeatin, from *Zea mays* (Chen 1998). Chemically, the naturally occurring cytokinins are *N*6-substituted adenine derivatives that have diverse effects, but cytokinin activity is also displayed by phenylurea derivatives (Frebort et al. 2011). Marine phytoplankton (macrophytes and microalgae) extracts have continually demonstrated cytokinin activity, which vary in content but includes both isoprenoid and aromatic cytokinins (Provasoli and Carlucci 1974). This suggests that algae have a complex hormonal system for metabolism and control (Tarakhovskaya et al. 2007). The mechanism of synthesis and physiological functions are not well understood in them. While in higher plants, cytokinins are produced via a biochemical modification of adenine and are subsequently transported to both the roots and shoots via the xylem (Osugi and Sakakibara 2015), and it regulate stomatal opening, enhance thermotolerance, and function antagonistically to abscisic acid (ABA) regulation (Pospíšilová 2003; Danilova et al. 2016).

8.2.3 Ethylene

Ethylene or ethene is a simple odorless, colorless, ignitable gaseous phytohormone of great significance in agriculture (Abhishek et al. 2016). Commercially, it is utilized in the form of ethephone (2-chloroethyl phosphonic acid), which produces ethylene gas upon hydrolysis. It has profound effect on growth, development, and stress response throughout the life cycle of photosynthetic organisms (Mattoo and Suttle 1991; Abeles et al. 2012). Ethylene consists of two carbon atoms linked by double bond. In seed plants, 1-aminocyclopropane-1-carboxylate (ACC) acts as substrate for ethylene production, while in non-seed plants, it is still unknown (Van de Poel and de Vries 2023). Application of low concentration of ethylene induces triple response (stem thickening, shoot elongation, and horizontal growth habit) during seed germination. It has profound roles in flower development, fruit ripening, and also stimulate plants to respond toward external stimuli, e.g., during flooding, it enhances the growth of adventitious roots, stimulate the epinasty, inhibit the shoot growth, and stomatal closure (Schaller and Kieber 2002). Ethylene also regulates the stress-responsive biochemical reactions in plants against various stresses such as drought, temperature fluctuations, chilling, wounding pathogen, and attack (Khan et al. 2017; Suhel et al. 2023). Ethylene receptors in plants have been extensively studied, with a common assumption that they are exclusive to plants. However, through an examination of sequenced genomes, it has been discovered that numerous bacterial species harbor potential ethylene receptors (Lacey and Binder 2016). The ethylene receptor has been identified from the cyanobacterium *Synechocystis* sp. PCC 6803, which regulate its physiology (Lacey and Binder 2016), and various findings suggested the presence of ethylene in cyanobacterial strains (Table 8.1) (Tsavkelova et al. 2006a, b).

8.2.4 Gibberellin

During research on the widespread rice disease "bakanae" (foolish seedling) (Kurosawa 1926), a class of phytohormones known as Gibberellin (GA) was identified (Stowe and Yamaki 1957). Bakanae disease causing fungus "*Fusarium fujikuroi*" were responsible for the production of gibberellins. Subsequently, gibberellin was also discovered from marine algae such as *Fucus* (Mowat 1965; Radley 1961), but still reports are low on the members of pheophyceae. The GAs are denoted as GA1, GA2, GA3, GA4, GA5, and so on according to their sequence of discovery (Sponsel and Hedden 2010). Although upto 136 different GAs have been discovered from the plants, few of them are biologically active. In *Arabidopsis,* increased concentration of GA associated with the longer hypocotyl, earlier flowering elongation of the stem, and inhibition of seed dormancy (Huang et al. 1998). While a compound resembling gibberellin has been identified in the blue-green alga *Phormidium foveolarum*. This substance exhibits activity in bioassays involving dwarf pea, dwarf maize, and cucumber hypocotyl (Gupta and Agarwal 1973).

8.2.5 Abscisic Acid

Initially, abscisic acid (ABA) was identified as dormin and abscission II from sycamore and cotton boll, respectively (Eagles and Wareing 1963; Cornforth et al. 1965; Dörffling 2015; Shu et al. 2018). Later, Addicott renamed these compounds as abscisic acid (ABA) (Addicott et al. 1968). ABA plays an essential role in a wide range of stress events, including low temperature, drought, and salinity (Rajasheker et al. 2019). The content of ABA enhanced when plant suffer wilting resulting in closure of stomata. Despite from that, ABA is also involved in the accumulation of reserve food materials in seeds, induce seed dormancy, and trigger abscission, which is also controlled by ethylene.

8.2.6 Brassinosteriods

Brassinosteroids (BRs) are ubiquitously distributed in plant kingdom (from algae to angiosperms), but not reported from cyanobacteria. Structurally, BRs are similar to animal steroids hormones and named as brassinosteroids because first time they have been isolated from pollen extract of *Brassica napus* (Mandava 1988; Clouse 2011). Till now, 69 BRs have been identified from 64 species of plants (Manghwar et al. 2022). Brassinolides are most common brassinosteroids and their precursor is campesterol (Yokota 1997). BRs play a pivotal role in diverse physiological processes as well as exert a major influence on the growth and development of photosynthetic organisms (Manghwar et al. 2022). This phytohormone, like other hormones, interacts with external environmental factors along with other phytohormones and affects the photomorphogenesis, pollen formation, flowering, and cellular dynamics in plants (Clouse 2011). Biochemical and genomics approaches have been used to trace the biosynthetic pathway and signal transduction mechanism of BRs.

8.2.7 Jasmonic Acid

Jasmonic acid (JA) is found from all photosynthetic organisms including cyanobacteria (Meyer et al. 1984; Manickavelu et al. 2006; Záveská Drábková et al. 2015; Zapata et al. 2021). Other than JA, methyl esters of jasmonic acid are also commonly found in these organisms. Initially, JA was discovered as a signaling molecule, but later it has been added in the list of phytohormones. JA is composed of pentacyclic ring structure (Table 8.2), derived from oxygenated octadecanoid fatty acids (Creelman and Mullet 1995). It is known as a stress hormone along with ABA and salicylic acid (Záveská Drábková et al. 2015). JAs regulate plant reproductive development, nutrient storage, and stress responses. A complex interaction of metabolites and jasmonic acid conjugates is involved in the plant responses to stress.

8.2.8 Salicylic Acid

Salicylic acid (SA), a phenolic molecule produced by the phenylpropanoid pathway in plants (Maruri-López et al. 2019), has been extensively investigated in plant physiological processes for its many functions (War et al. 2011; Li et al. 2022). SA is vital for pathogenesis-related gene expression, systemic acquired resistance, and hypersensitivity (Hayat et al. 2012). It also responds to ozone, salt, UV-B, drought, paraquat, heat, cold, and metal stress. SA affects stress-related flowering, tuberization, and senescence (Ananieva et al. 2002; Yadav et al. 2020; Miao et al. 2020; Maghsoudi et al. 2020; Shaukat et al. 2022). SA is a multifaceted molecule central to plant defense mechanisms and stress responses, impacting both gene expression and developmental processes.

Table 8.2 Major phytohormones with their physiological functions (modified from Tsavkelova et al. 2006a, b)

S. no.	Phytohormone	Site of synthesis	Synthetic analog	Precursor	Physiological functions
1.	Auxin 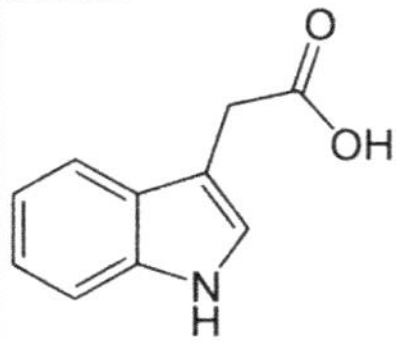	Young tissues, developing fruits	1-Naphthalene acetic acid, (NAA), 2,4-dichlorophenoxyaacetic acid (2,4-D)	Tryptophan	Apical dominance; cell elongation; tropisms; initiation of root formation; vascular tissue differentiation; promote parthenocarpy, etc.
2.	Cytokinin 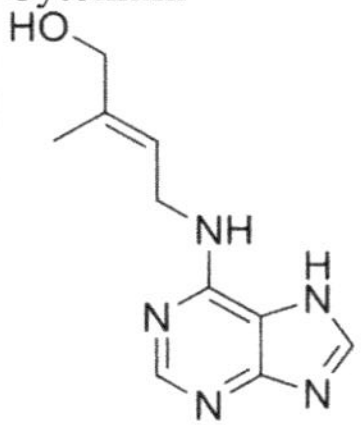	Root tips, young leaves, and developing seeds	Benzyladenine, (BAP), Forchlorfenuron	Biochemical modification of adenine	Control of cell division; lateral bud formation; delay leaf senescence, chloroplast maturation
3.	Gibberellin 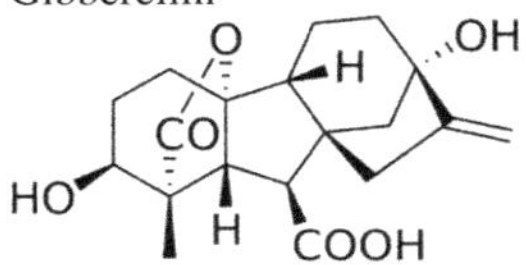	Young shoot tissues and developing seeds	Glucosyl esters	Glyceraldehyde-3-phosphate	Stem elongation; stimulation of flowering; initiation of seed germination; leaf expansion

(continued)

Table 8.2 (continued)

S. no.	Phytohormone	Site of synthesis	Synthetic analog	Precursor	Physiological functions
4.	Ethylene 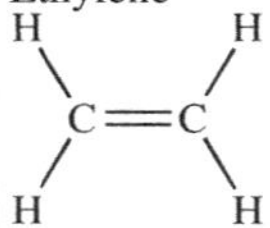	Stressed tissues and maturing fruits	–	Methionine	Senescence induction; stress response; initiation of defensive responses; fruit ripening
5.	ABA 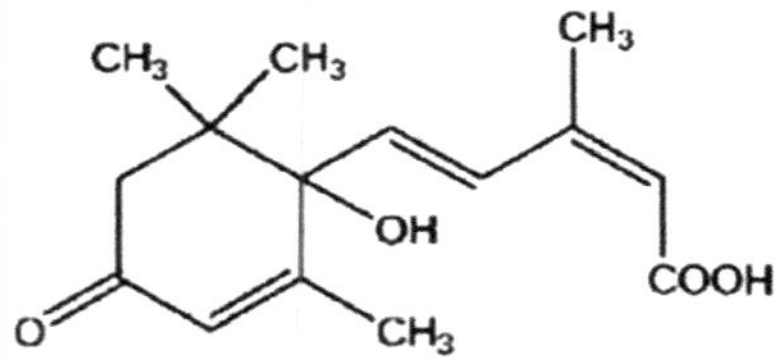	Roots and expanded leaves	3′-Phenyl alkynyl	Carotenoids	Senescence induction; stomatal regulation; control of root/shoot growth; desiccation tolerance
6.	Jasmonic acid 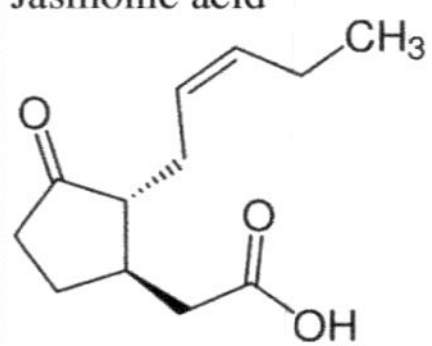	Various tissue	–	α-Linoleic acid	Provide defensive responses against insect herbivores; necrotrophic pathogen responses

7.	Salicylic acid 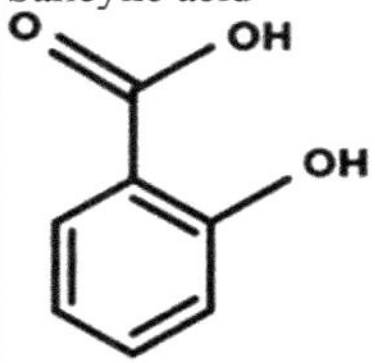	Various tissues	2,2,2,2′-tetra-fluroacetophenone (tetraFA)	Phenylalanine	Induction of the complex of defensive responses during pathogenesis
8.	Brassinosteriods 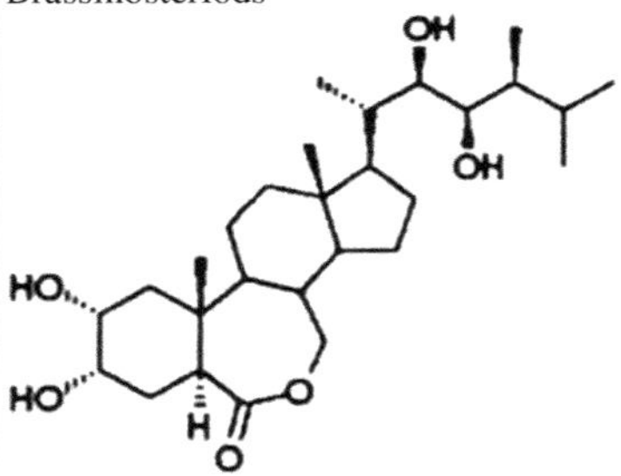	Various tissues	–	Mevalonic acid	Control of cell division; leaf senescence; stress response

8.3 Role of Phytohormones in Photosynthetic Resilience

Photosynthesis is the fundamental process for growth and development of plants (Simkin et al. 2020), and it is strongly affected by many endogenous (oxidative stress) and exogenous cues (high salinity, high light intensity, cold, heat, drought, and heavy metal toxicity) (Huang et al. 2019; Yang et al. 2020). These stressful conditions decrease the photosynthetic performance resuling in declination in growth of photosynthetic organisms. High ROS generation occurs due to stressful conditions in photosynthetic organisms, which may lead to further damage of thylakoid membranes of cyanobacteria and chloroplast, and inactivation of photosynthesis related proteins (Li and Kim 2022). In plants, stress destroys the ultrastructure of chloroplast and results in the declination of photosynthetic rate (Sidhu et al. 2017). Excess ROS can irreversibly damage the photosynthetic machinery if not mitigated by antioxidative enzymes and nonenzymatic antioxidants (Tung and Yuasa 2021). Continuous repair of D1 protein of damaged PSII occurs in the photosynthetic organisms and if repair process lag behind the damage PSII may get inactivated (Murata et al. 2007). In higher plants, drought and salinity lead to water deficiency, wilting of leaf and accumulation of ROS resulted in dissociation of protein complexes such as PSII, PSI along with water dissociation complex (Cruz de Carvalho 2008; Seleiman et al. 2021; dos Santos et al. 2022). To cope up this, plants deploy certain adaptive strategies for balancing growth and productivity. A significant role in such strategies is contributed by phytohormones (Kosakivska et al. 2022). Phytohormones are low molecular weight compound that influence the rate of photosynthesis directly or indirectly under optimal and stress conditions. These phytohormones interact with each other in cordial manner and respond to fluctuating environmental cues (Checker et al. 2018). Moreover, phytohormones also have predominant roles in protection of photosynthetic apparatus under various stress conditions (Müller and Munné-Bosch 2021). Role of the phytohormones in photosynthetic resilience can be understood by dividing phytohormones actions into two subsections: effects of exogenous applied and endogenously produced phytohormones. In the below sections, most of the information is on the higher plants in comparison to cyanobacteria and algae because most of the reports suggesting role of phytohormones on photosynthesis have been found only from the higher plants.

8.3.1 Exogenously Applied Phytohormones

Reponses of the exogenously applied phytohormones on stressed plants were seen in different studies to elucidate their role in photosynthetic resilience. It was reported that certain phytohormones (including auxin, cytokinin, and gibberellins), when applied exogenously, can positively affect various physiological processes such as glutathione cycle, transpiration rate, cytokinesis, nitrogen metabolism, which results in improved growth activity and photosynthesis in plants under abiotic stress conditions (Sytar et al. 2019). Directly or indirectly, auxin influences the photosynthetic rate by regulating the chloroplast development, leaf venation, and stomata development in control and stress conditions (Salazar-Iribe and De-la-Pena 2020). Exogenously applied 10^{-8} M concentration of auxin was most effective and maintains the rate of photosynthesis in *Brassica* sp. via improving PSII activity and electron transport rate under Cu stress (Mir et al. 2022). Several findings suggested that ABA acts as a positive regulator for re-establishing photosynthesis under light stress. Under high light, plants have found to accumulate nearly 20% transcripts of H_2O_2 and ABA-response related genes (Suzuki et al. 2015). It leads to stomatal closure and low CO_2 availability for plant (Raven 2014; Merilo et al. 2015). Moreover, exogenously applied ABA mediates the downregulation of *lhcb* gene promoter, which leads to reduction in antenna size followed by resilient the rate of net photochemical and carbon fixation reactions resulted into reduction in the impact of high light stress (Staneloni et al. 2008). The exogenously applied 12-OPDA (JA precursor) and ABA trigger the stomatal closure in *Arabidopsis*, *Brassica,* and tomato, and their combination was found more potent, as drought-responsive regulators, than individually application (Savchenko and Dehesh 2014). ABA also protects and improves the photosynthesis in algae during stress conditions. At low temperature stress in *Chlamydomonas reinhardtii*, ABA protects the photosynthetic system by repairing the photo-inactivated photosystem II complex (Saradhi et al. 2000). While in hypersaline conditions in *Neopyropia yezoensis* (red alga), small increment in F_v/F_m was found after exogenous application of ABA (Yang et al. 2022). Auxin, cytokinin, jasmonates, salicylic acid, and ethylene also involved in regulation of stomatal function (Miura and Tada 2014). Phytohormones such as brassinosteroids (BR), ethylene and salicylic acid (SA), and gibberellins have been reported to enhance the level of the antioxidative system, and decreased levels of ROS and lipid peroxidation resulted into the protection of photosynthesis in plants (Raza et al. 2022). Gibberellin (GA) mitigates the low-light conditions in plant (Djakovic-Petrovic et al. 2007). Exogenous application of GA3 on the maize plant enhanced the adaptability of plant toward low-light condition. Overexpression of key genes such as SOD, GR and downregulation of GA-related genes such as*GID1C1*, *GID1C2,* and *GID1C3* were found during exogenous application (Fu et al. 2023). Therefore, exogenously applied GA3 mitigates the abiotic stress in plants by regulating antioxidative metabolism and reducing the lipid peroxidation of cell membranes (Gilroy and Jones 1992; Jiang and Huang 2001), which finally improve the photosynthesis. Gibberellic acid (GA3) and auxin (IAA) ameliorated the photosynthesis, decreased due to copper toxicity in sunflower plant, by preserving the chlorophyll and carotenoids, and enhancement in the stability of light harvesting complex (LHC) of PSII core (Ouzounidou and Ilias 2005). When brassinosteroids were exogenously applied to the tomato plants, stressed by organic contamination and heavy metal, improvement in the F_v/F_m and Φ_{PSII} were seen along with the significant increment in photosynthetic pigments such as Chl a, Chl b, and carotenoids (Ahammed et al. 2013). In different cultivars of *Helianthus annuus* L., proline accumulation was seen followed by increment in net photosynthetic rate than Cu stressed treatment plants by addition of brassinosteroids (Filová et al. 2013). Salicylic acid application on the salt (NaCl) stressed plant (*Zea mays* L.) recuperated its photosynthetic performance by stimulating RubisCo activity and photosynthetic pigment production (Khodary 2004). Accumulation of proline and inhibition of ethylene production was occurred in drought-stressed mustard plant after exogenous application of salicylic acid. Proline accumulation leads to increase in the efficiency of PS II and stabilization of RubisCo so improvement in photosynthetic performance of stressed plant (Nazar et al. 2015). Exogenously applied jasmonic acid and methyl-jasmonate trigger the overexpression of PSII subunit genes such as *psbI, psbL, psbM*, in *Brassicaoleracea,* which leads to improved photosynthetic productivity and sugar accumulation (Sirhindi et al. 2020). In drought stress, ABA and strigolactone regulate the closing of stomata and make the plant adapted against the stress (Kim et al. 2010; Ha et al. 2014). Additionally, JA influences stomatal closure through the regulation of ABF1 and ABF4, contributing to plant resilience and growth maintenance (Lin et al. 2020). While exogenously applied salicylic acid in lemongrass alleviates drought stress by enhancing terpene citral synthesis (Idrees et al. 2010).

8.3.2 Endogenously Produced Phytohormones

Plants respond to photosynthesis decline under stress by initiating a series of reactions, along with the reactive oxygen species (ROS) generation (Sachdev et al. 2021) followed by activation of mitigation processes and promoting the production of phytometabolites and phytohormones. These cascading events induce physiological and biochemical changes, enhancing plant resilience and stabilizing growth.

Phytohormones are integral to the adaptive phase of plants, exerting endogenous influence even under optimal photosynthetic conditions (Müller and Munné-Bosch 2021). They play roles in both abiotic and biotic stress responses. For instance, when pathogens enter through stomata, salicylic acid undergoes an endogenous flux, leading to stomatal closure and preventing pathogen entry (Arnaud and Hwang 2015). This showcases the multifaceted involvement of phytohormones in regulating plant responses to various environmental challenges. While biotic stress (pathogen entry) induced ROS seems responsible for the above endogenous flux of salicylic acid (Melotto et al. 2006). During drought stress, salicylic acid accumulation occurs, which lead to ABA signaling and followed by stomata closing (Miura and Tada 2014). Salicylic acid triggers SHAM-sensitive peroxidase-mediated ROS signaling, followed by activation of CDPK1 and CDPK6, which subsequently activate SLAC1 (Prodhan et al. 2018).

8.4 Phytometabolites: In Photosynthetic Organisms

The term "phyto" in the word "phytochemical" originates from the Greek language that signifies plants (Adusei et al. 2019). There are a wide range of organic molecules produced by photosynthetic organisms that play a number of roles in the physiological processes in photosynthetic organisms, including in growth, reproduction, and defense against infections, herbivores, and environmental stresses, while their accumulation and synthesis is determined by environmental conditions (Ashraf

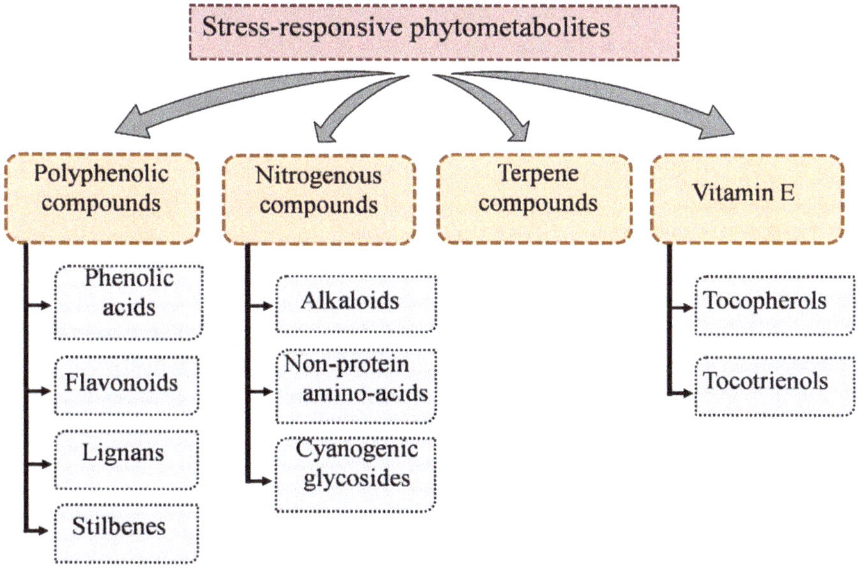

Fig. 8.2 Classification of various stress responsive phytometabolites

et al. 2018). There are more than 4000 different phytochemicals that differ from each other in chemical structure, physical nature, and functions (Malik et al. 2018; Mendoza and Silva 2018). There is no well-defined classification of phytometabolites. They are broadly divided into two categories, namely primary phytochemicals including chlorophyll, amino acids, sugars, proteins and secondary phytochemicals including saponins, glucosinolates, alkaloids, tannins, lignans, phenols, steroids, among them, phenols are the much-diversified category of secondary metabolites (Fig. 8.2) (Gregory 1998).

Polyphenols consist of over 8000 compounds (González-Sarrías et al. 2020). All synthesized through a shared pathway known as the shikimate/phenylpropanoid pathway (Sharma et al. 2019). This pathway serves as the common route for the production of these diverse molecules (Sharma et al. 2019). They display a wide range of structures, linked by the common feature of possessing one (in simple phenolics) or more (in polyphenols) hydroxyl substituents directly attached to aromatic or benzene rings. These compounds can be classified according to their structures into phenolic acids, flavonoids, stilbenoids, and lignans, highlighting the considerable diversity in their molecular arrangements and emphasizing the various forms these compounds can take in nature (Šamec et al. 2021).

8.4.1 Polyphenolic Compounds

8.4.1.1 Phenolic Acids

All phenolic acids share the common basic structure, i.e., a single hydroxy structure group attached to the benzene ring. Usually, they are found in conjugated form linked either to glycosides, amides, esters, and sometimes linked to structural components of plant cell wall through acetal, ether, and ester bonds. Further, they are divided into two subgroups, namely hydrocinnamic acids and hydrobenzoic acids. Hydroxycinnamic acids (HCAs) include caffeic, sinapic, *p*-coumaric, and ferulic acids and their esterified/etherified conjugates (e.g., chlorogenic acid) (Goleniowski et al. 2013). While hydroxybenzoic acids (HBAs) include salicylic acid (SA), gallic acid (GA), vanillic acid (VA), and variants of dihydroxybenzoic acid (2,3-DHBA; 2,5-DHBA; 3,4-DHBA; 3,5-DHBA) (Goleniowski et al. 2013). Functionally, phenolic acids act as powerful ROS scavengers under different stress conditions (Bistgani et al. 2019; Chen et al. 2019). Under such conditions, their biosynthesis is mediated by the activation of key genes including phenylalanine synthase (PAL) and chalcone synthase (CHS) of phenylpropanoid biosynthetic pathway and result in accumulation of various phenolic acids such as caffeic, cinnamic acid, gallic, ferulic, and vanillic acids (Commisso et al. 2016; Wang et al. 2019). Although phenolics play a multifaceted role in enhancing the photosynthetic performance of plants by protecting against oxidative stress (Linić et al. 2021; Hura et al. 2023), providing UV protection, reinforcing cell walls, regulating stomatal behavior, and interacting with signaling pathways. These adaptations contribute to the plant's ability to thrive in diverse and often challenging environmental conditions.

8.4.1.2 Flavonoids

These are extensively studied group of polyphenols. Structurally, the flavonoid molecules are made of two benzene rings (A and B) linked by three carbon atoms and one oxygen atom forming a central pyrene ring (C). Flavonoids share the isoalloxazine ring with flavins, yet they have more diverse structures (Panche et al. 2016). Depending on their structure, they are grouped into six classes namely, flavonols, flavones, flavan-3-ols, flavanonols, flavanones, isoflavones, and anthocyanins (Panche et al. 2016). They can be present in either free or conjugated form but mostly associated with sugar residues in conjugated forms as monoglycosides, diglycoslides which influences their extent of bioactivity (Wen et al. 2017) (Rauter et al. 2007). Furthermore, the antioxidant activity of flavonoids is the most widely investigated among biological samples. Their antioxidant activity is affected by a number of factors such as glycosylation. The increased number of sugar residues on ring B results in decrease activity, whereas the increased number of hydroxyl residues triggers the antioxidant activity. While the effect of exogenously applied flavonoids in plants is not well studied although they structurally mimic the flavins so it can be speculated that they show positive effects as exogenously applied flavins do.

8.4.1.3 Stilbenes

These are small subgroup of polyphenols composed of 14 carbon skeletons made up of 2 benzene rings connected by central ethylene moiety (Nagumo et al. 2019; Al-Khayri et al. 2023). Due to this central ethylene moiety, they exist in two forms: *cis* and *trans* form. More than 400 different stilbenes have been reported to date mainly belong to *trans* form (Valletta et al. 2021). Furthermore, they also possess antimicrobial, insecticidal and nematocidal activity and this property of stilbenes make them able to react them during biotic stress (herbivores, phytopathogenic microorganisms) on plants (Torres et al. 2003; Kim et al. 2022). Moreover stilbenes, due to their polyphenolic nature, also act as antioxidants during abiotic stress (UV radiations and tropospheric ozone). Therefore, stilbenes are important for plants in counteracting stress in two ways, first they act as antioxidants and scavenge ROS, produced during stress, and second, they act as toxic for pathogens.

Lignans are naturally occurring group of polyphenols largely contributed among higher plants. Structurally, they are made up of two phenylpropanoid units linked through a bond between carbons C8 and C8′ (Zálešák et al. 2019). While its formation is affected by its concentration, structure, and composition during developmental and stress conditions (Ménard et al. 2022).

8.4.2 Nitrogenous Compounds

8.4.2.1 Alkaloids

Nitrogenous compounds including alkaloids, cyanogenic glycosides, or glucosinolates and non-protein amino acids are widely distributed among higher plants (Das and Gezici 2018). Alkaloids are widespread in plants, bacteria, fungi, and animals (Thawabteh et al. 2019). Among the various alkaloid classes, indole alkaloids, characterized by a structural moiety of indole, are notable for their bioactive properties. While from cyanobacteria, more than 300 alkaloids have been reported from them (Tan 2007). In plants, these compounds serve as secondary metabolites triggered by biotic or abiotic stresses (Jan et al. 2021). Furthermore, under abiotic stress conditions, higher alkaloid content and its derivatives are produced such as poppy plants produce higher amounts of alkaloids during periods of drought as well as when they are exposed to salinity stress (Pandey et al. 2018). Drought and activated yeast extract treatment was influenced the content of alkaloids in lupins (*Lupinus termis*) (Taha et al. 2020).

8.4.3 Terpenes and Terpenoids

Terpenes are hydrocarbons, whereas terpenoids (isoprenoids) are oxygenated form of terpenes. Terpenoids are most diverse group of phytometabolites, and till date, more than 55,000 terpenoids have been discovered from diverse living organisms (Nandagopal et al. 2021). The precursor of terpenoids is isopentenyl pyrophosphate (IPP) (five carbon compound) and its functional isomer dimethylallyl pyrophosphate (DMAPP). Condensation of IPP and DMAPP in different proportion leads to synthesis of terpenoids having different chain length. On the basis of number of precursor units used in terpenoid synthesis, they can be conveniently classified as hemiterpenoid (C_5), monoterpenoids (C_{10}), sesquiterpenoids (C_{15}), diterpenoids (C_{20}), and so on (Pichersky and Raguso 2018). Most of the terpenoids are found in photosynthetic organisms especially in plants. There are two biosynthetic pathways of terpenoids are found in plants, algae, and cyanobacteria. First pathway is mevalonate (MEV) pathway, which uses Acetyl Co-A as substrate and second is methylerythritol-4-phosphate (MEP) pathway. The MEV pathway found in cytoplasm of eukaryotic phototrophs such as algae and plants, while MEP pathway runs in cyanobacteria and plastids of algae and plants (Rohmer 1999, 2003; Pattanaik and Lindberg 2015). Apart from the plant metabolite roles, various terpenoids are also valuable for human use such as, in pharmaceutical industry (Cox-Georgian et al. 2019). Terpenes or terpenoids are involved primary metabolism of the plant such as components of the electron transfer system and hormones. They also provide resistance against various biotic and abiotic stresses to plants. Terpenes take part in plant–insect communication during pollination and deterrent to herbivores (Breitmaier 2006). Moreover, they also provide resistance to the plant against plant pathogens, such as viruses, bacteria, and fungi (Toffolatti et al. 2021). Several reports point to the role of terpenes or terpenoids in maintaining photosynthesis under fluctuating environmental factors. Terpenes were seen to increase the thermotolerance of photosynthesis in some higher plants where it was hypothesized that due to membrane-stabilizing properties of terpenes they might increase thylakoid membrane stability at high temperature and so thermotolerance of photosynthesis (Singsaas et al. 1997; Singsaas 2000). During priming of plants for stress tolerance emission of terpenes increased from the plant which act as photosynthetic membrane stabilizer so provide tolerance to the plant for future stresses (Liu et al. 2021).

8.4.4 Vitamin E

Tocopherol and tocotrienols belong to the vitamin E family (Munné-Bosch and Alegre 2002). They can be found in various sources, including vegetable oils, wheat germ, barley, and specific nuts and grains, whereas animals are unable to synthesis this so depended on the photosynthetic organisms (Sharma et al. 2024). Tocotrienols come in four types: alpha, beta, gamma, and delta. In contrast to tocopherols, tocotrienols are unique in that they are unsaturated and possess an isoprenoid side chain (Munné-Bosch and Alegre 2002). Functionally, tocotrienols is known for its antioxidant properties. In photosynthetic organism, they might contribute to protecting the photosynthetic apparatus from oxidative damage (Maeda and DellaPenna 2007). Tocopherol is a well-known phenolic antioxidant conferring the mitigation of stress conditions by trapping reactive oxygen species (ROS) (Frankel 1989). It is exclusively synthesized by photosynthetic organisms including higher plants, thallophytes, and cyanobacteria. Its content and composition vary greatly among different groups (plants, cyanobacteria) and in different parts of plants (Lushchak and Semchuk 2012; Mudimu et al. 2017). These wide oscillations in tocopherol content are due to genetic and environmental factors. Similar to tocotrienols, tocopherol exist in four related derivatives (alpha, beta, gamma, and delta) with differing biological activities, where alpha tocopherol is most active and efficient form. Tocopherols are categorized relying on the number and position of the $-CH_3$ group on the chromanol ring (Górnaś et al. 2022). Tocopherol modulates the various physiological processes including preventing lipid peroxidation and hence maintain membrane integrity (Munné-Bosch and Alegre 2002). Apart from that, alpha tocopherol also involves in cellular osmotic adjustment. It enhances the activities of antioxidative enzymes such as APX, CAT, POD, and SOD, along with the accumulation of osmolytes such as glycine betaine, proline, and sugars (Shah et al. 2021). Since stress conditions affect tocopherol levels, stress-tolerant plants and cyanobacteria having higher levels and stress-sensitive species have lower levels of tocopherol. This increment under stress conditions is to provide extra shield against oxidative stress. Stress factors including drought, salt, and heat influence the tocopherol accumulation. Different nitrogen dosages also alter tocopherol accumulation (Hussain et al. 2014). Since rate of photosynthetic activity in the leaves is affected by light capturing capacity and number of photosynthetic pigments in plants, while the exogenous application of alpha tocopherol on plants increases the photosynthetic pigments (chlorophyll and carotenoids) (Taiz et al. 2015; Ali et al. 2019, 2020; Faizan et al. 2023), which suggest that alpha tocopherol has an important role in photosynthetic resilience in plants. Under drought conditions, tocopherol increases the plant water status and maintains the cellular water balance (Akram et al. 2023). This balancing of water content controls the stomatal movement, which finally leads to increase in net photosynthetic efficiency by leaves under water-limiting conditions.

In mid-1980s, the tocopherol biosynthetic pathway in photosynthetic organisms was elucidated. It involves the synthesis of homogentisic acid (HGA) from the cytosolic shikimate pathway and phytyl diphosphate (PDP) from the plastid methylerythritol phosphate pathway. HGA and PDP conjugate to produce tocopherols, with HGA providing the aromatic ring and PDP provides the tail. This process is necessary for plant tocopherol synthesis. Several genes were isolated from photosynthetic organisms to identified the enzymes that involved in tocopherol biosynthesis. Further, the key intermediates and rate-limiting reactions involved in its biosynthesis were identified via genetic manipulation and biochemical studies. Through genetic engineering, five key enzymes are often targeted for generating transgenic lines with particular type of tocopherol.

Two methyltransferases enzymes, gamma-tocopherol methyltransferase (γ-TMT) and 2-methyl-6-phytyl-1,4-benzoquinone methyltransferase (MPBQ MT), control the tocopherol composition in photosynthetic organisms. The γ-TMT is crucial and involve in crucial step for alpha tocopherol production (Fritsche et al. 2017). It transforms γ- and δ-tocopherols to α- and β-tocopherols. MPBQ MT, is involve in methylation of tocopherol. Although neither enzyme limits overall tocopherol flux, their activities determine tocopherol composition and vitamin E's biological efficiency.

γ-TMT governs the efficiency of tocopherol pool. This enzyme has been identified and characterized from many plant species but structural and functional characterization is still lacking in cyanobacteria. Here, bioinformatic analysis for this protein has been done to analyze its diversity in cyanobacteria. During search for similar proteins in cyanobacteria in NCBI database, "methyltransferase domain containing protein" annotated proteins were retrieved. Direct γ-TMT annotated proteins were absent in cyanobacteria so methyltransferase domain was used for further

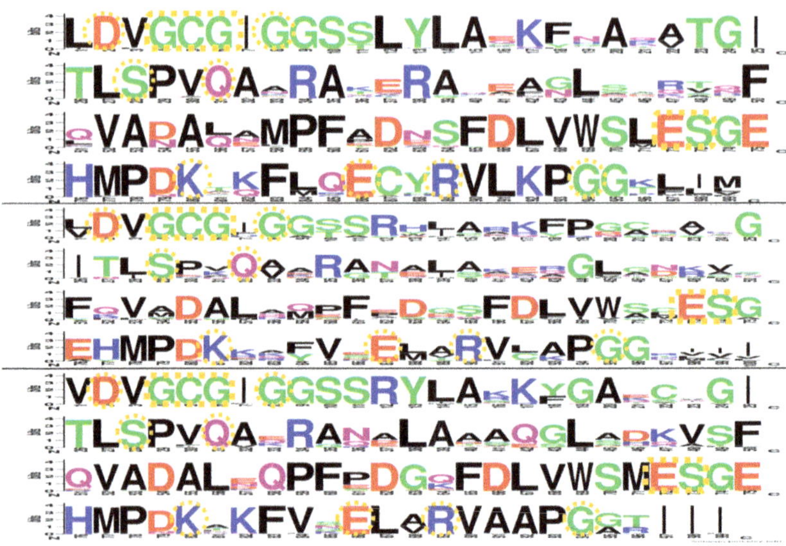
(a)
(b)
(c)

Fig. 8.3 Hidden Markov model (HMM) logo formed by the aligned amino acid sequences of methyltransferase domain. (**a**) Cyanobacteria, (**b**) algae, (**c**) plants

analysis. Amino acid sequence of the gamma tocopherol methyl transferase (γ-TMT) of *Arabidopsis* retrieved from the uniport. Domain search in this sequence was done at NCBI CD search portal. Extracted domain sequence was used as a query and DELTA BLAST was performed at NCBI BLAST server against the cyanobacteria, algae and plant sequences separately. Resulted sequences were downloaded and further aligned using MEGA X software. Three different Hidden Markov model (HMM) logos of the aligned amino acid sequences of methyltransferase domain of cyanobacteria, algae and plants were prepared by Weblogo server (Crooks et al. 2004). Conserved residues were analyzed among the aligned residues of methyltransferase domain through HMM logo and such conserved sites were encircled, as shown in Fig. 8.3. During comparison of logos of cyanobacteria, algae and plants it was seen that algae has least conserved residues. Moreover, some residues were found evolutionarily conserved, among all three groups i.e. cyanobacteria, algae and plants. These fully conserved residues were GCG at position fourth to sixth while G at position eighth. At second position an aspartic acid residue is also fully conserved. At 28th (29th position in algae) a conserved serine was found. Conserved glutamine was found at the 31st position (32nd in algae). Similarly, five more conserved sites were also found as shown in Fig. 8.3. This conservation in these residues suggested that these residues may have indispensable role during methyltransferase activity or may be vitally important for the native structure of the enzyme.

Further, phylogenetic tree of the methyltransferase domain sequences from plants, algae, and cyanobacteria was also prepared to analyze the diversity of this domain in different photosynthetic organism groups as shown in Fig. 8.4.

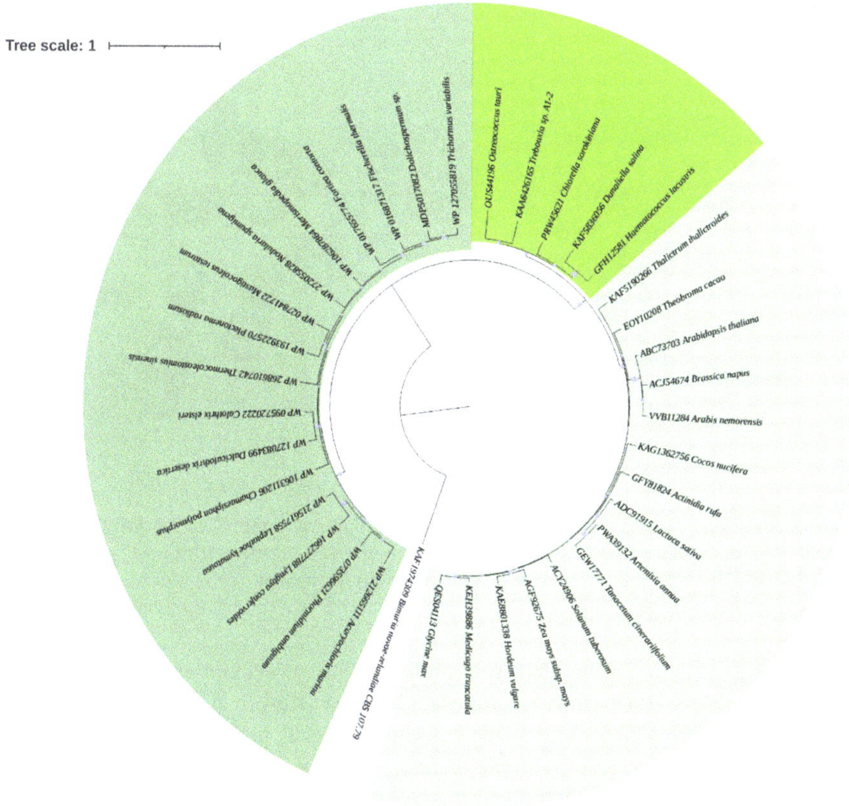

Fig. 8.4 Phylogenetic tree of the methyltransferase domain sequences from cyanobacteria, algae, and plants

Neighbor-Joining method was used for phylogenetic tree via MEGA 10 software (Saitou and Nei 1987; Kumar et al. 2018). Test of cluster togetherness was done by bootstrap value using 1000 replicates (Felsenstein 1985). Computation of evolutionary distances in tree was done using JTT matrix-based method (Jones et al. 1992). iTOL server was used for further editing of the tree (Letunic and Bork 2021). During phylogenetic analysis, it was shown that the methyltransferase domain of the cyanobacteria forms a big cluster, separated from the plants and algae. The domain sequence of plants and algae is somewhat similar to each other.

Additionally, molecular docking was conducted to test the interaction of gamma TMT with gamma tocopherol. Amino acid sequence of gamma tocopherol methyl transferase (gamma TMT), also known as *S*-adenosyl-L-methionine:gamma-tocopherol 5-*O*-methyltransferase, of *Nostocsphaeroides* CCNUC1 (id: 2653204) was retrieved from NCBI database. Homology modeling of this sequence was done using SWISS-MODEL server, and a PDB file was generated (id: A0A5P8VS37). Furthermore, the interaction between this enzyme and gamma tocopherol was

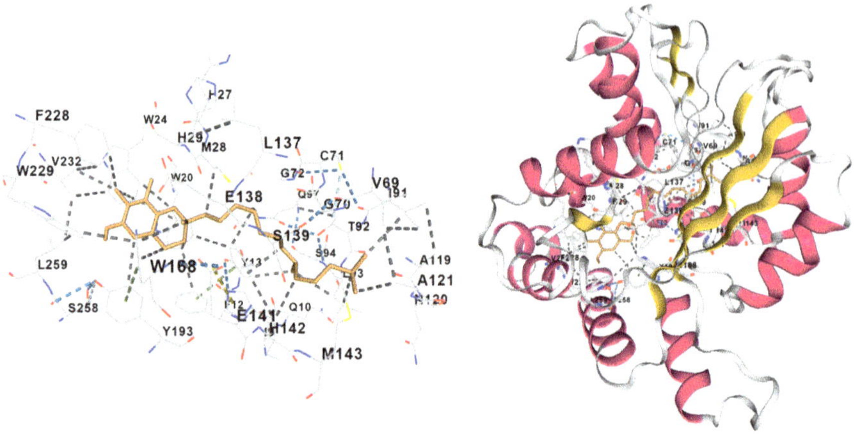

(a) (b)

Fig. 8.5 In silico interaction of gamma tocopherol (yellow) and gamma TMT domain-containing protein of *Nostoc sphaeroides* CCNUC1. (**a**) Interaction of gamma tocopherol with various amino acid residues, (**b**) fitting of gamma tocopherol in cavity formed in 3D model of gamma TMT domain containing protein

visualized in silico. Cavity detection-guided blind molecular docking was performed using CB-Dock (through AutoDock Vina) (Liu et al. 2020). The ligand (gamma tocopherol) was successfully docked (Fig. 8.5) into the topmost cavity of enzyme with showing cavity volume, center_x, center_y, center_z, size_x, size_y, size_z, and score; 3711, 12.834, 6.112, −2.956, 32, 30, 31, and −9.8, respectively. This in silico approach proved the presence of gamma TMT in cyanobacteria and should further confirmed with experimentation in laboratory. In Fig. 8.5a, only those amino acid residues of gamma TMT were visualized that bind with the gamma tocopherol (yellow color). Successful docking of ligand was visualized in Fig. 8.5b where ligand (yellow color) fitted into the center cavity of 3D model of gamma TMT. These hypotheses may be further proved using mutational studies in the future.

8.5 Role of Phytometabolites in Photosynthetic Resilience

A huge number of reports show the importance of exogenously applied antioxidants to improve plant growth and resilience under stress conditions (Rodrigues de Queiroz et al. 2023). Accumulation of secondary metabolites including total phenolics and flavonoids, allantoin, chlorogenic, and rosmarinic acids occurs under different stress conditions such as heavy metal stress (Dresler et al. 2017), UV radiation (Younis et al. 2010), salinity (Ksouri et al. 2007), and heat stress (Trabelsi et al. 2016) in phototrophs. Accumulated secondary metabolites scavenge reactive oxygen species under stress conditions, act as antioxidants, which ultimately maintained the photosynthetic components and growth of phototrophs. It was observed in the plant *Heracleum sosnowskyi* that the short-term heat exposer decreases photosynthetic efficiency of PSII (F_v/F_m). However, after long-term exposure (30 °C/7 days), the photosynthetic efficiency of PSII (F_v/F_m) reaches the same level as control plants along with accumulation of secondary metabolites which further suggested that accumulation of secondary metabolites help in resilience of photosynthesis (Rysiak et al. 2021). Several studies documented the role of secondary metabolites under stress conditions, i.e., heavy metal stress. (Dresler et al. (2017) documented that, under Zn/Pb stress conditions, increment of secondary metabolite concentration including total phenolics and flavonoids, allantoin, chlorogenic, and rosmarinic acids and enhance antioxidant activity. Accumulated secondary metabolites might help in scavenging reactive oxygen species under stress conditions. The exogenous application of vanillic acid on tomatoes may boost salt tolerance via the enhancement of osmolyte accumulation (proline), ion regulation (increased K, Ca, and Mg), increased antioxidant activities (SOD, CAT), and photosynthetic pigment amplification. It leads to increased growth in stressed plants (Parvin et al. 2020).

8.6 Crosstalk Between Phytohormones and Phytometabolites in Phototrophs in Photosynthetic Resilience

The interplay between phytohormones and phytometabolites in the plants is complex and dynamic process in context to photosynthetic resilience. In higher plants, these specific responses may vary depending on the type and intensity of the stress, the plant species, and the developmental stage of the plant. Among phytohormones, abscisic acid significantly contributes to plant responses to abiotic stress through interaction with salicylic acid, jasmonic acid, ethylene, and gibberellic acid (Fig. 8.6) (Gomez-Cadenas et al. 2015). Any environmental consequences regulate the downstream pathways by maintaining phytohormone homeostasis and signaling networks, promoting the production of phytometabolites.

Plant roots perceive the stress (drought, heavy metal and salinity) (Gorgues et al. 2022) and transmit signals to above-ground tissues, while plant leaf senses the temperature stress (Ding and Yang 2022) and then initiate systemic phytohormone signaling (Fig. 8.6). These signals then elicit the synthesis of phytohormones such as ABA and in turn result in closure of stomata to mitigate water loss. After a pathogen attack, there's an immediate response known as oxidative burst, characterized by swift ROS generation (Torres 2010). Phytometabolites especially secondary metabolites play a crucial role in scavenging reactive oxygen species (ROS), safeguarding plants from lipid peroxidation, and fulfilling other defense-related functions. Exogenous application of jasmonic acid induces defense genes, activating antioxidant enzymes including APX, dehydroascorbate reductase (DHAR), γ-glutamyl

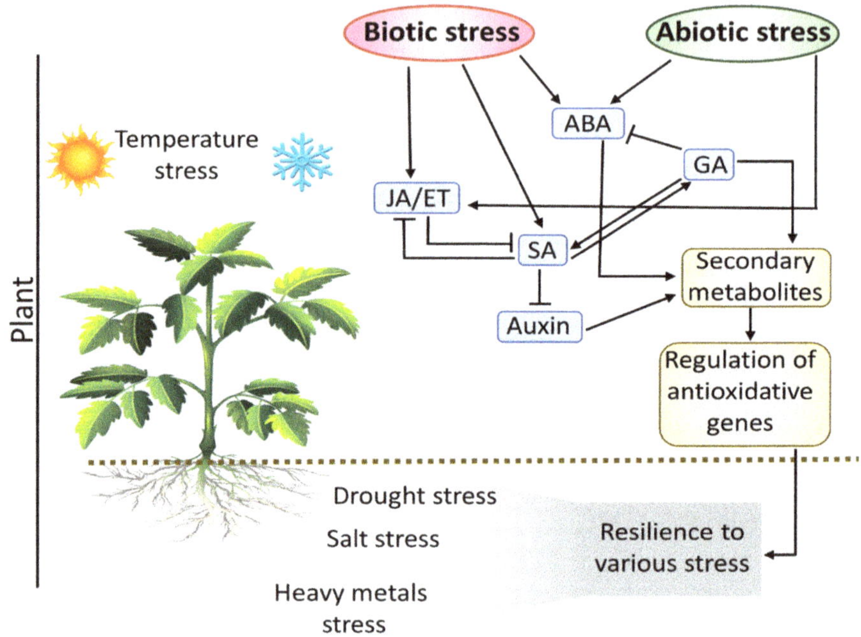

Fig. 8.6 Crosstalk within phytohormones and phytometabolites during photosynthetic resilience in higher plants

cysteine synthetase, glutathione reductase (GR), and L-galactono-1,4-lactone dehydrogenase (Shan and Liang 2010). Moreover, several plant species have been found that foster the production of stress-responsive metabolites, including alkaloids, taxanes, terpenoids, coumarins, and phenolic phytoalexins (Ramakrishna and Ravishankar 2011). A number of evidences are available that showed that ABA, SA, and JA are produced by plants in response to biotic and abiotic stress conditions, indicating potential cross-communication between biotic and abiotic signaling pathways via reactive oxygen species (ROS) regulation (Ku et al. 2018; Jogawat 2019; Singh and Roychoudhury 2023). While SA activation is crucial for defending against biotrophic and hemi biotrophic pathogens, as well as establishing systemic acquired resistance (SAR) in plants (Zhao et al. 2017). Conversely, jasmonic acid (JA) and ethylene (ET) are linked to necrotrophic pathogens and herbivorous insects, while abscisic acid (ABA) plays a more prominent role in abiotic stress signaling, indicating potential cross-communication between biotic and abiotic signaling pathways via reactive oxygen species (ROS) regulation (Zhao et al. 2017). Thus, in higher plants, it is clear that (Fig. 8.6) phytohormones communicate with each other under stress conditions either negatively or positively. While in case of algae, cyanobacteria, bryophytes, and pteriodophytes, this communication is not reported till now. Thus, understanding these interactions is crucial for developing strategies to enhance the resilience in the face of environmental challenges.

8.7 Conclusions

Photosynthesis is a crucial process for phototrophs and is extremely regulated by the various endogenously generated and exogenously provided phytohormones and phytometabolites. In nature, the phytohormones and phytometabolites help in resilience of photosynthetic process. Among all the major phytohormones, some are classified as growth-promoting hormones, while others (ABA, Jasmonic acid, SA, and brassinosteroids) are known as stress hormones because of their pivotal role during stress conditions. A wide distribution of phytohormones was generally reported in plant kingdom, whereas fewer reports were found among the cyanobacteria. It might be possible that cyanobacteria use other chemical substances such as phytometabolites, for the resiliency in their photosynthesis. An exogenous supply of phytohormones protected the photosynthetic machinery of the plants and recovered photosynthetic efficiency. During fluctuating environmental conditions, altered level of various phytohormones was also addressed for maintaining and stabilizing photosynthesis. Crosstalk between various phytohormones and phytometabolites is a common strategy during photosynthesis. An increasing trend has been observed in case of phytometabolites exploitation for human welfare. Some phytohormones are directly involved in primary metabolism such as terpenes for making accuracy in photosynthesis. Alpha tocopherol (vitamin E) is also an efficient phytohormone and natural antioxidant which involve in ROS scavenging in plants, produced during photosynthesis. Since alpha tocopherol is most active form among all tocopherols, and it has been already reported in plants and cyanobacteria. The bioinformatic analysis clearly indicated the presence of methyltransferase domain and also an involvement of the domain in the activity of gamma TMT in various cyanobacterial strains along with the elucidation of some conserved residues among the domain of cyanobacteria, algae, and plants. Also, the docking study elucidated the interaction of gamma tocopherol with gamma TMT in cyanobacteria.

8.8 Future Prospects

Although there are many reports suggesting the role of various phytohormones in photosynthetic resilience are found in the literature, but to fully elucidate the mechanism further research is needed in other photoautotrophs such as cyanobacteria and algae. Most of the phytohormones were found in the higher plants while being simple, short lifespan, and oxygenic photosynthetic organism, i.e. cyanobacteria can become a model organism for further understanding the mechanism of photosynthetic resilience. Since photosynthetic resilient crop plants can become future need, full elucidation of resilience mechanism and control of phytohormones and phytometabolites in it may help in engineering the crop plants for survival in the fluctuating environment. Among phytometabolites, the structural and functional characterization of enzymes involved in alpha tocopherol biosynthesis, such as gamma TMT, in cyanobacteria is essentially needed in future for further use of these enzymes in improving the photosynthetic resilience in crop plants during stress conditions.

Acknowledgement We acknowledged the Head of the Department of Botany, Banaras Hindu University, Varanasi, India, for providing necessary laboratory facilities. We are thankful to Institute of Eminence (IOE-6031), Banaras Hindu University for financial support. Vandana Sindhu is acknowledged to Department of Biotechnology (DBT), New Delhi for financial support in the form of Senior Research Fellowship.

Conflict of Interest The authors declare that they have no conflict of interest.

References

Abeles FB, Morgan PW, Saltveit ME Jr (2012) Ethylene in plant biology. Academic, New York

Abhishek RU, Venkatesh HN, Manjunath K, Mohana DC (2016) Artificial ripening of fruits-misleading ripe and health risk. Everyman's Sci 6(1):364–369

Addicott FT, Lyon JL, Ohkuma K, Thiessen WE, Carns HR, Smith OE, Cornforth JW, Milborrow BV, Ryback G, Wareing PF (1968) Abscisic acid: a new name for abscisin II (dormin). Science 159(3822):1493

Adusei S, Otchere JK, Oteng P, Mensah RQ, Tei-Mensah E (2019) Phytochemical analysis, antioxidant and metal chelating capacity of *Tetrapleura tetraptera*. Heliyon 5(11):e02762

Ahammed GJ, Choudhary SP, Chen S, Xia X, Shi K, Zhou Y, Yu J (2013) Role of brassinosteroids in alleviation of phenanthrene–cadmium co-contamination-induced photosynthetic inhibition and oxidative stress in tomato. J Exp Bot 64(1):199–213

Ahmad MR, Winter A (1969) Studies on the hormonal relationships of algae in pure culture: III. Tryptamine is an intermediate in the conversion of tryptophan to indole-3-acetic acid by the blue-green alga *Chlorogloea fritschii*. Planta 88:61–66

Akram NA, Bashir R, Ashraf G, Bashir S, Ashraf M, Alyemeni MN, Bajguz A, Ahmad P (2023) Exogenous α-tocopherol regulates the growth and metabolism of eggplant (*Solanum melongena* L.) under drought stress. Plants 12(2):237

Ali Q, Ali S, Iqbal N, Javed MT, Rizwan M, Khaliq R, Shahid S, Perveen R, Alamri SA, Alyemeni MN, Wijaya L, Ahmad P (2019) Alpha-tocopherol fertigation confers growth physio-biochemical and qualitative yield enhancement in field grown water deficit wheat (*Triticum aestivum* L.). Sci Rep 9(1):12924

Ali Q, Tariq Javed M, Haider MZ, Habib N, Rizwan M, Perveen R, Ali S, Alyemeni MN, El-Serehy HA, Al-Misned FA (2020) α-Tocopherol foliar spray and translocation mediates growth, photosynthetic pigments, nutrient uptake, and oxidative defense in maize (*Zea mays* L.) under drought stress. Agronomy 10(9):1235

Al-Khayri JM, Mascarenhas R, Harish HM, Gowda Y, Lakshmaiah VV, Nagella P, Al-Mssallem MQ, Alessa FM, Almaghasla FI, Rezk AAS (2023) Stilbenes, a versatile class of natural metabolites for inflammation—an overview. Molecules 28(9):3786

Ananieva EA, Alexieva VS, Popova LP (2002) Treatment with salicylic acid decreases the effects of paraquat on photosynthesis. J Plant Physiol 159(7):685–693

Arnaud D, Hwang I (2015) A sophisticated network of signaling pathways regulates stomatal defenses to bacterial pathogens. Mol Plant 8(4):566–581

Arnold TM, Targett NM, Tanner CE, Hatch WI, Ferrari KE (2001) Evidence for methyl jasmonate-induced phlorotannin production in *Fucus vesiculosus* (Phaeophyceae). J Phycol 37(6):1026–1029

Ashraf MA, Iqbal M, Rasheed R, Hussain I, Riaz M, Arif MS (2018) Environmental stress and secondary metabolites in plants: an overview. In: Plant metabolites and regulation under environmental stress. Elsevier, Amsterdam, pp 153–167

Bajguz A, Hayat S (2009) Effects of brassinosteroids on the plant responses to environmental stresses. Plant Physiol Biochem 47(1):1–8
Barciszewski J, Siboska GE, Pedersen BO, Clark BF, Rattan SI (1996) Evidence for the presence of kinetin in DNA and cell extracts. FEBS Lett 393(2–3):197–200
Bilgin DD, Zavala JA, Zhu JIN, Clough SJ, Ort DR, DeLucia EH (2010) Biotic stress globally downregulates photosynthesis genes. Plant Cell Environ 33(10):1597–1613
Bistgani ZE, Hashemi M, DaCosta M, Craker L, Maggi F, Morshedloo MR (2019) Effect of salinity stress on the physiological characteristics, phenolic compounds and antioxidant activity of *Thymus vulgaris* L. and *Thymus daenensis* Celak. Ind Crop Prod 135:311–320
Breitmaier E (2006) Terpenes: flavors, fragrances, pharmaca, pheromones. Wiley, New York
Checker VG, Kushwaha HR, Kumari P, Yadav S (2018) Role of phytohormones in plant defense: signaling and cross talk. In: Molecular aspects of plant-pathogen interaction. Springer, Singapore, pp 159–184
Chen CM (1998) The discovery of cytokinins. In: Discoveries in plant biology, vol I. World Scientific, Singapore, pp 1–15
Chen Z, Ma Y, Yang R, Gu Z, Wang P (2019) Effects of exogenous Ca^{2+} on phenolic accumulation and physiological changes in germinated wheat (*Triticum aestivum* L.) under UV-B radiation. Food Chem 288:368–376
Ciesielski T (1871) Untersuchungenüber die Abwärtskrümmung der Wurzel. éditeur non identifié
Clouse SD (2011) Brassinosteroids. In: The arabidopsis book, vol 9. American Society of Plant Biologists, Rockville, p e0151
Commisso M, Toffali K, Strazzer P, Stocchero M, Ceoldo S, Baldan B, Levi M, Guzzo F (2016) Impact of phenylpropanoid compounds on heat stress tolerance in carrot cell cultures. Front Plant Sci 7:1439
Cooke TJ, Poli D, Sztein AE, Cohen JD (2002) Evolutionary patterns in auxin action. Plant Mol Biol 49:319–338
Cornforth JW, Milborrow BV, Ryback G, Wareing PF (1965) Chemistry and physiology of 'dormins' in sycamore: identity of sycamore 'dormin' with abscisin II. Nature 205(4978):1269–1270
Cox-Georgian D, Ramadoss N, Dona C, Basu C (2019) Therapeutic and medicinal uses of terpenes. In: Medicinal plants: from farm to pharmacy. Springer, Berlin, pp 333–359
Creelman RA, Mullet JE (1995) Jasmonic acid distribution and action in plants: regulation during development and response to biotic and abiotic stress. Proc Natl Acad Sci 92(10):4114–4119
Crooks GE, Hon G, Chandonia JM, Brenner SE (2004) WebLogo: a sequence logo generator. Genome Res 14(6):1188–1190
Cruz de Carvalho MH (2008) Drought stress and reactive oxygen species: production, scavenging and signaling. Plant Signal Behav 3(3):156–165
Cruz CG, da Rosa APC, Costa JAV (2023) Identification of the phytohormones indole-3-acetic acid and trans-zeatin in microalgae. J Chem Technol Biotechnol 98(4):1048–1056
Danilova MN, Kudryakova NV, Doroshenko AS, Zabrodin DA, Vinogradov NS, Kuznetsov VV (2016) Molecular and physiological responses of *Arabidopsis thaliana* plants deficient in the genes responsible for ABA and cytokinin reception and metabolism to heat shock. Russ J Plant Physiol 63:308–318
Darwin C, Darwin F (1880) The power of movement in plants. John Murray, London
Das K, Gezici S (2018) Secondary plant metabolites, their separation and identification, and role in human disease prevention. Ann Phytomed 7:13–24
Ding Y, Yang S (2022) Surviving and thriving: how plants perceive and respond to temperature stress. Dev Cell 57(8):947–958
Djakovic-Petrovic T, Wit MD, Voesenek LA, Pierik R (2007) DELLA protein function in growth responses to canopy signals. Plant J 51(1):117–126
Dörffling K (2015) The discovery of abscisic acid: a retrospect. J Plant Growth Regul 34:795–808
dos Santos TB, Ribas AF, de Souza SGH, Budzinski IGF, Domingues DS (2022) Physiological responses to drought, salinity, and heat stress in plants: a review. Stresses 2(1):113–135
Dresler S, Wójciak-Kosior M, Sowa I, Stanisławski G, Bany I, Wójcik M (2017) Effect of short-term Zn/Pb or long-term multi-metal stress on physiological and morphological parameters of

metallicolous and nonmetallicolous *Echium vulgare* L. populations. Plant Physiol Biochem 115:380–389

Eagles CF, Wareing PF (1963) Dormancy regulators in woody plants: experimental induction of dormancy in *Betula pubescens*. Nature 199(4896):874–875

Fahad S, Hussain S, Matloob A, Khan FA, Khaliq A, Saud S, Hassan S, Shan D, Khan F, Ullah N, Faiq M, Khan MR, Tareen AK, Khan A, Ullah A, Ullah N, Huang J (2015) Phytohormones and plant responses to salinity stress: a review. Plant Growth Regul 75:391–404

Faizan M, Alam P, Rajput VD, Shareen, Kaur K, Faraz A, Minkina T, Ahmed SM, Rajpal VR, Hayat S (2023) Potential role of tocopherol in protecting crop plants against abiotic stresses. Physiol Mol Biol Plants 29:1–13

Farooqi AA, Shukla YN, Shukla A, Bhakuni DS (1990) Cytokinins from marine organisms. Phytochemistry 29(7):2061–2063

Felsenstein J (1985) Confidence limits on phylogenies: an approach using the bootstrap. Evolution 39(4):783–791

Filová A, Sytar O, Krivosudská E (2013) Effects of brassinosteroid on the induction of physiological changes in *Helianthus annuus* L. under copper stress. Acta Universitatis Agriculturae et Silviculturae Mendelianae Brunensis 61(3):623–629

Frankel EN (1989) The antioxidant and nutritional effects of tocopherols, ascorbic acid and beta-carotene in relation to processing of edible oils. Bibl Nutr Dieta 43:297–312

Frebort I, Kowalska M, Hluska T, Frébortová J, Galuszka P (2011) Evolution of cytokinin biosynthesis and degradation. J Exp Bot 62(8):2431–2452

Fritsche S, Wang X, Jung C (2017) Recent advances in our understanding of tocopherol biosynthesis in plants: an overview of key genes, functions, and breeding of vitamin E improved crops. Antioxidants 6(4):99

Fu J, Li L, Wang S, Yu N, Shan H, Shi Z, Li F, Zhong X (2023) Effect of gibberellic acid on photosynthesis and oxidative stress response in maize under weak light conditions. Front Plant Sci 14:1128780

Ge L, Yong JWH, Goh NK, Chia LS, Tan SN, Ong ES (2005) Identification of kinetin and kinetin riboside in coconut (*Cocos nucifera* L.) water using a combined approach of liquid chromatography–tandem mass spectrometry, high performance liquid chromatography and capillary electrophoresis. J Chromatogr B 829(1–2):26–34

Gilroy S, Jones RL (1992) Gibberellic acid and abscisic acid coordinately regulate cytoplasmic calcium and secretory activity in barley aleurone protoplasts. Proc Natl Acad Sci 89(8):3591–3595

Gjindali A, Herrmann HA, Schwartz JM, Johnson GN, Calzadilla PI (2021) A holistic approach to study photosynthetic acclimation responses of plants to fluctuating light. Front Plant Sci 12:668512

Goleniowski M, Bonfill M, Cusido R, Palazon J (2013) Phenolic acids. Nat Prod:1951–1973

Gomez-Cadenas A, Vives V, Zandalinas IS, Manzi M, Sanchez-Perez AM, Perez-Clemente RM, Arbona V (2015) Abscisic acid: a versatile phytohormone in plant signaling and beyond. Curr Protein Pept Sci 16(5):413–434

Gomez-Roldan V, Fermas S, Brewer PB, Puech-Pagès V, Dun EA, Pillot JP, Letisse F, Matusova R, Danoun S, Portais JC, Bouwmeester H, Bécard G, Beveridge CA, Rameau C, Rochange SF (2008) Strigolactone inhibition of shoot branching. Nature 455(7210):189–194

González-Sarrías A, Tomás-Barberán FA, García-Villalba R (2020) Structural diversity of polyphenols and distribution in foods. In: Dietary polyphenols: their metabolism and health effects. Wiley, New York, pp 1–29

Gorgues L, Li X, Maurel C, Martinière A, Nacry P (2022) Root osmotic sensing from local perception to systemic responses. Stress Biol 2(1):36

Gorham J (1977) Lunularic acid and related compounds in liverworts, algae and *Hydrangea*. Phytochemistry 16(2):249–253

Górnaś P, Baškirovs G, Siger A (2022) Free and esterified tocopherols, tocotrienols and other extractable and non-extractable tocochromanol-related molecules: compendium of knowledge, future perspectives and recommendations for chromatographic techniques, tools, and approaches used for tocochromanol determination. Molecules 27(19):6560

Gregory JF III (1998) Nutritional properties and significance of vitamin glycosides. Annu Rev Nutr 18(1):277–296

Grzegorczyk-Karolak I, Krzemińska M, Kiss AK, Owczarek-Januszkiewicz A, Olszewska MA (2023) Role of phytohormones in biomass and polyphenol accumulation in *Salvia bulleyana* in vitro culture. Biomol Ther 13(2):227

Gupta AB, Agarwal PR (1973) Extraction, isolation, and bioassay of a gibberellin-like substance from *Phormidium foveolarum*. Ann Bot 37(4):737–741

Ha CV, Leyva-González MA, Osakabe Y, Tran UT, Nishiyama R, Watanabe Y, Tanaka M, Seki M, Yamaguchi S, Dong NV, Yamaguchi-Shinozaki K, Shinozaki K, Herrera-Estrella L, Tran LSP (2014) Positive regulatory role of strigolactone in plant responses to drought and salt stress. Proc Natl Acad Sci 111(2):851–856

Hartung W (2010) The evolution of abscisic acid (ABA) and ABA function in lower plants, fungi and lichen. Funct Plant Biol 37(9):806–812

Hattab S, Dridi B, Chouba L, Kheder MB, Bousetta H (2009) Photosynthesis and growth responses of pea *Pisum sativum* L. under heavy metals stress. J Environ Sci 21(11):1552–1556

Hayat S, Irfan M, Wani AS, Nasser A, Ahmad A (2012) Salicylic acids: local, systemic or inter-systemic regulators? Plant Signal Behav 7(1):93–102

Huang S, Raman AS, Ream JE, Fujiwara H, Cerny RE, Brown SM (1998) Overexpression of 20-oxidase confers a gibberellin-overproduction phenotype in *Arabidopsis*. Plant Physiol 118(3):773–781

Huang R, Liu Z, Xing M, Yang Y, Wu X, Liu H, Liang W (2019) Heat stress suppresses *Brassica napus* seed oil accumulation by inhibition of photosynthesis and BnWRI1 pathway. Plant Cell Physiol 60(7):1457–1470

Hura T, Hura K, Ostrowska A, Urban K (2023) Toward resilient agriculture and environmental protection: the role of cell wall-bound phenolics. J Plant Physiol 287:154020

Hussain A, Krischke M, Roitsch T, Hasnain S (2010) Rapid determination of cytokinins and auxin in cyanobacteria. Curr Microbiol 61:361–369

Hussain N, Li H, Jiang YX, Jabeen Z, Shamsi IH, Ali E, Jiang LX (2014) Response of seed tocopherols in oilseed rape to nitrogen fertilizer sources and application rates. J Zhejiang Univ Sci B 15(2):181–193

Idrees M, Khan MMA, Aftab T, Naeem M, Hashmi N (2010) Salicylic acid-induced physiological and biochemical changes in lemongrass varieties under water stress. J Plant Interact 5(4):293–303

Jacobs WP (1993) A search for some angiosperm hormones and their metabolites in *Caulerpa Paspaloides* (Chlorophyta). J Phycol 29(5):595–600

Jan R, Asaf S, Numan M, Lubna, Kim KM (2021) Plant secondary metabolite biosynthesis and transcriptional regulation in response to biotic and abiotic stress conditions. Agronomy 11(5):968

Jennings JC (1967) The development of the gametophyte and young sporophyte of *Ecklonia radiata* (C. Ag.) J. Ag. (Laminariales). J R Soc Western Aust 50:93–96

Jennings RC (1968) Gibberellins as endogenous growth regulators in green and brown algae. Planta 80:34–42

Jiang Y, Huang B (2001) Effects of calcium on antioxidant activities and water relations associated with heat tolerance in two cool-season grasses. J Exp Bot 52(355):341–349

Jogawat A (2019) Crosstalk among phytohormone signaling pathways during abiotic stress. In: Molecular plant abiotic stress: biology and biotechnology. Wiley, New York, pp 209–220

Johnová P, Skalák J, Saiz-Fernández I, Brzobohatý B (2016) Plant responses to ambient temperature fluctuations and water-limiting conditions: a proteome-wide perspective. Biochim Biophys Acta 1864(8):916–931

Jones DT, Taylor WR, Thornton JM (1992) The rapid generation of mutation data matrices from protein sequences. Bioinformatics 8(3):275–282

Khan MIR, Fatma M, Per TS, Anjum NA, Khan NA (2015) Salicylic acid-induced abiotic stress tolerance and underlying mechanisms in plants. Front Plant Sci 6:462

Khan NA, Khan MIR, Ferrante A, Poor P (2017) Ethylene: a key regulatory molecule in plants. Front Plant Sci 8:1782

Khodary SEA (2004) Effect of salicylic acid on the growth, photosynthesis and carbohydrate metabolism in salt stressed maize plants. Int J Agric Biol 6(1):5–8

Kim TH, Böhmer M, Hu H, Nishimura N, Schroeder JI (2010) Guard cell signal transduction network: advances in understanding abscisic acid, CO_2, and Ca^{2+} signaling. Annu Rev Plant Biol 61:561–591

Kim YR, Han JY, Choi YE (2022) Production of nematicidalpinosylvin stilbenes in cell suspension cultures of *Pinus koraiensis* by fungal elicitation. Plants 11(21):2933

Kobayashi M, Hirai N, Kurimura Y, Ohigashi H, Tsuji Y (1997) Abscisic acid-dependent algal morphogenesis in the unicellular green alga *Haematococcus pluvialis*. Plant Growth Regul 22:79–85

Kögl F, Haagen-Smit AJ (1931) The chemistry of the growth substance. Proc Acad Sci Amsterdam 34:1411–1416

Kosakivska IV, Vedenicheva NP, Babenko LM, Voytenko LV, Romanenko KO, Vasyuk VA (2022) Exogenous phytohormones in the regulation of growth and development of cereals under abiotic stresses. Mol Biol Rep 49(1):617–628

Kreslavsky VD, Kobzar EF, Muzafarov EN (1997) Effect of red radiation, kinetin and linuron on growth and ethylene production in *Chlorella*. Biol Plant 39:427–430

Ksouri R, Megdiche W, Debez A, Falleh H, Grignon C, Abdelly C (2007) Salinity effects on polyphenol content and antioxidant activities in leaves of the halophyte *Cakile maritima*. Plant Physiol Biochem 45(3–4):244–249

Ku YS, Sintaha M, Cheung MY, Lam HM (2018) Plant hormone signaling crosstalks between biotic and abiotic stress responses. Int J Mol Sci 19(10):3206

Kul R, Ekinci M, Turan M, Ors S, Yildirim E (2020) How abiotic stress conditions affects plant roots. In: Plant roots. IntechOpen, London, pp 6–10

Kumar S, Stecher G, Li M, Knyaz C, Tamura K (2018) MEGA X: molecular evolutionary genetics analysis across computing platforms. Mol Biol Evol 35(6):1547

Kurosawa E (1926) Experimental studies on the nature of the substance secreted by the "bakanae" fungus. Nat Hist Soc Formos 16:213–227

Kusk KO, Nyholm N (1992) Toxic effects of chlorinated organic compounds and potassium dichromate on growth rate and photosynthesis of marine phytoplankton. Chemosphere 25(6):875–886

Lacey RF, Binder BM (2016) Ethylene regulates the physiology of the cyanobacterium *Synechocystis* sp. PCC 6803 via an ethylene receptor. Plant Physiol 171(4):2798–2809

Letunic I, Bork P (2021) Interactive tree of life (iTOL) v5: an online tool for phylogenetic tree display and annotation. Nucleic Acids Res 49(W1):W293–W296

Li M, Kim C (2022) Chloroplast ROS and stress signaling. Plant Commun 3(1):100264

Li A, Sun X, Liu L (2022) Action of salicylic acid on plant growth. Front Plant Sci 13:878076

Lin Q, Wang S, Dao Y, Wang J, Wang K (2020) Arabidopsis thaliana trehalose-6-phosphate phosphatase gene *TPPI* enhances drought tolerance by regulating stomatal apertures. J Exp Bot 71(14):4285–4297

Linić I, Mlinarić S, Brkljačić L, Pavlović I, Smolko A, Salopek-Sondi B (2021) Ferulic acid and salicylic acid foliar treatments reduce short-term salt stress in Chinese cabbage by increasing phenolic compounds accumulation and photosynthetic performance. Plants 10(11):2346

Liu Y, Grimm M, Dai WT, Hou MC, Xiao ZX, Cao Y (2020) CB-Dock: a web server for cavity detection-guided protein–ligand blind docking. Acta Pharmacol Sin 41(1):138–144

Liu B, Zhang L, Rusalepp L, Kaurilind E, Sulaiman HY, Püssa T, Niinemets Ü (2021) Heat priming improved heat tolerance of photosynthesis, enhanced terpenoid and benzenoid emission and phenolics accumulation in *Achillea millefolium*. Plant Cell Environ 44(7):2365–2385

Lushchak VI, Semchuk NM (2012) Tocopherol biosynthesis: chemistry, regulation and effects of environmental factors. Acta Physiol Plant 34(5):1607–1628

Maeda H, DellaPenna D (2007) Tocopherol functions in photosynthetic organisms. Curr Opin Plant Biol 10(3):260–265

Maghsoudi K, Arvin MJ, Ashraf M (2020) Mitigation of arsenic toxicity in wheat by the exogenously applied salicylic acid, 24-epi-brassinolide and silicon. J Soil Sci Plant Nutr 20:577–588
Malik S, Bano H, Rather RA, Ahmad S (2018) Cloud seeding; its prospects and concerns in the modern world—a review. Int J Pure Appl Biosci 6(5):791–796
Mandava NB (1988) Plant growth-promoting brassinosteroids. Annu Rev Plant Physiol Plant Mol Biol 39(1):23–52
Manghwar H, Hussain A, Ali Q, Liu F (2022) Brassinosteroids (BRs) role in plant development and coping with different stresses. Int J Mol Sci 23(3):1012
Manickavelu A, Nadarajan N, Ganesh SK, Ramalingam R, Raguraman S, Gnanamalar RP (2006) Organogenesis induction in rice callus by cyanobacterial extracellular product. Afr J Biotechnol 5(5):437–439
Maršálek B, Zahradníčková H, Hronková M (1992) Extracellular abscisic acid produced by cyanobacteria under salt stress. J Plant Physiol 139(4):506–508
Maruri-López I, Aviles-Baltazar NY, Buchala A, Serrano M (2019) Intra and extracellular journey of the phytohormone salicylic acid. Front Plant Sci 10:423
Mattoo AK, Suttle JC (1991) The plant hormone ethylene. CRC Press, Boca Raton
Mazhar S, Cohen JD, Hasnain S (2013) Auxin producing non-heterocystous cyanobacteria and their impact on the growth and endogenous auxin homeostasis of wheat. J Basic Microbiol 53(12):996–1003
McAdam SA, Brodribb TJ, Banks JA, Hedrich R, Atallah NM, Cai C, Geringer MA, Lind C, Nichols DS, Stachowski K, Geiger D, Sussmilch FC (2016) Abscisic acid-controlled sex before transpiration in vascular plants. Proc Natl Acad Sci 113(45):12862–12867
Melotto M, Underwood W, Koczan J, Nomura K, He SY (2006) Plant stomata function in innate immunity against bacterial invasion. Cell 126(5):969–980
Ménard D, Blaschek L, Kriechbaum K, Lee CC, Serk H, Zhu C, Lyubartsev A, Nuoendagula, Bacsik Z, Bergström L, Mathew A, Kajita S, Pesquet E (2022) Plant biomechanics and resilience to environmental changes are controlled by specific lignin chemistries in each vascular cell type and morphotype. Plant Cell 34(12):4877–4896
Mendoza N, Silva EME (2018) Introduction to phytochemicals: secondary metabolites from plants with active principles for pharmacological importance. In: Phytochemicals: source of antioxidants and role in disease prevention, vol 25. IntechOpen, London, pp 1–5
Merilo E, Jalakas P, Kollist H, Brosché M (2015) The role of ABA recycling and transporter proteins in rapid stomatal responses to reduced air humidity, elevated CO_2, and exogenous ABA. Mol Plant 8(4):657–659
Meyer A, Miersch O, Büttner C, Dathe W, Sembdner G (1984) Occurrence of the plant growth regulator jasmonic acid in plants. J Plant Growth Regul 3:1–8
Miao Y, Luo X, Gao X, Wang W, Li B, Hou L (2020) Exogenous salicylic acid alleviates salt stress by improving leaf photosynthesis and root system architecture in cucumber seedlings. Sci Hortic 272:109577
Miller CO, Skoog F, Von Saltza MH, Strong FM (1955) Kinetin, a cell division factor from deoxyribonucleic acid 1. J Am Chem Soc 77(5):1392
Mir AR, Alam P, Hayat S (2022) Auxin regulates growth, photosynthetic efficiency and mitigates copper induced toxicity via modulation of nutrient status, sugar metabolism and antioxidant potential in *Brassica juncea*. Plant Physiol Biochem 185:244–259
Miura K, Tada Y (2014) Regulation of water, salinity, and cold stress responses by salicylic acid. Front Plant Sci 5:4
Mowat JA (1965) A survey of results on the occurrence of auxins and gibberellins in algae. Bot Mar 8(1):149–155
Mudimu O, Koopmann IK, Rybalka N, Friedl T, Schulz R, Bilger W (2017) Screening of microalgae and cyanobacteria strains for α-tocopherol content at different growth phases and the influence of nitrate reduction on α-tocopherol production. J Appl Phycol 29:2867–2875
Müller M, Munné-Bosch S (2021) Hormonal impact on photosynthesis and photoprotection in plants. Plant Physiol 185(4):1500–1522

Munné-Bosch S, Alegre L (2002) The function of tocopherols and tocotrienols in plants. Crit Rev Plant Sci 21(1):31–57
Murata N, Takahashi S, Nishiyama Y, Allakhverdiev SI (2007) Photoinhibition of photosystem II under environmental stress. Biochim Biophys Acta Bioenerg 1767(6):414–421
Nagumo M, Ninomiya M, Oshima N, Itoh T, Tanaka K, Nishina A, Koketsu M (2019) Comparative analysis of stilbene and benzofuran neolignan derivatives as acetylcholinesterase inhibitors with neuroprotective and anti-inflammatory activities. Bioorg Med Chem Lett 29(17):2475–2479
Nandagopal P, Steven AN, Chan LW, Rahmat Z, Jamaluddin H, Mohd Noh NI (2021) Bioactive metabolites produced by cyanobacteria for growth adaptation and their pharmacological properties. Biology 10(10):1061
Nazar R, Umar S, Khan NA, Sareer O (2015) Salicylic acid supplementation improves photosynthesis and growth in mustard through changes in proline accumulation and ethylene formation under drought stress. S Afr J Bot 98:84–94
Nelson N, Ben-Shem A (2004) The complex architecture of oxygenic photosynthesis. Nat Rev Mol Cell Biol 5(12):971–982
Niemann DI, Dörffling K (1980) Growth-inhibitors and growth-promoters in enteromorpha compress a (CHLOROPHYTA). J Phycol 16(3):383–389
Nimura K, Mizuta H (2002) Inducible effects of abscisic acid on sporophyte discs from *Laminaria japonica* Areschoug (Laminariales, Phaeophyceae). J Appl Phycol 14:159–163
Ördög V, Stirk WA, Van Staden J, Novák O, Strnad M (2004) Endogenous cytokinins in three genera of microalgae from the Chlorophyta. J Phycol 40(1):88–95
Osugi A, Sakakibara H (2015) Q&A: how do plants respond to cytokinins and what is their importance? BMC Biol 13(1):1–10
Ouzounidou G, Ilias I (2005) Hormone-induced protection of sunflower photosynthetic apparatus against copper toxicity. Biol Plant 49:223–228
Ozturk M, Unal BT (2023) Exogenous application of phytohormones and phytometabolites to plants to alleviate the effects of drought stress. In: Phytohormones and stress responsive secondary metabolites. Academic, New York, pp 1–12
Panche AN, Diwan AD, Chandra SR (2016) Flavonoids: an overview. J Nutr Sci 5:E47
Pandey A, Jena SN, Shukla S (2018) Impact of abiotic stresses on metabolic adaptation in opium poppy (*Papaver somniferum* L.). In: Metabolic adaptations in plants during abiotic stress. CRC Press, Boca Raton, pp 369–380
Papazoglou EG, Karantounias GA, Vemmos SN, Bouranis DL (2005) Photosynthesis and growth responses of giant reed (*Arundo donax* L.) to the heavy metals Cd and Ni. Environ Int 31(2):243–249
Park SH, Han KS, Kim TW, Shim JK, Takatsuto S, Yokota T, Kim SK (1999) In vivo and in vitro conversion of teasterone to typhasterol in cultured cells of *Marchantia polymorpha*. Plant Cell Physiol 40(9):955–960
Parvin K, Nahar K, Hasanuzzaman M, Bhuyan MB, Mohsin SM, Fujita M (2020) Exogenous vanillic acid enhances salt tolerance of tomato: insight into plant antioxidant defense and glyoxalase systems. Plant Physiol Biochem 150:109–120
Pattanaik B, Lindberg P (2015) Terpenoids and their biosynthesis in cyanobacteria. Life 5(1):269–293
Pichersky E, Raguso RA (2018) Why do plants produce so many terpenoid compounds? New Phytol 220(3):692–702
Polevoi VV, Tarakhovskaya ER, Maslov YI, Polevoi AV (2003) Role of auxin in induction of polarity in *Fucus vesiculosus* zygotes. Russ J Dev Biol 34:360–364
Pospíšilová J (2003) Interaction of cytokinins and abscisic acid during regulation of stomatal opening in bean leaves. Photosynthetica 41:49–56
Prodhan MY, Munemasa S, Nahar MNEN, Nakamura Y, Murata Y (2018) Guard cell salicylic acid signaling is integrated into abscisic acid signaling via the Ca^{2+}/CPK-dependent pathway. Plant Physiol 178(1):441–450
Provasoli LA, Carlucci AF (1974) Vitamins and growth regulators. In: Algal physiology and biochemistry. Canadian Science Publishing, Ottawa, pp 741–787
Pryce R (1972) The occurrence of lunularic and abscisic acids in plants. Phytochemistry 11(5):1759–1761

Radley M (1961) Gibberellin-like substances in plants: leguminous root nodules. Nature 191(4789):684–685
Rajasheker G, Jawahar G, Jalaja N, Kumar SA, Kumari PH, Punita DL, Karumanchi AR, Reddy PS, Rathnagiri P, Sreenivasulu N, Kishor PBK (2019) Role and regulation of osmolytes and ABA interaction in salt and drought stress tolerance. In: Plant signaling molecules. Woodhead Publishing, Sawston, pp 417–436
Ramakrishna A, Ravishankar GA (2011) Influence of abiotic stress signals on secondary metabolites in plants. Plant Signal Behav 6(11):1720
Raman N, Elumalai S (1996) Presence of cytokinin in the root nodules of *Casuarina equisetifolia*. Indian J Exp Biol 34(6):577–579
Rauter AP, Lopes RG, Martins A (2007) C-Glycosyl flavonoids: identification, bioactivity and synthesis. Nat Prod Commun 2(11):1934578X0700201125
Raven JA (2014) Speedy small stomata? J Exp Bot 65(6):1415–1424
Raza A, Salehi H, Rahman MA, Zahid Z, Madadkar Haghjou M, Najafi-Kakavand S, Charagh S, Osman HS, Albaqami M, Zhuang Y, Siddique KHM, Zhuang W (2022) Plant hormones and neurotransmitter interactions mediate antioxidant defenses under induced oxidative stress in plants. Front Plant Sci 13:961872
Rodrigues de Queiroz A, Hines C, Brown J, Sahay S, Vijayan J, Stone JM, Bickford N, Wuellner M, Glowacka K, Buan NR, Roston RL (2023) The effects of exogenously applied antioxidants on plant growth and resilience. Phytochem Rev 22(2):407–447
Rohmer M (1999) The discovery of a mevalonate-independent pathway for isoprenoid biosynthesis in bacteria, algae and higher plants. Nat Prod Rep 16(5):565–574
Rohmer M (2003) Mevalonate-independent methylerythritol phosphate pathway for isoprenoid biosynthesis. Elucidation and distribution. Pure Appl Chem 75(2–3):375–388
Rohwer F, Bopp M (1985) Ethylene synthesis in moss protonema. J Plant Physiol 117(4):331–338
Rysiak A, Dresler S, Hanaka A, Hawrylak-Nowak B, Strzemski M, Kováčik J, Sowa I, Latalski M, Wójciak M (2021) High temperature alters secondary metabolites and photosynthetic efficiency in *Heracleum sosnowskyi*. Int J Mol Sci 22(9):4756
Sachdev S, Ansari SA, Ansari MI, Fujita M, Hasanuzzaman M (2021) Abiotic stress and reactive oxygen species: generation, signaling, and defense mechanisms. Antioxidants 10(2):277
Saitou N, Nei M (1987) The neighbor-joining method: a new method for reconstructing phylogenetic trees. Mol Biol Evol 4(4):406–425
Salazar-Iribe A, De-la-Pena C (2020) Auxins, the hidden player in chloroplast development. Plant Cell Rep 39(12):1595–1608
Šamec D, Karalija E, Šola I, Vujčić Bok V, Salopek-Sondi B (2021) The role of polyphenols in abiotic stress response: the influence of molecular structure. Plants 10(1):118
Saradhi PP, Suzuki I, Katoh A, Sakamoto A, Sharmila P, Shi DJ, Murata N (2000) Protection against the photo-induced inactivation of the photosystem II complex by abscisic acid. Plant Cell Environ 23(7):711–718
Savchenko T, Dehesh K (2014) Drought stress modulates oxylipin signature by eliciting 12-OPDA as a potent regulator of stomatal aperture. Plant Signal Behav 9(4):1151–1160
Schaller GE, Kieber JJ (2002) Ethylene. In: The Arabidopsis book, vol 1. American Society of Plant Biologists, Rockville, p e0071
Schiewer U (1967) Occurrence and metabolism of auxin in multicellular algae of the Baltic Sea: I. On the occurrence of indole-3-acetic acid. Planta 74:313–323
Seleiman MF, Al-Suhaibani N, Ali N, Akmal M, Alotaibi M, Refay Y, Dindaroglu T, Abdul-Wajid HH, Battaglia ML (2021) Drought stress impacts on plants and different approaches to alleviate its adverse effects. Plants 10(2):259
Sergeeva E, Liaimer A, Bergman B (2002) Evidence for production of the phytohormone indole-3-acetic acid by cyanobacteria. Planta 215:229–238
Shah W, Ullah S, Ali S, Idrees M, Khan MN, Ali K, Khan A, Ali M, Younas F (2021) Effect of exogenous alpha-tocopherol on physio-biochemical attributes and agronomic performance of lentil (*Lens culinaris* Medik.) under drought stress. PLoS One 16(8):e0248200
Shan C, Liang Z (2010) Jasmonic acid regulates ascorbate and glutathione metabolism in *Agropyron cristatum* leaves under water stress. Plant Sci 178(2):130–139

Sharma A, Shahzad B, Rehman A, Bhardwaj R, Landi M, Zheng B (2019) Response of phenylpropanoid pathway and the role of polyphenols in plants under abiotic stress. Molecules 24(13):2452

Sharma S, Joshi J, Kataria S, Verma SK, Chatterjee S, Jain M, Pathak K, Rastogi A, Brestic M (2020) Regulation of the Calvin cycle under abiotic stresses: an overview. In: Plant life under changing environment. Elsevier, Amsterdam, pp 681–717

Sharma S, Mishra A, Priyanka, Singh SS (2024) Redox cycle: signaling and metabolic cross-talks in cyanobacteria. In: Cyanobacteria. Academic, New York, pp 177–214

Shaukat K, Zahra N, Hafeez MB, Naseer R, Batool A, Batool H, Raza A, Wahid A (2022) Role of salicylic acid–induced abiotic stress tolerance and underlying mechanisms in plants. In: Emerging plant growth regulators in agriculture. Academic, New York, pp 73–98

Shu K, Luo X, Meng Y, Yang W (2018) Toward a molecular understanding of abscisic acid actions in floral transition. Plant Cell Physiol 59(2):215–221

Sidhu GPS, Singh HP, Batish DR, Kohli RK (2017) Alterations in photosynthetic pigments, protein, and carbohydrate metabolism in a wild plant *Coronopus didymus* L. (Brassicaceae) under lead stress. Acta Physiol Plant 39:1–9

Simkin AJ, Faralli M, Ramamoorthy S, Lawson T (2020) Photosynthesis in non-foliar tissues: implications for yield. Plant J 101(4):1001–1015

Singh A, Roychoudhury A (2023) Abscisic acid in plants under abiotic stress: crosstalk with major phytohormones. Plant Cell Rep 42:1–14

Singsaas EL (2000) Terpenes and the thermotolerance of photosynthesis. New Phytol 146(1):1–4

Singsaas EL, Lerdau M, Winter K, Sharkey TD (1997) Isoprene increases thermotolerance of isoprene-emitting species. Plant Physiol 115(4):1413–1420

Sirhindi G, Mushtaq R, Gill SS, Sharma P, Abd Allah EF, Ahmad P (2020) Jasmonic acid and methyl jasmonate modulate growth, photosynthetic activity and expression of photosystem II subunit genes in *Brassica oleracea* L. Sci Rep 10(1):9322

Sitnik KM, Musatenko LI, Vasyuk VA, Vedenicheva NP, Generalova VM, Martin GG, Nesterova AN (2003) Gormonal'nii Kompleksroslinigribiv (hormonal complex in plants and fungi). Akademperiodika, Kiiv

Skoog F, Armstrong DJ (1970) Cytokinins. Annu Rev Plant Physiol 21(1):359–384

Slattery RA, Walker BJ, Weber AP, Ort DR (2018) The impacts of fluctuating light on crop performance. Plant Physiol 176(2):990–1003

Sponsel VM, Hedden P (2010) Gibberellin biosynthesis and inactivation. In: Plant hormones: biosynthesis, signal transduction, action! Springer, Berlin, pp 63–94

Staneloni RJ, Rodriguez-Batiller MJ, Casal JJ (2008) Abscisic acid, high-light, and oxidative stress down-regulate a photosynthetic gene via a promoter motif not involved in phytochrome-mediated transcriptional regulation. Mol Plant 1(1):75–83

Stirk WA, Van Staden J (1997) Isolation and identification of cytokinins in a new commercial seaweed product made from *Fucus serratus* L. J Appl Phycol 9:327–330

Stirk WA, Ördög V, Van Staden J, Jäger K (2002) Cytokinin- and auxin-like activity in cyanophyta and microalgae. J Appl Phycol 14:215–221

Stirk WA, Novák O, Strnad M, Van Staden J (2003) Cytokinins in macroalgae. Plant Growth Regul 41:13–24

Stirk WA, Ördög V, Novák O, Rolčík J, Strnad M, Bálint P, van Staden J (2013) Auxin and cytokinin relationships in 24 microalgal strains1. J Phycol 49(3):459–467

Stowe BB, Yamaki T (1957) The history and physiological action of the gibberellins. Annu Rev Plant Physiol 8(1):181–216

Suhel M, Husain T, Prasad SM (2023) Ethylene as a modulator of redox reaction. In: The plant hormone ethylene. Academic, New York, pp 171–179

Suzuki N, Devireddy AR, Inupakutika MA, Baxter A, Miller G, Song L, Shulaev E, Azad RK, Shulaev V, Mittler R (2015) Ultra-fast alterations in mRNA levels uncover multiple players in light stress acclimation in plants. Plant J 84(4):760–772

Swaminathan S, Bock RM, Skoog F (1977) Subcellular localization of cytokinins in transfer ribonucleic acid. Plant Physiol 59(4):558–563

Sytar O, Kumari P, Yadav S, Brestic M, Rastogi A (2019) Phytohormone priming: regulator for heavy metal stress in plants. J Plant Growth Regul 38:739–752
Taha RS, Seleiman MF, Alhammad BA, Alkahtani J, Alwahibi MS, Mahdi AH (2020) Activated yeast extract enhances growth, anatomical structure, and productivity of *Lupinus termis* L. plants under actual salinity conditions. Agronomy 11(1):74
Taiz L, Zeiger E, Møller IM, Murphy A (2015) Plant physiology and development, 6th edn. Sinauer Associates Incorporated, Sunderland
Takahashi S, Murata N (2008) How do environmental stresses accelerate photoinhibition? Trends Plant Sci 13(4):178–182
Takezawa D, Komatsu K, Sakata Y (2011) ABA in bryophytes: how a universal growth regulator in life became a plant hormone? J Plant Res 124:437–453
Tan LT (2007) Bioactive natural products from marine cyanobacteria for drug discovery. Phytochemistry 68(7):954–979
Tarakhovskaya ER, Maslov YI, Shishova MF (2007) Phytohormones in algae. Russ J Plant Physiol 54:163–170
Teale WD, Paponov IA, Palme K (2006) Auxin in action: signalling, transport and the control of plant growth and development. Nat Rev Mol Cell Biol 7(11):847–859
Thawabteh A, Juma S, Bader M, Karaman D, Scrano L, Bufo SA, Karaman R (2019) The biological activity of natural alkaloids against herbivores, cancerous cells and pathogens. Toxins 11(11):656
Thomas RJ, Harrison MA, Taylor J, Kaufman PB (1983) Endogenous auxin and ethylene in *Pellia* (Bryophyta). Plant Physiol 73(2):395–397
Toffolatti SL, Maddalena G, Passera A, Casati P, Bianco PA, Quaglino F (2021) Role of terpenes in plant defense to biotic stress. In: Biocontrol agents and secondary metabolites. Woodhead Publishing, Sawston, pp 401–417
Tominaga N, Takahata M, Tominaga H (1993) Effects of NaCl and KNO_3 concentrations on the abscisic acid content of *Dunaliella* sp. (Chlorophyta). In: Saline Lakes V: proceedings of the Vth international symposium on Inland Saline Lakes, held in Bolivia, 22–29 March 1991. Springer, Dordrecht, pp 163–168
Torres MA (2010) ROS in biotic interactions. Physiol Plant 138(4):414–429
Torres P, Avila JG, de Vivar AR, García AM, Marín JC, Aranda E, Céspedes CL (2003) Antioxidant and insect growth regulatory activities of stilbenes and extracts from *Yucca periculosa*. Phytochemistry 64(2):463–473
Trabelsi L, Mnari A, Abdel-Daim MM, Abid-Essafi S, Aleya L (2016) Therapeutic properties in Tunisian hot springs: first evidence of phenolic compounds in the cyanobacterium *Leptolyngbya* sp. biomass, capsular polysaccharides and releasing polysaccharides. BMC Complement Altern Med 16(1):1–10
Tsavkelova EA, Klimova SY, Cherdyntseva TA, Netrusov AI (2006a) Hormones and hormone-like substances of microorganisms: a review. Appl Biochem Microbiol 42:229–235
Tsavkelova EA, Klimova SY, Cherdyntseva TA, Netrusov AI (2006b) Microbial producers of plant growth stimulators and their practical use: a review. Appl Biochem Microbiol 42:117–126
Tung T, Yuasa T (2021) Roles of autophagy in photosynthesis of soybean leaf damaged under chilling stress and light irradiation. Cryobiol Cryotechnol 67(1):41–45
Valletta A, Iozia LM, Leonelli F (2021) Impact of environmental factors on stilbene biosynthesis. Plants 10(1):90
Van de Poel B, de Vries J (2023) Evolution of ethylene as an abiotic stress hormone in streptophytes. Environ Exp Bot 214:105456
Vialet-Chabrand S, Matthews JS, Simkin AJ, Raines CA, Lawson T (2017) Importance of fluctuations in light on plant photosynthetic acclimation. Plant Physiol 173(4):2163–2179
Waaland SD, Watson BA (1980) Isolation of a cell-fusion hormone from *Griffithsia pacifica* Kylin, a red alga. Planta 149:493–497
Wang LI, Shan T, Xie B, Ling C, Shao S, Jin P, Zheng Y (2019) Glycine betaine reduces chilling injury in peach fruit by enhancing phenolic and sugar metabolisms. Food Chem 272:530–538

War AR, Paulraj MG, War MY, Ignacimuthu S (2011) Role of salicylic acid in induction of plant defense system in chickpea (*Cicer arietinum* L.). Plant Signal Behav 6(11):1787–1792

Wasternack C (2007) Jasmonates: an update on biosynthesis, signal transduction and action in plant stress response, growth and development. Ann Bot 100(4):681–697

Watanabe T, Kondo N (1976) Ethylene evolution in marine algae and a proteinaceous inhibitor of ethylene biosynthesis from red alga. Plant Cell Physiol 17(6):1159–1166

Watson BA, Waaland SD (1983) Partial purification and characterization of a glycoprotein cell fusion hormone from *Griffithsia pacifica*, a red alga. Plant Physiol 71(2):327–332

Wen L, Jiang Y, Yang J, Zhao Y, Tian M, Yang B (2017) Structure, bioactivity, and synthesis of methylated flavonoids. Ann N Y Acad Sci 1398(1):120–129

Went FW, Thimann KV (1937) Phytohormones. MacMillan, New York

Wenz J, Davis JG, Storteboom H (2019) Influence of light on endogenous phytohormone concentrations of a nitrogen-fixing *Anabaena* sp. cyanobacterium culture in open raceways for use as fertilizer for horticultural crops. J Appl Phycol 31:3371–3384

Yadav T, Kumar A, Yadav RK, Yadav G, Kumar R, Kushwaha M (2020) Salicylic acid and thiourea mitigate the salinity and drought stress on physiological traits governing yield in pearl millet-wheat. Saudi J Biol Sci 27(8):2010–2017

Yang Y, Zhang L, Huang X, Zhou Y, Quan Q, Li Y, Zhu X (2020) Response of photosynthesis to different concentrations of heavy metals in *Davidia involucrata*. PLoS One 15(3):e0228563

Yang J, Gu W, Feng Z, Yu B, Niu J, Wang G (2022) Synthesis of abscisic acid in *Neopyropia yezoensis* and its regulation of antioxidase genes expressions under hypersaline stress. Front Microbiol 12:775710

Yokota T (1997) The structure, biosynthesis and function of brassinosteroids. Trends Plant Sci 2(4):137–143

Yokota T, Kim SK, Fukui Y, Takahashi N, Takeuchi Y, Takematsu T (1987) Brassinosteroids and sterols from a green alga, *Hydrodictyon reticulatum*: configuration at C-24. Phytochemistry 26(2):503–506

Younis MEB, Hasaneen MNAG, Abdel-Aziz HMM (2010) An enhancing effect of visible light and UV radiation on phenolic compounds and various antioxidants in broad bean seedlings. Plant Signal Behav 5(10):1197–1203

Zahradníčková H, Maršálek B, Polišenská M (1991) High-performance thin-layer chromatographic and high-performance liquid chromatographic determination of abscisic acid produced by cyanobacteria. J Chromatogr A 555(1–2):239–245

Zálešák F, Bon DJYD, Pospíšil J (2019) Lignans and neolignans: plant secondary metabolites as a reservoir of biologically active substances. Pharmacol Res 146:104284

Zapata D, Arroyave C, Cardona L, Aristizábal A, Poschenrieder C, Llugany M (2021) Phytohormone production and morphology of *Spirulina platensis* grown in dairy wastewaters. Algal Res 59:102469

Záveská Drábková L, Dobrev PI, Motyka V (2015) Phytohormone profiling across the bryophytes. PLoS One 10(5):e0125411

Zažímalová E, Petrášek J, Benková E (eds) (2014) Auxin and its role in plant development. Springer, Vienna, pp 2675–2688

Zhang YB, Yang SL, Dao JM, Deng J, Shahzad AN, Fan X, Li R, Quan Y, Bukhari SAH, Zeng ZH (2020) Drought-induced alterations in photosynthetic, ultrastructural and biochemical traits of contrasting sugarcane genotypes. PLoS One 15(7):e0235845

Zhao P, Lu GH, Yang YH (2017) Salicylic acid signaling and its role in responses to stresses in plants. In: Mechanism of plant hormone signaling under stress, vol 1. Wiley, New York, pp 413–441

Zheng Y, Wang X, Cui X, Wang K, Wang Y, He Y (2023) Phytohormones regulate the abiotic stress: an overview of physiological, biochemical, and molecular responses in horticultural crops. Front Plant Sci 13:1095363

Stress Biology in Plant Tissue Culture: Recent Advances

9

Afreen Akhtar, Saad Bin Javed, and Altaf Ahmad

Abstract

Conditions or factors in the environment that disrupt normal physiological processes and growth of plants that can be arise through any abiotic or biotic source, alone or in combination, can be considered as stress. Stress biology deals with the different physiological and molecular modifications plants undergo during various stresses. Understanding how plants respond to stress is essential for agriculture, horticulture, and environmental conservation efforts. Not all plants show similar type of response in similar environment rather they respond differently; therefore, to study the effect of any particular type of stress on plants, a large sets of experimental setups are needed, with precise conditions which cannot be provided by growth chamber, field, and green house. Tissue culture provides a controlled environment where specific stress factors, such as drought, salinity, temperature extremes, or chemical pollutants, heavy metals can be applied accurately and consistently to plant tissues and can be monitored by measuring their effects on plant growth and health. In recent decades, a large number of studies have been conducted on various types of stress under in vitro conditions in plants. Stress-tolerant plant varieties are developed by analyzing the plants under heat, cold, drought, and salinity stress under in vitro conditions. Pathways that are triggered by the stress conditions are also identified to certain extent by the various studies. An increased production of secondary metabolites was also reported in several studies, under some heavy metal stress in vitro conditions. Efforts are also made in the direction of phytoremediation and environmental assessment using in vitro approaches. In this chapter, we will discuss the recent advancements that are taking place in stress biology using the plant tissue culture tech-

A. Akhtar · S. B. Javed (✉) · A. Ahmad
Department of Botany, Aligarh Muslim University, Aligarh, India
e-mail: javed.bt@amu.ac.in

A. K. Mishra (ed.), *Stress Biology in Photosynthetic Organisms*,
https://doi.org/10.1007/978-981-97-1883-2_9

niques and the future prospects that should be integrated into it for a better understanding.

Keywords

Stress biology · Secondary metabolites · Heavy metals · Light stress · Drought stress

9.1 Introduction

Plant stress is an adverse environmental condition that affects a plant's ability to grow and develop, which lowers agricultural productivity globally. Plants may be under biotic, abiotic, or both types of stress. According to Savary et al. (2012), biotic stress refers to the harm that living things like viruses, bacteria, fungus, insects, parasites, and weeds inflict, which lowers crop production by 25% globally. Abiotic stress refers to the combination of environmental conditions that negatively impact a plant's development and growth, resulting in a 50% reduction in agricultural output (Kayum et al. 2016). By 2050, more than 50% of arable land may be severely salinized due to a combination of increasing salinity and drought in several regions of the world (Wang et al. 2003). According to Mustafa et al. (2019), this causes an average 50% decline in global agricultural revenues. Abiotic stress management techniques include changing the timing of seeding, adding fertilizer to the soil, using high-quality water for irrigation, and more. Although conventional plant breeding has improved the resistance of many crops to major diseases, it is challenging to respond swiftly to the emergence of new virulent pathogen races because to the time-consuming procedures of progeny selection, back crosses, and cross-breeding. Furthermore, because breeders cannot access natural sources of resistance, plant breeding techniques cannot be used to combat many serious diseases (Dasgupta 1992; Melcher and Stuiver 2000). Tissue culture has emerged as a more practical and economical method for producing stress resistance in plants in recent times. Understanding the biochemistry and physiology of the plants cultured in a variety of environmental stress conditions is achieved using this technique that operates under checked parameters with less space as well as time. This technique has a very high potential for the production of stress-tolerant crop plants (Jan et al. 2018). Cell and plant tissue culture techniques enable the studying and screening of very large populations of regenerated plants and cells in a limited area and in a considerably extra controlled environment. The ability of plants to produce metabolites is thought to be an adaptive mechanism for dealing with demanding conditions in a challenging and changing environment (Naik and Al-Khayri 2016). Abiotic stressors also cause plants to produce secondary metabolites that have antioxidant and anti-inflammatory qualities that are beneficial to human health as well as defence systems for the plant (Sharma et al. 2022). Medicinal plants are thought to function as a biochemical factory that generates valuable secondary metabolites (SMs) (Neumann et al. 2009). The factors that restrict the industrial synthesis of these plant-based metabolites from field-grown plants include their overexploitation, availability and challenges in cultivating the source plant; seasonal and phytogeographical variation in productivity; low productivity; tissue/organ-specific production; variability of impurities; challenges in purification; and the financial inputs associated with choosing and implementing suitable screening bioassays. Furthermore, because to their stereospecificity and extremely complicated structures, chemical production is frequently not economically possible (Nandagopal et al. 2018). New methods based on plant tissue culture techniques offer advantages like easier source modification and more abundant and high-quality secondary chemical extraction. Callus and in vitro propagation are two plant tissue culture techniques that guarantee the production rate of secondary chemicals in larger amounts without limiting the influences of the environment (Beigmohamadi et al. 2019) Conflicting information has been reported regarding the influence of abiotic variables upon growth, productivity, and accumulation of secondary metabolites in medicinal plants. These parameters include temperature (hot or cold), exposure to temperature, increased CO_2, drought, and salinity (Mishra 2016). Higher proportions of different secondary metabolites (phenolics, alkaloids, terpenoids, and glycosides) have been found in plants treated with elevated CO_2 levels, such as *Catharanthus roseus* (Saravanan and Karthi 2014), *Elaeis guineensis* (Ibrahim and Jaafar 2012), *Ginkgo biloba* (Huang et al. 2010), *Zingiber officinale* (Ghasemzadeh et al. 2010), *Digitalis lanata* (Stuhlfauth and Fock 1990), and among others. Like this, in medicinal plant species stressed by heat, salinity, or drought, growth usually decreased while secondary metabolite content accumulated. It is now clear that secondary metabolite accumulation in plants and the stress response are correlated, either directly or indirectly. Using a plant tissue culture approach, this chapter elaborates on the various aspects of stress tolerance under different stresses and the production of various important metabolites.

9.2 Selection of Stress-Tolerant Plants Using Plant Tissue Culture

Plant cells, tissues, or organs can be cultured in vitro on a media containing selection agents to identify and regenerate plants with desired traits. This method is also successfully applied in order to develop tolerance, allowing for the survival and growth of desirable phenotypes through use of certain selection agents (Purohit et al. 1998). For in vitro selection, the selecting agents that are typically used are the causal organism itself (for disease resistance), certain phytotoxin (like fusaric acid) or fungal culture filtrate (FCF), PEG or mannitol (for drought tolerance), and NaCl (for salt tolerance). Diverse variety of these selective agents have been introduced to the growth media are exposed to the explants. The only explants that are chosen are those that can endure in these conditions over time. Two different approaches of selection have been proposed (a) shock treatment, where direct exposure of high concentration of shock are given to culture and only those that could withstand the treatment will survive and (b) sequential long-term treatment, where cultures are stressed and gradually the concentration of the selecting agent is increased (Purohit et al. 1998). Such techniques are based on the creation of genetic variation in cultured and regenerated plant cells, tissues, and/or organs (Mohamed et al. 2000). Somaclonal variation, which is induced in regenerated plants via tissue culture (Larkin and Scowcroft 1981), can produce a variety of genetically stable changes that are beneficial for agricultural development. Field selection can be complemented by in vitro selection, which can significantly reduce the amount of time required for selection and culturing of desired traits under selection pressure with little environmental harmony (Mohan Jain 2001). Epigenetic alterations, which are occasionally seen in tissue culture experiments, are mostly caused by altered DNA methylation, according to several studies (Guo et al. 2007; Li et al. 2007; Gao et al. 2010). Researchers have proposed that one-step selection or short-term could be used to overcome the issue regarding adaptation (epigenetic) during in vitro selection as it can hamper the growth of epigenetically adapted cells (Tal 1994).

9.3 Production of Secondary Metabolite Under Stress Conditions

Suggesting that plant tissue and cell cultures provide an additional edge for industry scale production of plant secondary metabolites (DiCosmo and Misawa 1985). This strategy achieved by modifying various growth parameters is self-sufficient and independent of seasonal or regional variation. Several methods were used with the cell culture system to produce SMs in large quantities. In comparison to mother plants, plant cell cultures synthesize SMs in varying quantities and with varying characteristics, which may vary over time (Tepe and Sokmen 2007). The employment of plant tissue culture methods offers a great avenue for the large-scale production of important secondary metabolites (Chattopadhyay et al. 2002). Now plant cell cultures are extensively used in the industrial production of SMs and are advantageous for study on biochemistry, cell biology, and genetics. Cell suspensions provide a fast alternative to produce large amounts of SMs for industrial use than tissue and organ culture. This process is the most trustworthy and effective way to synthesize secondary metabolites (Chattopadhyay et al. 2002; Vanisree et al. 2004). It is commonly known that numerous attempts have been undertaken to produce suspension culture on a commercial scale. Required metabolites are increased in suspension cultures, but eventually the capacity to synthesize declines because of inadequate nourishment or genetic differences. Thus, for suspension culture, choosing and maintaining a high-yielding cell line is crucial (Chattopadhyay et al. 2002). Suitable calluses are generated from valuable mother plants in an ideal culture medium. This suitable media aids in the mechanisms of differentiation and dedifferentiation. According to Barrales-Cureño et al. (2019), the calli formed by these plants are subcultured either for organogenesis, embryogenesis, or suspension culture, or for propagation.

Applying chemical or physical stressors to cell suspension cultures is another method of elicitation used in plant tissue culture. Stresses cause plants to produce secondary metabolites, which are not typically generated in them. These days, biotic elicitors like chitosan, different protein extracts, and pathogenic fungus mycelium that has been sterilized are used in elicitation, along with abiotic elements like heavy

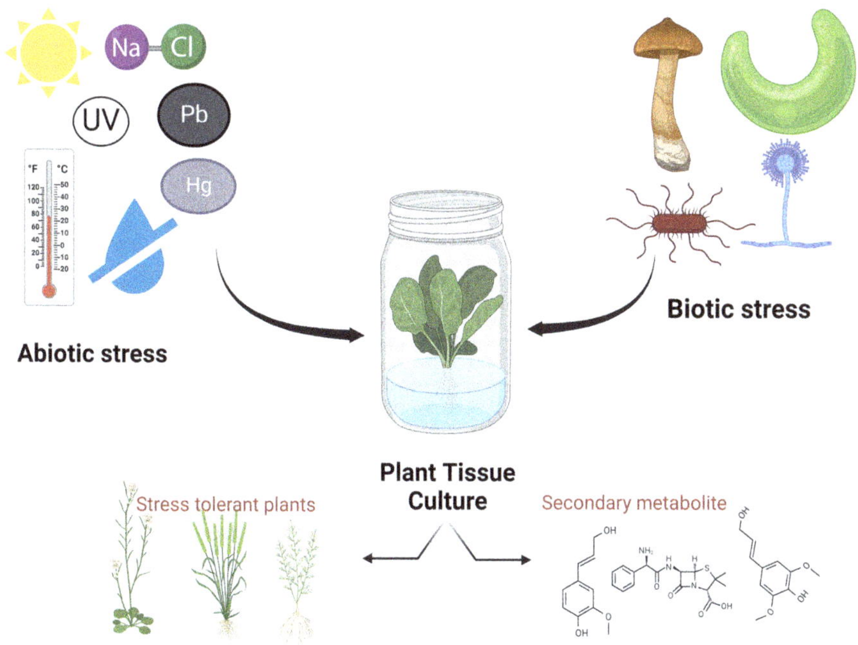

Fig. 9.1 Applications of plant tissue culture

metal salts, extremes in temperature, pH, UV light, and so forth. There are numerous findings that demonstrate how elicitors boost the number of SMs using tissue culture. Scholars have reported the use of diverse elicitor types to increase stem cell generation in in vitro systems (Karuppusamy 2009; Sudha and Ravishankar 2003). Applying chemical or physical stressors to cell suspension cultures is another method of elicitation used in plant tissue culture. Stresses cause plants to produce secondary metabolites, which are not typically generated in them. These days, biotic elicitors like chitosan, different protein extracts, and pathogenic fungus mycelium that has been sterilized are used in elicitation, along with abiotic elements like extremes in temperature, heavy metal salts, UV light, pH, and so forth (Fig. 9.1). There are numerous findings that demonstrate how elicitors boost the number of SMs in plant tissue culture.

9.4 Abiotic Stress

Physical and chemical agents are examples of nonbiological sources from which abiotic elicitors originate. High radiation (UV light and solar), salinity, nutrient shortage, flooding, water logging, drought, soil pH, extremely hot or cold temperatures, and man-made pollution are a few examples of different abiotic factors (Fig. 9.1) that can occur (Narayani and Srivastava 2017).

9.4.1 Salinity Stress

Some agricultural methods, like irrigation and inadequate drainage systems, have made the issue of soil salinity worse in the past few decades. According to estimates, salt affects 20% of the world's irrigated land. As salinization in agricultural fields increases, this is predicted to limit the amount of land suitable for agriculture by 30% over the next 25 years and by the year 2050 up to 50% (Rozema and Flowers 2008). Excessive concentration of salt in soil or tissue culture media might disrupt various physiological and biochemical functions, leading to issues like mineral shortage, ion imbalance, osmotic stress, and oxidative stress ion toxicity. These conditions eventually interact with various plant cellular components, such as lipids, proteins, DNA, a pigment, hindering the growth, and development of most crops (Zhu 2002). It is now a global challenge to preserve crops from damage caused by salinity. Osmotic/ionic stress is a result of high salinity, which is defined as elevated level of sodium and chloride ions in the soil solution (Hasegawa et al. 2000). Development of plants reacts to increased salinity levels in two ways: (a) a fast osmotic phase that begins as soon as the concentration of salt surrounding the roots reaches a threshold and causes a significant decline in shoot growth and (b) a steady ion-specific phase that begins when the concentration of salt in old leaves reaches toxic levels and causes them to perish (Munns and Tester 2008). A workable solution to this issue would be the creation of salt-tolerant crops that can uphold elevated saline concentration in the soil (Yamaguchi and Blumwald 2005). To create plants that can withstand NaCl various approaches are being developed. *Agrobacterium*-mediated transformation and in vitro selection techniques give a useful strategy for production of such resistant varieties (Hossain et al. 2007). Several attempts have been made to use in vitro methods to generate salt-tolerant plants. Many systems, including callus, somatic embryos, suspension cultures, shoot cultures, and others, are being utilized for this purpose. These systems are assessed for differences in their capacity to withstand relatively high concentrations of salt (NaCl) within culture media (Woodward and Bennett 2005). NaCl is the salt used in most salinity experiments. On the other hand, a number of researchers have examined how different SO_2^- and Cl^- salts, such as Na_2SO_4, $MgSO_4$, and KCl, react to in vitro screening. When *Nicotiana tabacum* was grown on seawater, artificial seawater, NaCl, mannitol, and other SO_2^- and Cl^- and salts, different reactions were seen (Chen et al. 1980). It could be wiser to use a different proportion of salts as a selection pressure since it will match the salinity in the field condition. Many stress-responsive mechanisms, such as the Na^+ translocation from tissue of leaf into the vacuoles, the exclusion of Na^+ from transpiration, and ROS detoxification mechanisms, result in the synthesis of plant SMs, determine a plant's sensitivity, or tolerance to salinity (Jorge et al. 2019; Yang and Guo 2018; Munns and Tester 2008). Many studies have shown that several organs and tissues can produce more phenolic compounds when subjected to salt stress (Table 9.1). For instance, *Hordeum vulgare* tissue culture showed a decrease in phenolic compounds in the shoots but an increase in the roots when NaCl was added (Ali and Abbas 2003).

Table 9.1 List of in vitro studies on salinity stress

Plant species	Culture type	Stressing agent	References
Durum wheat (*Triticum turgidum* var. durum)	Embryo culture	NaCl 0.3–2.1%	Arzani and Mirodjagh (1999)
Rice (*Oryza sativa* L.)	Embryo culture	NaCl 1–2%	Vajrabhaya et al. (1989)
Potato (*Solanum tuberosum*)	Callus culture	NaCl 0.25–0.5 M	Naik and Widholm (1993)
Sugarcane (*Saccharum* sp.)	Callus culture	NaCl 68 mM	Gandonou et al. (2006)
Tomato (*Solanum lycopersicum*)	Callus culture	NaCl 5–100 mM	Aazami et al. (2021)
Sweet potato (*Ipomoea batatas*)	Embryo culture	NaCl 25–200 mM	Anwar et al. (2010)
Carrot (*Daucus carota* subsp. sativus)	Protoplast culture	NaCl 10–400 mM	Kiełkowska et al. (2019)
Bamboo (*Dendrocalamus strictus*)	Embryo culture	NaCl 50–250 mM	Singh et al. (2003)
Sour Orange (*Citrus aurantium*)	Embryo culture	NaCl 100–300 mM	Koc et al. (2009)
Canola (*Brassica napus*)	Callus culture	NaCl 4000–16,000 ppm	Abdrabou et al. (2017)
Chrysanthemum (*Chrysanthemum morifolium*)	Callus culture	NaCl 50–100 mM	Hossain et al. (2007)
Aubergine (*Solanum melongena*)	Callus culture	NaCl 40–120 mM	Hannachi et al. (2021)
Bacopa monnieri	Nodal culture	NaCl 0–100 mM	Dogan (2020)
Limnophila aromatica	Nodal culture	NaCl 0–100 mM	Dogan (2020)

9.4.2 Temperature Stress

There exists a correlation between the optimum temperature range of a plant's natural habitat and its growth and development. Temperatures above or below a plant's normal range might negatively affect its ability to grow and produce (Li et al. 2020; Yadav et al. 2014). Research has demonstrated that signaling molecules are triggered in plants under abiotic stress conditions. Specifically, temperature upshifts initiate a lipid-dependent signaling cascade, which initiates a plant's stress adaption mechanism (Hou et al. 2016). The primary signaling lipids produced in response to temperature stress are phospholipids, which include lysophospholipids, oxylipins, phosphatidic acid, sphingolipids, and phosphoinositides (Zhu 2016). Salicylic acid (SA) is a key player in the signaling and tolerance generated by abiotic stress; in particular, it has been shown that SA may cause thermo-tolerance in a variety of genera (Zandalinas et al. 2018; Vishwakarma et al. 2017). Different temperatures in the in vivo and in vitro environments where plants are produced affect ontogeny and metabolic processes; greater temperatures can cause early leaf senescence, which can have an effect on plant metabolism (Morison and Lawlor 1999).

Altering the temperature of culture has the potential to impact the growth of plant and SMs production by altering the physiology and metabolism of the cultured cells. Generally, callus tissue induction and the development of cultured cells occur in the 17–25 °C temperature range. In the hairy root culture of *Panax ginseng*, temperature upshifts and quality of light affect the ginsenoside production (Yu et al. 2005). According to Chan et al. (2010), *Melastoma malabathricum* cell cultured at 20 ± 2 °C developed better and produced more anthocyanin as compared to those grown at 29 ± 2 and 26 ± 2 °C. The anthocyanin yield is maximized at the optimal temperature of 25 °C, as seen in *Perilla frutescens* (Zhong and Yoshida 1993) as well as in strawberry cell cultures (Zhang et al. 1997). Lower temperatures slow down cell proliferation but promote the accumulation of anthocyanins (Zhang et al. 1997). Anthocyanin synthesis was significantly decreased at 28 °C, in suspension cultures of *Perilla frutescens*, whereas for the pigment's productivity 25 °C was ideal (Zhong and Yoshida 1993). Analogous findings regarding the ideal anthocyanin productivity in cell suspensions of *Daucus carota* have been documented (Narayan et al. 2005). It was observed that pigment was released from hairy roots of *Beta vulgaris* when its root cultures were exposed to changing temperatures (Thimmaraju et al. 2003). In the callus cultures of *Helicteres isora,* light and temperature had a synergistic effect on the formation of SMs (Joshi 2015). It has been demonstrated that when beans and tobacco are exposed to water stress and heat shock, polyamines are produced and then phenylamides are formed (Edreva et al. 1995, 1998, 2008). In a similar way, heat treatments marginally reduced the amount of carotenoids generated by members of the Brassicaceae family (Morison and Lawlor 1999), whereas high concentrations encouraged root SM and leaf senescence in *Panax quinquefolius* (Jochum et al. 2007). While tree adaptation to the cold temperature was linked to the synthesis of high amounts of chlorogenic acid (Pérez-Ilzarbe et al. 1997), cold stress encouraged the formation of phenolics and their subsequent inclusion in plant cell walls as suberin or lignin (Griffith and Yaish 2004). For instance, oxidative stress, which was more pronounced at higher temperatures than at lower ones, was demonstrated to have an impact on the formation of SMs in *Eleutherococcus senticosus* somatic embryos (Shohael et al. 2006). It affected the amount of flavolignans that accumulated in the *Silybum marianum* hairy root cultures; when paired with the treatments, acid pH proved to be the most effective (Rahimi and Hasanloo 2016). An increase in incubation temperature had an impact on the growth of *Stevia rebaudiana* hairy root cultures, but it also partially increased stevioside synthesis (Kumari and Chandra 2016). Therefore, for physiological functions, which include the biosynthesis of the SMs, each plant species, cultivar, or genotype has certain ideal temperature ranges; deviation from these ranges may have an influence on biomass and SM biosynthesis. Because of this, different seasons and geographical locations may experience variations in the metabolite yield during the in vivo growth of a particular plant species, cultivar, or genotype. These variations can then be mimicked in in vitro cultures by manipulating the media and cultural conditions, which has served as the foundation for the development of numerous in vitro-based SM enhancement strategies (Isah 2019).

9.4.3 Drought Stress

Drought, or inadequate water availability, is one of the most common stressors that affects terrestrial plants. It will interfere with essential metabolic processes and turgor pressure at the cellular level, which will impact the creation of cell walls and cell expansion—two processes that are unavoidably more sensitive to water scarcity. When there is strong radiation and heat at the same time, the damage caused by oxidative stress is increased. Therefore, elevated reactive oxygen species (ROS) concentrations are a common characteristic of almost all abiotic stressors, and strengthening the anti-oxidative system typically accompanies stress adaptation (Tippmann et al. 2006). For example, research on maize has demonstrated a clear correlation between improved plant drought tolerance and elevated antioxidative enzyme activity (Malan et al. 1990). Abscisic acid (ABA), a plant growth regulator (PGR), is one of the key messengers under drought stress (Ingram and Bartels 1996). A connection holds between cellular and in vivo plant responses, drought-tolerant plants have been produced through in vitro culture (Mohamed et al. 2000). There have been reports of in vitro selection in recent years for cells that show enhanced resistance to water stress or drought stress (Table 9.2). Several researchers have employed mannitol, sorbitol, polyethylene glycol (PEG), or sucrose as agents of osmotic stress for in vitro selection of plants (Hassan et al. 2004; Mohamed et al. 2000). PEG, on the other hand, has been applied in plants to induce water stress most significantly. This high molecular weight molecule reduces the water potential of nutrient-rich solutions without getting absorbed by plants or becoming phytotoxic. It is an inert osmoticum that does not penetrate (Hassan et al. 2004). Water is removed from the cell wall as well as the cell itself since PEG does not penetrate the apoplast. PEG solutions therefore more closely resemble dry soil than do solutions of low molecular weight chemicals, which allow solute to penetrate the cell wall (Verslues et al. 1998). In a study conducted on *Carum copticum* in in vitro cultures under drought stress, 36 distinct essential oil components (terpenes) production was enhanced (Razavizadeh and Komatsu 2018). By enhancing the activity and expression levels of important enzymes involved in biosynthesis of baicalin—phenylalanine chalcone synthase, cinnamate-4-hydroxylase, and ammonia-lyase—*Scutellaria baicalensis* can produce more baicalin (Cheng et al. 2018). Drought stress can enhance the quality of SMs, artemisinin in *Artemisia annua* and betulinic acid, rutin and quercetin in *Hypericum brassilians* (Chen et al. 2004). Under drought conditions, *Camellia sinensis* leaves showed an increase in α-tocopherol, phytol, and trans-2-hexenal (Podda et al. 2019); levels of flavone, isoflavone, anthocyanins, and flavonol also increased (Gai et al. 2020).

9.4.4 Light Stress

UV-A, UV-B, photosynthetic active, and infrared radiation all make up the solar radiation that penetrates the Earth's surface; however, plants only use a small portion of UV-B, which is the most energetic component of day light, for growth and

Table 9.2 List of in vitro studies on drought stress

Plant species	Culture type	Stressing agent	References
Soybean (*Glycine max*)	Callus culture	20% PEG 6000	Mishra et al. (2021)
Soybean (*Glycine max*)	Embryo culture	15% PEG 6000	Sunaryo et al. (2016)
Durum wheat (*Triticum durum*)	Embryo culture	10% and 20% PEG 10000	Bouiamrine and Diouri (2012)
Durum wheat (*Triticum durum*)	Embryo culture	10% and 20% PEG 6000	Kacem et al. (2017)
Sorghum (*Sorghum bicolor*)	Callus culture	Range of 0–15% PEG 8000	Duncan et al. (1995)
Wheat (*Triticum aestivum*)	Embryo culture	Mannitol	Abdel-Ghany et al. (2004)
Wheat (*Triticum aestivum*)	Embryo culture	PEG 6000	Mahmood et al. (2012)
Tagetes (*Tagetes minuta*)	Callus culture	Range of mannitol	Mohamed et al. (2000)
Potato (*Solanum tuberosum*)	Nodal culture	Range of sorbitol	Gelmesa et al. (2017)
Potato (*Solanum tuberosum*)	Microtubers	Range of sorbitol and PEG 6000	Gopal and Iwama (2007)
Rice (*Oryza sativa*)	Callus culture	Range of PEG	Biswas et al. (2002)
Rice (*Oryza sativa*)	Embryo culture	Range of (0–2%) PEG 6000	Wani et al. (2010)
Sugarcane (*Saccharum* sp.)	Callus culture	Range of mannitol	Errabii et al. (2006)
Coconut (*Cocos nucifera*)	Embryo culture	Range of PEG 6000 (1–5%)	Karunaratne et al. (1991)
Brown mustard (*Brassica juncea*)	Callus culture	Range of mannitol	Gangopadhyay et al. (1997)
Ground nut (*Arachis hypogaea*)	Callus culture	Range of PEG 6000	Purushotham et al. (1998)

development (Caldwell et al. 2007). This is dependent on the exposure wavelength and how the radiation interacts with environmental signals. In vivo and in vitro plant development and growth are influenced by a number of limiting factors, including light. Depending on the species or genotype, developmental stage, kind of light, and duration of exposure, light can have an impact on the production of small mass plants (Ghosh et al. 2018; Isah and Umar 2020). In *Plantago lanceolata* progeny, phenotypic plasticity related to light and nutritional conditions affected biomass and iridoid glycoside accumulation (Miehe-Steier et al. 2015). The quality of the light had an impact on the growth and production of flavonoids in *Hyptis marrubioides* seedlings grown in vitro. Red light was shown to be the most effective for plant growth and leaf formation, while blue and white light promoted the accumulation of rutin (Pedroso et al. 2017). The duly expression of genes involved in *Camptotheca acuminata* seedling biomass and camtothecin (CPT) content production was found to be regulated by the intensity and duration of light exposure (Hu et al. 2016). It is clear that light has a complex impact on plant development and SM, depending on the species studied at the in vivo or in vitro growth stage(s), the physiological condition of the tissue or organ assessed, by the light source's spectrum level (Ghosh et al. 2018).

UV-B radiation has been used to study a number of terpenes. Important pharmacological compounds called monoterpenes, like linalool, limonene, borneol, camphor, myrcene, geraniol, nerol, sabinene, 1,8-cineole (eucalyptol), z-citral, and pinene, increase in *Ocimum basilicum* (Ioannidis et al. 2002; Chang et al. 2009; Bertoli et al. 2013; Takshak and Agrawal 2019), *Cymbopogon citratus* (Kumari et al. 2009), and *Mentha piperita* (Maffei and Scannerini 2000; Dolzhenko et al. 2010), Culture of *Vitis vinifera* exposed to UV-B is reported to create diterpenes such as phytol and α- and γ-tocopherol along with triterpenes including sitosterol, stigmasterol, and lupeol (Gil et al. 2012). About 700 tetraterpenoid (40C) lipophilic pigments make to the category known as carotenoids. Studies have shown that UV-B radiation can affect their concentration (Takshak and Agrawal 2019; Jansen et al. 2008). *Picea asperata* was discovered to have higher carotenoid concentrations when exposed to UV-B radiation (Han et al. 2009). Exposure of ultraviolet radiation causes *Arachis hypogea* to accumulate stilbene, and *V. vinifera* callus cultures to produce secondary metabolites (Cetin 2014).

9.4.5 Heavy Metals

The majority of plants lack specific defence systems against the active influx of heavy metals from the soil. Increased uptake of heavy metals impairs nutrition absorption, enzyme function, and membrane integrity. Mercury (Hg), cadmium (Cd), and lead (Pb) are examples of metals with low redox capacity that interfere with the activation of the antioxidative glutathione pool, iron-mediated mechanisms, and Ca^{2+}-dependent systems (Pinto et al. 2003). It is assumed that the signal transduction pathways are mimicked by the O_3-mediated generation of ROS prior to oxidative burst during an avirulent causal organism attack (Jan et al. 2018). Still, not every plant reacts to the stress of metal and mineral ions in the same manner. Plant development retardation in barley can result from the accumulation of phenolic chemicals, which stimulate indoleacetic acid oxidase and peroxidases. This may lead to a reduced auxin levels in the tissues (Ali and Abbas 2003). The accumulation of steroidal sapogenins was positively impacted by the addition of metal ions individually; however, *Agave amanuensis* sapogenin content dramatically decreases upon addition of metal ions in the culture (Sri Andrijany et al. 1998). According to Ajungla et al. (2009), certain metal ions, such Ca^{2+}, facilitate the induction of peroxidases, which are employed in the breakdown of secondary metabolites in *Datura metel*. According to Tauqeer et al. (2016), *Alternanthera bettzickiana* photosynthetic pigment content rose in medium with low Cd and Pb concentrations but reduced in photosynthetic pigments with greater Cd and Pb applications. These findings demonstrate that the impact of heavy metals on pigments involved in photosynthetic processes varies based on the kind of metal and the type of plant. Inhibiting photosynthetic pigment concentration and activity are the most common signs of lead toxicity (Singh et al. 2010). When added in the proper amounts, heavy metal ions such as silver, cadmium and cobalt are powerful inducers that may promote the formation of phenolic acid in *V. vinifera* cell culture without causing a reduction in biomass (Cai et al. 2013). In tissue culture these have been used to boost the surpluous production of SMs, advance plant growth, and boost the effectiveness of germination (Wang et al. 2016; Ruttkay-Nedecky et al. 2017). Several important metal oxide nanoparticles, including as copper, iron, zinc, and titanium oxide, have been reported to benefit plants by advancing the production of plant secondary metabolites (Table 9.3) (Mohammed 2015). As reported, in Jasim et al. (2017), treatment with silver nanoparticles raised the concentration of diosgenin in fenugreek, the concentration of artemisinin in *Artemisia annua* L. hairy root cultures (Zhang et al. 2013), and the upregulation of biosynthetic genes for the formation of anthocyanins and flavonoids in *Arabidopsis thaliana* (García-Sánchez et al. 2015). When AgNPs were added to hairy root cultures of *Brassica rapa* (turnip), the glucosinolates (glucobrassicanapin, glucoallysin, gluconasturtin, sinigrin, 4-methoxy-glucobrassicin, progoitrin, gluconapin, 4-hydroxyglucobrassicin, glucobrassicin and neoglucobrassicin) were found to be elevated (Chung et al. 2018). The production of scopolamine, hyoscyamine, tropane and alkaloids were effectively boosted in *Hyoscyamus reticulatus* L. hairy root cultures when nano-iron oxide (FeNPs) was added to it (Moharrami et al. 2017). In *Withania somnifera* WcAgNPs elicitation increased the synthesis of withanolide A and withaferin A by 22.8 and 11.15 times, respectively. (Tripathi et al. 2021).

Table 9.3 List of in vitro studies on heavy metal stress

Plant species	Culture type	Stressing agent	References
Tobacco (*Nicotiana tabacum*)	Callus culture	Copper	Rout and Sahoo (2007)
Potato (*Solanum tuberosum*)	In vitro cell line	Cadmium	Ashrafzadeh and Leung (2017)
Mustard (*Brassica juncea*)	Callus culture	Cadmium	Gasic and Korban (2007)
Rice (*Oryza sativa* L.)	Callus culture	Aluminium	Roy and Mandal (2005)
Brassica campestris, *Brassica juncea*	Callus culture Callus culture	Zinc Zinc	Rout et al. (1999) Rout et al. (1999)
Foxtail millet (*Setaria italica*)	Callus culture	Zinc	Samantaray et al. (1999)
Jungle rice (*Echinochloa colona*)	Callus culture	Chromium, nickel	Samantaray et al. (2001)
Flax (*Linum usitatissimum*)	Callus/shoot tip culture	Cadmium	Berková et al. (2022)
Silver poplar (*Populus alba*)	Microshoot culture	Cadmium, copper	Marzilli et al. (2018)
Silver poplar (*Populus alba*)	Microshoot culture	Cadmium, arsenic, zinc, copper,	Di Lonardo et al. (2011)
Downy oak (*Quercus pubescens*)	Shoot culture	Cadmium, copper	Di Santo et al. (2016)
Prunella vulgaris	Shoot-tip culture	Cadmium, zinc	Çetin and Ay (2020)

9.5 Biotic Stress

In addition to environmental stressors, pathogenic pathogens like as bacteria, fungus, viruses, nematodes, and insects (Fig. 9.1) are constant latent foes that plants must contend with. The pathogen's effect is determined by the type of attack. The amount of soluble carbohydrates that are present in the plant's diseased sections changes when it becomes infected with diseases. According to Brodmann et al. (2002) and Abood and Lösel (2003), biotrophs typically cause intracellular osmotic stress by accumulating larger amounts of carbohydrates, which manipulates the host's entire metabolism to their advantage. Plant stress is caused by the activation of pathogen stress responses, which in turn leads to disease and defence mechanisms that also lower yield (Heil and Bostock 2002). Many commercially significant crops continue to have very low yields because of their vulnerability to a variety of bacterial, viral, and fungal diseases. Chemical control of these infections is frequently challenging, expensive, resource, and labour intensive (Bézier et al. 2002). Furthermore, some fungicides that are chemically manufactured have the potential to pollute the environment because they are not biodegradable and can accumulate to high quantities in the soil, which lowers its productivity, and in the water table, which puts plants and animals at risk of illness. Therefore, research on creating disease-tolerant plants using current or new techniques has taken on greater significance. To produce stress-tolerant lines, an appealing alternative method is in vitro selection (Jayasankar et al. 2000; Ganesan and Jayabalan 2006). A significant strategy for desirable plant selection is in vitro selection using increased expression of antifungal peptides, pathogenesis-related (PR) proteins, or phytoalexin production (Ganesan and Jayabalan 2006; Vijaya Kumar et al. 2008). When it comes to enhanced disease tolerance, this technology is simpler and less expensive than the transgenic approach (Jayasankar et al. 2000). Embryogenic or organogenic calli, somatic embryos, shoots, or cell suspensions can be used in the in vitro selection for pathogen resistance by subjecting them to toxins generated in the filtrate of the pathogen culture, or by the pathogen itself (Vijaya Kumar et al. 2008).

9.5.1 Bacterial

Bacterial elicitation comprises of using bacterial cells, its components, and cell extracts to cause a reaction in plants. They are also employed to stimulate the in vitro synthesis of SMs in plants. Different secondary metabolites can be produced by different bacterial species in varying amounts (Table 9.4). According to Zhao et al. (2005) and Biswas et al. (2016), the differences result from reactive oxygen species formation, cytoplasmic acidification, G protein activation, and elicitor-membrane receptor binding. To elicitate the synthesis of SMs, a variety of bacterial components are used, including culture filtrates (Biswas et al. 2016, 2018), live cells (Park et al. 2006; Awad et al. 2014), cellular extracts (Gandi et al. 2012; Chodisetti et al. 2013), and cell homogenate (Buitelaar et al. 1992; Jung et al. 2003). Glycyrrhizic acid was extracted from the roots of *Taverniera cuneifolia* (Roth) Arn

Table 9.4 List of in vitro studies on secondary metabolite production using bacterial stress

Plant species	Culture type	Bacterial elicitor	Targeted metabolite(s)	References
Hypericum perforatum	Cell suspension	*Agrobacterium rhizogenes*	Xanthone	Tusevski et al. (2015)
Taverniera cuneifolia	Root		Glycyrrhizic acid	Awad et al. (2014)
Gymnema sylvestre	Cell suspension		Gymnemic acid	Chodisetti et al. (2013)
Scutellaria laterifora	Hairy root		Acteoside, Wogonin, Baicalin, Scutellarin, and Wogonoside	Wilczańska-Barska et al. (2012)
Andrographis paniculata	Cell suspension		Andrographolide	Gandi et al. (2012)
Taverniera cuneifolia	Root	*Agrobacterium tumefaciens*	Glycyrrhizic acid	Awad et al. (2014)
Hypericum perforatum	Cell suspension		Xanthone	Tusevski et al. (2015)
Datura metel	Hairy root	*Bacillus cereus*	Atropine	Shakeran et al. (2015)
Panax ginseng	Root	*Bacillus* sp. CWJ-1	Ginsenoside	Le et al. (2018)
Helicteres isora	Cell suspension	*Bacillus subtilis*	Diosgenin	Shaikh et al. (2020)
Gymnema sylvestre	Cell suspension		Gymnemic acid	Chodisetti et al. (2013)

using cultures of *Rhizobium leguminosarum, Bacillus cereus*, *Bacillus aminovorans*, *Agrobacterium tumefaciens*, and *Agrobacterium rhizogenes*. Tanshinone production increased 13.5 times when *Bacillus cereus* and *D. metel* L. hairy root cultures were co-cultured; additions of cellular extract and culture supernatant increased tanshinone synthesis by 2 and 4 times, respectively. *Bacillus circulans* and *Pseudomonas monteilii* culture filtrates were used to induce ginsenoside synthesis in *Panax quinquefolius* cell suspension culture. *P. monteilii* demonstrated an increase of 2.5-fold in total ginsenoside production, while *B. circulans* showed a slight decline. Extended elicitor exposure has been linked to a 20–30% decrease in overall ginsenoside synthesis (Biswas et al. 2016). In root extracts of *Panax ginseng* C.A. Meyer, ginsenosides have been produced using the culture filtrates of gram-positive bacteria such as *Bacillus* sp., *Leuconostoc* sp., *and Lactobacillus plantarum* as well as gram-negative bacteria such as *Azotobacter beijerinckii*, *Bradyrhizobium ganzhouense, Mesorhizobium amorphae*, *Mesorhizobium huakuii*, ginsenoside synthesis increased in all types of bacteria, although gram-negative bacteria produced the most (Le et al. 2018). Synthesis of diosgenin was induced in the cell suspensions of *Helicteres isora* L. using *Escherichia coli and Bacillus subtilis* culture filtrates, the production of diosgenin increased 9.1-fold in *E. coli* (1.5%) and 6.1-fold in *B. subtilis* (2%) (Shaikh et al. 2020).

9.5.2 Fungal

According to Vijaya Kumar et al. (2008), oxidative stress was brought on by use of fungal culture filtrate during in vitro selection. The expression levels of numerous antioxidative defence enzymes, including superoxide dismutase, ascorbate peroxidase, catalase, and glutathione reductase, are indicative of oxidative damage. In the defensive system against the pathogen or the pathogen culture filtrate, antioxidant enzymes are essential. When a pathogen attacks, receptor-induced signalling triggers the production of superoxide radicals (O_2^-), which are extremely deadly and can aid in the destruction of the invasive pathogen, by apoplast or membrane localized oxidases (amine oxidase or NAPDH oxidase). Superoxide dismutase (SOD) quickly converts this O_2^- into hydrogen peroxide (H_2O_2). H_2O_2, as opposed to O_2, has the ability to penetrate into cells and triggers a variety of plant defence mechanisms, such as programmed cell death (PCD) (Apel and Hirt 2004; Dangl and Jones 2001). A study by Gayatri et al. (2005) found that in vitro selection increased the number of resistant plants compared to tissue culture without selection. This was achieved by administering host-specific phytotoxins like fusaric acid and pathogen-produced nonspecific phytotoxins like crude pathogen culture filtrate, deoxynivalenol (DON), and occasionally the pathogen itself to the culture (Gayatri et al. 2005). Herbaceous plants, such as maize, potatoes, alfalfa, barley, and rice, have been successfully selected in vitro for disease resistance lines utilizing pathogen culture filtrate. Peaches, lemons, mangoes, and grapes are among the woody species wherein in vitro selection for disease resistance has been successfully reported (Table 9.5). Two polyembryonic mango cultivars that were selected against the culture of *Colletotrichum gloeosporioides* showed tolerance to the fungus in vitro in their nucellar embryogenic cultures (Jayasankar and Litz 1998). Similarly, Jayasankar et al. (2000) found that the resistance to the pathogen was increased in regenerated plants after screening the proembryogenic mass of grape plant against filtered culture made by *Elsinoe ampelina*, the causal organism of anthracnose disease. Tests for resistance in sugarcane (Sengar et al. 2009), safflower (Vijaya Kumar et al. 2008), cotton (Ganesan and Jayabalan 2006), turmeric (Gayatri et al. 2005), flax (Rutkowska-Krause et al. 2003), tomatoes (Fiume and Fiume 2003), wheat (Yang et al. 1998), and other crops have also demonstrated the utility of these studies as assays. Gentile et al. (1992, 1993) in in vitro culture used deoxynivalenol as a selection agent to successfully screen *Triticum aestivum* L. plantlets tolerant to *Fusarium graminearum* (Yang et al. 1998). Fungal stress, including both endophytic and free-living strains, is the most important and frequently employed biotic stressor in the production of industrial chemicals. Plant–fungus interaction causes hypersensitive reactions that in turn trigger defence mechanisms in the plant, leading to an increase in phytoalexins and more effectively enhance secondary metabolites production (Baldi et al. 2009; Zhai et al. 2017). Typically, the hyphal tip culture yields pure fungal cultures (Salehi et al. 2019). Instead of other non-biological elicitation, *Aspergillus niger* produced more menthol when employed as an elicitor in *Mentha*

Table 9.5 List of in vitro studies on secondary metabolite production using fungal stress

Plant species	Culture type	Fungal elicitor	Targeted metabolite(s)	References
Panax ginseng	Adventitious root	*Alternaria panax*	Ginsenoside	Hao et al. (2020)
Corylus avellana	Cell suspension	*Chaetomium globosum* and *Paraconiothyrium brasiliense*	Paclitaxel	Salehi et al. (2019)
Panax sikkimensis	Cell suspension	*Trichoderma atroviride* and *Trichoderma harzianum*	Ginsenoside and anthocyanin	Biswas et al. (2018)
Artemisia annua	Callus	*Penicillium oxalicum*	Artemisinin	Zheng et al. (2016)
Panax quinquefolius	Cell suspension	*Trichoderma atroviride* and *Trichoderma harzianum*	Ginsenoside	Biswas et al. (2016)
Ocimum basilicum	Hairy root	*Rhizophagus irregularis*	Rosmarinic acid	Srivastava et al. (2016)
Hypericum perforatum	Cell suspension	*Phoma exigua, Fusarium oxysporum,* and *Botrytis cinerea*	Phenylpropanoid and naphtodianthrone	Gadzovska Simic et al. (2015)
Centella asiatica	Axillary shoot	*Colletotrichum lindemuthianum*	Asiaticoside	Prasad et al. (2013)
Abrus precatorius	Cell suspension	*Aspergillus niger* and *Rhizopus stolonifer*	Glycyrrhizin	Karwasara et al. (2011)
Gymnema sylvestre	Cell suspension	*Aspergillus niger*	Gymnemic acids	Devi and Srinivasan (2011)

piperita L. cell cultures (Chakraborty and Chattopadhyay 2008). The production of gymnemic acid in *Aspergillus niger* cultures of *Gymnema sylvestre* (Retz.) increased nine times (Devi and Srinivasan 2011). The cultures of *Abrus precatorius* L. and *Rhizopus stolonifera* showed significant increase in glycyrrhizin production, respectively (Karwasara et al. 2011). One plant that produces dyes that is essential to the economy is *Oldenlandia umbellate* L. to promote its growth, elicitors such as *Trichoderma viride*, *Mucor prayagensis* and *Aspergillus niger* were employed. *P. quinquefolius* L. cell suspensions were treated with the culture filtrates of *Trichoderma harzianum and Trichoderma atroviride* to extract ginsenoside. The largest amount of ginsenoside was produced by *T. atroviride* (3.2 times greater than control) (Biswas et al. 2016). In *Primula sikkimensis* R. N. Banerjee cell suspension cultures, they were also utilized to induce the dual metabolites anthocyanin and ginsenoside (Biswas et al. 2018). Using *Centella asiatica* L. cultures, the effects of fungal elicitors were investigated, *T. harzianum* filtrate was added, and asiaticoside production increased 2.53-fold, while asiaticoside production dropped but biomass increased (Prasad et al. 2013). Endophytic fungi such as *Paraconiothyrium brasiliense* and *Chaetomium globosum*, which were isolated from *Corylus avellane* L. and *Taxus baccata* L., respectively, were added to cell suspensions of *Corylus avellane*

L. The addition of 10% (v/v) *C. globosum* resulted in maximum elicitation, with 4.1 times greater paclitaxel production (Salehi et al. 2019).

9.6 Discussion and Future Perspectives

The utilization of intensive in vitro cellular and tissue culture-based systems provides an exceptional instrument for analyzing the molecular, physiological, and biochemical regulation involved in plant development and stress management. Utilizing in vitro methods, significant strides have been made in the creation and selection of stress-tolerant genotypes in recent years. The generation of transgenic genotypes using various techniques, in vitro selection of plants resistant to various abiotic challenges, and induction of somaclonal variation have shown to be the most effective applied technologies. It is feasible to shorten the time needed to establish commercial crops and other plant species lines that are resistant to disease and abiotic stress thanks to in vitro selection. To ensure the genetic stability of the chosen features in the field, field testing should be conducted on the in vitro variations that were chosen. The advancement of in vitro selection technology, in conjunction with functional genomics and molecular methods, presents a novel prospect for enhancing plant stress tolerance that is pertinent to food production and ecological sustainability. In plant-stress biology, the creation of transgenic plants with the use of biotechnological instruments has gained significance. Previous studies using molecular and genetic methods have demonstrated that the majority of characteristics resistant to abiotic stress are multigenic. Thus, it is necessary to transfer many stress-related genes in order to increase stress tolerance. In recent times, modifying individual transcription factors has had comparable results to manipulating numerous genes. This strategy has shown promise in producing crops that can withstand abiotic stress. With rare exceptions, genetic manipulation techniques necessitate regeneration of transformed plants using in vitro plant tissue culture equipment, which poses a barrier to the technology's wider use. Despite recent increases in the number of species susceptible to *Agrobacterium*-mediated transformation, there are still numerous genotypes for which regeneration techniques are lacking. However, because plant tissue culture allows the cultivation of many plants in a confined area without the influence of other stress factors (biotic or abiotic), it is also a priceless laboratory tool for studying fundamental aspects of growth and development as well as for manipulating these processes. The time needed to create lines of significant plant species that are resistant to abiotic stress and disease will be reduced by in vitro selection. To verify the genetic stability of the chosen characteristic, field testing should be conducted on the in vitro variations that were chosen (Mohan Jain 2001). Explants with high morphogenic potential can be used to circumvent the ability to regenerate which might be lost during in vitro selections, perhaps ensuring effective regeneration. Another barrier to the identification and selection of uncommon mutants displaying tolerance, that is, meiotically inherited, is epigenetic adaptation, which can be avoided by employing one-step or short-term selection (Tal 1994). Despite these issues, many reports

(Winicov 1996; Patnaik and Debata 1997; Singh et al. 2003; Hossain et al. 2007) describe the positive regeneration of plants from specific cell/callus lines demonstrating stability in stress tolerance by the whole plant. Additionally, we know very little about the molecular pathways that plants use to tolerate stress in vitro. The plant microbiome is an intriguing avenue for future investigation. Recent findings indicate that microorganisms can regulate the systemic and local responses of plants to external abiotic stressors (Meena et al. 2017). In vitro co-cultivation of plants along with their endophytic microorganisms under abiotic stressor may be a useful technique to increase the production of plant secondary metabolites that mitigate the detrimental effects of stress on plant growth, although more research is needed to determine the role the microbiome plays in helping plants overcome abiotic stress and how it influences the synthesis of plant secondary metabolites. When combined with functional and molecular genomics, in vitro selection can offer a fresh way to enhance plant stress tolerance that is important for both food production and environmental sustainability.

Conflict of Interest The authors declare no conflict of interest.

References

Aazami MA, Rasouli F, Ebrahimzadeh A (2021) Oxidative damage, antioxidant mechanism and gene expression in tomato responding to salinity stress under in vitro conditions and application of iron and zinc oxide nanoparticles on callus induction and plant regeneration. BMC Plant Biol 21(1):597

Abdel-Ghany HM, Nawar AA, Ibrahim ME, El-Shamarka A, Selim MM, Fahmi AI (2004) Using tissue culture to select for drought tolerance in bread wheat. In: Proceedings of the 4th international crop science congress Brisbane, Australia, vol 26

Abdrabou R, Fergani MA, Azzam CR, Morsi N (2017) Devolopment of some canola genotypes to salinity tolerance using tissue culture technique. Egypt J Agron 39(3):431–448

Abood JK, Lösel DM (2003) Changes in carbohydrate composition of cucumber leaves during the development of powdery mildew infection. Plant Pathol 52(2):256–265

Ajungla L, Patil PP, Barmukh RB, Nikam TD (2009) Influence of biotic and abiotic elicitors on accumulation of hyoscyamine and scopolamine in root cultures of Datura metel L. Indian J Biotechnol 8:317–322

Ali RM, Abbas HM (2003) Response of salt stressed barley seedlings to phenylurea. Plant Soil Environ 49(4):158–162

Anwar N, Kikuchi A, Watanabe KN (2010) Assessment of somaclonal variation for salinity tolerance in sweet potato regenerated plants. Afr J Biotechnol 9(43):7256–7265

Apel K, Hirt H (2004) Reactive oxygen species: metabolism, oxidative stress, and signal transduction. Annu Rev Plant Biol 55:373–399

Arzani A, Mirodjagh SS (1999) Response of durum wheat cultivars to immature embryo culture, callus induction and in vitro salt stress. Plant Cell Tissue Organ Cult 58:67–72

Ashrafzadeh S, Leung DW (2017) Novel potato plants with enhanced cadmium resistance and antioxidative defence generated after in vitro cell line selection. PLoS One 12(10):e0185621

Awad V, Kuvalekar A, Harsulkar A (2014) Microbial elicitation in root cultures of Taverniera cuneifolia (Roth) Arn. for elevated glycyrrhizic acid production. Ind Crop Prod 54:13–16

Baldi A, Srivastava AK, Bisaria VS (2009) Fungal elicitors for enhanced production of secondary metabolites in plant cell suspension cultures. In: Symbiotic fungi: principles and practice. Springer, Berlin, pp 373–380

Barrales-Cureño HJ, Reyes CR, García IV, Valdez LGL, De Jesús AG, Ruíz JAC et al (2019) Alkaloids of pharmacological importance in Catharanthus roseus. In: Alkaloids—their importance in nature and human. IntechOpen, London

Beigmohamadi M, Movafeghi A, Sharafi A, Jafari S, Danafar H (2019) Cell suspension culture of Plumbago europaea L. towards production of plumbagin. Iran J Biotechnol 17(2):46

Berková V, Berka M, Griga M, Kopecká R, Prokopová M, Luklová M et al (2022) Molecular mechanisms underlying flax (Linum usitatissimum L.) tolerance to cadmium: a case study of proteome and metabolome of four different flax genotypes. Plants 11(21):2931

Bertoli A, Lucchesini M, Mensuali-Sodi A, Leonardi M, Doveri S, Magnabosco A, Pistelli L (2013) Aroma characterisation and UV elicitation of purple basil from different plant tissue cultures. Food Chem 141(2):776–787

Bézier A, Lambert B, Baillieul F (2002) Study of defense-related gene expression in grapevine leaves and berries infected with Botrytis cinerea. Eur J Plant Pathol 108:111–120

Biswas J, Chowdhury B, Bhattacharya A, Mandal AB (2002) In vitro screening for increased drought tolerance in rice. In Vitro Cell Dev Biol Plant 38:525–530

Biswas T, Kalra A, Mathur AK, Lal RK, Singh M, Mathur A (2016) Elicitors' influenced differential ginsenoside production and exudation into medium with concurrent Rg3/Rh2 panaxadiol induction in Panax quinquefolius cell suspensions. Appl Microbiol Biotechnol 100:4909–4922

Biswas T, Pandey SS, Maji D, Gupta V, Kalra A, Singh M et al (2018) Enhanced expression of ginsenoside biosynthetic genes and in vitro ginsenoside production in elicited Panax sikkimensis (Ban) cell suspensions. Protoplasma 255:1147–1160

Bouiamrine EH, Diouri M (2012) Response of durum wheat (Triticum durum Desf.) callus culture to osmosis-induced drought stress caused by polyethylene glycol (PEG). Ann Biol Res 3(9):4555–4563

Brodmann D, Schuller A, Ludwig-Müller J, Aeschbacher RA, Wiemken A, Boller T, Wingler A (2002) Induction of trehalase in Arabidopsis plants infected with the trehalose-producing pathogen Plasmodiophora brassicae. Mol Plant Microbe Interact 15(7):693–700

Buitelaar RM, Casário MT, Tramper J (1992) Elicitation of thiophene production by hairy roots of Tagetes patula. Enzym Microb Technol 14(1):2–7

Cai Z, Kastell A, Speiser C, Smetanska I (2013) Enhanced resveratrol production in Vitis vinifera cell suspension cultures by heavy metals without loss of cell viability. Appl Biochem Biotechnol 171:330–340

Caldwell MM, Bornman JF, Ballaré CL, Flint SD, Kulandaivelu G (2007) Terrestrial ecosystems, increased solar ultraviolet radiation, and interactions with other climate change factors. Photobiol Sci 6(3):252–266

Cetin ES (2014) Induction of secondary metabolite production by UV-C radiation in Vitis vinifera L. Öküzgözü callus cultures. Biol Res 47:1–7

Çetin B, Ay Ç (2020) Determination of CD (˙I ˙I) and ZN (˙I ˙I) tolerance and phytoremediation ability of Prunella Vulgaris L. In: Plant tissue culture conditions. Taylor & Francis, Abingdon

Chakraborty A, Chattopadhyay S (2008) Stimulation of menthol production in Mentha piperita cell culture. In Vitro Cell Dev Biol Plant 44:518–524

Chan LK, Koay SS, Boey PL, Bhatt A (2010) Effects of abiotic stress on biomass and anthocyanin production in cell cultures of Melastoma malabathricum. Biol Res 43(1):127–135

Chang X, Alderson PG, Wright CJ (2009) Enhanced UV-B radiation alters basil (Ocimum basilicum L.) growth and stimulates the synthesis of volatile oils. J Hortic For 12:27–31

Chattopadhyay S, Farkya S, Srivastava AK, Bisaria VS (2002) Bioprocess considerations for production of secondary metabolites by plant cell suspension cultures. Biotechnol Bioprocess Eng 7:138–149

Chen Y, Zahavi E, Barak P, Umiel N (1980) Effects of salinity stresses on tobacco: I. The growth of Nicotiana tabacum callus cultures under seawater, NaCl, and mannitol stresses. Z Pflanzenphysiol 98(2):141–153

Chen G, Chunchao L, Hechun Y, Guofeng L (2004) Effect of temperature on growth and artemisinin biosynthesis in hairy root cultures of *Artemisia annua*. Acta Botan Boreali-Occiden Sin 24(10):1828–1831

Cheng L, Han M, Yang LM, Li Y, Sun Z, Zhang T (2018) Changes in the physiological characteristics and baicalin biosynthesis metabolism of Scutellaria baicalensis Georgi under drought stress. Ind Crop Prod 122:473–482

Chodisetti B, Rao K, Gandi S, Giri A (2013) Improved gymnemic acid production in the suspension cultures of Gymnema sylvestre through biotic elicitation. Plant Biotechnol Rep 7:519–525

Chung IM, Rekha K, Rajakumar G, Thiruvengadam M (2018) Influence of silver nanoparticles on the enhancement and transcriptional changes of glucosinolates and phenolic compounds in genetically transformed root cultures of Brassica rapa ssp. rapa. Bioprocess Biosyst Eng 41:1665–1677

Dangl JL, Jones JD (2001) Plant pathogens and integrated defence responses to infection. Nature 411(6839):826–833

Dasgupta MK (1992) Rice sheath blight: the challenge continues. In: Plant diseases of international importance, vol 1. The University of Chicago, Chicago, pp 130–157

Devi CS, Srinivasan VM (2011) In vitro studies on stimulation of gymnemic acid production using fungal elicitor in suspension and bioreactor-based cell cultures of Gymnema sylvestre R.BR. Recent Res Sci Technol 3:101–104

Di Lonardo S, Capuana M, Arnetoli M, Gabbrielli R, Gonnelli C (2011) Exploring the metal phytoremediation potential of three Populus alba L. clones using an in vitro screening. Environ Sci Pollut Res 18:82–90

Di Santo P, Cocozza C, Tognetti R, Palumbo G, Iorio ED, Paura B (2016) A quick screening to assess the phytoextraction potential of cadmium and copper in Quercus pubescens plantlets. iForest Biogeosci For 10(1):93

DiCosmo F, Misawa M (1985) Eliciting secondary metabolism in plant cell cultures. Trends Biotechnol 3(12):318–322

Dogan M (2020) Effect of salt stress on in vitro organogenesis from nodal explant of Limnophila aromatica (Lamk.) Merr. and Bacopa monnieri (L.) Wettst. and their physio-morphological and biochemical responses. Physiol Mol Biol Plants 26:803–816

Dolzhenko Y, Bertea CM, Occhipinti A, Bossi S, Maffei ME (2010) UV-B modulates the interplay between terpenoids and flavonoids in peppermint (Mentha × piperita L.). J Photochem Photobiol B Biol 100(2):67–75

Duncan RR, Waskom RM, Nabors MW (1995) In vitro screening and field evaluation of tissue-culture-regenerated sorghum (Sorghum bicolor (L.) Moench) for soil stress tolerance. In: The methodology of plant genetic manipulation: criteria for decision making: proceedings of the Eucarpia plant genetic manipulation section meeting held at Cork, Ireland from September 11 to September 14, 1994. Springer, Dordrecht, pp 373–380

Edreva A, Yordanov I, Kardjieva R, Hadjiiska E, Gesheva E (1995) Expression of phenylamides in abiotic stress conditions. Bulg J Plant Physiol 21(2–3):15–23

Edreva A, Yordanov I, Kardjieva R, Gesheva E (1998) Heat shock responses of bean plants: involvement of free radicals, antioxidants and free radical/active oxygen scavenging systems. Biol Plant 41(2):185–191

Edreva A, Velikova V, Tsonev T, Dagnon S, Gürel A, Aktaş L, Gesheva E (2008) Stress-protective role of secondary metabolites: diversity of functions and mechanisms. Gen Appl Plant Physiol 34(1–2):67–78

Errabii T, Gandonou CB, Essalmani H, Abrini J, Idaomar M, Skali-Senhaji N (2006) Growth, proline and ion accumulation in sugarcane callus cultures under drought-induced osmotic stress and its subsequent relief. Afr J Biotechnol 5(16)

Fiume F, Fiume G (2003) Use of culture filtrates of Pyrenochaeta lycopersici in tests for selecting tolerant varieties of tomato. J Plant Pathol:131–133

Gadzovska Simic S, Tusevski O, Maury S, Delaunay A, Lainé E, Joseph C, Hagège D (2015) Polysaccharide elicitors enhance phenylpropanoid and naphtodianthrone production in cell suspension cultures of Hypericum perforatum. Plant Cell Tissue Organ Cult 122:649–663

Gai Z, Wang YU, Ding Y, Qian W, Qiu C, Xie H et al (2020) Exogenous abscisic acid induces the lipid and flavonoid metabolism of tea plants under drought stress. Sci Rep 10(1):12275

Gandi S, Rao K, Chodisetti B, Giri A (2012) Elicitation of andrographolide in the suspension cultures of Andrographis paniculata. Appl Biochem Biotechnol 168:1729–1738

Gandonou CB, Errabii T, Abrini J, Idaomar M, Senhaji NS (2006) Selection of callus cultures of sugarcane (Saccharum sp.) tolerant to NaCl and their response to salt stress. Plant Cell Tissue Organ Cult 87:9–16
Ganesan M, Jayabalan N (2006) Isolation of disease-tolerant cotton (Gossypium hirsutum L. cv. SVPR 2) plants by screening somatic embryos with fungal culture filtrate. Plant Cell Tissue Organ Cult 87:273–284
Gangopadhyay G, Basu S, Gupta S (1997) In vitro selection and physiological characterization of NaCl-and mannitol-adapted callus lines in Brassica juncea. Plant Cell Tissue Organ Cult 50:161–169
Gao X, Yang D, Cao D, Ao M, Sui X, Wang Q et al (2010) In vitro micropropagation of Freesia hybrida and the assessment of genetic and epigenetic stability in regenerated plantlets. J Plant Growth Regul 29:257–267
García-Sánchez S, Bernales I, Cristobal S (2015) Early response to nanoparticles in the Arabidopsis transcriptome compromises plant defence and root-hair development through salicylic acid signalling. BMC Genomics 16(1):1–17
Gasic K, Korban SS (2007) Expression of Arabidopsis phytochelatin synthase in Indian mustard (Brassica juncea) plants enhances tolerance for Cd and Zn. Planta 225:1277–1285
Gayatri MC, Darshini VR, Kavyashree R (2005) Selection of turmeric callus tolerant to culture filtrate of Pythium Graminicolum and regeneration of plants. Plant Cell Tissue Organ Cult 83:33–40
Gelmesa D, Dechassa N, Mohammed W, Gebre E, Monneveux P, Bündig C, Winkelmann T (2017) In vitro screening of potato genotypes for osmotic stress tolerance. Open Agric 2:308–316
Gentile A, Tribulato E, Continella G, Vardi A (1992) Differential responses of citrus calli and protoplasts to culture filtrate and toxin of Phoma tracheiphila. Theor Appl Genet 83:759–764
Gentile A, Tribulato E, Deng ZN, Galun E, Fluhr R, Vardi A (1993) Nucellar callus of 'Femminello' lemon, selected for tolerance to Phoma tracheiphila toxin, shows enhanced release of chitinase and glucanase into the culture medium. Theor Appl Genet 86:527–532
Ghasemzadeh A, Jaafar HZ, Rahmat A (2010) Elevated carbon dioxide increases contents of flavonoids and phenolic compounds, and antioxidant activities in Malaysian young ginger (Zingiber officinale Roscoe.) varieties. Molecules 15(11):7907–7922
Ghosh S, Watson A, Gonzalez-Navarro OE, Ramirez-Gonzalez RH, Yanes L, Mendoza-Suárez M et al (2018) Speed breeding in growth chambers and glasshouses for crop breeding and model plant research. Nat Protoc 13(12):2944–2963
Gil M, Pontin M, Berli F, Bottini R, Piccoli P (2012) Metabolism of terpenes in the response of grape (Vitis vinifera L.) leaf tissues to UV-B radiation. Phytochemistry 77:89–98
Gopal J, Iwama K (2007) In vitro screening of potato against water-stress mediated through sorbitol and polyethylene glycol. Plant Cell Rep 26:693–700
Griffith M, Yaish MW (2004) Antifreeze proteins in overwintering plants: a tale of two activities. Trends Plant Sci 9(8):399–405
Guo WL, Wu R, Zhang YF, Liu XM, Wang HY, Gong L et al (2007) Tissue culture-induced locus-specific alteration in DNA methylation and its correlation with genetic variation in Codonopsis lanceolata Benth. et Hook. f. Plant Cell Rep 26:1297–1307
Han C, Liu Q, Yang Y (2009) Short-term effects of experimental warming and enhanced ultraviolet-B radiation on photosynthesis and antioxidant defense of Picea asperata seedlings. J Plant Growth Regul 58:153–162
Hannachi S, Werbrouck S, Bahrini I, Abdelgadir A, Siddiqui HA, Van Labeke MC (2021) Obtaining salt stress-tolerant eggplant somaclonal variants from in vitro selection. Plants 10(11):2539
Hao YJ, An XL, Sun HD, Piao XC, Gao R, Lian ML (2020) Ginsenoside synthesis of adventitious roots in Panax ginseng is promoted by fungal suspension homogenate of Alternaria panax and regulated by several signaling molecules. Ind Crop Prod 150:112414
Hasegawa PM, Bressan RA, Zhu JK, Bohnert HJ (2000) Plant cellular and molecular responses to high salinity. Annu Rev Plant Biol 51(1):463–499

Hassan NM, Serag MS, El-Feky FM (2004) Changes in nitrogen content and protein profiles following in vitro selection of NaCl resistant mung bean and tomato. Acta Physiol Plant 26:165–175

Heil M, Bostock RM (2002) Induced systemic resistance (ISR) against pathogens in the context of induced plant defences. Ann Bot 89(5):503–512

Hossain Z, Mandal AKA, Datta SK, Biswas AK (2007) Development of NaCl-tolerant line in Chrysanthemum morifolium Ramat. through shoot organogenesis of selected callus line. J Biotechnol 129(4):658–667

Hou Q, Ufer G, Bartels D (2016) Lipid signalling in plant responses to abiotic stress. Plant Cell Environ 39(5):1029–1048

Hu Y, Yu W, Song L, Du XH, Ma X, Liu Y et al (2016) Effects of light on production of camptothecin and expression of key enzyme genes in seedlings of Camptotheca acuminate Decne. Acta Physiol Plant 38:1–9

Huang W, He XY, Liu CB, Li DW (2010) Effects of elevated carbon dioxide and ozone on foliar flavonoids of Ginkgo biloba. Adv Mater Res 113:165–169

Ibrahim MH, Jaafar HZ (2012) Impact of elevated carbon dioxide on primary, secondary metabolites and antioxidant responses of Eleais guineensis Jacq. (oil palm) seedlings. Molecules 17(5):5195–5211

Ingram J, Bartels D (1996) The molecular basis of dehydration tolerance in plants. Annu Rev Plant Biol 47(1):377–403

Ioannidis D, Bonner L, Johnson CB (2002) UV-B is required for normal development of oil glands in Ocimum basilicum L. (sweet basil). Ann Bot 90(4):453–460

Isah T (2019) Stress and defense responses in plant secondary metabolites production. Biol Res 52:39

Isah T, Umar S (2020) Influencing in vitro clonal propagation of Chonemorpha fragrans (moon) Alston by culture media strength, plant growth regulators, carbon source and photo periodic incubation. J For Res 31:27–43

Jan N, Qazi HA, Ramzan S, John R (2018) Developing stress-tolerant plants through in vitro tissue culture: family Brassicaceae. In: Biotechnologies crop improvement, volume 1: cellular approaches. Springer, Berlin, pp 327–372

Jansen MA, Hectors K, O'Brien NM, Guisez Y, Potters G (2008) Plant stress and human health: do human consumers benefit from UV-B acclimated crops? Plant Sci 175(4):449–458

Jasim B, Thomas R, Mathew J, Radhakrishnan EK (2017) Plant growth and diosgenin enhancement effect of silver nanoparticles in fenugreek (Trigonella foenum-graecum L.). Saudi Pharm J 25(3):443–447

Jayasankar S, Litz RE (1998) Characterization of embryogenic mango cultures selected for resistance to Colletotrichum gloeosporioides culture filtrate and phytotoxin. Theor Appl Genet 96:823–831

Jayasankar S, Li Z, Gray DJ (2000) In-vitro selection of Vitis vinifera 'Chardonnay' with Elsinoe ampelina culture filtrate is accompanied by fungal resistance and enhanced secretion of chitinase. Planta 211:200–208

Jochum GM, Mudge KW, Thomas RB (2007) Elevated temperatures increase leaf senescence and root secondary metabolite concentrations in the understory herb Panax quinquefolius (Araliaceae). Am J Bot 94(5):819–826

Jorge TF, Tohge T, Wendenburg R, Ramalho JC, Lidon FC, Ribeiro-Barros AI et al (2019) Salt-stress secondary metabolite signatures involved in the ability of Casuarina glauca to mitigate oxidative stress. Environ Exp Bot 166:103808

Joshi N (2015) Influence of light and temperature on secondary metabolite accumulation in callus cultures of Helicteres isora L. IOSR J Environ Sci Toxicol Food Technol 1(1):2319–2399

Jung HY, Kang SM, Kang YM, Kang MJ, Yun DJ, Bahk JD et al (2003) Enhanced production of scopolamine by bacterial elicitors in adventitious hairy root cultures of Scopolia parviflora. Enzym Microb Technol 33(7):987–990

Kacem NS, Delporte F, Muhovski Y, Djekoun A, Watillon B (2017) In vitro screening of durum wheat against water-stress mediated through polyethylene glycol. J Genet Eng Biotechnol 15(1):239–247
Karunaratne S, Santha S, Kovoor A (1991) An in vitro assay for drought-tolerant coconut germplasm. Euphytica 53:25–30
Karuppusamy S (2009) A review on trends in production of secondary metabolites from higher plants by in vitro tissue, organ and cell cultures. J Med Plants Res 3(13):1222–1239
Karwasara VS, Tomar P, Dixit VK (2011) Infuence of fungal elicitation on glycyrrhizin production in transformed cell cultures of Abrus precatorius Linn. Pharmacog Mag 7:307–314
Kayum MA, Kim HT, Nath UK, Park JI, Kho KH, Cho YG, Nou IS (2016) Research on biotic and abiotic stress related genes exploration and prediction in Brassica rapa and B. oleracea: a review. Plant Breed Biotechnol 4(2):135–144
Kiełkowska A, Grzebelus E, Lis-Krzyścin A, Maćkowska K (2019) Application of the salt stress to the protoplast cultures of the carrot (Daucus carota L.) and evaluation of the response of regenerants to soil salinity. Plant Cell Tissue Organ Cult 137:379–395
Koc NK, Baş B, Koc M, Küsek M (2009) Investigations of in vitro selection for salt tolerant lines in sour orange (Citrus aurantium L.). Biotechnology 8(1):155–159
Kumari M, Chandra S (2016) Secondary metabolite production in transformed cultures: Stevioside glycosides production from Stevia rebaudiana hairy root cultures. In: Kumari M, Chandra S (eds) Transgenesis and secondary metabolism: part of the series reference series in phytochemistry. Springer International Publishing, Berlin, pp 1–19
Kumari R, Agrawal SB, Sarkar A (2009) Evaluation of changes in oil cells and composition of essential oil in lemongrass (Cymbopogon citratus (DC) Stapf.) due to supplemental ultraviolet-B irradiation. Curr Sci 97(8):1137–1142
Larkin PJ, Scowcroft WR (1981) Somaclonal variation—a novel source of variability from cell cultures for plant improvement. Theor Appl Genet 60:197–214
Le KC, Im WT, Paek KY, Park SY (2018) Biotic elicitation of ginsenoside metabolism of mutant adventitious root culture in Panax ginseng. Appl Microbiol Biotechnol 102:1687–1697
Li X, Yu X, Wang N, Feng Q, Dong Z, Liu L et al (2007) Genetic and epigenetic instabilities induced by tissue culture in wild barley (Hordeum brevisubulatum (Trin.) Link). Plant Cell Tissue Organ Cult 90:153–168
Li Y, Kong D, Fu Y, Sussman MR, Wu H (2020) The effect of developmental and environmental factors on secondary metabolites in medicinal plants. Plant Physiol Biochem 148:80–89
Maffei M, Scannerini S (2000) UV-B effect on photomorphogenesis and essential oil composition in peppermint (Mentha piperita L.). J Essent Oil Res 12(5):523–529
Mahmood I, Razzaq A, Ashraf M, Ahmad Hafiz I, Kaleem S, Qayyum A, Ahmad M (2012) In vitro selection of tissue culture induced somaclonal variants of wheat for drought tolerance. J Agric Res 50(2):03681157
Malan C, Greyling MM, Gressel J (1990) Correlation between CuZn superoxide dismutase and glutathione reductase, and environmental and xenobiotic stress tolerance in maize inbreds. Plant Sci 69(2):157–166
Marzilli M, Di Santo P, Palumbo G, Maiuro L, Paura B, Tognetti R, Cocozza C (2018) Cd and Cu accumulation, translocation and tolerance in Populus alba clone (Villafranca) in autotrophic in vitro screening. Environ Sci Pollut Res 25:10058–10068
Meena KK, Sorty AM, Bitla UM, Choudhary K, Gupta P, Pareek A et al (2017) Abiotic stress responses and microbe-mediated mitigation in plants: the omics strategies. Front Plant Sci 8:172
Melchers LS, Stuiver MH (2000) Novel genes for disease resistance breeding. Curr Opin Plant Biol 3:147–152
Miehe-Steier A, Roscher C, Reichelt M, Gershenzon J, Unsicker SB (2015) Light and nutrient dependent responses in secondary metabolites of Plantago lanceolata offspring are due to phenotypic plasticity in experimental grasslands. PLoS One 10(9):e0136073
Mishra T (2016) Climate change and production of secondary metabolites in medicinal plants: a review. Int J Herb Med 4(4):27–30

Mishra N, Tripathi MK, Tiwari S, Tripathi N, Sapre S, Ahuja A, Tiwari S (2021) Cell suspension culture and in vitro screening for drought tolerance in soybean using poly-ethylene glycol. Plants 10(3):517
Mohamed MH, Harris PJC, Henderson J (2000) In vitro selection and characterisation of a drought tolerant clone of Tagetes minuta. Plant Sci 159(2):213–222
Mohammed AE (2015) Green synthesis, antimicrobial and cytotoxic effects of silver nanoparticles mediated by Eucalyptus camaldulensis leaf extract. Asian Pac J Trop Biomed 5(5):382–386
Mohan Jain S (2001) Tissue culture-derived variation in crop improvement. Euphytica 118:153–166
Moharrami F, Hosseini B, Sharafi A, Farjaminezhad M (2017) Enhanced production of hyoscyamine and scopolamine from genetically transformed root culture of Hyoscyamus reticulatus L. elicited by iron oxide nanoparticles. In Vitro Cell Dev Biol Plant 53:104–111
Morison JIL, Lawlor DW (1999) Interactions between increasing CO2 concentration and temperature on plant growth. Plant Cell Environ 22(6):659–682
Munns R, Tester M (2008) Mechanisms of salinity tolerance. Annu Rev Plant Biol 59:651–681
Mustafa G, Akhtar MS, Abdullah R (2019) Global concern for salinity on various agro-ecosystems. In: Salt stress, microbes, and plant interactions: causes and solution, vol 1. Springer, Berlin, pp 1–19
Naik PM, Al-Khayri JM (2016) Abiotic and biotic elicitors-role in secondary metabolites production through in vitro culture of medicinal plants. In: Abiotic and biotic stress in plants—recent advances and future perspectives. IntechOpen, Rijeka, pp 247–277
Naik PS, Widholm JM (1993) Comparison of tissue culture and whole plant responses to salinity in potato. Plant Cell Tissue Organ Cult 33:273–280
Nandagopal K, Halder M, Dash B, Nayak S, Jha S (2018) Biotechnological approaches for production of anti-cancerous compounds resveratrol, podophyllotoxin and zerumbone. Curr Med Chem 25(36):4693–4717
Narayan MS, Thimmaraju R, Bhagyalakshmi N (2005) Interplay of growth regulators during solid-state and liquid-state batch cultivation of anthocyanin producing cell line of Daucus carota. Process Biochem 40(1):351–358
Narayani M, Srivastava S (2017) Elicitation: a stimulation of stress in in vitro plant cell/tissue cultures for enhancement of secondary metabolite production. Phytochem Rev 16:1227–1252
Neumann KH, Kumar A, Imani J (2009) Plant cell and tissue culture: a tool in biotechnology, vol 12. Springer, Berlin
Park SY, Lee WY, Park Y, Ahn JK (2006) Effects of nitrogen source and bacterial elicitor on isoflavone accumulation in root cultures of Albizzia kalkora (Roxb.) Prain. J Integr Plant Biol 48(9):1108–1114
Patnaik J, Debata BK (1997) Regeneration of plantlets from NaCl tolerant callus lines of Cymbopogon martinii (Roxb.) wats. Plant Sci 128(1):67–74
Pedroso RCN, Branquinho NAA, Hara AC, Costa AC, Silva FG, Pimenta LP et al (2017) Impact of light quality on flavonoid production and growth of Hyptis marrubioides seedlings cultivated in vitro. Rev Bras 27:466–470
Pérez-Ilzarbe J, Hernández T, Estrella I, Vendrell M (1997) Cold storage of apples (cv. Granny Smith) and changes in phenolic compounds. Zeitschrift für Lebensmitteluntersuchung und-Forschung A 204:52–55
Pinto E, Sigaud-kutner TC, Leitao MA, Okamoto OK, Morse D, Colepicolo P (2003) Heavy metal–induced oxidative stress in algae. J Phycol 39(6):1008–1018
Podda A, Pollastri S, Bartolini P, Pisuttu C, Pellegrini E, Nali C et al (2019) Drought stress modulates secondary metabolites in Brassica oleracea L. convar. acephala (DC) Alef, var. sabellica L. J Sci Food Agric 99(12):5533–5540
Prasad A, Mathur A, Kalra A, Gupta MM, Lal RK, Mathur AK (2013) Fungal elicitor-mediated enhancement in growth and asiaticoside content of Centella asiatica L. shoot cultures. J Plant Growth Regul 69:265–273
Purohit M, Srivastava S, Srivastava PS (1998) Stress tolerant plants through tissue culture. In: Plant tissue culture and molecular biology: application and prospects. Narosa Publishing House, New Delhi, pp 554–578

Purushotham MG, Patil V, Chandrashekra Raddey P, Prasad TG, Vajranabhaiah SN (1998) Development of in vitro PEG stress tolerant cell lines in two groundnut (Arachis hypogaea L.) genotypes. Indian J Plant Physiol 3(1):49–51

Rahimi S, Hasanloo T (2016) The effect of temperature and pH on biomass and bioactive compound production in Silybum marianum hairy root cultures. Res J Pharmacogn 3(2):53–59

Razavizadeh R, Komatsu S (2018) Changes in essential oil and physiological parameters of callus and seedlings of Carum copticum L. under in vitro drought stress. J Food Meas Charact 12:1581–1592

Rout GR, Sahoo S (2007) In vitro selection and plant regeneration of copper-tolerant plants from leaf explants of Nicotiana tabacum L. cv. 'Xanthi'. Plant Breed 126(4):403–409

Rout GR, Samantaray S, Das P (1999) In vitro selection and biochemical characterisation of zinc and manganese adapted callus lines in brassica spp. Plant Sci 146:89–100

Roy B, Mandal AB (2005) Towards development of Al-toxicity tolerant lines in indica rice by exploiting somaclonal variation. Euphytica 145(3):221–227

Rozema J, Flowers T (2008) Crops for a salinized world. Science 322(5907):1478–1480

Rutkowska-Krause I, Mankowska G, Lukaszewicz M, Szopa J (2003) Regeneration of flax (Linum usitatissimum L.) plants from anther culture and somatic tissue with increased resistance to Fusarium oxysporum. Plant Cell Rep 22:110–116

Ruttkay-Nedecky B, Krystofova O, Nejdl L, Adam V (2017) Nanoparticles based on essential metals and their phytotoxicity. J Nanobiotechnol 15(1):1–19

Salehi M, Moieni A, Safaie N, Farhadi S (2019) Elicitors derived from endophytic fungi Chaetomium globosum and Paraconiothyrium brasiliense enhance paclitaxel production in Corylus avellana cell suspension culture. Plant Cell Tissue Organ Cult 136:161–171

Samantaray S, Rout GR, Das P (1999) In vitro selection and regeneration of zinc tolerant calli from Setaria italica L. Plant Sci 143(2):201–209

Samantaray S, Rout GR, Das P (2001) Induction, selection and characterization of Cr and Ni-tolerant cell lines of Echinochloa colona (L.) link in vitro. J Plant Physiol 158(10):1281–1290

Saravanan S, Karthi S (2014) Effect of elevated CO2 on growth and biochemical changes in Catharanthus roseus—an valuable medicinal herb. World J Pharm Pharm Sci 3(11):411–422

Savary S, Ficke A, Aubertot JN, Hollier C (2012) Crop losses due to diseases and their implications for global food production losses and food security. Food Secur 4(4):519–537

Sengar AS, Thind KS, Kumar B, Pallavi M, Gosal SS (2009) In vitro selection at cellular level for red rot resistance in sugarcane (Saccharum sp.). J Plant Growth Regul 58:201–209

Shaikh S, Shriram V, Khare T, Kumar V (2020) Biotic elicitors enhance diosgenin production in Helicteres isora L. suspension cultures via up-regulation of CAS and HMGR genes. Physiol Mol Biol Plants 26:593–604

Shakeran Z, Keyhanfar M, Asghari G, Ghanadian M (2015) Improvement of atropine production by different biotic and abiotic elicitors in hairy root cultures of Datura metel. Turk J Biol 39(1):111–118

Sharma D, Shree B, Kumar S, Kumar V, Sharma S, Sharma S (2022) Stress induced production of plant secondary metabolites in vegetables: functional approach for designing next generation super foods. Plant Physiol Biochem 192:252

Shohael AM, Ali MB, Yu KW, Hahn EJ, Paek KY (2006) Effect of temperature on secondary metabolites production and antioxidant enzyme activities in Eleutherococcus senticosus somatic embryos. Plant Cell Tissue Organ Cult 85:219–228

Singh M, Jaiswal U, Jaiswal VS (2003) In vitro selection of NaCl-tolerant callus lines and regeneration of plantlets in a bamboo (Dendrocalamus strictus Nees). In Vitro Cell Dev Biol. Plant 39:229–233

Singh A, Agrawal M, Marshall FM (2010) The role of organic vs. inorganic fertilizers in reducing phytoavailability of heavy metals in a wastewater-irrigated area. Ecol Eng 36(12):1733–1740

Sri Andrijany V, Indrayanto G, Adi Soehono L (1998) Simultaneous effect of calcium, magnesium, copper and cobalt ions on sapogenin steroids content in callus cultures of Agave amaniensis. Plant Cell Tissue Organ Cult 55:103–108

Srivastava S, Conlan XA, Cahill DM, Adholeya A (2016) Rhizophagus irregularis as an elicitor of rosmarinic acid and antioxidant production by transformed roots of Ocimum basilicum in an in vitro co-culture system. Mycorrhiza 26:919–930

Stuhlfauth T, Fock HP (1990) Effect of whole season CO2 enrichment on the cultivation of a medicinal plant, Digitalis lanata. J Agron Crop Sci 164(3):168–173

Sudha G, Ravishankar GA (2003) Elicitation of anthocyanin production in callus cultures of Daucus carota and involvement of calcium channel modulators. Curr Sci:775–779

Sunaryo W, Widoretno W, Nurhasanah N, Sudarsono S (2016) Drought tolerance selection of soybean lines generated from somatic embryogenesis using osmotic stress simulation of polyethylene glycol (PEG). Nusantara Biosci 8(1)

Takshak S, Agrawal SB (2019) Defense potential of secondary metabolites in medicinal plants under UV-B stress. J Photochem Photobiol B Biol 193:51–88

Tal M (1994) In vitro selection for salt tolerance in crop plants: theoretical and practical considerations. In Vitro Plant 30:175–180

Tauqeer HM, Ali S, Rizwan M, Ali Q, Saeed R, Iftikhar U et al (2016) Phytoremediation of heavy metals by Alternanthera bettzickiana: growth and physiological response. Ecotoxicol Environ Saf 126:138–146

Tepe B, Sokmen A (2007) Production and optimisation of rosmarinic acid by Satureja hortensis L. callus cultures. Nat Prod Res 21(13):1133–1144

Thimmaraju R, Bhagyalakshmi N, Narayan MS, Ravishankar GA (2003) Kinetics of pigment release from hairy root cultures of Beta vulgaris under the influence of pH, sonication, temperature and oxygen stress. Process Biochem 38(7):1069–1076

Tippmann HF, Schlüter U, Collinge DB (2006) Common themes in biotic and abiotic stress signalling in plants. Floric Ornam Biotechnol:52–67

Tripathi D, Rai KK, Pandey-Rai S (2021) Impact of green synthesized WcAgNPs on in-vitro plant regeneration and withanolides production by inducing key biosynthetic genes in Withania coagulans. Plant Cell Rep 40:283–299

Tusevski O, Petreska Stanoeva J, Stefova M, Gadzovska Simic S (2015) Agrobacterium enhances xanthone production in Hypericum perforatum cell suspensions. J Plant Growth Regul 76:199–210

Vajrabhaya M, Thanapaisal T, Vajrabhaya T (1989) Development of salt tolerant lines of KDML and LPT rice cultivars through tissue culture. Plant Cell Rep 8(7):411–414

Vanisree M, Lee CY, Lo SF, Nalawade SM, Lin CY, Tsay HS (2004) Studies on the production of some important secondary metabolites from medicinal plants by plant tissue cultures. Bot Bull Acad Sin 45(1):1–22

Verslues PE, Ober ES, Sharp RE (1998) Root growth and oxygen relations at low water potentials. Impact of oxygen availability in polyethylene glycol solutions. Plant Physiol 116(4):1403–1412

Vijaya Kumar J, Ranjitha Kumari BD, Sujatha G, Castano E (2008) Production of plants resistant to Alternaria carthami via organogenesis and somatic embryogenesis of safflower cv. NARI-6 treated with fungal culture filtrates. Plant Cell Tissue Organ Cult 93:85–96

Vishwakarma K, Upadhyay N, Kumar N, Yadav G, Singh J, Mishra RK et al (2017) Abscisic acid signaling and abiotic stress tolerance in plants: a review on current knowledge and future prospects. Front Plant Sci 8:161

Wang W, Vinocur B, Altman A (2003) Plant responses to drought, salinity and extreme temperatures: towards genetic engineering for stress tolerance. Planta 218:1–14

Wang P, Lombi E, Zhao FJ, Kopittke PM (2016) Nanotechnology: a new opportunity in plant sciences. Trends Plant Sci 21(8):699–712

Wani SH, Sofi PA, Gosal SS, Singh NB (2010) In vitro screening of rice (Oryza sativa L.) callus for drought tolerance. Commun Biometry Crop Sci 5(2):108–115

Wilczańska-Barska A, Królicka A, Głód D, Majdan M, Kawiak A, Krauze-Baranowska M (2012) Enhanced accumulation of secondary metabolites in hairy root cultures of Scutellaria lateriflora following elicitation. Biotechnol Lett 34:1757–1763

Winicov I (1996) Characterization of rice (Oryza sativa L.) plants regenerated from salt-tolerant cell lines. Plant Sci 113(1):105–111

Woodward AJ, Bennett IJ (2005) The effect of salt stress and abscisic acid on proline production, chlorophyll content and growth of in vitro propagated shoots of Eucalyptus camaldulensis. Plant Cell Tissue Organ Cult 82:189–200

Yadav RK, Sangwan RS, Sabir F, Srivastava AK, Sangwan NS (2014) Effect of prolonged water stress on specialized secondary metabolites, peltate glandular trichomes, and pathway gene expression in Artemisia annua L. Plant Physiol Biochem 74:70–83

Yamaguchi T, Blumwald E (2005) Developing salt-tolerant crop plants: challenges and opportunities. Trends Plant Sci 10(12):615–620

Yang Y, Guo Y (2018) Elucidating the molecular mechanisms mediating plant salt-stress responses. New Phytol 217(2):523–539

Yang Z, Yang X, Huang D (1998) Studies on somaclonal variants for resistance to scab in bread wheat (Triticum aestivum L.) through in vitro selection for tolerance to deoxynivalenol. Euphytica 101(2):213–219

Yu KW, Murthy HN, Hahn EJ, Paek KY (2005) Ginsenoside production by hairy root cultures of Panax ginseng: influence of temperature and light quality. Biochem Eng J 23(1):53–56

Zandalinas SI, Mittler R, Balfagón D, Arbona V, Gómez-Cadenas A (2018) Plant adaptations to the combination of drought and high temperatures. Physiol Plant 162(1):2–12

Zhai X, Jia M, Chen L, Zheng CJ, Rahman K, Han T, Qin LP (2017) The regulatory mechanism of fungal elicitor-induced secondary metabolite biosynthesis in medical plants. Crit Rev Microbiol 43(2):238–261

Zhang W, Seki M, Furusaki S (1997) Effect of temperature and its shift on growth and anthocyanin production in suspension cultures of strawberry cells. Plant Sci 127(2):207–214

Zhang B, Zheng LP, Yi Li W, Wen Wang J (2013) Stimulation of artemisinin production in Artemisia annua hairy roots by Ag-SiO2 core-shell nanoparticles. Curr Nanosci 9(3):363–370

Zhao J, Davis LC, Verpoorte R (2005) Elicitor signal transduction leading to production of plant secondary metabolites. Biotechnol Adv 23(4):283–333

Zheng LP, Tian H, Yuan YF, Wang JW (2016) The influence of endophytic Penicillium oxalicum B4 on growth and artemisinin biosynthesis of in vitro propagated plantlets of Artemisia annua L. J Plant Growth Regul 80:93–102

Zhong JJ, Yoshida T (1993) Effects of temperature on cell growth and anthocyanin production in suspension cultures of Perilla frutescens. J Ferment Bioeng 76(6):530–531

Zhu JK (2002) Salt and drought stress signal transduction in plants. Annu Rev Plant Biol 53(1):247–273

Zhu JK (2016) Abiotic stress signaling and responses in plants. Cell 167(2):313–324

Photosynthetic Organisms in Extreme Environments

10

Sanjay Sharma, Tameshwar Jaiswal, and Satya Shila Singh

Abstract

Various extreme habitats such as thermal springs, deserts, hypersaline environment, polar regions, and acidic pools are present in the Earth, and these habitats harbor extremophilic oxygenic photosynthetic organisms. Photosynthetic organisms play key roles in sustaining the ecosystems of extremophilic habitats by acting as primary producers. Phototrophs thriving in these environments may face multiple extreme factors at the same time, e.g., in acidic thermal springs, both acidic conditions and high temperature are available. Phototrophs inhabiting extreme habitats exhibit a wide range of diversity, i.e., from prokaryotic archaea, bacteria, and cyanobacteria to eukaryotic algae and plants, which have been discussed in the present chapter. To sustain the photosynthetic processes at extreme habitats, these phototrophs possess some special adaptation strategies including morphological alterations, modulations in light absorption machinery, changes in carbon fixation enzymes, and structural and compositional modifications in chloroplasts and cyanobacteria. While the morphological changes, especially in the leaf structure and canopy architecture, govern the survival of higher plants in extreme conditions, adaptations in the composition and structure of phycobiliproteins form the basis of cyanobacterial existence in such conditions. Other notable adaptive responses of cyanobacteria thriving in the extreme environments include substitution of some amino acids in the proteins of photosystem II (PSII) complex and plastocyanin. Sequence analysis of D1 proteins of PSII from different groups of extremophilic phototrophs reflected the occurrence of many isoforms of this protein. Nevertheless, such analysis of plastocyanin synthesized by extremophilic phototrophs has not been performed yet. In this chapter, phylogenetic analysis of plastocyanin showed that extremophilic plants

S. Sharma · T. Jaiswal · S. S. Singh (✉)
Laboratory of Cyanobacterial Systematics and Stress Biology, Department of Botany, Institute of Science, Banaras Hindu University, Varanasi, India

A. K. Mishra (ed.), *Stress Biology in Photosynthetic Organisms*,
https://doi.org/10.1007/978-981-97-1883-2_10

and eukaryotic algae share a similar type of plastocyanin, while the plastocyanin of extremophilic cyanobacteria made separate clade. Besides, the chapter also encompasses elaborate discussion on the pellucid role of compatible solutes and alkanes for unabating photosynthesis in extremophilic phototrophs. Although many research articles, related to photosynthesis behavior of organisms in extreme conditions, have been found in the literature, comprehensive compilation of these studies is still needed to further elucidate the existing lacunae. So, this chapter inclusively has diversity of extremophilic photosynthetic organisms and their adaptation features which make their photosynthesis sustainable in extreme conditions.

Keywords

Photosynthetic organisms · Extreme habitats · Photosystem II · Plastocyanin · Thermophiles

10.1 Introduction

Cyanobacteria (prokaryotes), some protists, algae, bryophytes, and tracheophytes are oxygenic photosynthetic organisms or photoautotrophs. They evolve oxygen during photosynthesis and use light and water as the source of energy and electrons, respectively (Barber 1998, 2017; Haumann and Junge 1999). These photoautotrophs are found in various habitats on the Earth's biosphere including extreme habitats, where they called extremophilic photoautotrophs. The successful establishment of such organisms in extreme habitats is achieved by means of some adaptative alterations in their structures and metabolisms. Photosynthesis, the foundation for plant survival (Zhang et al. 2023), is one of the important metabolism processes of photoautotrophs which exhibits high sensitivity to environmental stresses (heat, cold, salt, light intensity, etc.) (Apte et al. 1998; Hu et al. 2020). So, adaptations in photosynthetic machinery including morphological, structural, physiological, and biochemical adaptations are necessary for survival in extreme environment. In a particular habitat, one or more extreme abiotic factors can be present, e.g., deep sea has light scarcity, high pressure, and high salinity. Level of adaptations in photosynthesis-related structures is varying among different organisms, i.e., from prokaryotes to eukaryotes, unicellular to multicellular organisms, and thallophytes to higher plants. In cyanobacteria, which is a prokaryote and either unicellular or filamentous, morphological adaptations present as cell envelope modifications, thylakoid membrane orientations, chromatic adaptations, and modulations in the shape and size of cells. Mucilaginous layers also protect the cyanobacteria from stress conditions. However, in higher plants, complex morphological structures are present to combat photosynthesis from stress in extreme conditions, e.g., presence of wax layers and tuft of small filaments on the leaf, and variations in the shape and size of leaves and in the canopy structure of plants. Microorganisms like cyanobacteria and microalgae also make biofilms to enhance the survivability. Biochemical changes are generally similar in all forms of photosynthetic organisms. Accumulation of compatible solutes and their role in the protection of photosynthetic structures have been seen in the cells of most photoautotrophs. Basic structure of photosynthetic machinery and their way of functioning in all oxygenic photoautotrophs are universally conserved except some differences (Mulo et al. 2012), and these small variations make these organisms able to thrive in extreme conditions (Fig. 10.1). Photosynthesis-related proteins of thermophiles and psychrophiles have variations in their activity and flexibility at different temperatures. Ferredoxins from the cold-adapted cyanobacteria showed some conserved sequences which might be responsible for their stability and optimum activity at cold temperatures. Phycocyanin, a phycobiliprotein in cyanobacteria, isolated from thermophilic cyanobacteria also showed thermostable nature and are stable up to temperature 50 °C even in the acidic medium (Liang et al. 2018). Moreover, thermostable water splitting complex has also been found in the thermophilic cyanobacteria. Plastocyanin of thermophilic cyanobacteria also showed thermostable nature. Phylogenetic analysis of plastocyanin has also been presented in this chapter. Some extremophilic photoautotrophs, apart from the modifications in the structures and metabolisms, also show peculiarities in their behavior. Mosses in cold environment of Antarctic have the photosynthetic temperature optima (above 20 °C) above the mean temperature of their habitat which suggested that they are not psychrophiles but their survival and growth at such low temperature suggested their specific strategies of photosynthesis. These mosses have the ability to maximize their photosynthesis in the summer, for short period, and minimize carbon loss during respiration under cold conditions which resulted into survival in such extreme cold environment (Perera-Castro et al. 2020; Zhang et al. 2023). Different physiological and biochemical adaptations have been

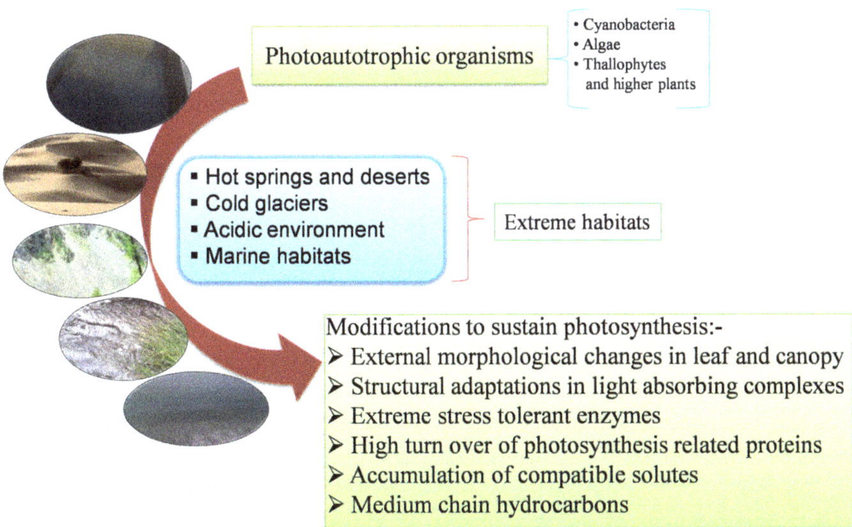

Fig. 10.1 Schematic representation of modifications found in photoautotrophic organisms found in various extreme habitats

found in the photoautotrophs which minimize the photosynthetic damages due to extreme environmental factors.

Although there are many research articles and reviews on the effects of stress conditions on the photosynthesis found in the scientific literature, a detailed discussion on photosynthesis-related adaptations in extremophilic photoautotrophs is still missing. This chapter has the comprehensive explanation of strategies adapted by photoautotrophs to sustain photosynthetic reactions in the extreme habitats, and a brief scenario of photoautotrophic organisms' diversity has also been presented.

10.2 Diversity of Photosynthetic Organisms in Extreme Habitats

There are various extreme habitats on the Earth which have single or combined environmental factors in utmost state such as extremely high and low temperature, high salinity, pressure, acidic conditions, etc. Photosynthetic organisms are found in various extreme habitats. Those organisms which found in thermal springs are known as thermophiles or thermotolerant, while low-temperature inhabitants are called psychrophiles. Similarly, organisms found in high salinity, acidity, and at high pressure are called halophiles, acidophiles, and piezophiles, respectively. When the organisms live in extreme habitats, which have more than one extreme factor, they are known as polyextremophiles, e.g., deep-sea inhabitants deal with the very high or very low temperature, high salinity, and high pressure simultaneously (Fang et al. 2010; Poli et al. 2017). Extreme habitats harbor diverse range of photosynthetic organisms, i.e., from cyanobacteria to bryophytes and higher plants (Seckbach and Oren 2007; Szabados et al. 2011; Zhang et al. 2023). Some habitats have photosynthetic organisms from all the groups, i.e., from cyanobacteria to plants, e.g., marine habitats, while some have representatives from little type of photoautotrophic organisms such as extreme acidic environments. Here, in the following sections, diversity of photosynthetic organisms in different extreme conditions has been discussed.

10.2.1 Hot Springs and Deserts

Thermal springs or hot springs are the most ancient ecosystem on the Earth and are still inhabitant by vast microbiota (Reysenbach and Shock 2002). Thermal springs mainly harbor the prokaryotic microbes. Factors such as temperature and level of H_2S and sodium affect the microbial diversity in these habitats. Along with the archaebacteria, oxygenic photoautotrophic microbes such as cyanobacteria are also the inhabitants of thermal springs. Cyanobacteria which are found in hot springs may be thermophiles or thermotolerants. Worldwide, all forms of cyanobacteria, i.e., from unicellular (*Thermosynechococcus* spp.) to filamentous branched (*Mastigocladus ambikapurensis*) forms, are found in the thermal springs (Jaiswal et al. 2021, 2023; Prondzinsky et al. 2021;). Thermal springs are either extremely acidic or alkaline. Due to these two extreme factors (temperature and pH) of thermal springs, the organisms thrive in these habitats may be polyextremophiles, e.g., *Chlorogloeopsis* and *Cyanidioschyzon*. However, cyanobacteria were not found from the extreme acidic thermal springs having pH below 4, but eukaryotic algae are common in the acidic thermal springs, as shown in Table 10.1. Although prokaryotic organisms are the main inhabitants of many hot springs, diversity of diatoms was also reported from some thermal springs. Recently, 49, 46, and 23 species of diatoms were recorded from the hot springs of Kamchatka, northern Thailand, and southwest region of Saudi Arabia, respectively (Abdelwahab and Amin 2017; Fazlutdinova et al. 2020; Pumas et al. 2018). Members of chlorophyceae and rhodophyceae were also reported from the thermal springs (Nikulina 2010; Jonker et al. 2013; Bhakta et al. 2016; van Etten et al. 2023), as shown in Table 10.1. Bryophytes and tracheophytes are also present in soils heated by the thermal spring, around these springs, and face the high temperature (above 40 °C) (Stout and Al-Niemi 2002; Convey and Smith 2006). In parallel with the thermal springs, one more extreme habitat is the hot deserts, which possess a common stress factor with hot springs which is high temperature. But apart from the high temperature, the photosynthetic organisms in hot deserts also deal with scarcity of water, due to low precipitation and high sunlight resulted into acquiring of some adaptations by these organisms to thrive in such conditions. A wide range of photosynthetic organisms are found in the deserts. Many cyanobacterial strains have also been reported from various deserts of the world such as *Leptolyngbya*, *Chroococcidiopsis*, *Desertifilum*, *Phormidium*, *Trichocoleus*, *Schizothrix*, *Plectonema* etc. (Wynn-Williams 2000; Dadheech et al. 2012; Etemadi-Khah et al. 2017; Lacap-Bugler et al. 2017). Recently, an extreme light tolerant cyanobacterial strain, *Cyanobacterium aponinum* 0216, has been isolated from Sonoran Desert (Dobson et al. 2021). Eukaryotic algae are also present diversely in the desert habitats. Some common desert algae are *Chlorosarcinopsis*, *Hormidium*, *Trebouxia*, *Radiosphaera*, *Chlamydomonas*, *Chlorococcum*, *Desertella*, *Eremochloris*, *Xerochlorella* etc. (Friedmann et al. 1967; Fučíková et al. 2014). Other than thallophytes, the bryophytes and tracheophytes show a vast diversity in deserts (McCleary 1959; Harten and Eickmeier 1987; Stark and Whittemore 2000). *Welwitschia*, a gymnosperm, is known as the most wonderful and extraordinary plant and is found in the African deserts (Hooker 1863; Jürgens et al. 2021; Wan et al. 2021). Similarly, a range of angiosperms are found in the deserts which have special adaptations to protect themselves from hot arid and high solar radiations, which have been discussed in the present chapter in the latter sections.

10.2.2 Cold Glaciers

Around more than 80% of the Earth's biosphere is perennially cold and majorly represented by Antarctic, Arctic, and Alpine regions (Margesin and Häggblom 2007; Zakhia et al. 2008; Goel et al. 2022). These cold regions harbor the wide range of photosynthetic organisms. Cyanobacteria are found in these habitats where

Table 10.1 Diversity of cyanobacteria, algae, and plants in various extreme habitats

Habitat	Phototrophic organisms		
	Cyanobacteria	Algae	Bryophytes and tracheophyta
Thermal springs	*Synechococcus*, *Spirulina*, *Calothrix*, *Fischerella*, *Leptolyngbya*, *Thermosynechococcus*, *Microcoleus*, *Phormidium*, *Mastigocladus*, *Chroococcus* (Peary and Castenholz 1964; Stewart 1970; Brock 1978; Hayashi et al. 1994; Papke et al. 2003; Klatt et al. 2011; Siangbood and Ramanujam 2011)	*Chlorella*, *Spirogyra*, *Cosmarium* (Bhakta et al. 2016); *Oedogonium* (Nikulina 2010); *Coelastrum* (Jonker et al. 2013); *Cyanidioschyzon*, *Galdieria* (van Etten et al. 2023)	*Gnaphalium*, *Heterotheca*, *Euphorbia*, *Rumex*, *Juncus*, *Agrostis*, *Dichanthelium*, *Panicum* (Stout and Al-Niemi 2002); *Campylopus* (Convey and Smith 2006)
Cold	*Synechococcus*, *Leptolyngbya*, *Gloeocapsa*, *Chroococcus*, *Chlorogloea*, *Aphanocapsa*, *Merismopedia*, *Nostoc*, *Dichothrix*, *Tolypothrix*, *Pseudanabaena*, *Phormidium*, *Oscillatoria* (Jungblut et al. 2010)	*Chlamydomonas* (Cvetkovska et al. 2017); *Chloromonas*, *Raphidonema* (Segawa et al. 2018)	*Bryum*, *Ceratodon*, *Chorisodontium*, *Polytrichastrum*, *Sanionia*, *Schistidium* (Perera-Castro et al. 2020); *Pohlia* (Zhang et al. 2023)
Acidic	*Nostoc*, *Aphanothece*, *Oscillatoria*, *Chroococcus*, *Gloeothece* (Dominic and Madhusoodanan 1999); *Chlorogloeopsis* (Fecteau et al. 2022)	*Pseudococcomyxa*, *Stichococcus*, *Viridiella*, *Cyanidioschyzon*, *Galdieria* (Gross 2000); *Chlamydomonas*, *Chlorella*, *Dunaliella*, *Cyanidium*, *Zygnemopsis*, *Klebsormidium* (Aguilera 2013); *Coccomyxa* (Hrdinka et al. 2013); *Ulothrix*, *Microthamnion*, *Hormidium*, *Zygogonium*, *Stichococcus* (Hargreaves et al. 1975)	*Typha*, *Juncus*, *Drepanocladus*, *Dicranella* (Hargreaves et al. 1975); *Sphagnum* (Bragina et al. 2012)
Marine	*Synechococcus*, *Prochlorococcus* (Chisholm et al. 1992); *Vermifilum*, *Leptochromothrix*, *Ophiophagus* (Berthold et al. 2021); *Tigrinifilum*, *Sirenicapillaria*, *Capilliphycus* (Berthold et al. 2022); *Acaryochloris* (Miyashita et al. 2003); *Prochloron*, *Katagnymene*, *Trichodesmium*, *Richelia*, *Lyngbya*, *Aphanocapsa*, *Synechocystis*, *Borzia*, *Calothrix*, *Oscillatoria*, *Phormidium*, *Rivularia*, *Microcoleus*, *Scytonema*, *Schizothrix* (Hoffman 1999); *Spirulina*, *Plectonema* (Silambarasan et al. 2012)	Members of rhodophyceae, phaeophyceae, ulvaceae, and siphonales. Some members of other chlorophycean orders e.g. *Dunaliella* (Preetha et al. 2012)	Many angiosperms

sometimes temperature reaches below −20 °C, which is a metabolically lowest thermal limit to prokaryotic life on Earth (Clarke et al. 2013). Recently, proteins and rRNA sequences of 53 cyanobacteria genomes from these cold habitats were collected, and phylogenetic analysis has been done. The 20 lineages (clades) of cyanobacterial strains have been predicted, which likely to be ancestors of cold-tolerant cyanobacteria or most recent common ancestors (MRCAs). These lineages consisted of one clade of *Gloeobacter*, four sub clades of *Pseudanabaena*, two clades of *Aphanocapsa*, one subclade of *Chamaesiphon subglobosus*, two clades of *Leptolyngbya*, two clades of *Phormidesmis*/*Plectolyngbya*, one clade of *Microcoleus*, one clade of *Wilmottia murrayi* (*Phormidium murray*), one clade of *Nodularia* and one of *Tolypothrix*, two clades of *Chroococcidiopsis,* and two clades of *Nostoc* (Chrismas et al. 2015). Cyanobacteria are the principal primary producers in Antarctic ecosystems where they act as a source of fixed carbon, nitrogen, and energy for the other living organisms and their survivability in such extreme conditions is may be due to their biofilm producing nature (Singh and Elster 2007; Komárek and Komárek 2010; Bowman et al. 2012; Convey 2013; Chong et al. 2015; Chrismas et al. 2015; Prieto-Barajas et al. 2018; Velichko et al. 2021). Many cyanobacterial strains have been isolated from different places of Antarctic. Some of the most studied genera from Antarctic are *Aphanocapsa*, *Chamaesiphon*, *Gloeocapsa*, *Leptolyngbya*, *Microcoleus*, *Phormidium*, and *Nostoc* (Velichko et al. 2021). Some species of cyanobacteria such as *Phormidium priestleyi* Fritsch, *Leptolyngbya frigida* (Fritsch) Anagnostidis and Komarek, *Leptolyngbya antarctica* (West and West) Anagnostidis and Komarek, *Phormidium pseudopriestleyi* Anagnostidis and Komarek, *Geitlerinema deflexum* (West & West) Anagnostidis, and *Oscillatoria subproboscidea* West & West were found to be endemic in Antarctic (Komárek 1999; Taton et al. 2006a, b). However, according to recent studies on Arctic microbiota, three species viz. *Phormidium priestleyi* Fritsch, *Leptolyngbya frigida* (Fritsch) Anagnostidis and Komarek, and *Leptolyngbya antarctica* (West and West) Anagnostidis and Komarek showed >99% similarity to the sequences from the Canadian High Arctic (Jungblut et al. 2010). Arctic, in comparison to Antarctic, was less studied for cyanobacterial diversity. Cyanobacterial strains of the order Chroococcales, Oscillatoriales, and Nostocales were found from the microbial mats of Arctic in association with the chlorophytes and chromophytes. Although both Arctic and Antarctic have the same environment, i.e., cold, but differences in the cyanobacterial diversity are seen between the Arctic and Antarctic, e.g., *Gloeobacter* was not reported from the Antarctic region, while it is found in the Arctic. The one reason behind this distribution is geographical location of these habitats. Antarctic cyanobacteria are comparatively less connected to the temperate latitudes, due to the presence of Southern Ocean, than the Arctic cyanobacteria (Jungblut et al. 2010). Algal species such as *Chlamydomonas* spp., *Chloromonas*, *Asterochloris*, *Chloroidium*, *Chlorella*, *Raphidonema*, *Himantothallus*, *Plocamium*, *Phaeurus*, and *Kallymenia* were reported from the polar environments (Cvetkovska et al. 2017; Segawa et al. 2018; Ko et al. 2022). Various species of bryophytes such as *Schistidium antarctici*, *Sanionia uncinata*, *Polytrichastrum alpinum*, *Ceratodon purpureus*, *Chorisodontium aciphyllum,* and *Bryum pseudotriquetrum* were also collected from Antarctic region, as shown in Table 10.1 (Perera-Castro et al. 2020). Angiosperms such as *Saxifraga oppositifolia* were also reported from the cold habitats, but they generally need 5 °C or above temperature for growth and reproduction and no any higher plant found to grow below 0 °C till now (Körner 2011). Although one-fourth of the higher plant families have plants found in alpine regions, the predominant families are Asteraceae, Brassicaceae, Caryophyllaceae, Cyperaceae, Poaceae, Ranunculaceae, and Rosaceae (Körner 2003).

10.2.3 Acidic Environment

The acidic habitats are sparsely distributed on the Earth. Naturally, these habitats are results of volcanic activities, while anthropogenic activities like mining of coal and metals are also responsible for the creation of such habitats (Sharma et al. 2012). Characteristics of acidic habitat are low dissolved organic carbon and high heavy metal concentration due to low pH (Gross 2000). These habitats harbor many autotrophic and heterotrophic microorganisms. Autotrophic organisms may be either chemolithoautotrophs or photoautotrophs. Most of the photoautotrophic organisms are eukaryotic algae such as *Chlamydomonas acidophila*, *Klebsormidium fluitans*, *Ulothrix zonata,* and *Galdieria sulphuraria*. Generally, cyanobacteria are found at neutral or slightly alkaline pH, but some cyanobacteria were also reported from the lower pH ranges, as shown in Table 10.1 (Dominic and Madhusoodanan 1999). Two mosses *Drepanocladus* and *Dicranella* were reported from the sites having pH 2.6–3.0 (Hargreaves et al. 1975). Some aquatic angiosperms were also found in acid water. Submerged species *Juncus bulbosus* (pH 2.5–3.0) and *Sparganium emersum* (pH 3.1), and emergent species *Carex rostrata*, *Eriophorum angustifolium*, *Juncus* spp., *Phragmites australis*, *Schoenoplectus lacustris*, and *Typha* spp. (pH 2.1–3.0) are some angiosperms grow in acidic habitats (Fyson 2000).

10.2.4 Marine Habitat

The major marine habitats are the oceans. Around 71% surface of the earth is covered by the oceans, and around 16% of the total named living species are reported from these marine habitats (Costello et al. 2010, 2013; Costello and Chaudhary 2017). Seawater contains nearly 3.5% salt which is very high and creates osmotic stress. Floral biodiversity of the marine habitats ranges from prokaryotes to tracheophyta. Three phytoplankton, i.e., diatoms, dinoflagellates, and coccolithophorids, are the main primary producers in oceans (Falkowski et al. 2004; Knoll et al. 2007). Apart from these three phytoplankton, cyanobacteria also play an important role in the primary production of oceans. Some common genera of cyanobacteria found in oceans are *Prochlorococcus*, *Synechococcus*, *Trichodesmium*, *Richelia*, *Microcystis*, *Cylindrospermopsis*, *Aphanizomenon,* and *Anabaena* (Hoffman 1999; Khalifa et al. 2021). Members of green algae such as *Dunaliella*, *Enteromorpha*, *Ulva*, and most of the members of phaeophyceae and rhodophyceae are exclusively marine (Preetha et al. 2012; Rashmi and Mona 2021). Till March 2017, around 312 species of tracheophytes have also been reported from oceans which are found as mangroves and seagrasses (Costello and Chaudhary 2017). However, no bryophyte has been reported from marine habitats till now.

10.3 Protective Measures for Photosynthesis at Extreme Habitats

10.3.1 Modifications in External Morphology

Photoautotrophs that live in extreme habitats adapted some special morphological features in their body organization. These adaptations help photoautotrophs in maintaining their photosynthetic reaction in extreme conditions. In higher plants, the adaptations are seen in the growth pattern, canopy structure, leaf morphology, and anatomy. Plants found in extreme environments tend to maintain their leaf microenvironment optimum using some strategies, e.g., rosette and cushion leaf canopy are frequently found in the many plants of polar regions. Some plants of water-deficient regions have hygroscopic plant parts such as leaf and scales which directly absorb the water from fog or dew to complement the water absorb through roots, e.g., *Trianthema hereroensis*, fairy duster, desert marigold, and twist flower (Von Hase et al. 2006; Gurera and Bhushan 2020; Simberloff and Rejmánek 2011).

Plants found in dry warm deserts have reduced leaf size to avoid a high transpiration rate. Leaves of these plants have small surface area, so less heat is absorbed from the sunlight and they remain cooler resulted in reduction in photorespiration (Gibson 1998; Mohanta et al. 2023). To cool the leaves, some modifications are found on the leaf surface of desert plants such as coating of reflective waxes and thick layer of trichomes, which reflect the direct solar radiations (Serna and Martin 2006; Liu et al. 2021; Mohanta et al. 2023). Plants found in these climates have leaves with highly dissected margins along with a higher perimeter/area ratio of leaves leading to dissipation of heat occur efficiently (Vogel 1970, 2009; Sisó et al. 2001; Peguero-Pina et al. 2020). Although there is high scarcity of water in deserts, some plants use transpiration to keep the leaf surface cool during lethal high temperatures but this only occurs for the short time. Heat dissipation of leaves occurs at the expense of high stomatal conductance resulting in high water loss. However, this type of mechanism is generally found in plants which have deep root system, so they can compensate water loss from high transpiration rates (Smith 1978; Aparecido et al. 2020; Peguero-Pina et al. 2020). Apart from these modifications, sometimes leaves are modified into entirely different structures such as phyllodes, photosynthetic organs with reduced blade and broadened petiole, which reduce the water loss due to their higher stomatal control. When water deficit occurs, the stomata on phyllodes respond more quickly in comparison to the true leaves (Brodribb and Hill 1993; Dong and He 2017). Unlike plants, lower forms such as algae and cyanobacteria of extreme habitats do not have complex morphological adaptations but some modifications at cellular level are present. Among unicellular and colonial forms, latter are more stress resistant because surface cells of the colony protect the inner cells (Xiong et al. 1996; Kvíderová et al. 2018). Predictions of cold adaptations in polar cyanobacteria through exopolysaccharide secretion around cells are also present in the literature (Chrismas et al. 2016).

10.3.2 Modifications in Photosynthetic Machinery

10.3.2.1 Phycobilisomes

Phycobilisomes (PBS) are found on the thylakoid membranes of cyanobacteria and in the red algae (Sidler 1994; Chen et al. 2022). These are the supramolecular complexes of water soluble phycobiliproteins or chromoproteins attached to the outer surface of thylakoid membrane above the photosystems. These act as energy funnels as they capture the light energy and transfer it to the reaction center chlorophyll (David et al. 2014; Dagnino-Leone et al. 2022). A phycobiliprotein contains alpha and beta polypeptide subunits in 1:1 ratio, and each polypeptide subunit may contain 1–3 chromophores (Stadnichuk and Kusnetsov 2023). There are mainly four types of phycobiliproteins: phycocyanin, allophycocyanin, phycoerythrin, and phycoerythrocyanin. During phycobilisome assembly, allophycocyanin, in hexameric structure, always occupy the core or central portion and rest of the phycobiliproteins form hexameric rods attached to each other through linker polypeptides and radiate from the core part, and this whole phycobilisome is attached to the reaction center via anchor polypeptides (Liu et al. 2005). Phycocyanin and allophycocyanin are two most prevalent phycobilins found in most of the cyanobacterial species, while phycoerythrin and phycoerythrocyanin are also found in cyanobacteria (Stadnichuk et al. 2015). The phycoerythrin was not observed in thermophiles which grow above 60 °C (Edwards et al. 1997). Crystal structures of phycobilins have been determined from different mesophilic and thermophilic cyanobacteria (Schirmer et al. 1986; Brejc et al. 1995; Adir et al. 2001; Nield et al. 2003) as well as their structural comparison has also been done for elucidation of thermophilic, psychrophilic, and halophilic behavior (Chen and Berns 1978; Adir et al. 2001). It was found that thermophilic cyanobacteria possess thermostable phycobilins which are not denatured up to 70 °C, while denaturing temperature for phycobilins of mesophilic cyanobacteria was much below 70 °C. When the phycobilisomes are energetically coupled and firmly attached to reaction center, high photochemical quenching occurs which leads to high photosynthetic efficiency. Fluorescent emission measurement is one of the methods to know the stability of phycobilisomes. When phycobilisomes and PSII are energetically coupled, the excitation at 605 nm gives fluorescence emission at 680 nm. With increase in temperature, the continuous increase in fluorescence occurs which shows the uncoupling of PSII from phycobilisome and finally a sudden decrease in fluorescence suggests the denaturation of phycobilisome. The *Synechococcus* sp. OH28 with highly thermostable phycobilisome showed fluorescent maxima at 80 °C which was higher than other strains (Pedersen and Miller 2017). Thermophilicity may not be the result of single significant factor (Adir et al. 2001), but it could involve more than one factor responsible for protein thermostability like amino acid composition, weak bond interactions, ionic pairs, secondary structures, and cavities in final structure of protein subunits (Szilágyi and Závodszky 2000). Several conserved cysteine residues have been found in the phycobiliprotein of thermophiles. Moreover, the alanine amino acid was found as most abundant amino acid in the phycobiline proteins of these organisms, and concentration of serine and glycine was lesser, while glutamate and lysine were found more in thermophiles than their mesophilic counterparts (Tang et al. 2023). Role of aggregations of monomeric phycobilin subunits or involvement of some unknown cellular factor would be there in the thermostability of these phycobilins in the cell because at higher temperature, i.e., 70 °C, monomers of phycobilins of thermophiles also get degenerate in same way as monomers of mesophilic origin. This monomeric instability at higher temperature could be understood on the basis that variations may occur at some zones in thermophiles such as chromophore binding, monomer formation (alpha and beta), trimer formation, hexamer formation, rod formation, and phycobilisome assembly by rod to rod, and rod to core attachment, and these may increase the stability of phycobilisomes (Adir et al. 2001). Phycocyanin is the most studied phycobilin, and most of the researches used this pigment for their studies on thermophilic aspects of phycobilins. Recently, phycocyanin isolated from *Thermosynechococcus* PCC6715 was found to be one of the most stable phycocyanin as it remained stable in acidic medium at 50 °C for 4 h incubation (Liang et al. 2018). Thermostability of phycobilisomes of cold-adapted cyanobacteria is lower than the mesophilic cyanobacteria. Decrease in the number of salt bridges in the phycobilin part is mainly responsible for the lower thermostability of these proteins (Luo et al. 2017). Although decrease in thermostability and increase in the flexibility are found in the cold-adapted enzymes, here phycobilins do not act as enzymes so flexibility is not the limiting factor. In marine inhabitant *Synechococcus* strains, the rod and core regions of phycobilisomes are evolved separately. The core region having allophycocyanin evolved along with the core genome, while rods evolved likely through horizontal gene transfer according to the existing light conditions in the environment (Six et al. 2007). *Phyllophora antarctica*, a benthic marine red alga, has a novel type of r-phycoerythrin, i.e., R-phycoerythrine IV, which has lowest ratio of phycoerythrobilin to phycourobilin. These features make these algae able to absorb light in blue region, which is generally found in the light-deprived deep-sea regions (MacColl et al. 1996; Samsonoff and MacColl 2001).

10.3.2.2 Photosystem II

Photosystem II (PSII) is a large protein-pigment complex found in all oxygenic photosynthetic organisms. In plants and eukaryotic algae, these complexes are found in chloroplast, while in cyanobacteria, these are found on the thylakoid membranes of the cells. However, these are absent in the mature heterocytes of cyanobacteria. In all oxygenic photoautotrophs, the basic structure of the PS II is the same except some minor variations and these small variations are responsible for the adaptation of these organisms in different habitats. First time, Rhee et al. (1998) reported the 8 Å structure of CP47 RC PSII and studied the 2D crystals of PSII. The

PSII particles with active water oxidation activity have been crystallized from a thermophilic cyanobacteria *Thermosynechococcus elongatus* (Zouni et al. 2000) and their X-ray structure at 3.8 Å resolution was reported (Zouni et al. 2001). Subsequently, many studies related to crystallization and structure elucidation of PSII from different organisms using different techniques have been reported (Loll et al. 2005; Guskov et al. 2009; Umena et al. 2011; Ago et al. 2016; Su et al. 2017). Most of the reports were related to the PSII of thermophilic cyanobacteria due to their thermostable nature. Generally, in cyanobacteria, the active PSII is found in dimeric form and each monomer consisted of 20 protein subunits: 17 are intrinsic and 3 are extrinsic (Nickelsen and Rengstl 2013). Besides, each monomer also has 35 chlorophyll a molecules, 20–25 lipids, 12 β-carotenes, 2–3 plastoquinones, 2 pheophytins, 2 hemes, 1 non-heme iron, water oxidizing complex (Mn_4O_5Ca), 4 Ca^{2+} ions, and 3 Cl^- ions (Nixon et al. 2010; Heinz et al. 2016). The overall structure of PS II of plants is similar to the cyanobacterial PS II. Moreover, the types of cofactors are also same in PS II of both cyanobacteria and plants, but numbers may vary such as in C_2S_2 PSII-LHCII super-complex of spinach, there are 105 chlorophylls, 28 β-carotenes and xanthophylls, 1 heme, 1 Mn_4CaO_5 cluster, and 1 plastoquinone (Gao et al. 2018). During photosynthesis, antenna complexes harvest the light and transferred it to reaction center chlorophyll which leads to excitation of an electron from reaction center and get transferred to pheophytin (primary electron acceptor), and further this electron pass through transport chain. Due to loss of an electron, a positive charge is developed on the other side of PSII resulted into charge separation which leads to oxidation of water with the help of oxygen evolving complex (OEC) and electrons generated from water get transfer to PSII reaction center so PSII again comes in its original state. Pathway between the oxidation of water to electron transfer to pheophytin molecules involved many proteins and co-factors of PSII which make this crucial process efficient in all conditions, even at extreme conditions. PSII is continuously degraded due to stresses such as light, heat etc. and consecutive ROS generation but simultaneous repair also persists. During this degradation and repair cycle, whole PSII is not degraded, rather D1 subunit of PSII is sacrificed itself to protect other subunits and proteolytic degradation of same is occurred by an enzyme FtsH (Aro et al. 1993; Lindahl et al. 2000; Tikkanen et al. 2014). The photosynthetic organisms found in thermal springs such as thermophilic cyanobacteria pointed toward the possibility of thermostable PSII and an efficient repair system because PSII of the photosynthetic machinery is known as the most temperature-sensitive site and its stability may become a limiting factor for their growth (Inoue et al. 2000).

Since most of PSII particles are crystallized from thermophilic cyanobacteria and their comparative studies with mesophiles are less, so interpretation of thermophilic properties of PSII is still not fully elucidated so for understanding the mechanism of thermophilicity in PSII compilation of physiological, biochemical, and genetic studies with these crystal structures is utmost needed. At high temperature, two sites in PSII: (1) D1 and nearby proteins of oxygen evolving complex (OEC) and (2) Thylakoid membrane lipids nearby PSII, are most probably heat-affected sites which may either decrease photosynthesis or fully inhibit it and for thriving at higher temperature of hot springs, thermophiles have to combat these sites from heat stress. D1 protein is one of the core proteins of PSII, while D1 and D2 proteins carry most of the cofactors, such as four chlorophylls, pheophytin, QA, QB, and involve in the electron transport reactions in the PSII (Sugiura et al. 2010). PsbA gene family is responsible for encoding the D1 protein and in cyanobacteria 1–5 copies of *psbA* gene found which encodes 1–3 unique D1 isoform per species differentially express according to environmental conditions (Vinyard et al. 2013). However, some species may have more D1 types such as *Leptolyngbya* sp. PCC 7375 has seven, and *Leptolyngbya* sp. Heron Island J has eight D1 proteins (Cardona et al. 2015). In general, D1 protein is continuously damaged and repaired to protect the photosystem from photoinhibition and when repair process is lag behind the damage, then photoinhibition is become apparent (Nishiyama et al. 2006), and recent studies suggested that moderate stresses other than light negatively affect the repair process of D1 rather than direct inactivation of PSII (Nishiyama and Murata 2014). At high temperature, cyanobacteria feels oxidative stress due to the generation of ROS especially at photosynthetic sites where $^1O_2$ is formed at acceptor side of PSII due to lipid peroxidation and H_2O_2 is generated at donor side due to incomplete oxidation of H_2O which later convert into OH· through Fenton reaction (Pospíšil 2016). D1 protein may either directly inactivate due to change in conformation by heat (Yamamoto et al. 2008) or inactivation may occur due to ROS, generated due to heat stress, which may directly inhibit D1 protein (Kale et al. 2017) or inhibit its translation (Nishiyama et al. 2004). As D1 protein bears many cofactors, necessary for electron transport, its conformational change due to heat may become unsuitable for cofactors binding so during heat stress, QB plastoquinone was destabilized and released from its binding pocket. This unbinding might be due to conformational change in the DE loop of D1 protein; however, it may occur due to chemical modification in the loop (Nishiyama and Murata 2014). ROS globally downregulates the protein synthesis in cell (Allakhverdiev and Murata 2004), so also to D1 protein, and oxidation of elongation factors by ROS is one of the possible reasons for this (Kojima et al. 2007). There are two studies on cyanobacteria which revealed about the elongation factor oxidation and D1 protein translation inhibition. First suggested due to oxidation via ROS, an intramolecular disulfide bond formation between two specific cysteine residues, cys105 and cys242, of elongation factor EF-G occurs which inhibit the translocation of peptidyl tRNA during translation of *psbA* mRNA (Kojima et al. 2009). Second study suggest that elongation factor EF-Tu is also involved in the oxidative inhibition of translation where cysteine residue, Cys 82, on this factor oxidized by ROS and consecutively intermolecular disulfide bond and sulfenic acid is formed (Yutthanasirikul et al. 2016). Recent in vivo study on EG-Tu in *Synechocystis* reported that after replacing cys82 by the serine increases de novo protein synthesis, including D1, as well as decreases the photoinhibition, suggesting an increase in repair process of PSII (Jimbo et al. 2018). Therefore, to protect PSII from damage due to heat and its related stresses like ROS, thermophilic cyanobacteria used several plausible mechanisms related to D1 protein. The preliminary way is the differential expression of D1 variants in different environmental conditions. Many researches related to *psbA* gene variants in *Synechococcus* (Clarke et al. 1993; Kulkarni and Golden 1994) and *Thermosynechococcus* (Loll et al. 2008; Kós et al. 2008; Hughes et al. 2010; Sugiura et al. 2010; Sugiura and Boussac 2014) were found in literature where different types as well as copies of this gene were found which expressed in different conditions. In *Thermosynechococcus elongatus,* three D1 forms, *psbA1*, *psbA2*, *psbA3*, were found where *psbA1* is dominant and constitutively expressed, while *psbA2* and *psbA3* were expressed under conditions like microaerobic and high light, respectively (Fig. 10.2). In *Synechocystis* 6803 genome also, three D1 isoforms, *psbA1*, *psbA2*, *psbA3*, were found but here *psbA2* and *psbA3* were dominant forms and similar in sequence and *psbA1* expressed under microaerobic conditions. In the same way, many cyanobacterial strains contain different number of D1 variants and these all D1 isoforms from different strains classified into four groups, D1m, D1:1, D1:2, D1′ (Mulo et al. 2009), on the basis of expression pattern in different stress and acclimated conditions. In D1′ two conserved modifications at position, Phe158Leu and Thr286Al are located in the binding pockets of chlorophyll

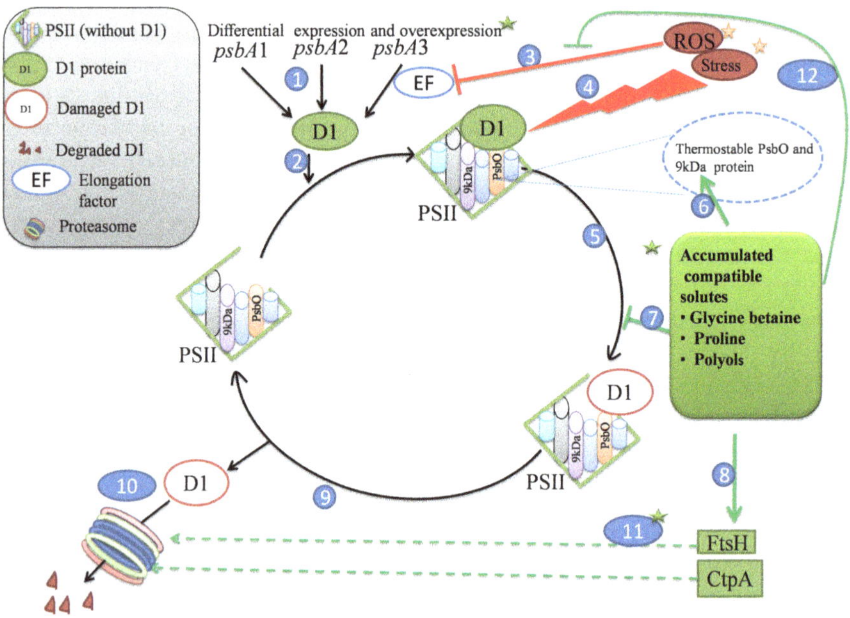

Fig. 10.2 Strategies used by photosynthetic organisms to protect photosystem II (PSII) from different stresses prevail in extreme habitats. Here red stars showing the stress conditions in extreme habitats, while green stars denoting the modifications to sustain in these conditions. Step 1 in the small blue circle showing the translation of D1 protein, step 2 is the assembly of D1 in new PSII, step 3 is inhibition of elongation factor via ROS, step 4 is the inhibitory effects of abiotic stresses on PSII, step 5 is the inhibition of PSII, step 6 is the protection of 9 kDa protein of PSII via compatible solutes, step 7 is the protection of PSII via compatible solutes, step 8 is the protection of FtsH via compatible solutes, step 9 is the sacrifice of D1, step 10 is the proteasomal degradation of D1, step 11 is the involvement of FtsH and CtpA in D1 degradation, and step 12 shows the protection of elongation factor by compatible solutes

residues, Chl D1 and Chl PD1, the places where initial charge separation and location of the stable cation $P680^+$ found, respectively, but function of these modifications and selective superiority of D1′ over normal D1 are still to be elucidated (Mulo et al. 2009). Moreover, there may be possibility that expression of D1′ help to cope with heat stress to thermophilic cyanobacteria because it was seen that at high temperature in aquatic environment, microaerobic conditions get created due to decrease in gaseous solubility at high temperature. Recently, phylogenetic analysis of all cyanobacterial D1 proteins has been done (Cardona et al. 2015), where the origin and evolution of oxygenic photosynthesis have been described and classified the D1 protein into five types on the basis of their sequences. Other adaptive strategy of thermophilic cyanobacteria regarding D1 protein may be high turnover number of D1 protein (Fig. 10.2). It is found that in mesophilic cyanobacteria, D1\D2 turnover gets increased at high temperature during the acclimation process, which was proved by the elevated expression of *ctpA*, an enzyme involved in processing of D1 protein into catalytically active form, at fully acclimated conditions (Rowland et al. 2010). In cyanobacteria, there are three extrinsic proteins PsbU, PsbO, and Cytochrome c550, which are known to stabilize oxygen evolving complex (OEC) (Nishiyama et al. 1994, 1999) and among them, 33 kDa extrinsic protein PsbO stabilizes the Mn of Mn cluster so known as Mn stabilizing protein and dissociation of this protein from its binding site is the main reason behind the heat inactivation of oxygen evolving complex (Enami et al. 1994). Mn stabilizing protein of thermophilic *Synechococcus elongatus* has been found more thermostable than spinach 33 kDa protein which also proved by an experiment in which substitution of 33 kDa protein from spinach by 34 kDa thermostable homologue protein from thermophilic cyanobacteria *Phormidium laminosum* and *Synechococcus elongatus* provides resistance toward heat inactivation and restore oxygen evolution of the complex (Pueyo et al. 2002).

10.3.2.3 Proteins Involved in Electron Transport Between PSII and PSI

Plastoquinone (PQ), cytochrome b6f complex (Cyt b6f), and plastocyanin (PC) or cytochrome c553 are the components of the photosynthetic electron transport chain which connect the PSII to the PSI; however, in cyanobacteria, these components are also shared with the respiratory electron transport chain due to lack of compartmentalization in cyanobacteria cell and occurrence of photosynthesis and respiration machinery on the same thylakoid membrane (Nagarajan and Pakrasi 2001). However, the advantage of this sharing is that the terminal oxidases of respiration electron transport chain work as sink for excess electrons generated by PSII therefore prevent the cell from ROS formation (Lea-Smith et al. 2016a). Plastoquinone (PQ), plastocyanin (PC), and Cyt c553 are the mobile carriers of electron where PQ is found soluble in thylakoid membrane, while PC and Cyt c553 are present toward the luminal side of the membrane. Selection of PC or Cyt c553 as electron carrier from Cyt b6f complex to PSI depends on the species of cyanobacteria and copper availability in the cyanobacterial medium, whereas only PC is found as electron carrier in higher plants (De la Cerda et al. 2007). Plastoquinones are small lipophilic molecules, and their movement and electron transport activity are affected by physical state of membrane lipids, which further affected by temperature, so the process of electron transport from PSII to PSI through PQ is temperature-affected process (Hirano et al. 1981). Plastocyanin is a copper-containing protein which reduced at Cyt b_6f complex by accepting an electron from it and again oxidized at PSI. The reduced form of PC has been found more thermostable (Tm = 71 °C) than oxidized form (Tm = 61 °C) in spinach as well as in mesophilic *Synechocystis* (Gross et al. 1992; Feio et al. 2006), while contradictory to this, the oxidized form of the PC has been found to be more thermostable than reduced form in thermophilic *Phormidium laminosum* and this contradiction might be due to the structural differences between PCs of these species (Feio et al. 2004). A comparative study found that PC from thermophilic *Phormidium laminosum* was more thermostable than PCs found in mesophilic *Anabaena*, *Pseudanabaena*, and *Synechocystis,* further suggested that PC is the limiting factor in thermostability of PC/PSI couple (Balme et al. 2001). Differences in the stability of PCs of *Synechocystis* and thermophilic *Phormidium* might be due to the factors like ion strength inside thylakoid membrane and molecular dynamics. It was speculated that hydrophobic interactions, which are more thermostable, seems more important than electrostatic interactions in *Phormidium* (Feio et al. 2006). Cyanobacteria use PC and Cyt c553 (Cyt c6) as alternate electron carriers, and in vitro study suggested that both PC and Cyt c6 interact with PSI through simple singe collision mechanism (Balme et al. 2001).

Here in this chapter, phylogenetic analysis of plastocyanin from cyanobacteria, algae, and plants has been done by taking amicyanin from *Paracoccus* as an outgroup (Fig. 10.3). The plastocyanin amino acid sequences of higher plants, algae, and cyanobacteria were retrieved from NCBI database. Alignment of the sequences was done using MEGA X software, and phylogenetic tree was prepared using the Neighbor-Joining method in the same software (Saitou and Nei 1987; Kumar et al. 2018). During replication of tree, the cluster togetherness was tested through bootstrap value using 1000 replicates (Felsenstein 1985). The evolutionary distances in tree were computed using the JTT matrix-based method which was shown through branch length in the tree (Jones et al. 1992). Final editing of tree was done using iTOL server (Letunic and Bork 2021). Initially, the phylogenetic tree was found to contain two clades: first one was of cyanobacteria and second was of eukaryotic photoautotrophs. The eukaryotic clade further contained algae and higher plants. The sequence similarity between plastocyanin of cyanobacteria and eukaryotic photoautotrophs was also not high (around 50%). In the cyanobacterial clade, the strains were grouped according to their taxonomical position and not purely on the basis of their habitat, which further suggested that substitution of amino acids, as an adaptation for extreme conditions, was not enough to categorize them into distinct taxonomical positions. Further, *Leptolyngbya* sp. NIES-2104 (GAP94489) and *Thermoleptolyngbya oregonensis* (WP_316788584) have been selected for comparative study of plastocyanin from these mesophilic and thermophilic strains (Fig. 10.4). Homology modeling was done, and 3D models of plastocyanin were retrieved using SWISS-MODEL server, and PDBsum server was used for 2D structures and topology drawing (Laskowski et al. 2018). Differences between the

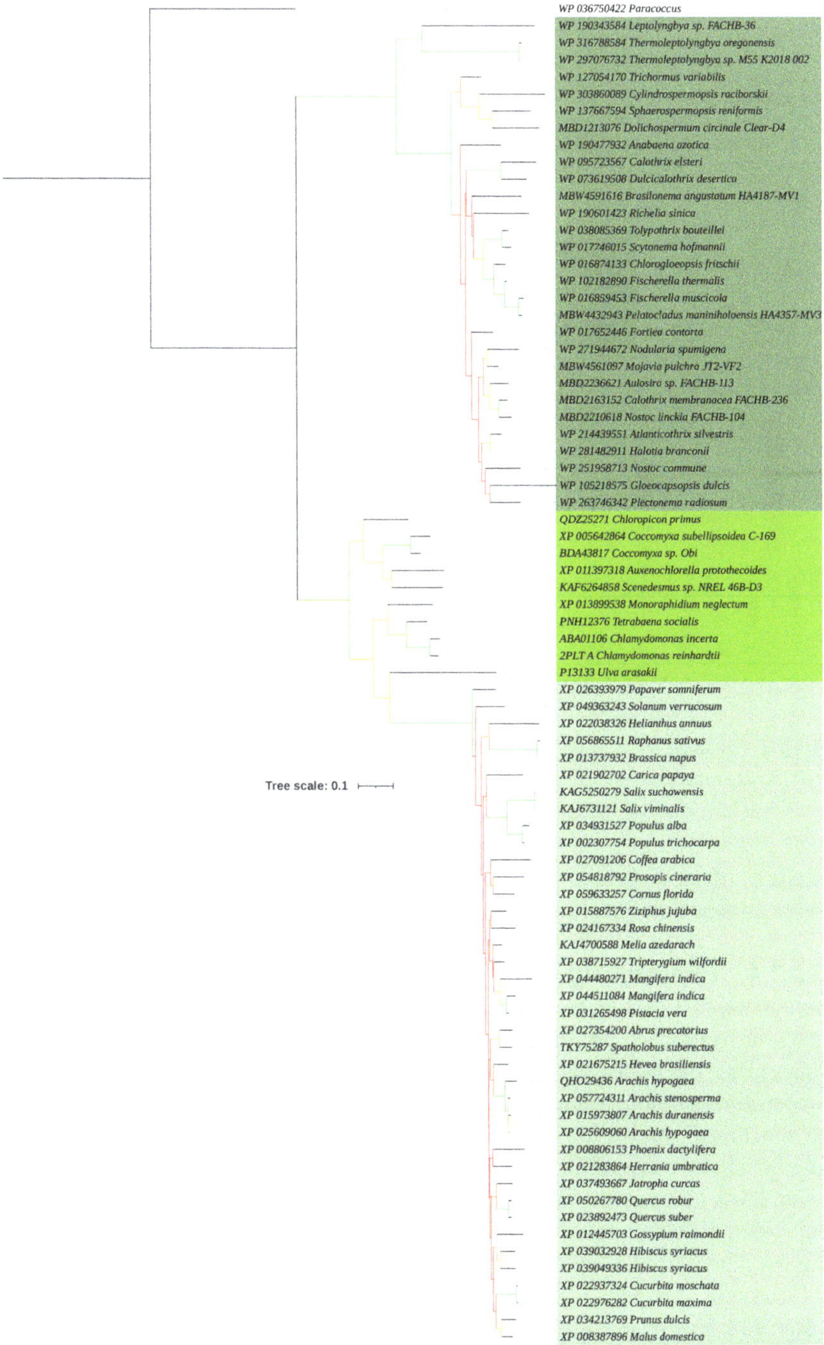

Fig. 10.3 Phylogenetic tree of plastocyanin from cyanobacteria, algae and plants

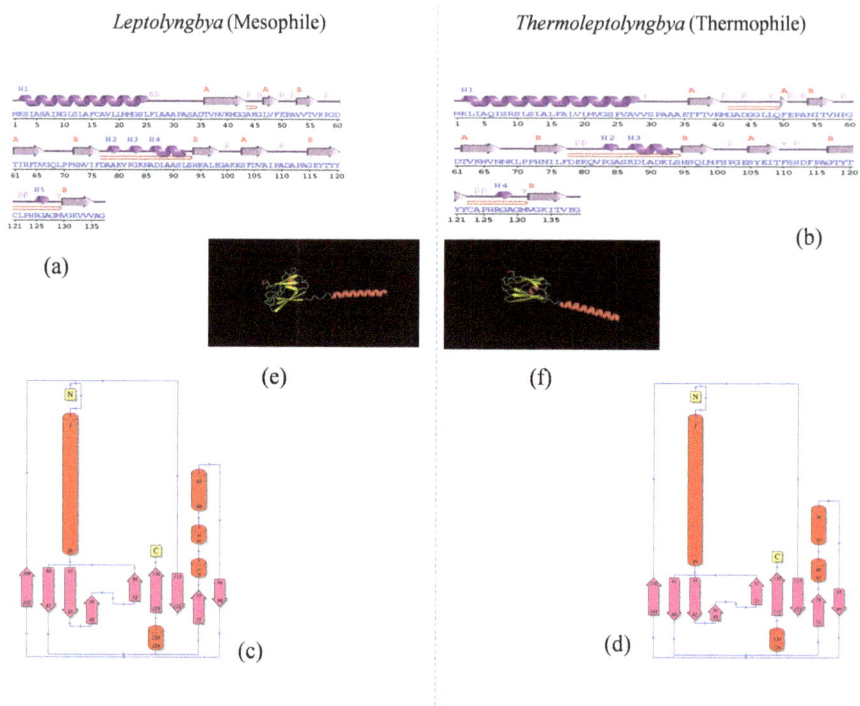

Fig. 10.4 Comparison of structural modification in *Leptolyngbya* and *Thermoleptolyngbya*. Here, (**a**, **b**) are 2D representation of plastocyanin structures, (**c**, **d**) denoted the topologies, while (**e**, **f**) show the 3D models of plastocyanin of *Leptolyngbya* and *Thermoleptolyngbya*

structures of mesophilic and thermophilic plastocyanin were seen in the 2D diagrams, as shown in Fig. 10.4.

10.3.2.4 Photosystem I

Photosystem I (PSI) is an integral protein-pigment complex of thylakoid membranes in cyanobacteria and chloroplasts and involved in the electron transport chain during the photosynthesis. PSI takes part in both linear and cyclic electron transfer during light reactions of photosynthesis (Yamori et al. 2015). Structural and functional elucidation of PSI through various molecular techniques such as mutation studies, crystallization, and X-ray structural analysis at resolution 6, 4.5, 4, and 2.5 Å (Witt et al. 1988; Koike et al. 1989; Ikeuchi et al. 1991; Mühlenhoff et al. 1993; Krauß et al. 1996; Jordan et al. 2001; El-Mohsnawy et al. 2010) suggested a generalized view of composition of PSI complex. PSI monomer consists of 12 protein subunits, 128 cofactors (96 chlorophyl a, 2 phylloquinones, 3 FeS cluster, 22 carotenoids, 4 lipid molecules, 1 Ca ion), and 201 water molecules. Out of 12 subunits, 2 subunits, i.e., PsaA and PsaB, are central subunits and make heterodimer which carry P700 reaction center of PSI. PsaX subunit is only found in PSI of cyanobacteria and isolated from thermophilic cyanobacteria *Synechococcus vulcanus* and *Anabaena variabilis,* and its homologous sequence was also found in the 2.5 Å resolution structure of PSI from thermophilic *Synechococcus elongatus* (Jordan et al. 2001; Saenger et al. 2002). Most of the studies related to PS I in photoautotrophs of extreme conditions were found on the cyanobacteria. Photosystem I in thermophilic cyanobacteria is comparatively less sensitive toward heat and more thermostable than PSII. Intact as well as isolated thylakoid membranes of a thermophilic cyanobacterium, *Synechococcus* sp., showed the similar optimum temperature for PSI reaction (disproportionation of diphenylcarbazone) which suggested the inbuilt thermostable nature of PSI in the thermophilic cyanobacteria (Yamaoka et al. 1978). In cyanobacteria, both monomeric and trimeric type of PS I have been reported, while in plants only monomeric PS I was found; however, in some cyanobacteria, dimeric and tetrameric complex was also present (Li et al. 2014; Semchonok et al. 2016). PsaL and PsaI subunits are known to be responsible for the trimerization of monomeric PSI (Chitnis and Chitnis 1993; Schluchter et al. 1996; Aspinwall et al. 2004). In mesophilic cyanobacteria *Synechocystis* PCC6803, trimer to monomer ratio increases with increase in temperature, further suggesting the superiority of trimer PSI on monomer in temperature stress (Kłodawska et al. 2015). PSI subunits majorly consist of alpha subunits; hence, FTIR spectroscopy was used to compare the thermal denaturation of monomeric and trimeric PSI from thermophilic cyanobacteria in temperature range from 20 to 100 °C. During this study, it was found that monomeric PSI has one more absorption declination phase at >60 °C before final irreversible denaturation aggregation of both monomer and trimeric PSI at >80 °C and speculated that this high thermal stability of trimeric PSI than monomeric PSI might be due to shielding of the denaturation sensitive sides at the monomer-monomer interface within a trimer (Shubin et al. 2017). Tetrameric PSI is found in most of the heterocyte-forming cyanobacteria, and its expression gets increase under high light condition with novel PSI-bound carotenoids such as canthaxanthin, echinenone, and myxoxanthophyll which inferred that tetrameric PSI is the adaptation to grow at high light intensity (Li et al. 2019). In vivo study using different mutants and circular dichroism (CD) suggested that xanthophylls and polyunsaturated lipid deficiencies in the thylakoid membrane affect the stability of PSI trimer at moderately high temperature and high light conditions; however, the exact location of xanthophylls in photosynthetic complex is still unknown (Zakar et al. 2017). Although the functional significance of trimeric PSI is still not fully understood, but on the basis of these studies, it can be speculated that trimeric as well as tetrameric organization of cyanobacterial PSI is an adaptational feature that make cyanobacteria thrive in optimal as well as extreme conditions. High temperature inhibits the functioning of PSII, but some reports regarding effect of heat on PSI shown that electron transport rate get increased with increasing temperature in higher plants (Mohanty et al. 1987; Havaux 1993) as well as in intact cyanobacterial cells (Venkataramanaiah et al. 2003; Zhang and Liu 2016); however, this increment may be due to increase in permeability of DCPIP into cell as increase in electron transport rate in PSI of isolated thylakoid membranes of cyanobacteria found insignificant (Venkataramanaiah et al. 2003). Researches related to effect of heat on PSI of thermophilic cyanobacteria are lower than PSII. It might be due to the high stability of PSI in thermophilic cyanobacteria, the half denaturation temperature of P700 in thermophilic *Synechococcus elongatus* has been found near 100 °C (Koike et al. 1982), high sensitivity of PSII site toward heat, and also the techniques to analyze the PSI activities are limited which attracted the attention of most of the researchers towards PSII of thermophilic cyanobacteria. Although PSI is comparatively heat stable, its damage by other stresses, such as extreme light, is more devastating for cell due to the absence of its repair system unlike PSII (Dobson et al. 2021).

10.3.2.5 Ferredoxin (Fd) and Ferredoxin:NADP$^+$ Oxidoreductase (FNR)

Ferredoxins are small iron–sulfur proteins found in all domains of life. In photosynthetic organisms during the light reaction of the photosynthesis, PSI transfer electrons to ferredoxins and these electrons are finally involved in the reduction of NADP$^+$ to NADPH via FNR (Fukuyama 2004; Cassier-Chauvat and Chauvat 2014; Sharma et al. 2024). Plant type Ferredoxin (Fd) (present from cyanobacteria to higher vascular plants) contains single (2Fe-2S) cluster, and it is one of the strongest reducing agents in nature having redox potential (E_m) around −0.42 eV (Hall et al. 1973). Thermostable Fds were isolated from thermophilic *Mastigocladus laminosus* and *Synechococcus elongatus* (mesophilic) to analyze their amino acid sequences and found that at high temperature (65 °C) *Mastigocladus* Fd shown highest activity (Hase et al. 1978; Baumann et al. 1996; Hatanaka et al. 1997). In comparative study of the thermophilic *Mastigocladus* Fd with mesophilic Fds of three mesophilic cyanobacteria, i.e., *Anabaena* sp. PCC 7119, *Spirulina platensis*, *Synechocystis* sp. PCC 6803, and one higher plant *Spinacia oleracea*, it was found that two structural differences in *Mastigocladus* Fd might be responsible for differences in thermostability. These two modifications were a salt bridge network in the C-terminal region, and the flexible L1, 2 loop that increases hydrophobic accessible surface area (Fish et al. 2005). Similarly, the ferredoxins from cold environment also have some adaptations in their structure. Like other psychrophilic proteins, Fds of cold-adapted photosynthetic organisms also have high flexibility at low temperature. Substitutions of some amino acids are responsible for this change in flexibility (Cvetkovska et al. 2018). This increased flexibility leads to an increase in activity at cold temperatures but consequently resulted in low stability of Fds at moderate temperatures. However, Fds from *Chlamydomonas* sp. UWO241 have both high activity at lower temperatures and high stability at moderate temperatures (40 °C). Moreover, two copies of Fd gene were found in *Chlamydomonas* sp. UWO241 and could be responsible for cold adaptation of this organism because it is believed that harboring of multiple copies of genes is a strategy to adapt in harsh environmental conditions (Kondrashov 2012; Cvetkovska et al. 2018). Ferredoxin: NADP$^+$ oxidoreductase (FNR) is the enzyme of flavoenzymes group which catalyzes the reduction of NADP$^+$. FNR from thermophilic *Synechococcus* sp. showed stability till 65 °C and suggested that thermostability of FNR correlates with the photosynthetic growth of these cyanobacteria at high temperature (Koike and Katoh 1980). FNR is encoded by petH gene. CpcD-like domain at N–terminal of

FNR was isolated from thermophilic *Synechococcus elongatus* which was responsible for the association of FNR with phycocyanin. Further it was inferred that this association provides stability to the FNR in above thermophilic cyanobacteria (Nakajima et al. 2002).

10.3.2.6 RubisCo Enzyme

Ribulose-1,5-bisphosphate carboxylase/oxygenase (RubisCo) (EC 4.1.1.39) is an enzyme involved in the so-called dark reaction of photosynthesis in photoautotrophic organisms. Here dark reactions do not mean that they occur in dark but because they do not require light directly during this part of photosynthesis. In Calvin–Benson pathway, RubisCo catalyzes the carboxylation of the ribulose-1,5-bisphosphate (RuBP) by fixing atmospheric CO_2 followed by conversion of this into two molecules of 3-phosphoglycerate (3PGA) (Raven 2013). Although RubisCo is a crucial enzyme for photoautotrophs, it is also an inefficient enzyme due to its slow turnover (Erb and Zarzycki 2018). Apart from carboxylation, RubisCo also has oxygen binding activity which leads to photorespiration. During photorespiration RubisCo binds with the oxygen molecule, instead of CO_2, and catalyzes the oxygenation of RuBP which further leads to the production of single molecule of 3PG and one molecule of 2-phosphoglycolate as byproduct (Shi and Bloom 2021). This 2-phosphoglycolate recycles into 3PG molecule by using ATP. Since 3PG is a toxic compound and needs recycling on the expense of ATP, while photosynthetic productivity also decreases, photorespiration is known as wasteful process. At equal concentrations, RubisCo binds strongly to CO_2 than O_2 but binding of RubisCo to O_2 occurs when CO_2 concentration goes down (van Lun et al. 2014). RubisCo of oxygenic photoautotrophic organisms is a hexadecameric structure composed of eight large and eight small subunits, where small subunits are known as CO_2 reservoirs (Knight et al. 1990; Tabita 1999; van Lun et al. 2014). Since deactivation of RubisCo, along with the reduction in chloroplast electron transport rate, is mainly responsible for the reduction in photosynthetic CO_2 fixation during high temperature in plants (Scafaro et al. 2023), adaptation in the RubisCo is very necessary for growing in extreme hot temperatures. Photoautotrophs of extreme habitats have modifications in their RubisCo structure which make them able to efficiently fix CO_2 at those conditions. Thermophilic *Synechococcus* has thermostable RubisCo in comparison to its mesophilic relatives. Reason behind the thermostability of RubisCo in thermophiles was substitutions of some amino acids in polypeptides of RubisCo (Miller et al. 2013). Adaptations are also found in the RubisCo of psychrophiles. Comparative study on activity, thermolability, and amount of RubisCo from psychrophilic *Chloromonas*, mesophilic *Chlamydomonas reinhardtii,* and mesophilic *Chloromonas* showed adaptation features in psychrophiles. RubisCo of psychrophilic *Chloromonas* was found temperature-sensitive in comparison to mesophilic counterpart. But the relative amount of RubisCo was higher in psychrophilic algae, which may be an adaptational feature in psychrophiles where this increased amount may offset the decreased enzyme activity at low temperatures (Devos et al. 1998). Engineering of extremophilic RubisCo subunits into crop plants can make stress resistant crop varieties.

10.3.3 Metabolic Modifications

10.3.3.1 Compatible Solutes

Compatible solutes are of low molecular weight, water soluble organic osmolytes present in variety of organisms from three domains of life and can be accumulated in the cells at higher concentrations without interfering with cellular processes (Kurz 2008). These are either synthesized by organisms itself or take up from outside. De novo synthesis of these compounds is mainly preferred by the photoautotrophs, while heterotrophs favor the absorption of these solutes from outer environment. Compatible solutes act in two ways to adapt the cell in extreme conditions; first they act as osmolarity regulator of the cell and second are the stabilization of enzymes, proteins, and nucleic acids (Brown 1976; Santos and Da Costa 2002; Kurz 2008). Compatible solute accumulation in cyanobacteria mainly occurs in marine and high temperature conditions. In marine conditions, cyanobacteria accumulate various compatible solutes in differential manner. When salt concentration is low cyanobacteria generally accumulate sucrose and trehalose, at moderate salt conditions glucosylglycerol accumulation occurs while at high salt concentrations cyanobacterial strains accumulate glycine betaine for osmoregulation (Klähn and Hagemann 2011). Apart from the saline conditions, the importance of these compounds in thermo tolerance of cyanobacteria has also been studied but still full mechanism is still to be elucidated. Glycine betaine, an osmolyte, has been found to be involved in protection of oxygen evolution complex by inhibiting the dissociation of 9 kDa protein and stabilizing the coordination of Mn with protein cleft against thermal stress in cyanobacteria (Mamedov et al. 1991; Stamatakis and Papageorgiou 1993; Papageorgiou and Murata 1995). As heat lowers PSII activity by inhibiting its repair cycle via inhibition of D1 translation and glycine, betaine may protect PSII possibly by preventing the inhibition of translational machinery during heat stress (Fig. 10.2). It was also evident in a study conducted on *Synechococcus* sp. PCC 7942 where transformed cells having high glycine betaine content shown more de nova synthesis of D1 protein than control cells at 40 °C (Allakhverdiev et al. 2007). Although in vitro studies of certain model enzymes from various organisms revel that compatible solutes can inhibit the thermal inactivation of these enzymes (Diamant et al. 2001; Borges et al. 2002; Klähn and Hagemann 2011), but more molecular and in vivo studies are needed to fully elucidate the mechanism of thermotolerance. Marine and halotolerant species of eukaryotic algae also accumulate various compatible solutes. Members of chlorophyceae such as *Dunaliella* sp. and *Chlamydomonas* sp. accumulate glycerol, while phaeophyceae members generally accumulate mannitol (Borowitzka and Brown 1974; Wegmann 1986). In plants, glycine betaine, proline, and polyols are the main compatible solutes accumulated under high salinity where these compounds provide tolerance to plants against salt and drought stress (Bohnert et al. 1995; Gharsallah et al. 2016; Giri 2011; Zhu et al. 2022). FtsH protease is responsible for the degradation of inactive D1 protein during PSII repair cycle and glycine betaine may act as protectant of FtsH in abiotic stresses (Fig. 10.2) such as in salt stress (Ohnishi and Murata 2006) and likewise this osmolyte may be responsible for the protection of

FtsH in thermophilic cyanobacteria, which grow at extreme temperature. Attempts to engineer these compatible solute synthesis pathway genes into crop plants are continued for increasing the yield and stress tolerance of these plants (Chen and Murata 2011; Wani et al. 2013).

10.3.3.2 Lipids and Hydrocarbons of Thylakoid Membranes

Lipids and hydrocarbons are the integral components of the thylakoid membranes of cyanobacteria and chloroplasts. There are four main classes of lipids found in the thylakoid membranes: monogalactosyldiacylglycerol (MGDG) (~40%), digalactosyldiacylglycerol (DGDG) (~32%), sulfoquinovosyldiacylglycerol (SQDG) (~15%), and phosphatidylglycerol (PG) (~13%) (Sakurai et al. 2006; van Eerden et al. 2015). These lipids consisted of fatty acid chains which may be saturated or unsaturated at varying extent. Change in growth temperature modulates the level of unsaturation in the fatty acids. Hydrocarbons are also found in the membranes where they provide flexibility and curvature to the thylakoid membranes which ultimately affect the cell size, division, and growth of the cyanobacteria (Lea-Smith et al. 2016b). Apart from the unsaturations in hydrocarbons and lipids of thylakoid membranes under cold conditions, certain medium chain hydrocarbons also play a crucial role in maintaining photosynthesis. Accumulation of medium chain hydrocarbons (heptadecane) was seen in the wild type cyanobacteria (*Synechocystis* sp. PCC 6803), while slower growth was found in "no alkane" mutant (unable to synthesis alkanes) of the same cyanobacterial strain during cold stress (Berla et al. 2015). This study suggested that medium chain alkanes have some role in the regulation of redox poise and reductant partitioning on the thylakoid membranes probably via affecting cyclic electron flow under stress conditions.

10.4 Conclusion

Extreme habitats harbor diverse range of photoautotrophs ranging from prokaryotes to eukaryotic algae and higher plants which sustain the light-driven ecosystem in such conditions. Extremophilic photoautotrophs found in these habitats sometimes may face more than one extreme abiotic factor where they are called polyextremophiles. The extreme habitats possess differential diversity of photoautotrophs, and dominance or absence of any group can be seen in a particular habitat, e.g., cyanobacteria are rarely found in extreme acidic conditions and absent from thermal acidic springs. Since extreme habitats are usually isolated from the rest of the environment, the photoautotrophs found in these habitats evolved some special features in their morphology and metabolism to maintain the photosynthesis. Morphological adaptations are mainly found in leaves and canopies of higher plants at deserts and polar regions. Metabolic adaptations were seen in the physiological and biochemical behavior along with the photosynthetic machinery especially in the proteins of light absorbing antennas and electron transport chain. During phylogenetic analysis of plastocyanin protein, it was seen that it can be grouped into prokaryotic, which found in cyanobacteria, and eukaryotic types, which found in both eukaryotic algae and plants. Adaptations in the behavior of plants also maintain the photosynthetic process at extreme conditions, e.g., some polar plants do photosynthesis only in the summers and they remain dormant in colder conditions. Compatible solutes accumulation and synthesis of medium chain length alkanes are also used by the extremophilic photoautotrophs as strategies to the sustain photosynthesis.

10.5 Future Prospects

Globally many extreme habitats are still untouched even in the terms of diversity analysis and require rapid exploration to fully understand the species composition of photoautotrophs and further metabolic studies. Generally, the cyanobacteria were used as model organisms to study the photosynthesis in extreme conditions perhaps due to its easy laboratory culturing, short life cycle, and presence of whole photosynthetic machinery, and its related metabolism, in a single cell. Although these studies on extremophilic cyanobacteria elucidated many adaptational features found in the reactions of photosynthesis, more researches are required on the higher taxa of extremophilic photoautotrophs to make generalize species-independent framework of extremophilic photosynthetic metabolism.

Acknowledgments We are thankful to the Head of the Department of Botany, Banaras Hindu University, Varanasi, India for providing necessary laboratory facilities. We are also thankful to Institute of Eminence (IOE-6031), Banaras Hindu University, for financial support. Sanjay Sharma is gratefully acknowledged to the Council of Scientific and Industrial Research (CSIR), New Delhi, for financial support in the form of Senior Research Fellowship. Tameshwar Jaiswal is thankful to the UGC, New Delhi, for Senior Research Fellowship.

Conflict of Interest The authors declare that they have no conflict of interest.

References

Abdelwahab H, Amin A (2017) Diatoms diversity of thermal springs in the southwest region, Saudi Arabia. Egypt Acad J Biol Sci H Bot 8(2):59–74

Adir N, Dobrovetsky Y, Lerner N (2001) Structure of C-phycocyanin from the thermophilic cyanobacterium *Synechococcus vulcanus* at 2.5 Å: structural implications for thermal stability in phycobilisome assembly. J Mol Biol 313(1):71–81

Ago H, Adachi H, Umena Y, Tashiro T, Kawakami K, Kamiya N, Tian L, Han G, Kuang T, Liu Z, Wang F, Zou H, Enami I, Miyano M, Shen JR (2016) Novel features of eukaryotic photosystem II revealed by its crystal structure analysis from a red alga. J Biol Chem 291(11):5676–5687

Aguilera A (2013) Eukaryotic organisms in extreme acidic environments, the Río Tinto case. Life 3(3):363–374

Allakhverdiev SI, Murata N (2004) Environmental stress inhibits the synthesis *de novo* of proteins involved in the photodamage–repair cycle of photosystem II in *Synechocystis* sp. PCC 6803. Biochim Biophys Acta Bioenerg 1657(1):23–32

Allakhverdiev SI, Los DA, Mohanty P, Nishiyama Y, Murata N (2007) Glycinebetaine alleviates the inhibitory effect of moderate heat stress on the repair of photosystem II during photoinhibition. Biochim Biophys Acta Bioenerg 1767(12):1363–1371

Aparecido LMT, Woo S, Suazo C, Hultine KR, Blonder B (2020) High water use in desert plants exposed to extreme heat. Ecol Lett 23(8):1189–1200

Apte SK, Fernandes T, Badran H, Ballal A (1998) Expression and possible role of stress-responsive proteins in *Anabaena*. J Biosci 23(4):399–406

Aro EM, Virgin I, Andersson B (1993) Photoinhibition of photosystem II. Inactivation, protein damage and turnover. Biochim Biophys Acta Bioenerg 1143(2):113–134

Aspinwall CL, Duncan J, Bibby T, Mullineaux CW, Barber J (2004) The trimeric organisation of photosystem I is not necessary for the iron-stress induced CP43′ protein to functionally associate with this reaction centre. FEBS Lett 574(1–3):126–130

Balme A, Hervás M, Campos LA, Sancho J, De la Rosa MA, Navarro JA (2001) A comparative study of the thermal stability of plastocyanin, cytochrome c_6 and photosystem I in thermophilic and mesophilic cyanobacteria. Photosynth Res 70:281–289

Barber J (1998) Photosystem two. Biochim Biophys Acta Bioenerg 1365(1–2):269–277

Barber J (2017) A mechanism for water splitting and oxygen production in photosynthesis. Nat Plants 3(4):1–5

Baumann B, Sticht H, Schärpf M, Sutter M, Haehnel W, Rösch P (1996) Structure of *Synechococcus elongatus* [Fe_2S_2] ferredoxin in solution. Biochemistry 35(39):12831–12841

Berla BM, Saha R, Maranas CD, Pakrasi HB (2015) Cyanobacterial alkanes modulate photosynthetic cyclic electron flow to assist growth under cold stress. Sci Rep 5(1):14894

Berthold DE, Lefler FW, Laughinghouse HD IV (2021) Untangling filamentous marine cyanobacterial diversity from the coast of South Florida with the description of Vermifilaceae fam. nov. and three new genera: *Leptochromothrix* gen. nov., *Ophiophycus* gen. nov., and *Vermifilum* gen. nov. Mol Phylogenet Evol 160:107010

Berthold DE, Lefler FW, Laughinghouse HD (2022) Recognizing novel cyanobacterial diversity in marine benthic mats, with the description of Sirenicapillariaceae fam. nov., two new genera, *Sirenicapillaria* gen. nov. and *Tigrinifilum* gen. nov., and seven new species. Phycologia 61(2):146–165

Bhakta S, Das SK, Adhikary SP (2016) Algal diversity in hot springs of Odisha. Nelumbo 58:157–173

Bohnert HJ, Nelson DE, Jensen RG (1995) Adaptations to environmental stresses. Plant Cell 7(7):1099–1111

Borges N, Ramos A, Raven NDH, Sharp RJ, Santos H (2002) Comparative study of the thermostabilizing properties of mannosylglycerate and other compatible solutes on model enzymes. Extremophiles 6(3):209–216

Borowitzka LJ, Brown AD (1974) The salt relations of marine and halophilic species of the unicellular green alga, *Dunaliella*—the role of glycerol as a compatible solute. Arch Microbiol 96(1):37–52

Bowman JS, Rasmussen S, Blom N, Deming JW, Rysgaard S, Sicheritz-Ponten T (2012) Microbial community structure of Arctic multiyear sea ice and surface seawater by 454 sequencing of the 16S RNA gene. ISME J 6(1):11–20

Bragina A, Berg C, Cardinale M, Shcherbakov A, Chebotar V, Berg G (2012) *Sphagnum* mosses harbour highly specific bacterial diversity during their whole lifecycle. ISME J 6(4):802–813

Brejc K, Ficner R, Huber R, Steinbacher S (1995) Isolation, crystallization, crystal structure analysis and refinement of allophycocyanin from the cyanobacterium *Spirulina platensis* at 2.3 Å resolution. J Mol Biol 249(2):424–440

Brock TD (1978) The thermophilic blue-green algae. In: Thermophilic microorganisms and life at high temperatures. Springer, Berlin, pp 217–254

Brodribb T, Hill RS (1993) A physiological comparison of leaves and phyllodes in *Acacia melanoxylon*. Aust J Bot 41(3):293–305

Brown AD (1976) Microbial water stress. Bacteriol Rev 40(4):803–846

Cardona T, Murray JW, Rutherford AW (2015) Origin and evolution of water oxidation before the last common ancestor of the cyanobacteria. Mol Biol Evol 32(5):1310–1328

Cassier-Chauvat C, Chauvat F (2014) Function and regulation of ferredoxins in the cyanobacterium, *Synechocystis* PCC6803: recent advances. Life 4(4):666–680

Chen CH, Berns DS (1978) Comparison of the stability of phycocyanins from thermophilic, mesophilic, psychrophilic and halophilic algae. Biophys Chem 8(3):203–213

Chen THH, Murata N (2011) Glycinebetaine protects plants against abiotic stress: mechanisms and biotechnological applications. Plant Cell Environ 34(1):1–20

Chen H, Qi H, Xiong P (2022) Phycobiliproteins—a family of algae-derived biliproteins: productions, characterization and pharmaceutical potentials. Mar Drugs 20(7):450

Chisholm SW, Frankel SL, Goericke R, Olson RJ, Palenik B, Waterbury JB, West-Johnsrud L, Zettler ER (1992) *Prochlorococcus marinus* nov. gen. nov. sp.: an oxyphototrophic marine prokaryote containing divinyl chlorophyll a and b. Arch Microbiol 157:297–300

Chitnis VP, Chitnis PR (1993) PsaL subunit is required for the formation of photosystem I trimers in the cyanobacterium *Synechocystis* sp. PCC 6803. FEBS Lett 336(2):330–334

Chong CW, Pearce DA, Convey P (2015) Emerging spatial patterns in Antarctic prokaryotes. Front Microbiol 6:1058

Chrismas NAM, Anesio AM, Sánchez-Baracaldo P (2015) Multiple adaptations to polar and alpine environments within cyanobacteria: a phylogenomic and Bayesian approach. Front Microbiol 6:1070

Chrismas NAM, Barker G, Anesio AM, Sánchez-Baracaldo P (2016) Genomic mechanisms for cold tolerance and production of exopolysaccharides in the Arctic cyanobacterium *Phormidesmis priestleyi* BC1401. BMC Genomics 17(1):1–14

Clarke AK, Hurry VM, Gustafsson P, Oquist G (1993) Two functionally distinct forms of the photosystem II reaction-center protein D1 in the cyanobacterium *Synechococcus* sp. PCC 7942. Proc Natl Acad Sci 90(24):11985–11989

Clarke A, Morris GJ, Fonseca F, Murray BJ, Acton E, Price HC (2013) A low temperature limit for life on earth. PLoS One 8(6):e66207

Convey P (2013) Antarctic ecosystems. In: Levin SA (ed) Encyclopedia of biodiversity. Elsevier, San-Diego, pp 179–188

Convey P, Smith RIL (2006) Geothermal bryophyte habitats in the South Sandwich Islands, maritime Antarctic. J Veg Sci 17(4):529

Costello MJ, Chaudhary C (2017) Marine biodiversity, biogeography, deep-sea gradients, and conservation. Curr Biol 27(11):R511–R527

Costello MJ, Cheung A, De Hauwere N (2010) Surface area and the seabed area, volume, depth, slope, and topographic variation for the world's seas, oceans, and countries. Environ Sci Technol 44(23):8821–8828

Costello MJ, Bouchet P, Boxshall G, Fauchald K, Gordon D, Hoeksema BW, Poore GCB, van Soest RWM, Stöhr S, Walter TC, Vanhoorne B, Decock W, Appeltans W (2013) Global coordination and standardisation in marine biodiversity through the World Register of Marine Species (WoRMS) and related databases. PLoS One 8(1):e51629

Cvetkovska M, Hüner NPA, Smith DR (2017) Chilling out: the evolution and diversification of psychrophilic algae with a focus on Chlamydomonadales. Polar Biol 40(6):1169–1184

Cvetkovska M, Szyszka-Mroz B, Possmayer M, Pittock P, Lajoie G, Smith DR, Hüner NPA (2018) Characterization of photosynthetic ferredoxin from the Antarctic alga *Chlamydomonas* sp. UWO241 reveals novel features of cold adaptation. New Phytol 219(2):588–604

Dadheech PK, Abed RMM, Mahmoud H, Krishna Mohan M, Krienitz L (2012) Polyphasic characterization of cyanobacteria isolated from desert crusts, and the description of *Desertifilum tharense* gen. et sp. nov. (Oscillatoriales). Phycologia 51(3):260–270

Dagnino-Leone J, Figueroa CP, Castañeda ML, Youlton AD, Vallejos-Almirall A, Agurto-Muñoz A, Pérez JP, Agurto-Muñoz C (2022) Phycobiliproteins: structural aspects, functional characteristics, and biotechnological perspectives. Comput Struct Biotechnol J 20:1506–1527

David L, Prado M, Arteni AA, Elmlund DA, Blankenship RE, Adir N (2014) Structural studies show energy transfer within stabilized phycobilisomes independent of the mode of rod-core assembly. Biochim Biophys Acta Bioenerg 1837(3):385–395

De la Cerda B, Castielli O, Durán RV, Navarro JA, Hervás M, De la Rosa MA (2007) A proteomic approach to iron and copper homeostasis in cyanobacteria. Brief Funct Genomics Proteomics 6(4):322–329

Devos N, Ingouff M, Loppes R, Matagne RF (1998) RUBISCO adaptation to low temperatures: a comparative study in psychrophilic and mesophilic unicellular algae. J Phycol 34(4):655–660
Diamant S, Eliahu N, Rosenthal D, Goloubinoff P (2001) Chemical chaperones regulate molecular chaperones *in vitro* and in cells under combined salt and heat stresses. J Biol Chem 276(43):39586–39591
Dobson Z, Ahad S, Vanlandingham J, Toporik H, Vaughn N, Vaughn M, Williams D, Reppert M, Fromme P, Mazor Y (2021) The structure of photosystem I from a high-light tolerant cyanobacteria. elife 10:e67518
Dominic TK, Madhusoodanan PV (1999) Cyanobacteria from extreme acidic environments. Curr Sci 77(8):1021–1023
Dong Z, He H (2017) Phyllode anatomy and histochemistry of four *Acacia* species (Leguminosae: Mimosoideae) in the Great Sandy Desert, north-Western Australia. J Arid Environ 139:110–120
Edwards MR, Hauer C, Stack RF, Eisele LE, MacColl R (1997) Thermophilic C-phycocyanin: effect of temperature, monomer stability, and structure. Biochim Biophys Acta Bioenerg 1321(2):157–164
El-Mohsnawy E, Kopczak MJ, Schlodder E, Nowaczyk M, Meyer HE, Warscheid B, Karapetyan NV, Rögner M (2010) Structure and function of intact photosystem 1 monomers from the cyanobacterium *Thermosynechococcus elongatus*. Biochemistry 49(23):4740–4751
Enami I, Kitamura M, Tomo T, Isokawa Y, Ohta H, Katoh S (1994) Is the primary cause of thermal inactivation of oxygen evolution in spinach PS II membranes release of the extrinsic 33 kDa protein or of Mn? Biochim Biophys Acta Bioenerg 1186(1–2):52–58
Erb TJ, Zarzycki J (2018) A short history of RubisCO: the rise and fall (?) of nature's predominant CO_2 fixing enzyme. Curr Opin Biotechnol 49:100–107
Etemadi-Khah A, Pourbabaee AA, Alikhani HA, Noroozi M, Bruno L (2017) Biodiversity of isolated cyanobacteria from desert soils in Iran. Geomicrobiol J 34(9):784–794
Falkowski PG, Katz ME, Knoll AH, Quigg A, Raven JA, Schofield O, Taylor FJR (2004) The evolution of modern eukaryotic phytoplankton. Science 305(5682):354–360
Fang J, Zhang L, Bazylinski DA (2010) Deep-sea piezosphere and piezophiles: geomicrobiology and biogeochemistry. Trends Microbiol 18(9):413–422
Fazlutdinova A, Gabidullin Y, Allaguvatova R, Gaysina L (2020) Diatoms in Kamchatka's hot spring soils. Diversity 12(11):1–16
Fecteau KM, Boyd ES, Lindsay MR, Amenabar MJ, Robinson KJ, Debes RV, Shock EL (2022) Cyanobacteria and algae meet at the limits of their habitat ranges in moderately acidic hot springs. J Geophys Res Biogeosci 127:e2021JG006446
Feio MJ, Navarro JA, Teixeira MS, Harrison D, Karlsson BG, De la Rosa MA (2004) A thermal unfolding study of plastocyanin from the thermophilic cyanobacterium *Phormidium laminosum*. Biochemistry 43(46):14784–14791
Feio MJ, Díaz-Quintana A, Navarro JA, De la Rosa MA (2006) Thermal unfolding of plastocyanin from the mesophilic cyanobacterium *Synechocystis* sp. PCC 6803 and comparison with its thermophilic counterpart from *Phormidium laminosum*. Biochemistry 45(15):4900–4906
Felsenstein J (1985) Confidence limits on phylogenies: an approach using the bootstrap. Evolution 39(4):783–791
Fish A, Danieli T, Ohad I, Nechushtai R, Livnah O (2005) Structural basis for the thermostability of ferredoxin from the cyanobacterium *Mastigocladus laminosus*. J Mol Biol 350(3):599–608
Friedmann I, Lipkin Y, Ocampo-Paus R (1967) Desert algae of the Negev (Israel). Phycologia 6(4):185–200
Fučíková K, Lewis PO, Lewis LA (2014) Widespread desert affiliation of trebouxiophycean algae (Trebouxiophyceae, Chlorophyta) including discovery of three new desert genera. Phycol Res 62(4):294–305
Fukuyama K (2004) Structure and function of plant-type ferredoxins. Photosynth Res 81(3):289–301
Fyson A (2000) Angiosperms in acidic waters at pH 3 and below. Hydrobiologia 433:129–135
Gao J, Wang H, Yuan Q, Feng Y (2018) Structure and function of the photosystem supercomplexes. Front Plant Sci 9:357

Gharsallah C, Fakhfakh H, Grubb D, Gorsane F (2016) Effect of salt stress on ion concentration, proline content, antioxidant enzyme activities and gene expression in tomato cultivars. AoB Plants 8:plw055

Gibson AC (1998) Photosynthetic organs of desert plants: structural designs of nonsucculent desert plants cast doubt on the popular view that saving water is the key strategy. Bioscience 48(11):911–920

Giri J (2011) Glycinebetaine and abiotic stress tolerance in plants. Plant Signal Behav 6(11):1746–1751

Goel R, Soni R, Suyal DC, Khan M (eds) (2022) Survival strategies in cold-adapted microorganisms. Springer, Berlin

Gross W (2000) Ecophysiology of algae living in highly acidic environments. Hydrobiologia 433:31–37

Gross EL, Draheim JE, Curtiss AS, Crombie B, Scheffer A, Pan B, Chiang C, Lopez A (1992) Thermal denaturation of plastocyanin: the effect of oxidation state, reductants, and anaerobicity. Arch Biochem Biophys 298(2):413–419

Gurera D, Bhushan B (2020) Passive water harvesting by desert plants and animals: lessons from nature. Philos Trans R Soc A Math Phys Eng Sci 378(2167):20190444

Guskov A, Kern J, Gabdulkhakov A, Broser M, Zouni A, Saenger W (2009) Cyanobacterial photosystem II at 2.9-Å resolution and the role of quinones, lipids, channels and chloride. Nat Struct Mol Biol 16(3):334–342

Hall DO, Cammack RC, Rao KK (1973) The plant ferredoxins and their relationship to the evolution of ferredoxins from primitive life. Pure Appl Chem 34(3–4):553–578

Hargreaves JW, Lloyd EJH, Whitton BA (1975) Chemistry and vegetation of highly acidic streams. Freshw Biol 5(6):563–576

Harten JB, Eickmeier WG (1987) Comparative desiccation tolerance of three desert pteridophytes: response to long-term desiccation. Am Midl Nat 118(2):337–347

Hase T, Wakabayashi S, Matsubara H, Rao KK, Hall DO, Widmer H, Gysi J, Zuber H (1978) The amino acid sequence of ferredoxin from the alga *Mastigocladus laminosus*. Phytochemistry 17(11):1863–1867

Hatanaka H, Tanimura R, Katoh S, Inagaki F (1997) Solution structure of ferredoxin from the thermophilic cyanobacterium *Synechococcus elongatus* and its thermostability. J Mol Biol 268(5):922–933

Haumann M, Junge W (1999) Photosynthetic water oxidation: a simplex-scheme of its partial reactions. Biochim Biophys Acta Bioenerg 1411(1):86–91

Havaux M (1993) Characterization of thermal damage to the photosynthetic electron transport system in potato leaves. Plant Sci 94(1–2):19–33

Hayashi NR, Peerapornpisal Y, Nishihara H, Ishii M, Igarashi Y, Kodama T (1994) Isolation and cultivation of thermophilic cyanobacteria from hot springs of northern Thailand. J Ferment Bioeng 78(2):179–181

Heinz S, Liauw P, Nickelsen J, Nowaczyk M (2016) Analysis of photosystem II biogenesis in cyanobacteria. Biochim Biophys Acta Bioenerg 1857(3):274–287

Hirano M, Satoh K, Katoh S (1981) The effect on photosynthetic electron transport of temperature-dependent changes in the fluidity of the thylakoid membrane in a thermophilic blue-green alga. Biochim Biophys Acta Bioenerg 635(3):476–487

Hoffman L (1999) Marine cyanobacteria in tropical regions: diversity and ecology. Eur J Phycol 34(4):371–379

Hooker JD (1863) I. On *Welwitschia*, a new genus of Gnetaceæ. Trans Linn Soc Lond 24(1):1–48

Hrdinka T, Šobr M, Fott J, Nedbalová L (2013) The unique environment of the most acidified permanently meromictic lake in The Czech Republic. Limnologica 43(6):417–426

Hu S, Ding Y, Zhu C (2020) Sensitivity and responses of chloroplasts to heat stress in plants. Front Plant Sci 11:375

Hughes JL, Cox N, Rutherford AW, Krausz E, Lai TL, Boussac A, Sugiura M (2010) D1 protein variants in photosystem II from *Thermosynechococcus elongatus* studied by low temperature optical spectroscopy. Biochim Biophys Acta Bioenerg 1797(1):11–19

Ikeuchi M, Nyhus KJ, Inoue Y, Pakrasi HB (1991) Identities of four low-molecular-mass subunits of the photosystem I complex from *Anabaena variabilis* ATCC 29413: evidence for the presence of the psaI gene product in a cyanobacterial complex. FEBS Lett 287(1–2):5–9
Inoue N, Emi T, Yamane Y, Kashino Y, Koike H, Satoh K (2000) Effects of high-temperature treatments on a thermophilic cyanobacterium *Synechococcus vulcanus*. Plant Cell Physiol 41(4):515–522
Jaiswal TP, Chakraborty S, Singh P, Mishra AK, Singh SS (2021) Description of hot spring dwelling *Mastigocladus ambikapurensis* sp. nov., using a polyphasic approach. Plant Syst Evol 307(3):1–12
Jaiswal TP, Chakraborty S, Sharma S, Mishra A, Mishra AK, Singh SS (2023) Prospects of a hot spring–originated novel cyanobacterium, *Scytonema ambikapurensis*, for wastewater treatment and exopolysaccharide-enriched biomass production. Environ Sci Pollut Res 30(18):53424–53444
Jimbo H, Yutthanasirikul R, Nagano T, Hisabori T, Hihara Y, Nishiyama Y (2018) Oxidation of translation factor EF-Tu inhibits the repair of photosystem II. Plant Physiol 176(4):2691–2699
Jones DT, Taylor WR, Thornton JM (1992) The rapid generation of mutation data matrices from protein sequences. Bioinformatics 8(3):275–282
Jonker CZ, van Ginkel C, Olivier J (2013) Association between physical and geochemical characteristics of thermal springs and algal diversity in Limpopo Province, South Africa. Water SA 39(1):95–104
Jordan P, Fromme P, Witt HT, Klukas O, Saenger W, Krauß N (2001) Three-dimensional structure of cyanobacterial photosystem I at 2.5 Å resolution. Nature 411(6840):909–917
Jungblut AD, Lovejoy C, Vincent WF (2010) Global distribution of cyanobacterial ecotypes in the cold biosphere. ISME J 4(2):191–202
Jürgens N, Oncken I, Oldeland J, Gunter F, Rudolph B (2021) *Welwitschia*: phylogeography of a living fossil, diversified within a desert refuge. Sci Rep 11(1):1–14
Kale R, Hebert AE, Frankel LK, Sallans L, Bricker TM, Pospíšil P (2017) Amino acid oxidation of the D1 and D2 proteins by oxygen radicals during photoinhibition of photosystem II. Proc Natl Acad Sci 114(11):2988–2993
Khalifa SAM, Shedid ES, Saied EM, Jassbi AR, Jamebozorgi FH, Rateb ME, Du M, Abdel-Daim MM, Kai GY, Al-Hammady MAM, Xiao J, Guo Z, El-Seedi HR (2021) Cyanobacteria—from the oceans to the potential biotechnological and biomedical applications. Mar Drugs 19(5)
Klähn S, Hagemann M (2011) Compatible solute biosynthesis in cyanobacteria. Environ Microbiol 13(3):551–562
Klatt CG, Wood JM, Rusch DB, Bateson MM, Hamamura N, Heidelberg JF, Grossman AR, Bhaya D, Cohan FM, Kühl M, Bryant DA, Ward DM (2011) Community ecology of hot spring cyanobacterial mats: predominant populations and their functional potential. ISME J 5(8):1262–1278
Kłodawska K, Kovács L, Várkonyi Z, Kis M, Sozer Ö, Laczkó-Dobos H, Kóbori O, Domonkos I, Strzałka K, Gombos Z, Malec P (2015) Elevated growth temperature can enhance photosystem I trimer formation and affects xanthophyll biosynthesis in cyanobacterium *Synechocystis* sp. PCC6803 cells. Plant Cell Physiol 56(3):558–571
Knight S, Andersson I, Brändén CI (1990) Crystallographic analysis of ribulose 1,5-bisphosphate carboxylase from spinach at 2.4 Å resolution: subunit interactions and active site. J Mol Biol 215(1):113–160
Knoll AH, Summons RE, Waldbauer JR, Zumberge JE (2007) The geological succession of primary producers in the oceans. In: Evolution of primary producers in the sea. Academic, New York, pp 133–163
Ko SH, Lim YH, Kim EJ, Ko YW, Hong IS, Kim S, Jung YJ (2022) Antarctic marine algae extracts as a potential natural resource to protect epithelial barrier integrity. Mar Drugs 20(9):562
Koike H, Katoh S (1980) Thermal properties of NADP: ferredoxin oxidoreductase and ferredoxin isolated from a thermophilic blue-green alga. Photosynth Res 1:163–170
Koike H, Satoh K, Katoh S (1982) Heat-stabilities of electron transport related to photosystem I in a thermophilic blue-green alga, *Synechococcus* sp. Plant Cell Physiol 23(2):293–299

Koike H, Ikeuchi M, Hiyama T, Inoue Y (1989) Identification of photosystem I components from the cyanobacterium, *Synechococcus vulcanus* by N-terminal sequencing. FEBS Lett 253(1–2):257–263

Kojima K, Oshita M, Nanjo Y, Kasai K, Tozawa Y, Hayashi H, Nishiyama Y (2007) Oxidation of elongation factor G inhibits the synthesis of the D1 protein of photosystem II. Mol Microbiol 65(4):936–947

Kojima K, Motohashi K, Morota T, Oshita M, Hisabori T, Hayashi H, Nishiyama Y (2009) Regulation of translation by the redox state of elongation factor G in the cyanobacterium *Synechocystis* sp. PCC 6803. J Biol Chem 284(28):18685–18691

Komárek J (1999) Diversity of cyanoprokaryotes (cyanobacteria) of King George Island, maritime Antarctica—a survey. Algological Studies/Archiv Für Hydrobiologie 94:181–193

Komárek O, Komárek J (2010) Diversity and ecology of cyanobacterial microflora of Antarctic seepage habitats: comparison of King George Island, Shetland Islands, and James Ross Island, NW Weddell Sea, Antarctica. In: Microbial mats: modern and ancient microorganisms in stratified systems. Springer, Berlin, pp 515–539

Kondrashov FA (2012) Gene duplication as a mechanism of genomic adaptation to a changing environment. Proc R Soc B Biol Sci 279(1749):5048–5057

Körner C (2003) Alpine plant life: functional plant ecology of high mountain ecosystems. Springer, Berlin

Körner C (2011) Coldest places on earth with angiosperm plant life. Alp Bot 121:11–22

Kós PB, Deák Z, Cheregi O, Vass I (2008) Differential regulation of *psbA* and *psbD* gene expression, and the role of the different D1 protein copies in the cyanobacterium *Thermosynechococcus elongatus* BP-1. Biochim Biophys Acta Bioenerg 1777(1):74–83

Krauß N, Schubert WD, Klukas O, Fromme P, Witt HT, Saenger W (1996) Photosystem I at 4 Å resolution represents the first structural model of a joint photosynthetic reaction centre and core antenna system. Nat Struct Biol 3(11):965–973

Kulkarni RD, Golden SS (1994) Adaptation to high light intensity in *Synechococcus* sp. strain PCC 7942: regulation of three *psbA* genes and two forms of the D1 protein. J Bacteriol 176(4):959–965

Kumar S, Stecher G, Li M, Knyaz C, Tamura K (2018) MEGA X: molecular evolutionary genetics analysis across computing platforms. Mol Biol Evol 35(6):1547

Kurz M (2008) Compatible solute influence on nucleic acids: many questions but few answers. Saline Syst 4(1):1–14

Kvíderová J, Elster J, Komárek J (2018) Ecophysiology of cyanobacteria in the polar regions. In: Cyanobacteria: from basic science to applications. Academic, London, pp 277–302

Lacap-Bugler DC, Lee KK, Archer S, Gillman LN, Lau MCY, Leuzinger S, Lee CK, Maki T, McKay CP, Perrott JK, de los Rios-Murillo A, Warren-Rhodes KA, Hopkins DW, Pointing SB (2017) Global diversity of desert hypolithic cyanobacteria. Front Microbiol 8:867

Laskowski RA, Jabłońska J, Pravda L, Vařeková RS, Thornton JM (2018) PDBsum: structural summaries of PDB entries. Protein Sci 27(1):129–134

Lea-Smith DJ, Bombelli P, Vasudevan R, Howe CJ (2016a) Photosynthetic, respiratory and extracellular electron transport pathways in cyanobacteria. Biochim Biophys Acta Bioenerg 1857(3):247–255

Lea-Smith DJ, Ortiz-Suarez ML, Lenn T, Nürnberg DJ, Baers LL, Davey MP, Parolini L, Huber RG, Cotton CAR, Mastroianni G, Bombelli P, Ungerer P, Stevens TJ, Smith AG, Bond PJ, Mullineaux CW, Howe CJ (2016b) Hydrocarbons are essential for optimal cell size, division, and growth of cyanobacteria. Plant Physiol 172(3):1928–1940

Letunic I, Bork P (2021) Interactive Tree Of Life (iTOL) v5: an online tool for phylogenetic tree display and annotation. Nucleic Acids Res 49(W1):W293–W296

Li M, Semchonok DA, Boekema EJ, Bruce BD (2014) Characterization and evolution of tetrameric photosystem I from the thermophilic cyanobacterium *Chroococcidiopsis* sp TS-821. Plant Cell 26(3):1230–1245

Li M, Calteau A, Semchonok DA, Witt TA, Nguyen JT, Sassoon N, Boekema EJ, Whitelegge J, Gugger M, Bruce BD (2019) Physiological and evolutionary implications of tetrameric photosystem I in cyanobacteria. Nat Plants 5(12):1309–1319

Liang Y, Kaczmarek MB, Kasprzak AK, Tang J, Shah MMR, Jin P, Klepacz-Smółka A, Cheng JJ, Daroch M (2018) Thermosynechococcaceae as a source of thermostable C-phycocyanins: properties and molecular insights. Algal Res 35:223–235

Lindahl M, Spetea C, Hundal T, Oppenheim AB, Adam Z, Andersson B (2000) The thylakoid FtsH protease plays a role in the light-induced turnover of the photosystem II D1 protein. Plant Cell 12(3):419–431

Liu LN, Chen XL, Zhang YZ, Zhou BC (2005) Characterization, structure and function of linker polypeptides in phycobilisomes of cyanobacteria and red algae: an overview. Biochim Biophys Acta Bioenerg 1708(2):133–142

Liu LB, Bai WP, Li HJ, Tian Y, Yuan HJ, Garant TM, Liu HS, Zhang J, Bao AK, Rowland O, Wang SM (2021) ZxABCG11 from the xerophyte *Zygophyllum xanthoxylum* enhances drought tolerance in *Arabidopsis thaliana* through modulating cuticular wax accumulation. Environ Exp Bot 190:104570

Loll B, Kern J, Saenger W, Zouni A, Biesiadka J (2005) Towards complete cofactor arrangement in the 3.0 Å resolution structure of photosystem II. Nature 438(7070):1040–1044

Loll B, Broser M, Kos PB, Kern J, Biesiadka J, Vass I, Saenger W, Zouni A (2008) Modeling of variant copies of subunit D1 in the structure of photosystem II from *Thermosynechococcus elongatus*. Biol Chem 389:609–617

Luo W, Chen B, Li CY, Li K, Zhang XY, Qin QL, Zhou BC, Chen XL, Zhang YZ, Xie BB (2017) Structural insights into the cold adaptation of the photosynthetic pigment-protein C-phycocyanin from an Arctic cyanobacterium. Biochim Biophys Acta Bioenerg 1858(4):325–335

MacColl R, Eisele LE, Williams EC, Bowser SS (1996) The discovery of a novel R-phycoerythrin from an antarctic red alga. J Biol Chem 271(29):17157–17160

Mamedov MD, Hayashi H, Wada H, Mohanty PS, Papageorgiou GC, Murata N (1991) Glycinebetaine enhances and stabilizes the evolution of oxygen and the synthesis of ATP by cyanobacterial thylakoid membranes. FEBS Lett 294(3):271–274

Margesin R, Häggblom MM (2007) Thematic issue: microorganisms in cold environments. FEMS Microbiol Ecol 59(2):215–216

McCleary JA (1959) The bryophytes of a Desert Region in Arizona. Bryologist 62(1):58–62

Miller SR, McGuirl MA, Carvey D (2013) The evolution of RuBisCO stability at the thermal limit of photoautotrophy. Mol Biol Evol 30(4):752–760

Miyashita H, Ikemoto H, Kurano N, Miyachi S, Chihara M (2003) *Acaryochloris marina* gen. et sp. nov. (cyanobacteria), an oxygenic photosynthetic prokaryote containing Chl d as a major pigment. J Phycol 39(6):1247–1253

Mohanta TK, Mohanta YK, Kaushik P, Kumar J (2023) Physiology, genomics, and evolutionary aspects of desert plants. J Adv Res

Mohanty N, Murthy SDS, Mohanty P (1987) Reversal of heat-induced alterations in photochemical activities in wheat primary leaves. Photosynth Res 14:259–267

Mühlenhoff U, Haehnel W, Witt H, Herrmann RG (1993) Genes encoding eleven subunits of photosystem I from the thermophilic cyanobacterium *Synechococcus* sp. Gene 127(1):71–78

Mulo P, Sicora C, Aro EM (2009) Cyanobacterial *psbA* gene family: optimization of oxygenic photosynthesis. Cell Mol Life Sci 66:3697–3710

Mulo P, Sakurai I, Aro EM (2012) Strategies for *psbA* gene expression in cyanobacteria, green algae and higher plants: from transcription to PSII repair. Biochim Biophys Acta 1817(1):247–257

Nagarajan A, Pakrasi HB (2001) Membrane-bound protein complexes for photosynthesis and respiration in cyanobacteria. eLS, pp 1–8

Nakajima M, Sakamoto T, Wada K (2002) The complete purification and characterization of three forms of ferredoxin-NADP+ oxidoreductase from a thermophilic cyanobacterium *Synechococcus elongatus*. Plant Cell Physiol 43(5):484–493

Nickelsen J, Rengstl B (2013) Photosystem II assembly: from cyanobacteria to plants. Annu Rev Plant Biol 64:609–635

Nield J, Rizkallah PJ, Barber J, Chayen NE (2003) The 1.45 Å three-dimensional structure of C-phycocyanin from the thermophilic cyanobacterium *Synechococcus elongatus*. J Struct Biol 141(2):149–155

Nikulina TV (2010) Biodiversity of algae of hot springs from the Kuril Islands (Russia). Int J Algae 12(4):299–320
Nishiyama Y, Murata N (2014) Revised scheme for the mechanism of photoinhibition and its application to enhance the abiotic stress tolerance of the photosynthetic machinery. Appl Microbiol Biotechnol 98:8777–8796
Nishiyama Y, Hayashi H, Watanabe T, Murata N (1994) Photosynthetic oxygen evolution is stabilized by cytochrome c550 against heat inactivation in *Synechococcus* sp. PCC 7002. Plant Physiol 105(4):1313–1319
Nishiyama Y, Los DA, Murata N (1999) PsbU, a protein associated with photosystem II, is required for the acquisition of cellular thermotolerance in *Synechococcus* species PCC 7002. Plant Physiol 120(1):301–308
Nishiyama Y, Allakhverdiev SI, Yamamoto H, Hayashi H, Murata N (2004) Singlet oxygen inhibits the repair of photosystem II by suppressing the translation elongation of the D1 protein in *Synechocystis* sp. PCC 6803. Biochemistry 43(35):11321–11330
Nishiyama Y, Allakhverdiev SI, Murata N (2006) A new paradigm for the action of reactive oxygen species in the photoinhibition of photosystem II. Biochim Biophys Acta Bioenerg 1757(7):742–749
Nixon PJ, Michoux F, Yu J, Boehm M, Komenda J (2010) Recent advances in understanding the assembly and repair of photosystem II. Ann Bot 106(1):1–16
Ohnishi N, Murata N (2006) Glycine betaine counteracts the inhibitory effects of salt stress on the degradation and synthesis of D1 protein during photoinhibition in *Synechococcus* sp. PCC 7942. Plant Physiol 141(2):758–765
Papageorgiou GC, Murata N (1995) The unusually strong stabilizing effects of glycine betaine on the structure and function of the oxygen-evolving photosystem II complex. Photosynth Res 44:243–252
Papke RT, Ramsing NB, Bateson MM, Ward DM (2003) Geographical isolation in hot spring cyanobacteria. Environ Microbiol 5(8):650–659
Peary JA, Castenholz RW (1964) Temperature strains of a thermophilic blue-green alga. Nature 202(4933):720
Pedersen D, Miller SR (2017) Photosynthetic temperature adaptation during niche diversification of the thermophilic cyanobacterium *Synechococcus* A/B clade. ISME J 11(4):1053–1057
Peguero-Pina JJ, Vilagrosa A, Alonso-Forn D, Ferrio JP, Sancho-Knapik D, Gil-Pelegrín E (2020) Living in drylands: functional adaptations of trees and shrubs to cope with high temperatures and water scarcity. Forests 11(10):1–23
Perera-Castro AV, Waterman MJ, Turnbull JD, Ashcroft MB, Mckinley E, Watling JR, Bramley-Alves J, Casanova-Katny A, Zuniga G, Flexas J, Robinson SA (2020) It is hot in the sun: Antarctic mosses have high temperature optima for photosynthesis despite cold climate. Front Plant Sci 11:1178
Poli A, Finore I, Romano I, Gioiello A, Lama L, Nicolaus B (2017) Microbial diversity in extreme marine habitats and their biomolecules. Microorganisms 5(2):1–30
Pospíšil P (2016) Production of reactive oxygen species by photosystem II as a response to light and temperature stress. Front Plant Sci 7:1950
Preetha K, John L, Subin CS, Vijayan KK (2012) Phenotypic and genetic characterization of *Dunaliella* (Chlorophyta) from Indian aalinas and their diversity. Aquat Biosyst 8(1):10–12
Prieto-Barajas CM, Valencia-Cantero E, Santoyo G (2018) Microbial mat ecosystems: structure types, functional diversity, and biotechnological application. Electron J Biotechnol 31:48–56
Prondzinsky P, Berkemer SJ, Ward LM, McGlynn SE (2021) The thermosynechococcus genus: wide environmental distribution, but a highly conserved genomic core. Microbes Environ 36(2):1–9
Pueyo JJ, Alfonso M, Andrés C, Picorel R (2002) Increased tolerance to thermal inactivation of oxygen evolution in spinach photosystem II membranes by substitution of the extrinsic 33-kDa protein by its homologue from a thermophilic cyanobacterium. Biochim Biophys Acta Bioenerg 1554(1–2):29–35

Pumas C, Pruetiworanan S, Peerapornpisal Y (2018) Diatom diversity in some hot springs of northern Thailand. Botanica 24(1):69–86

Rashmi P, Mona K (2021) Marine macroalgae diversity at Bhuigaon Coast, District Palghar, Maharashtra. J Sci Res 65(6):126–131

Raven JA (2013) Rubisco: still the most abundant protein of Earth? New Phytol 198(1):1–3

Reysenbach AL, Shock E (2002) Merging genomes with geochemistry in hydrothermal ecosystems. Science 296(5570):1077–1082

Rhee KH, Morris EP, Barber J, Kühlbrandt W (1998) Three-dimensional structure of the plant photosystem II reaction centre at 8 Å resolution. Nature 396(6708):283–286

Rowland JG, Pang X, Suzuki I, Murata N, Simon WJ, Slabas AR (2010) Identification of components associated with thermal acclimation of photosystem II in *Synechocystis* sp. PCC6803. PLoS One 5(5):e10511

Saenger W, Jordan P, Krauß N (2002) The assembly of protein subunits and cofactors in photosystem I. Curr Opin Struct Biol 12(2):244–254

Saitou N, Nei M (1987) The neighbor-joining method: a new method for reconstructing phylogenetic trees. Mol Biol Evol 4(4):406–425

Sakurai I, Shen JR, Leng J, Ohashi S, Kobayashi M, Wada H (2006) Lipids in oxygen-evolving photosystem II complexes of cyanobacteria and higher plants. J Biochem 140(2):201–209

Samsonoff WA, MacColl R (2001) Biliproteins and phycobilisomes from cyanobacteria and red algae at the extremes of habitat. Arch Microbiol 176(6):400–405

Santos H, Da Costa MS (2002) Compatible solutes of organisms that live in hot saline environments. Environ Microbiol 4(9):501–509

Scafaro AP, Posch BC, Evans JR, Farquhar GD, Atkin OK (2023) Rubisco deactivation and chloroplast electron transport rates co-limit photosynthesis above optimal leaf temperature in terrestrial plants. Nat Commun 14(1):2820

Schirmer T, Huber R, Schneider M, Bode W, Miller M, Hackert ML (1986) Crystal structure analysis and refinement at 2.5 Å of hexameric C-phycocyanin from the cyanobacterium *Agmenellum quadruplicatum*: the molecular model and its implications for light-harvesting. J Mol Biol 188(4):651–676

Schluchter WM, Shen G, Zhao J, Bryant DA (1996) Characterization of psaI and psaL mutants of *Synechococcus* sp. strain PCC 7002: a new model for state transitions in cyanobacteria. Photochem Photobiol 64(1):53–66

Seckbach J, Oren A (2007) Oxygenic photosynthetic microorganisms in extreme environments: possibilities and limitations. In: Algae and cyanobacteria in extreme environments. Springer Netherlands, Dordrecht, pp 3–25

Segawa T, Matsuzaki R, Takeuchi N, Akiyoshi A, Navarro F, Sugiyama S, Yonezawa T, Mori H (2018) Bipolar dispersal of red-snow algae. Nat Commun 9(1)

Semchonok DA, Li M, Bruce BD, Oostergetel GT, Boekema EJ (2016) Cryo-EM structure of a tetrameric cyanobacterial photosystem I complex reveals novel subunit interactions. Biochim Biophys Acta Bioenerg 1857(9):1619–1626

Serna L, Martin C (2006) Trichomes: different regulatory networks lead to convergent structures. Trends Plant Sci 11(6):274–280

Sharma A, Kawarabayasi Y, Satyanarayana T (2012) Acidophilic bacteria and archaea: acid stable biocatalysts and their potential applications. Extremophiles 16(1):1–19

Sharma S, Mishra A, Priyanka, Singh SS (2024) Redox cycle: signaling and metabolic cross-talks in cyanobacteria. In: Cyanobacteria. Academic, New York, pp 177–214

Shi X, Bloom A (2021) Photorespiration: the futile cycle? Plants 10(5):908

Shubin VV, Terekhova IV, Bolychevtseva YV, El-Mohsnawy E, Rögner M, Mäntele W, Kopczak MJ, Džafić E (2017) Thermostability of photosystem I trimers and monomers from the cyanobacterium *Thermosynechococcus elongatus*. Spectrochim Acta A Mol Biomol Spectrosc 179:17–22

Siangbood H, Ramanujam P (2011) A report on thermophilic cyanophyta (cyanobacteria) from Jakrem Hotspring, Meghalaya. Int J Algae 13(2):178–185

Sidler WA (1994) Phycobilisome and phycobiliprotein structures. In: The molecular biology of cyanobacteria. Springer Netherlands, Dordrecht, pp 139–216

Silambarasan G, Ramanathan T, Kathiresan K (2012) Diversity of marine cyanobacteria from three mangrove environment in Tamil Nadu coast, South East coast of India. Curr Res J Biol Sci 4(3):235–238

Simberloff D, Rejmánek M (eds) (2011) Encyclopedia of biological invasions (No. 3). University of California Press, Berkeley

Singh SM, Elster J (2007) Cyanobacteria in Antarctic lake environments: a mini-review. In: Algae and cyanobacteria in extreme environments. Springer, Berlin, pp 303–320

Sisó S, Camarero J, Gil-Pelegrín E (2001) Relationship between hydraulic resistance and leaf morphology in broadleaf *Quercus* species: a new interpretation of leaf lobation. Trees Struct Funct 15(6):341–345

Six C, Thomas JC, Garczarek L, Ostrowski M, Dufresne A, Blot N, Scanlan DJ, Partensky F (2007) Diversity and evolution of phycobilisomes in marine *Synechococcus* spp.: a comparative genomics study. Genome Biol 8(12):R259

Smith WK (1978) Temperatures of desert plants: another perspective on the adaptability of leaf size. Science 201(4356):614–616

Stadnichuk IN, Kusnetsov VV (2023) Phycobilisomes and phycobiliproteins in the pigment apparatus of oxygenic photosynthetics: from cyanobacteria to tertiary endosymbiosis. Int J Mol Sci 24(3)

Stadnichuk IN, Krasilnikov PM, Zlenko DV (2015) Cyanobacterial phycobilisomes and phycobiliproteins. Microbiology 84:101–111

Stamatakis C, Papageorgiou GC (1993) Stabilization of photosystem II particles isolated from the thermophilic cyanobacterium *Phormidium laminosum* with glycinebetaine and glycerol. Biochim Biophys Acta Bioenerg 1183(2):333–338

Stark LR, Whittemore AT (2000) Bryophytes from the Northern Mojave Desert. Southwest Nat 45(2):226–232

Stewart WD (1970) Nitrogen fixation by blue-green algae in Yellowstone thermal areas. Phycologia 9(3–4):261–268

Stout RG, Al-Niemi TS (2002) Heat-tolerant flowering plants of active geothermal areas in Yellowstone National Park. Ann Bot 90(2):259–267

Su X, Ma J, Wei X, Cao P, Zhu D, Chang W, Liu Z, Zhang X, Li M (2017) Structure and assembly mechanism of plant $C_2S_2M_2$-type PSII-LHCII supercomplex. Science 357(6353):815–820

Sugiura M, Boussac A (2014) Some photosystem II properties depending on the D1 protein variants in *Thermosynechococcus elongatus*. Biochim Biophys Acta Bioenerg 1837(9):1427–1434

Sugiura M, Iwai E, Hayashi H, Boussac A (2010) Differences in the interactions between the subunits of photosystem II dependent on D1 protein variants in the thermophilic cyanobacterium *Thermosynechococcus elongatus*. J Biol Chem 285(39):30008–30018

Szabados L, Kovács H, Zilberstein A, Bouchereau A (2011) Plants in extreme environments: importance of protective compounds in stress tolerance. In: Advances in botanical research, vol 57. Academic, New York, pp 105–150

Szilágyi A, Závodszky P (2000) Structural differences between mesophilic, moderately thermophilic and extremely thermophilic protein subunits: results of a comprehensive survey. Structure 8(5):493–504

Tabita FR (1999) Microbial ribulose 1,5-bisphosphate carboxylase/oxygenase: a different perspective. Photosynth Res 60:1–28

Tang J, Zhou H, Yao D, Du L, Daroch M (2023) Characterization of molecular diversity and organization of phycobilisomes in thermophilic cyanobacteria. Int J Mol Sci 24(6):5632

Taton A, Grubisic S, Balthasart P, Hodgson DA, Laybourn-Parry J, Wilmotte A (2006a) Biogeographical distribution and ecological ranges of benthic cyanobacteria in East Antarctic lakes. FEMS Microbiol Ecol 57(2):272–289

Taton A, Grubisic S, Ertz D, Hodgson DA, Piccardi R, Biondi N, Tredici MR, Mainini M, Losi D, Marinelli F, Wilmotte A (2006b) Polyphasic study of Antarctic cyanobacterial strains. J Phycol 42(6):1257–1270

Tikkanen M, Mekala NR, Aro EM (2014) Photosystem II photoinhibition-repair cycle protects photosystem I from irreversible damage. Biochim Biophys Acta Bioenerg 1837(1):210–215

Umena Y, Kawakami K, Shen JR, Kamiya N (2011) Crystal structure of oxygen-evolving photosystem II at a resolution of 1.9 Å. Nature 473(7345):55–60

van Eerden FJ, De Jong DH, De Vries AH, Wassenaar TA, Marrink SJ (2015) Characterization of thylakoid lipid membranes from cyanobacteria and higher plants by molecular dynamics simulations. Biochim Biophys Acta Biomembr 1848(6):1319–1330

van Etten J, Cho CH, Yoon HS, Bhattacharya D (2023) Extremophilic red algae as models for understanding adaptation to hostile environments and the evolution of eukaryotic life on the early earth. In: Seminars in cell and developmental biology, vol 134. Academic, New York, pp 4–13

van Lun M, Hub JS, van der Spoel D, Andersson I (2014) CO_2 and O_2 distribution in Rubisco suggests the small subunit functions as a CO_2 reservoir. J Am Chem Soc 136(8):3165–3171

Velichko N, Smirnova S, Averina S, Pinevich A (2021) A survey of Antarctic cyanobacteria. Hydrobiologia 848(11):2627–2652

Venkataramanaiah V, Sudhir P, Murthy SDS (2003) Effect of high temperature on photosynthetic electron transport activities of the cyanobacterium *Spirulina platensis*. Photosynthetica 41:331–334

Vinyard DJ, Ananyev GM, Charles Dismukes G (2013) Photosystem II: the reaction center of oxygenic photosynthesis. Annu Rev Biochem 82:577–606

Vogel S (1970) Convective cooling at low airspeeds and the shapes of broad leaves. J Exp Bot 21(1):91–101

Vogel S (2009) Leaves in the lowest and highest winds: temperature, force and shape: Tansley review. New Phytol 183(1):13–26

Von Hase A, Cowling RM, Ellis AG (2006) Petal movement in cape wildflowers protects pollen from exposure to moisture. Plant Ecol 184(1):75–87

Wan T, Liu Z, Leitch IJ, Xin H, Maggs-Kölling G, Gong Y, Li Z, Marais E, Liao Y, Dai C, Liu F, Wu Q, Song C, Zhou Y, Huang W, Jiang K, Wang Q, Yang Y, Zhong Z, Yang M, Yan X, Hu G, Hou C, Su Y, Feng S, Yang J, Yan J, Chu JF, Ran J, Wang X, de Peer YV, Leitch AR, Wang Q (2021) The *Welwitschia* genome reveals a unique biology underpinning extreme longevity in deserts. Nat Commun 12(1):4247

Wani SH, Brajendra Singh N, Haribhushan A, Iqbal Mir J (2013) Compatible solute engineering in plants for abiotic stress tolerance—role of glycine betaine. Curr Genomics 14(3):157–165

Wegmann K (1986) Osmoregulation in eukaryotic algae. FEMS Microbiol Lett 39(1–2):37–43

Witt I, Witt HT, Di Fiore D, Rögner M, Hinrichs W, Saenger W, Granzin J, Betzel C, Dauter Z (1988) X-ray characterization of single crystals of the reaction center I of water splitting photosynthesis. Ber Bunsenges Phys Chem 92(12):1503–1506

Wynn-Williams DD (2000) Cyanobacteria in deserts—life at the limit? In: The ecology of cyanobacteria: their diversity in time and space. Springer, Dordrecht, pp 341–366

Xiong F, Lederer F, Lukavský J, Nedbal L (1996) Screening of freshwater algae (Chlorophyta, Chromophyta) for ultraviolet-B sensitivity of the photosynthetic apparatus. J Plant Physiol 148(1–2):42–48

Yamamoto Y, Aminaka R, Yoshioka M, Khatoon M, Komayama K, Takenaka D, Yamashita A, Nijo N, Inagawa K, Morita N, Sasaki T, Yamamoto Y (2008) Quality control of photosystem II: impact of light and heat stresses. Photosynth Res 98:589–608

Yamaoka T, Satoh K, Katoh S (1978) Photosynthetic activities of a thermophilic blue-green alga. Plant Cell Physiol 19(6):943–954

Yamori W, Shikanai T, Makino A (2015) Photosystem I cyclic electron flow via chloroplast NADH dehydrogenase-like complex performs a physiological role for photosynthesis at low light. Sci Rep 5(1):13908

Yutthanasirikul R, Nagano T, Jimbo H, Hihara Y, Kanamori T, Ueda T, Haruyama T, Konno H, Yoshida K, Hisabori T, Nishiyama Y (2016) Oxidation of a cysteine residue in elongation factor EF-Tu reversibly inhibits translation in the cyanobacterium *Synechocystis* sp. PCC 6803. J Biol Chem 291(11):5860–5870

Zakar T, Herman E, Vajravel S, Kovacs L, Knoppová J, Komenda J, Domonkos I, Kis M, Gombos Z, Laczko-Dobos H (2017) Lipid and carotenoid cooperation-driven adaptation to light and temperature stress in *Synechocystis* sp. PCC6803. Biochim Biophys Acta Bioenerg 1858(5):337–350

Zakhia F, Jungblut AD, Taton A, Vincent WF, Wilmotte A (2008) Cyanobacteria in cold ecosystems. In: Psychrophiles: from biodiversity to biotechnology. Springer, Berlin, pp 121–135

Zhang L, Liu J (2016) Effects of heat stress on photosynthetic electron transport in a marine cyanobacterium *Arthrospira* sp. J Appl Phycol 28:757–763

Zhang L, Zhang Z, Cao J, Wang K, Qin L, Sun Y, Ju W, Qu C, Miao J (2023) Extreme environmental adaptation mechanisms of Antarctic bryophytes are mainly the activation of antioxidants, secondary metabolites and photosynthetic pathways. BMC Plant Biol 23(1):399

Zhu M, Li Q, Zhang Y, Zhang M, Li Z (2022) Glycine betaine increases salt tolerance in maize (*Zea mays* L.) by regulating Na^+ homeostasis. Front Plant Sci 13:978304

Zouni A, Jordan R, Schlodder E, Fromme P, Witt HT (2000) First photosystem II crystals capable of water oxidation. Biochim Biophys Acta Bioenerg 1457(3):103–105

Zouni A, Witt HT, Kern J, Fromme P, Krauss N, Saenger W, Orth P (2001) Crystal structure of photosystem II from *Synechococcus elongatus* at 3.8 Å resolution. Nature 409(6821):739–743

Underwater Photosynthesis in Cyanobacteria: Challenges and Adaptations

11

Vinod Kumar, Vaishnavi Kedia, Satyam Jaiswal, and Surbhi Kharwar

Abstract

Within the class of photosynthetic organisms, cyanobacteria are among the most prevalent and diverse groups. They can multiply under a wide range of conditions which can be attributed to the adaptability of their light-harvesting system. Compared to terrestrial ecosystems, oceans receive different wavelengths of solar energy that are needed for photosynthesis. Red light cannot pass below ~15 m because water quickly attenuates longer light wavelengths. Blue and green wavelengths dominate the illumination of near-surface waters (those above 30 m) in the open ocean, but blue light (400–500 nm) is more abundant in deeper waters (100–200 m). Since seawater in coastal places contains higher amounts of humic compounds and phytoplankton, green wavelengths typically prevail in these environments. Via light-harvesting antennae, oxygen-consuming photosynthetic organisms transform light energy into electron flow within the reaction center of their photosystems. They handle low-gas exchange underwater circumstances in addition to light. Gas penetration, which has a 10^4-fold lower diffusion coefficient in water than in air, is a major barrier for cyanobacteria. As a result, aquatic organisms are far more likely than terrestrial ones to have inorganic carbon limitation in photosynthesis. Underwater, limited gas exchange has led to the evolution of a variety of adaptive traits. The ability to absorb and utilize the particular wavelengths that enrich a particular area and depth of the ocean would provide a photosynthetic organism an advantage over other organisms in that habitat. Many pigment-protein complexes have evolved in photosynthetic organisms to absorb light and transfer excited-state energy to reaction centers where

V. Kumar
Department of Botany, CollegePandit Prithi Nath (PG) College, Kanpur, Uttar Pradesh, India

V. Kedia · S. Jaiswal · S. Kharwar (✉)
Department of Botany, University of Lucknow, Lucknow, Uttar Pradesh, India

A. K. Mishra (ed.), *Stress Biology in Photosynthetic Organisms*,
https://doi.org/10.1007/978-981-97-1883-2_11

photochemical and electron transfer activities occur. These light-harvesting antennas boost the effective absorption cross section of the reaction centers. Hence, the present chapter provides insight into the adaptive features of cyanobacteria to improve light consumption underwater, as these acclimations affect underwater photosynthesis.

Keywords

Cyanobacteria · Adaptation · Underwater · Photosynthesis · Light · Light-harvesting system

11.1 Introduction

Earth and life on it have transformed as a result of photosynthetic organisms converting sunlight into energy. Early in Earth's history, photosynthetic life first appeared and it is most likely that these first forms were anoxygenic (non-oxygen evolving). The enzymatic and energetic foundations of life underwent a radical change when oxygenic photosynthesis was developed and atmospheric oxygen levels rose, about 2.4 billion years ago. This revolution has resulted in new biosynthetic routes for photosynthetic cofactors and changes to electron carriers, pigments, and both conventional and non-conventional photosynthetic carbon fixation methods.

Both endosymbiotic events and lateral gene transfers involving photosynthetic components complicate the evolutionary history of photosynthetic organisms. Given that all recognized instances of the process are located in the bacterial domain, it is highly likely that the bacterial domain is where the process originated. Before 1900, cyanobacteria, purple bacteria, and green sulfur bacteria were classified; however, the most recent type of photosynthetic organisms has just lately been identified (Bryant et al. 2007). Currently, we are aware of six major categories, or phyla, that comprise species that contain chloroplasts. They include filamentous anoxygenic phototrophs, purple bacteria, green sulfur bacteria, cyanobacteria, heliobacteria, and acidobacteria. The distribution of bacterial groups is highly uneven, which is probably caused by a mix of lateral gene transfer events involving photosynthetic components and vertical inheritance of the primary metabolic capacities (Raymond et al. 2002).

Stanier and Cohen-Bazire (Stanier and Cohen-Bazire 1977) reviewed the cyanobacterial cell's overall structure. The majority of cyanobacteria have two photosystems (photosystem I and II) on their typical thylakoid membranes (Whittson and Potts 2007). But in contrast to the majority of photosynthetic eukaryotes, they lack chlorophyll *b* and instead obtain their light energy from special multimolecular structures called phycobilisome (PBS). The only other known living things with such structures are cyanobacteria, except for eukaryotic red algae. PBS are found uniformly on the stromal surface of the thylakoid membrane and are exceptionally efficient at about 100% when it comes to light harvesting and energy transmission (Glazer 1989). The internal inclusions of cyanobacteria are also frequently very diverse; most of these have reserve roles and accumulate in unfavorable nutritional imbalance situations. A non-unit membrane may encircle some of these inclusions (Allen 1984; Jensen 1984; Codd 1988) possibly used as a source of nitrogen under some conditions. According to Fogg et al. (1973), there are two major biological categories of aquatic cyanobacteria: planktonic cyanobacteria, which float freely in the water column, and benthic cyanobacteria, which stick to submerged solid surfaces such as sediments, rocks, algae, and aquatic plants. One known element that affects the growth and dispersal of phytoplankton in the spring is temperature. However, temperature and light have different effects on phytoplankton growth and dispersal, and these effects are frequently mingled, making it difficult to distinguish between them in field investigations.

In addition to variations in solar elevation and waves, Water bodies have variable photophotic environments due to complex depth-dependent relationships between light and suspended particulates and immersed matter, including light absorption and scattering. Because the supply of light has an immediate influence on the improvement, existence, and subsistence approaches of autotrophs like cyanobacteria, it is critical for primary production. Over short periods and spatial scales, changes in the phytoplankton's depth distribution, for example, due to vertical movement, may result in significant variations in the obtainability of light and food.

As such, nothing is known about how light affects the succession of spring phytoplankton groups. Light has a direct impact on the growth and production of phytoplankton due to its influence on photosynthesis. Light intensity has a major impact on photosynthetic efficiency and can also have an impact on phytoplankton growth and activity (Gao et al. 2019). Furthermore, eutrophication-induced increases in phytoplankton biomass. In the spring, Bacillariophytes or filamentous cyanobacteria may exacerbate light attenuation and further decrease water clarity (Zhang et al. 2018). Because of a progressive decline in incoming total radiation and water clarity, aquatic life has much-reduced access to undersea light (Zhang et al. 2020). The production of chlorophyll and photosynthetic pigments in Bacillariophyta and Chlorophyta is influenced by decreasing light intensity, which lowers their light usage efficiency. All aquatic photoautotrophs require fast acclimation mechanisms concerning the variability of the underwater light. Photoinhibition occurs above a certain irradiance level, which is one reason why photosynthesis is nonlinearly light-dependent.

Research on the photosynthesis of aquatic phytoplankton has enabled the study of underwater photosynthesis, the most significant metabolic activity for microorganisms (Smith and Walker 1980). Underwater organisms eventually run out of sugar as a result of continuous carbohydrate use during respiration or fermentation. By using underwater photosynthesis, several underwater microorganisms can generate carbohydrates. Underwater photosynthesis is crucial for the survival and proliferation of underwater organisms since it can benefit them by giving them carbohydrates and enhancing internal aeration.

The environment of the water has a direct impact on underwater photosynthesis (Schwarzenbach and Meier 1958). Because of the effects of an aquatic body's dissolved substances, suspended parts, and depth, light attenuation is the most sensitive phenomenon (Raven and Beardall 2016). Aquatic organisms developed specific strategies to cope with such adverse circumstances.

Light is commonly thought to be the primary limiting factor for the proliferation of aquatic microorganisms. China has recently experienced significant light attenuation as a result of human air pollution and aerosol loadings (Li et al. 2017; Wang et al. 2012). According to the researchers, eutrophication reduces water clarity in freshwater ecosystems and increases the concentration of suspended particle matter (Fang et al. 2022). Climate change has made floods more prevalent in China, resulting in a drop in submerged photosynthesis-active radiation (PAR) and a rise in water visibility in habitats containing freshwater (Wu and Ding 2019). According to Zhang et al., this darkening of underwater has several negative repercussions, including the ability to substantially diminish species abundances, change in the energy and material transport dynamics of freshwater ecosystems, and disrupt food chains (Zhang et al. 2022). Therefore, microbes alter their photopigments to acclimatize environmental change.

In addition to light, underwater photosynthesis is also stressed by the lack of CO_2 because CO_2 is constrained by the extracellular diffusion layer and the pace at which underwater gas diffusion occurs (Colmer et al. 2011; Yang and Liu 2015). In cyanobacteria, carbonic anhydrase catalyzes the transition from HCO_3^- to CO_2, indicating a carbon-concentrated mechanism. It might effectively raise the rate of CO_2 utilization and improve the carbon uptake of submerged organisms. According to the reports, pH levels affect the inorganic carbon sources that submerged organisms employ for photosynthesis in freshwater habitats (Pedersen et al. 2013). Furthermore, the temperature of the water may influence underwater photosynthesis. According to earlier research, the ideal temperature is between 25 °C and 30 °C, but when temperatures were substantially lower (10–15 °C), there were noticeable changes in the photosynthetic capability (Yang et al. 2015). Salinity additionally exerts an impact on the intensity of aquatic photosynthesis in seawater systems in conjunction with the above-mentioned factors (Pedersen et al. 2013).

Moreover, several researchers are focused on underwater photosynthesis, but complete knowledge about the challenges and adaptations in cyanobacterial photosynthesis under aquatic conditions is unavailable in the literature. Thus, this chapter serves to offer the underwater photosynthesis challenges and adaptations of cyanobacteria.

11.2 Cyanobacterial Underwater Photosynthesis

The biological transformation of light energy into chemical energy, which is then harnessed as organic carbon molecules, is known as photosynthesis. Each year, aquatic environments account for around 45% of all photosynthesis on Earth (Falkowski and Raven 2007; Field et al. 1998). Photosynthesis serves as the main source of organic matter for every organism in an aquatic ecosystem. As a result, the rate of photosynthesis limits the total biomass and ecosystem productivity and the total biological flow of energy on the surface of this planet.

The fixation of carbon dioxide (or HCO_3^- in water; Eq. 11.2), which is driven by light and produces organic matter, O_2, and OH, is a common way to summarize the net process of photosynthesis (Eq. 11.1). This process is catalyzed by several enzymes, including Rubisco.

$$CO_2 + H_2O \rightarrow CH_2O + O_2 \quad (11.1)$$

$$HCO_3^- + H_2O \rightarrow CH_2O + O_2 + OH^- \quad (11.2)$$

The amount of new organic matter and oxygen produced, as well as the amount of carbon dioxide (CO_2) and bicarbonate (HCO_3) that are consumed, can be used to assess the rate of the process.

Photosynthesis in general is an alkalinization process, as shown by the formation of OH in Eq. 11.2 and the corresponding utilization of CO_2 and protons in Eq. 11.1 (i.e., CO_2 + H_2O + H^+ + HCO_3^-). Because carbonate alkalinity is the primary cause of buffer capacity, it can be assessed by taking into account the pH increases that result from converting to dissolved inorganic carbon (DIC) consumption.

Cyanobacteria are Gram-negative oxygenic photosynthetic organisms that are cosmopolitan in their distribution. They are recognized by evolved aerobic forms of life on Earth's surface because of their innate ability for oxygen evolution driven by photosynthesis (Olson 2006). Over the past 3.5 billion years, cyanobacteria have evolved through adaptive diversification that maximizes their photosynthesis in various ecological niches (Fischer 2008). As a result, a wide variety of photopigments have evolved, each with unique characteristics that enable them to successfully compete on the planet (Singh et al. 2015).

The cyanobacteria include two different reaction centers (RCs) of P700, i.e., photosystem I (PSI), and P680, i.e., photosystem II (PSII). Each RC has a light-harvesting protein-pigment complex (LHC) antenna attached to it (Bearden and Malkin 1974). In the RC, excited state energy promotes primary charge separation thereby drives the linear transport of electrons between PSII and PSI (Duysens et al. 1961). RC contains the universal pigment biomolecule referred as chlorophyll *a* (Chl *a*) (Ferreira et al. 2004). In oxygenic photosynthesis, P700 and P680 chlorophyll molecules are excited at the RCs of PS I and II (which are embedded in the thylakoid membrane) by photons of light that are captured by protein complexes termed PBS. With the release of oxygen as a byproduct, this enables the extraction of electrons from water. PSII, cytochrome b6f (Cyt b6f), plastocyanin (PC), and PSI transport electrons to ferredoxin (Fd), which serves as the center for the transportation of photosynthetic electrons. The proton motive force (pmf) that is subsequently generated across the thylakoid membrane works as an electrochemical potential difference that is produced by the photosynthetic electron transfer chain (PETC) and drives adenosine triphosphate (ATP) synthases for the generation of ATP from ADP and Pi. Electrons are thereby transferred from Fd via Fd-NAD(P)H oxidoreductase (FNR), reducing $NADP^+$ to NADPH, primarily used for CO_2 reduction and fixation in the Calvin–Benson–Bassham cycle known as dark reactions (also called as light-independent reactions) to build up carbohydrate (Bassham et al. 1954). This process depends on the environmental cues and the metabolic status of the cell.

Light-harvesting complexes (LHCs) absorb light photons as the initial stage in the process of photosynthesis. Depending on their many origins, photosynthetic antenna complexes exhibit an extensive variety of spectrum characteristics and macromolecular conformations. The main light-harvesting antennae complexes in cyanobacteria are called PBS. They act as external antenna complexes connected to the stromal surfaces (Liu et al. 2008; Arteni et al. 2008). They are made up of linker polypeptides and phycobiliproteins (PBPs). PBPs that are found in the Chlorophyta and Rhodophyta are water-soluble (Liu et al. 2005).

Three cylinders grouped in a triangle PBS core make up the inner domain of PBS, which accounts for around up to 60% of the total protein content in cyanobacteria. The peripheral domain, on the other hand, consists of six rod-like structures that originate from the core (Casella 2017). PBS transmits energy to RCs while utilizing 30–50% of the total light-harvesting capacity (Singh et al. 2009).

PBSs are made up of hydrophobic linker proteins, which are non-chromophoric basic polypeptides, and acidic polypeptides are water-soluble and linked with chromophores known as PBPs, and these PBPs serve as main photoreceptors (Chaiklahan et al. 2011). According to Scheer and Zhao (2008), PBPs have a covalent interaction with the apoprotein through one or two thioether linkages, which bind the characteristic tetrapyrrole chromophores known as bilins. Bilins give them their bright colors, they absorb light of various colors. These bilins are in charge of changing the color of light by absorbing specifically visible light spectra of one color and producing light of another (Ghosh and Mishra 2017).

PBPs comprise ~60% of the total soluble cellular protein (Soni et al. 2008). PBPs are first chromophorylated to begin the assembly of PBS, which is then followed by methylation, oligomerization, and the insertion of a linker protein (Singh et al. 2015). PBS architecture and assembly are designed to make it easier to transfer energy from higher to lower energy states. The PBS core and rods are both made up of stacked PBP hexamers (Casella 2017).

The PBPs are collection of brightly colored proteins, and based on their color, energy, and absorption maxima, they are split into three major categories, namely:

(a) Pinkish-red, high-energy phycoerythrin (PE; max 540–570 nm), which is in charge of absorbing the green wavelengths of sunlight.
(b) Intermediate energy-phycocyanin (PC; max 610–620 nm), which is deep blue and absorbs the orange-red component of sunlight.
(c) Low-energy allophycocyanin (APC; max 650–655 nm), which is light blue and absorbs red wavelengths of sunlight (Ghosh and Mishra 2017).

The absorption spectrum of biliproteins is influenced by the presence of the open-chain tetrapyrrole chromophore and the surrounding in which it is located. According to Singh et al. (2015), PBS is particularly well suited for energy absorption, transport, and directing to the RCs because of interactions between chromophore, apoprotein, non-chromophoric linker, and degree of organization. APC consists of the majority of the domain in typical PBSs, while PC and PE make up the majority of the peripheral rods. In the PBS of *Synechocystis* PCC 6803 and *Synechococcus elongatus* 7942, APC is the main PBP of the core, while rods are made up of PC only (Casella 2017).

11.3 How Does the Light Energy Transfer Take Place in Phycobiliproteins?

The fluorescence process causes the light to turn from green to red. All have the fascinating PBP characteristic of emitting visible light that is a different color than the light that hits them. It is anticipated that energy transfer in PBSs will gradually move from PE to PC, APC, and then to chlorophyll.

The absorbance ranges that PSII and PSI cover are increased by the presence of PBSs. PC picks up the yellow-orange light after PE has converted it to yellow-orange and then transforms it to light red. For Chl, APC adopts this light-red light and transforms it into a deep red light. As a result, the green light has been transformed into red, which is the wavelength of light that Chl was designed by nature to absorb. The entire procedure resembles a relay race (Ghosh and Mishra 2017).

Olson and Blankenship (2004) described that color-protein complexes known as photosynthetic RCs convert excitation energy into chemical potential energy. The primary photosynthesis reactions take place in two homologous PSI and PSII types of RCs. The most prevalent photosystem in cyanobacteria, PSI, transfers electrons to NADPH as part of the linear electron transport process. 96, 22, 2, 4, 3, chlorophyll *a* molecules, phylloquinones, lipids, L-carotenes, and iron–sulfur proteins, respectively, act as intermediate redox modulators. These all are included in the multi-subunit (12 subunit) pigment-protein complex known as PSI. The names of polypeptide subunits that make up PSI are PsaA, PsaF, PsaI, PsaM, and PsaX. PsaB and PsaA, both are the basic elements of the RC core, constitute the primary electron donor P700, which is a dimer of Chl *a* molecule (Xu et al. 2001). PSI trimer formation has been shown to require the PsaL component. Compared to higher plants and algae, which exclusively contain PSI as a monomer, cyanobacteria mostly contain PSI complexes in the form of monomers and trimers (MacGregor-Chatwin et al. 2017; Casella 2017).

The release of molecular oxygen from the oxidation of water is catalyzed by PSII. The translated products of *psbC* and *psbB* are CP43 and CP47, respectively, whereas *psbE* and *psbF* encode cytochrome b-559. The smaller subunits composed of PsbI, PsbH, PsbJ, PsbM, PsbK, PsbN, PsbL, and PsbX are found in cyanobacteria and green plants. The major electron donor P680, Chl *a*, is positioned in the middle of PSII RC, which is constituted by subunits D1 and D2 (Ferreira et al. 2004).

The fundamental process of charge separation in RCs is to release electrons that are mostly transferred via different cytochromes. The transmembrane is home to the dimeric membrane-intrinsic Cyt b_6f complex. It is essential for the electron transport chains involved in respiration and photosynthesis processes. Cyt b_6f collects electrons from PSII via plastoquinol and sends them to PSI via reducing plastocyanin in the linear electron transfer route of oxygenic photosynthesis. As a result, H^+ is released in the lumen, creating an electrochemical gradient of H^+ across the membrane. According to Shikanai (2014), Cyt b_6f is capable of transitioning from a cyclic form of electron transfer around PSI to a linear mode of electron transport across both PSs. Electron transport is carried out by four major subunits of the Cyt b_6f complex: Rieske iron–sulfur protein, subunit IV, cytochrome f, and cytochrome

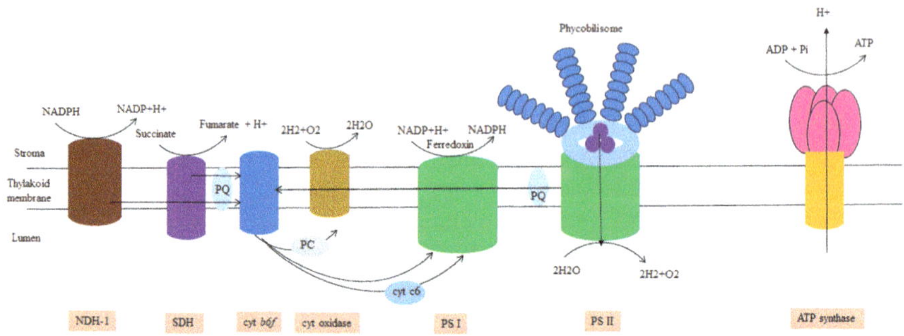

Fig. 11.1 A summary of photosynthesis and transfer of electron pathway during photosynthetic pathway in cyanobacteria. *SDH* succinate dehydrogenase, *NDH-1* type 1 NADPH dehydrogenase, *NADP(H)* nicotinamide-adenine dinucleotide phosphate (reduced form), *Cyt* b_6f cytochrome b6f, *ADP* adenosine diphosphate, *ATP* adenosine triphosphate, *PC* plastocyanin, *PQ* plastoquinone

b6. Four tiny subunits, namely PetG, PetM, PetL, and PetN, aid in stabilizing the complex (Rexroth et al. 2014). To catalyze the conversion of ADP and Pi into ATP, a protein complex known as ATP synthase uses proton flux across the membrane. The proton gradient produced by electron transfer drives the transfer of protons (Fig. 11.1).

Although cyanobacterial photosynthesis operates as described above in the open terrestrial environment, it faces various hurdles while performing in the aquatic system. Some of these challenges are discussed below.

11.4 Challenges in Underwater Cyanobacterial Photosynthesis

11.4.1 Light Attenuation and Penetration

Light can also drastically restrict photosynthesis (Kirk 1994). According to the equation,

$$I_Z = I_0 \left(1 - f\right) e^{-Z\epsilon}$$

where Iz and I_0 are the available irradiance at a given depth (z) and irradiance at the surface, respectively. ϵ is the attenuation coefficient. The proportion of downwelling irradiance is 0.9 due to the relative reflection and backscattering at the water surface (f), which is normally around 0.1 (Kirk 1994). Pure water has an average attenuation value of 0.03 m^{-1} across the photosynthetic spectrum.

When compared to the conditions above the sea, light quantity and quality are different underwater. Aquatic environments are typically thought of as dark surroundings since light is diminished by back scattering, surface reflection, and absorption by water molecules and suspended particles (Holmes and Klein 1987; Sand-Jensen 1989).

11.4.2 Reduced Gas Diffusion

Because gases in water have a diffusion coefficient that is 10^4 times lower than that of air, there can be problems when immersed (Armstrong 1980; Smith and Walker 1980; Maberly and Madsen 2002) leading to restricted photosynthesis and respiration. Due to the limited CO_2 intake, underwater photosynthesis declines. In addition, total inorganic carbon (TIC)—the raw material for photosynthesis exists in the form of CO_2 in the terrestrial environment, but when carbon dioxide is dissolved in water, bicarbonate (HCO_3^-) and carbonate (HCO_3^{2-}) ions are also produced. Ionic strength, temperature, and pH all affect the relative ratios of the three components of dissolved inorganic carbon (DIC), which are CO_2, HCO_3^-, and HCO_3^{2-} (Schwarzenbach and Meier 1958). CO_2 and HCO_3^- are the predominant forms at neutral pH and ambient temperatures; some plants' ability to adapt to submerged environments depends on their ability to acquire and assimilate HCO_3^- (Raven and Hurd 2012).

Due to the restricted gas exchange beneath the water, macroalgal thalli have evolved a range of adaptive features, such as the increased usage of HCO_3^-, to decrease the impact of diffuse boundary layers (DBLs) upon the exchange of O_2 and CO_2 (Colmer et al. 2011). The pool of HCO_3^- ion is many times more than that of CO_2 in seawater and many freshwaters, making it an appealing substitute for CO_2. In marine as well as freshwater systems, HCO_3 use has undergone numerous evolutionary changes among unicellular algae, macroalgae, and angiosperms (Raven and Hurd 2012). This use could entail protons being extruded in acid bands or direct absorption in the cells, or external conversion to CO_2 in the DBL mediated by surface-bound carbonic anhydrase (Lucas and Smith 1973).

The implications of illumination and DIC on aquatic photosynthesis are not only mutually exclusive but also highly dependent on the geographical scale. Several mechanisms, including both metabolic and morphological flexibility, have evolved in cyanobacteria that are thought to lessen the detrimental consequences of submersion. Such mechanisms, known as the physiological acclimation phenomenon of photosynthetic apparatus in cyanobacteria, are unparallel in the terrestrial environment. Hence, the present chapter describes the adaptive traits of cyanobacteria in response to light variations, like the adjustment of pigment contents and morphological changes.

11.5 Adaptations in Cyanobacterial Photosynthetic Process Under Changing Light Conditions

The solar radiation wavelengths that can be used to power photosynthesis in aquatic systems are different from those in land-based ecosystems. At varying depths in aquatic ecosystems, there are notable differences in both the quantity and quality of light. The surfaces of the aquatic environment receive high irradiance of light enhanced with red light (RL) wavelengths, while their bottoms receive low irradiance of light loaded with green light (GL) wavelengths. Babin et al. (2003) demonstrated that the primary causes of this variation in light quantity and quality throughout the water column are light absorption by photosynthetic microbes at the surface and light scattering by dissolved organic matter in the water. Light is a vital component that influences photosynthesis and, hence, impacts phytoplankton growth and productivity. The light quality and quantity have a major role in determining the photosynthetic efficiency of phytoplankton, and they can also impact algae biomass and biological activity (Ooms et al. 2016).

Red light does not penetrate below approximately 15 meters in the aquatic system because water quickly attenuates longer wavelengths of light (Kirk 1994). In the open ocean, blue and green wavelengths of light dominate in waters close to the surface (i.e., above 30 m), but blue light (400–500 nm) is more abundant in deeper waters (100–200 m) (Kirk 1994). Because seawater in coastal regions contains higher amounts of humic compounds and phytoplankton, green wavelengths typically prevail there (Kirk 1994). Hence, cyanobacteria would have a competitive advantage in that environment if they could efficiently absorb and use the particular wavelengths of light that enrich a given area and depth of the ocean (Bennett and Bogorad 1973; Bhaya 2016).

According to Ganf and Oliver (1982), cyanobacteria may move vertically in water by changing their buoyancy and choosing the right stratum of water to live in. Moreover, cyanobacteria have unique chloroplast architecture and arrangement, a greater light absorption cross-sectional area, flexible photosynthetic pigment composition, and the capacity to adapt to various light environments.

11.5.1 Adjustment of Pigment Proteins: As an Adaptive Feature

Light quantity is a dynamic factor; thus, light quantity may vary according to the time and depth of the water. Under varying light conditions, cyanobacteria control the amount and/or quality of pigment-protein complexes in their PBS as well as the interactions between them. Proteins that comprise light-harvesting antennas absorb different wavelengths of light. By allocating collected light energy to photosystems (PSs) in the right proportion, oxygen-based photosynthetic organisms, i.e., cyanobacteria, efficiently carry out photosynthesis. It is necessary to modify the mechanisms involved in energy transfer and light harvesting to maintain photosynthetic efficiency under fluctuating light circumstances.

Reduced light intensity causes cyanobacteria to produce more pigments, including phycocyanin and phycoerythrin. This enables them to perform photosynthetic processes even in low light by absorbing light more effectively (Falkowski and Raven 1997). Zhang et al. (2012) reported that they also enlarge the size of their antenna, which allows them to absorb more light energy.

The ability of *Prochlorococcus* to absorb blue light wavelengths is especially effective. Despite being smaller than *Synechococcus*, it can theoretically reach a higher photosynthetic rate per unit biomass due to its similar absorption cross section (~0.1 µm2, blue maximum) (Morel et al. 1993). Furthermore, due to its small size and unique antenna pigment concentration (divinyl chlorophyll), *Prochlorococcus* cells are more likely to absorb blue wavelength photons rather than scatter them; this is demonstrated by the fact that *Prochlorococcus* cells have an average ratio of energy absorbed to energy striking their geometrical cross section that is almost twice that of *Synechococcus* (Morel et al. 1993). *Prochlorococcus* population seems to have an edge in blue light thereby makes up for a dominant part of the community in deep waters at depths of 100–200 m (Partensky et al. 1999; Moore et al. 1998), which are rich in blue light.

In particular, *Prochlorococcus* communities are found deeper in the water column than *Synechococcus* populations, with a maximum depth of 230 m below the surface in the Atlantic, Pacific, Mediterranean, and Arabian seas. Due to the PEB pigments in its PBS, While *Prochlorococcus* is more adept at absorbing blue light, *Synechococcus* may be able to absorb green wavelengths (~500–550 nm) (Ong and Glazer 1991; Morel et al. 1993; Moore et al. 1995). Marine forms like *Synechococcus* are therefore suited to grow in green light, which is primarily found in coastal waters and has a wavelength of 500–550 nm. Green light penetrates the open ocean's upper water column (~0–30 m) to a similar degree as blue light (Kirk 1994), which may help explain why *Synechococcus* can thrive at these depths (Morel et al. 1993).

The synthesis of PBPs offers cyanobacteria considerable competitive advantages in such an environment. The process through which cyanobacteria respond to light quality by changing the formation of phycobiliproteins is called chromatic adaptation (CA). Tandeau de Marsac (1977) described that this process results in the favored synthesis of phycobiliprotein elements (phycocyanin in the presence of red light and phycoerythrin in the presence of green light, respectively), which efficiently takes in the available light quality. Stated differently, CA is a mechanism that modifies the structure of a pigment that harvests light in response to variations in the color of the surrounding light. Furthermore, the color of the light that irradiates cyanobacteria determines their color phenotype, for example; upon irradiance of red light the cyanobacterial cells turns bluish-green, while upon irradiance of green light, the cells turns red. This procedure is reversible. Recent years have seen an increase in interest in the study of CA due to its discovery, which allows photosynthetic microorganisms worldwide to evolve in response to an extensive spectrum of light wavelengths. Consequently, the selective use of different chromophores that maximize the absorption of the dominant colors (wavelengths) of light in the vicinity is the outcome of CA-regulated PBS changes. When developing photosynthesis-supporting structures (PBSs) in environments with fluctuating light colors, this technique can collect significantly more sunlight than building PBS with unchanging chromophore types, where absorption capabilities do not match any given ambient light color (Sanfilippo et al. 2019).

By adding a variety of PBSs and linker proteins to their PBP, cyanobacterial species can adapt to changes in the quality of the surrounding light. Growth is most selectively aided when light availability is limited, as this reduces energy losses caused by the "wasteful" synthesis of a redundant phycobiliprotein. By precisely adjusting the makeup of its pigments that capture light to closely resemble the spectrum of light, an organism can improve photosynthetic efficiency across a range of lighting conditions (Fig. 11.2). This process is referred to as complementary chromatic adaptation, or CCA (Kehoe and Grossman 1994).

Based on the amount of phycocyanin and phycoerythrin in cyanobacterial species that react to red and green light, CCA is classified into the following categories: Group I species, comprising 27% of all cyanobacterial species, including *Oscillatoria agardhii* 7821 (Wyman and Fay 1986). Under red/green light conditions, their PC and PE levels remain unchanged. Levels of PE are higher in green light and lower in the presence of red light in Group II species. There is a decrease in PE synthesis and an increase in PC synthesis in the third group (Group III) under red light. Group IV strains, in addition, are very new and distinct from groups I, II, and III in that PC and PE remains same; instead, a bilin that absorbs green light is added to PE in the presence of green light, and a bilin that absorbs blue light is added to PE in the presence of blue light (Kehoe 2010), for example, in the *Synechococcus* strain. There is no correlation between light quality and the quantity of allophycocyanin units linked to certain PBS (Glazer 1982). Glazer (1982) reported that while strains belonging to group III exhibit stable phycobilisome sizes regardless of light quality, group II organisms exhibit lower phycobilisome sizes under red light. Consequently, CCA is a phenomena which promotes comprehensive rebuilding of the light-harvesting antennae complexes, or PBSs, which primarily provide light energy to the reaction centers of the photosystem II (Kehoe and Gutu 2006). Changes in PC and PE synthesis, which is primarily regulated at the transcriptional level of the *cpcBAC* and *cpeBAC* genes encoding PC and PE structural units, respectively, are the main cause of CCA.

Apart from that, some of the cyanobacterial species that are present in underneath coral-reef in the shaded niche possess Chl *d* (near-infrared light) as the principal light-harvesting complex, which has its absorption maxima at 700–720 nm (Kühl et al. 2005) (Fig. 11.3). Moreover, few cyanobacteria hosted by stromatolites enriched in near-infrared light contain Chl *f* which has an absorption maxima of 750 nm (Chen et al. 2010) (Fig. 11.3).

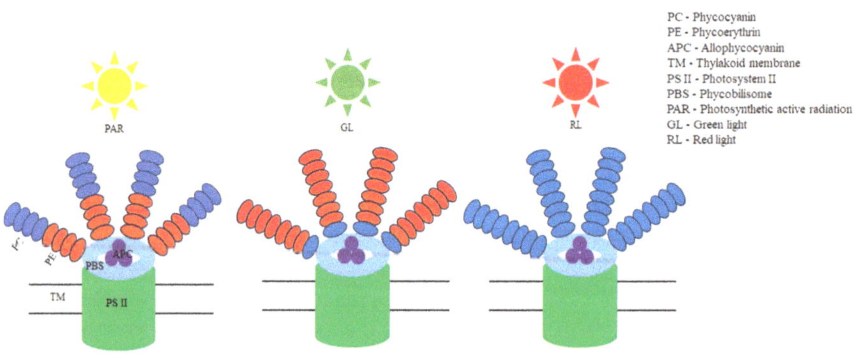

Fig. 11.2 An overview of the anatomy of the phycobilisome and the response of the various components of the phycobilisome to light quality undergoing CCA in cyanobacteria

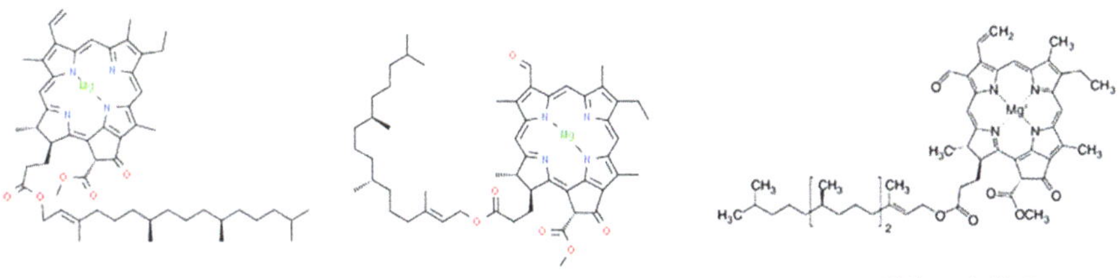

Chlorophyll *a* Chlorophyll *d* Chlorophyll *f*

Fig. 11.3 Structure of chlorophyll present in cyanobacteria

11.5.2 Morphological Alterations: As an Adaptive Feature

An ecologically significant phenomenon is called CCA, in which cyanobacteria can change their morphology in response to light signals to maximize their ability to harvest light energy and, as a result, maximize their fitness in the presence of dominant light in any environment (Singh and Montgomery 2011). Cells and filaments are shaped during CCA in order to maximize the capture of radiant energy by altering the exposed surface area of the cell and/or filament (Singh and Montgomery 2014). As mentioned earlier, when the cells detect a low-light environment that is abundant in green light, they adopt a rectangular shape that is linked to an increase in the organism's filament length. In contrast, a high-light environment that is enriched in red light causes spherical cell shape and shorter filament lengths (Bordowitz and Montgomery 2008). Several authors demonstrated that *Fremyella diplosiphon* change their morphological forms under the fluctuating light environment to optimize the fitness and absorption of the accessible photon (Campbell 1996; Pattanaik et al. 2012; Singh and Montgomery 2014; Montgomery 2016). According to Campbell et al. (1993), the process behind phenotypic alteration in *F. diplosiphon* is asymmetric stimulation of PSI and PSII that is dependent on RL and GL (Fig. 11.4). This results in an impaired redox state. A recent study shows redox regulated, i.e., multiple ways that accumulate reactive oxygen species (ROS) under GL and RL in *F. diplosiphon* also leads to shift in cell and filament phenotype (Singh and Montgomery 2012). Studies show that the amount and quality of light have an impact on the levels of ROS in cells. Furthermore, spherical cell shape is associated with quality-dependent generation of high ROS levels and light intensity (Pattanaik et al. 2012; Walters et al. 2013). The RL features of spherical cells and short filaments are associated with increased ROS levels. *F. diplosiphon* uses long filaments and rectangular cellular architecture at low GL intensities in the natural environment to optimize photon absorption (Singh and Montgomery 2012). This alteration in the morphological form of cyanobacteria is associated with modulations in *bolA* morphogenes and cytoskeleton protein-encoding genes such as *mreB*, *mreC*, and *mreD* that modify the biosynthesis of peptidoglycan scaffold (Singh and Montgomery 2014, 2015). Figure 11.2. represents the mechanistic summary of *mreB* in the cyanobacterial cell morphology.

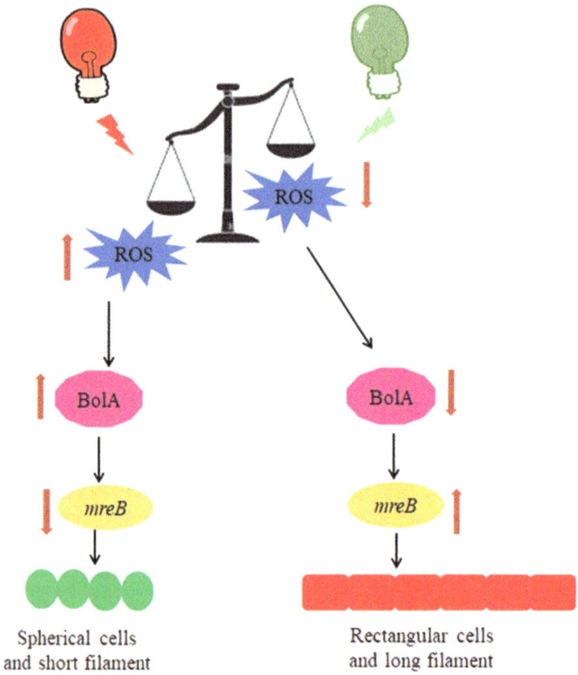

Fig. 11.4 The light-dependent alterations in *F. diplosiphon* cellular morphology are depicted schematically in the diagram. While GL-dependent downregulation of BolA due to its less transcription results in higher accumulation of mreB and associated rectangular cell morphology and long filament, RL-dependent increased accumulation of BolA leads to downregulation of *mreB* transcription and related spherical morphology

11.6 Conclusion and Future Prospectives

In conclusion, this chapter provides a solid foundation of theory for an extensive study of aquatic photosynthesis, which helps in the precise delineation of scientific domains and hubs. The aquatic environment has a direct impact on underwater photosynthesis. Light attenuation is the most vulnerable state because of the consequences of suspended particulates, dissolved ingredients in the water body, and the depth of the water. To cope with this unfavorable environment, cyanobacteria evolved various tactics. From a morphological perspective, they exhibit a spherically shaped, short filament. Furthermore, the topics of complementary chromatic adaptation and adjustments to photosynthetic pigments were covercd. Although physiological levels are important for comprehending the underwater photosynthetic process, future studies should concentrate more on the molecular mechanics of underwater photosynthesis.

Acknowledgments The Head, Department of Botany, University of Lucknow, Lucknow is deeply acknowledged for their invaluable support and encouragement. Vinod Kumar is thankful to

Pandit Prithi Nath (PG) College, Kanpur for providing laboratory facilities. Surbhi Kharwar and Vinod Kumar express their sincere gratitude to the editor Prof. A.K. Mishra.

Author Contribution The study was conceptualized by SK and VK. Together with SJ, SK, VK, and VK found relevant literature and contributed to the manuscript's writing. SK and VK reviewed and made edits to the manuscript's final draft.

Conflict of Interest Statement Every author state that there were no financial or commercial ties that would raise the possibility of a conflict of interest during the research.

References

Allen MM (1984) Cyanobacterial cell inclusions. Ann Rev Microbiol 38:1–25
Armstrong W (1980) Aeration in higher plants. In: Advances in botanical research, vol 7. Academic Press, San Diego, pp 225–332
Arteni AA, Liu LN, Aartsma TJ, Zhang YZ, Zhou BC, Boekema EJ (2008) Structure and organization of PBS on membranes of the red alga *Porphyridium cruentum*. Photosynth Res 95:169–174
Babin M, Stramski D, Ferrari GM, Claustre H, Bricaud A, Obolensky G, Hoepffner N (2003) Variations in the light absorption coefficients of phytoplankton, nonalgal particles, and dissolved organic matter in coastal waters around Europe. J Geophys Res Oceans 108(C7):111579
Bassham JA, Benson AA, Kay LD, Harris AZ, Wilson AT, Calvin M (1954) The path of carbon in photosynthesis. XXI. The cyclic regeneration of carbon dioxide acceptor1. J Am Chem Soc 76(7):1760–1770
Bearden A, Malkin R (1974) Primary photochemical reactions in chloroplast photosynthesis. Q Rev Biophys 7:131–177
Bennett A, Bogorad L (1973) Complementary chromatic adaptation in a filamentous blue-green alga. J Cell Biol 58(2):419–435
Bhaya D (2016) In the limelight: photoreceptors in cyanobacteria. MBio 7(3):10–1128
Bordowitz JR, Montgomery BL (2008) Photoregulation of cellular morphology during complementary chromatic adaptation requires sensor-kinase-class protein RcaE in *Fremyella diplosiphon*. J Bacteriol 190(11):4069–4074
Bryant DA, Costas AMG, Maresca JA, Chew AGM, Klatt CG, Bateson MM, Ward DM (2007) Candidatus *Chloracidobacterium thermophilum*: an aerobic phototrophic acidobacterium. Science 317(5837):523–526
Campbell D (1996) Complementary chromatic adaptation alters photosynthetic strategies in the Cyanobacterium calothrix. Microbiology 142(5):1255–1263
Campbell D, Houmard J, De Marsac NT (1993) Electron transport regulates cellular differentiation in the filamentous Cyanobacterium calothrix. Plant Cell 5(4):451–463
Casella S (2017) Organization, dynamics and adaptation of the photosynthetic machinery in Cyanobacteria. Liver 1:4–24
Chaiklahan R, Chirasuwan N, Loha V, Tia S, Bunnag B (2011) Separation and purification of phycocyanin from *Spirulina* sp. using a membrane process. Bioresour Technol 102:7159–7164
Chen M, Schliep M, Willows RD, Cai ZL, Neilan BA, Scheer H (2010) A red-shifted chlorophyll. Science 329(5997):1318–1319
Codd GA (1988) Carboxysomes and ribulose bisphosphate carboxylase/oxygenase. Adv Microb Physiol 29:115–164
Colmer TD, Winkel A, Pedersen O (2011) A perspective on underwater photosynthesis in submerged terrestrial wetland plants. AoB Plants plr030
Duysens LN, Amesz J, Kamp BM (1961) Two photochemical systems in photosynthesis. Nature 190(4775):510–511
Falkowski PG, Raven JA (1997) Aquatic photosynthesis, vol 375. Blackwell Science, Malden, p 525

Falkowski PG, Raven JA (2007) Aquatic photosynthesis, 2nd edn. Princeton Univ. Press, Princeton, NJ
Fang C, Jacinthe P, Song C, Zhang C, Song K (2022) Climate-driven variations in suspended particulate matter dominate water clarity in shallow lakes. Opt Express 30:4028–4045
Field CB, Behrenfeld MJ, Randerson JT, Falkowski P (1998) Primary production of the biosphere: integrating terrestrial and oceanic components. Science 281(5374):237–240
Ferreira KN, Iverson TM, Maghlaoui K, Barber J, Iwata S (2004) Architecture of the photosynthetic oxygen-evolving center. Science 303:1831–1838
Fischer WW (2008) Life before the rise of oxygen. Nature 455:1051
Fogg GE, Stewart WDP, Fay P, Walsby AE (1973) The blue-green algae. Academic Press, London
Ganf GG, Oliver RL (1982) Vertical separation of light and available nutrients as a factor causing replacement of green algae by blue-green algae in the plankton of a stratified lake. J Ecol:829–844
Gao JS, Zhu J, Dong WY (2019) Influence mechanism of light on common algae and its application. Environ Eng 37(5):111–116
Ghosh T, Mishra S (2017) How does photosynthesis take place. In: Frontiers for young mind, p 5
Glazer AN (1982) PBS: structure and dynamics. Annu Rev Microbiol 36(1):173–198
Glazer AN (1989) Light guides: directional energy transfer in a photosynthetic antenna. J Biol Chem 264(1):1–4
Holmes MG, Klein WH (1987) The light and temperature environments. Spec Publ Br Ecol Soc 5:3–22
Jensen TE (1984) Cyanobacterial cell inclusions of irregular occurence: systematic and evolutionary implications. Cytobios 39(153):35–62
Kehoe DM (2010) Chromatic adaptation and the evolution of light color sensing in cyanobacteria. Proc Natl Acad Sci 107(20):9029–9030
Kehoe DM, Grossman AR (1994) Complementary chromatic adaptation: photoperception to gene regulation. In: Seminars in cell biology, vol 5. Academic Press, London, pp 303–313
Kehoe DM, Gutu A (2006) Responding to color: the regulation of complementary chromatic adaptation. Annu Rev Plant Biol 57:127–150
Kirk JTO (1994) The nature of the underwater light field. In: Light and photosynthesis in aquatic ecosystems. Cambridge University Press, Cambridge, p 509
Kühl M, Chen M, Ralph PJ, Schreiber U, Larkum AWD (2005) Ecology: a niche for cyanobacteria containing chlorophyll d. Nature 433:820
Li X, Wagner F, Peng W, Yang J, Mauzerall DL (2017) Reduction of solar photovoltaic resources due to air pollution in China. Proc Natl Acad Sci USA 114:11867–11872
Liu L-N, Chen X-L, Zhang Y-Z, Zhou B-C (2005) Characterization, structure and function of linker polypeptides in PBS of cyanobacteria and red algae: an overview. Biochim Biophys Acta 1708:133–142
Liu L-N, Elmalk AT, Aartsma TJ, Thomas J-C, Lamers GE, Zhou B-C, Zhang Y-Z (2008) Light-induced energetic decoupling as a mechanism for phycobilisome-related energy dissipation in red algae: a single molecule study. PloS One 3:1
Lucas WJ, Smith FA (1973) The formation of alkaline and acid regions at the surface of *Chara corallina* cells. J Exp Bot 24(1):1–14
Maberly SC, Madsen TV (2002) Freshwater angiosperm carbon concentrating mechanisms: processes and patterns. Funct Plant Biol 29(3):393–405
MacGregor-Chatwin C, Sener M, Barnett SF, Hitchcock A, Barnhart-Dailey MC, Maghlaoui K, Hunter CN (2017) Lateral segregation of photosystem I in cyanobacterial thylakoids. Plant Cell 29(5):1119–1136
Montgomery BL (2016) Mechanisms and fitness implications of photomorphogenesis during chromatic acclimation in cyanobacteria. J Exp Bot 67(14):4079–4090
Moore LR, Goericke R, Chisholm SW (1995) Comparative physiology of *Synechococcus* and *Prochlorococcus*: influence of light and temperature on growth, pigments, fluorescence and absorptive properties. Mar Ecol Prog Ser 116:259–275

Moore LR, Rocap G, Chisholm SW (1998) Physiology and molecular phylogeny of coexisting *Prochlorococcus* ecotypes. Nature 393(6684):464–467
Morel A, Ahn YH, Partensky F, Vaulot D, Claustre H (1993) *Prochlorococcus* and *Synechococcus*: a comparative study of their optical properties in relation to their size and pigmentation. J Mar Res 51(3):617–649
Olson JM (2006) Photosynthesis in the Archean era. Photosynth Res 88:109–117
Olson JM, Blankenship RE (2004) Thinking about the evolution of photosynthesis. Photosynth Res 80:373–386
Ong LJ, Glazer AN (1991) Phycoerythrins of marine unicellular cyanobacteria. I. Bilin types and locations and energy transfer pathways in *Synechococcus* spp. phycoerythrins. J Biol Chem 266(15):9515–9527
Ooms MD, Dinh CT, Sargent EH, Sinton D (2016) Photon management for augmented photosynthesis. Nat Commun 7(1):12699
Partensky F, Hess WR, Vaulot D (1999) *Prochlorococcus*, a marine photosynthetic prokaryote of global significance. Microbiol Mol Biol Rev 63(1):106–127
Pattanaik B, Whitaker MJ, Montgomery BL (2012) Light quantity affects the regulation of cell shape in *Fremyella diplosiphon*. Front Microbiol 3:170
Pedersen O, Colmer TD, Sand-Jensen K (2013) Underwater photosynthesis of submerged plants–recent advances and methods. Front Plant Sci 4:140
Raven JA, Beardall J (2016) The ins and outs of CO2. J Exp Bot 67(1):1–13
Raven JA, Hurd CL (2012) Ecophysiology of photosynthesis in macroalgae. Photosynth Res 113:105–125
Raymond J, Zhaxybayeva O, Gogarten JP, Gerdes SY, Blankenship RE (2002) Whole-genome analysis of photosynthetic prokaryotes. Science 298(5598):1616–1620
Rexroth S, Rexroth D, Veit S, Plohnke N, Cormann KU, Nowaczyk MM, Rögner M (2014) Functional characterization of the small regulatory subunit PetP from the cytochrome b6f complex in *Thermosynechococcus elongatus*. Plant Cell 26:3435–3448
Sand-Jensen KAJ (1989) Environmental variables and their effect on photosynthesis of aquatic plant communities. Aquat Bot 34(1–3):5–25
Sanfilippo JE, Garczarek L, Partensky F, Kehoe DM (2019) Chromatic acclimation in cyanobacteria: a diverse and widespread process for optimizing photosynthesis. Ann Rev Microbiol 73:407–433
Scheer H, Zhao K-H (2008) Biliprotein maturation: the chromophore attachment. Mol Microbiol 68:263
Schwarzenbach G, Meier J (1958) Formation and investigation of unstable protonation and deprotonation products of complexes in aqueous solution. J Inorg Nucl Chem 8:302–312
Shikanai T (2014) Central role of cyclic electron transport around photosystem I in the regulation of photosynthesis. Curr Opin Biotechnol 26:25–30
Singh SP, Montgomery BL (2011) Determining cell shape: adaptive regulation of cyanobacterial cellular differentiation and morphology. Trends Microbiol 19(6):278–285
Singh SP, Montgomery BL (2012) Reactive oxygen species are involved in the morphology-determining mechanism of *Fremyella diplosiphon* cells during complementary chromatic adaptation. Microbiology 158(9):2235–2245
Singh SP, Montgomery BL (2014) Morphogenes bolA and mreB mediate the photoregulation of cellular morphology during complementary chromatic acclimation in *Fremyella diplosiphon*. Mol Microbiol 93(1):167–182
Singh SP, Montgomery BL (2015) Regulation of BolA abundance mediates morphogenesis in *Fremyella diplosiphon*. Front Microbiol 6:1215
Singh NK, Parmar A, Madamwar D (2009) Optimization of medium components for increased production of C-phycocyanin from *Phormidium ceylanicum* and its purification by single step process. Bioresour Technol 100:1663–1669
Singh NK, Sonani RR, Rastogi RP, Madamwar D (2015) The PBS: an early requisite for efficient. Excli J 1:268–289

Smith FA, Walker NA (1980) Photosynthesis by aquatic plants: effects of unstirred layers in relation to assimilation of CO2 and HCO3− and to carbon isotopic discrimination. New Phytol 86(3):245–259

Soni B, Trivedi U, Madamwar D (2008) A novel method of single step hydrophobic interaction chromatography for the purification of phycocyanin from *Phormidium fragile* and its characterization for antioxidant property. Bioresour Technol 99:188–194

Stanier RY, Cohen-Bazire G (1977) Phototrophic prokaryotes: the cyanobacteria. Ann Rev Microbiol 31(1):225–274

Tandeau de Marsac N (1977) Occurrence and nature of chromatic adaptation in cyanobacteria. J Bacteriol 130(1):82–91

Walters KJ, Whitaker MJ, Singh SP, Montgomery BL (2013) Light intensity and reactive oxygen species are centrally involved in photoregulatory responses during complementary chromatic adaptation in *Fremyella diplosiphon*. Commun Integr Biol 6(5):e25005

Wang KC, Dickinson RE, Wild M, Liang S (2012) Atmospheric impacts on climatic variability of surface incident solar radiation. Atmos Chem Phys 12:9581–9592

Whitton BA, Potts M (eds) (2007) The ecology of cyanobacteria: their diversity in time and space. Springer, Dordrecht

Wu H, Ding J (2019) Global change sharpens the double-edged sword effect of aquatic alien plants in China and beyond. Front Plant Sci 10:787

Wyman M, Fay P (1986) Underwater light climate and the growth and pigmentation of planktonic blue-green algae (cyanobacteria) I. The influence of light quantity. Proc Royal Soc B Biol Sci 227(1248):367–380

Xu W, Tang H, Wang Y, Chitnis PR (2001) Proteins of the cyanobacterial photosystem I. Biochim Biophys Acta 1507:32–40

Yang T, Liu X (2015) Comparing photosynthetic characteristics of *Isoetes sinensis* palmer under submerged and terrestrial conditions. Sci Rep 5(1):17783

Yang R, Chen B, Liu H, Liu Z, Yan H (2015) Carbon sequestration and decreased CO2 emission caused by terrestrial aquatic photosynthesis: insights from diel hydrochemical variations in an epikarst spring and two spring-fed ponds in different seasons. Appl Geochem 63:248–260

Zhang M, Duan H, Shi X, Yu Y, Kong F (2012) Contributions of meteorology to the phenology of cyanobacterial blooms: implications for future climate change. Water Res 46(2):442–452

Zhang M, Shi X, Yang Z, Yu Y, Shi L, Qin B (2018) Long-term dynamics and drivers of phytoplankton biomass in eutrophic Lake Taihu. Sci Total Environ 645:876–886

Zhang Y, Qin B, Shi K, Zhang Y, Deng J, Wild M, Li L, Zhou Y, Yao X, Liu M, Zhu G, Zhang L, Gu B, Brookes JD (2020) Radiation dimming and decreasing water clarity fuel underwater darkening in lakes. Sci Bull 65:1675–1684

Zhang M, Yang Z, Shi X, Yu Y (2022) The synergistic effect of rising temperature and declining light boosts the dominance of bloom-forming cyanobacteria in spring. Harmful Algae 116:102252

Adaptation of Photoautotrophs in Extraterrestrial Environments: Responses and Mechanisms of Survival

12

Sindhunath Chakraborty and Ranjan Kumar Tiwari

Abstract

In the last few decades, impeccable progress has been made in the field of astrobiological research whose purpose was to search for efficient Earth organisms capable of surviving in space environment followed by using them as tools for human space exploration. Among the investigated organisms, photoautotrophs constitute the most rigorously studied group. In the present chapter, responses and survival strategies of photoautotrophs in space and/or simulated extraterrestrial conditions have been discussed in detail. Space exposure and survival-related studies on photoautotrophs have mainly been performed using cyanobacteria, algae, lichens and higher plants. Reports accumulated over 60 years reflected that many species belonging to these photoautotrophic groups have an outstanding ability to survive in space. However, the exact mechanisms of their survival in space have not yet been completely explored. In order to unveil the survival strategies of photoautotrophs in space environment, various plant species i.e. *Arabidopsis thaliana*, *Brassica rapa var. nipposinica*, *Ceratopteris richardii* and *Oryza sativa,* have recently been studied thoroughly through mRNAseq and microarray-based transcriptomic analyses. Results obtained from these studies denoted that plants exhibit a myriad of survival responses like cell wall remodelling, oxidative stress, activation of antioxidative machineries, lipid utilization, cell cycle regulation, increased telomerase activity, reduced genome oxidation, altered hormone signalling, protection of photosynthetic apparatus etc. Few identical responses like oxidative stress, activation of

S. Chakraborty (✉)
Laboratory of Microbial Genetics, Department of Botany, Banaras Hindu University, Varanasi, India

R. K. Tiwari
Laboratory of Cyanobacterial Systematics and Stress Biology, Department of Botany, Banaras Hindu University, Varanasi, India

A. K. Mishra (ed.), *Stress Biology in Photosynthetic Organisms*,
https://doi.org/10.1007/978-981-97-1883-2_12

antioxidant machineries, increased lipid metabolism and protection of photosynthetic apparatus have also been noticed in cyanobacteria and algae when they have been exposed to space like stratospheric and high-altitude environments respectively, followed by being subjected to gene expression analysis. Interestingly, complete absence of such studies has been recorded for lichens, hence the space survival mechanisms of this photoautotrophic group still remain absolutely unknown. Despite a large number of researches on this topic, only a few review articles have been written so far and those reviews have either focussed on only plants' survival responses in space or that of cyanobacteria and algae. Therefore, the present chapter will be very useful to the upcoming researchers in this field as it meticulously elaborates the survival responses and strategies of all groups of photoautotrophs in space conditions.

Keywords

Photoautotrophs · Astrobiology · Extraterrestrial environment · Space flight · LEO (Low Earth Orbit) · Fv/Fm · Gravitational force · International Space Station · Microgravity · Cosmic radiation

12.1 Introduction

In the past few decades, astronomical researches explored many aspects of our solar system. One of which is the future engulfment of the planet Earth by the Sun (Schröder and Smith, 2008). Apart from this, satellite-based imaging of the Earth from space disclosed the event of climate change which unveiled its possible detrimental impact on the environment (Abbass et al. 2022). These two facts raised a serious question on the existence of humans and all living beings on the Earth. As a result, scientists from different parts of the world have become interested in searching for strategies to increase our longevity. One such strategy may be to explore another planet from the solar system or interstellar space with either habitable conditions or the possibility of being redesigned in order to support life. Since no planet with habitable conditions for humans has yet been found, researchers have directed their focus on designing a planet artificially to mimic Earth. The environment outside the earth or its atmosphere is called as extraterrestrial environment (Cleland 2019) which represents extremely harsh conditions like lack of oxygen, highly variable temperature range, exposure to cosmic rays, etc. Due to such conditions, life in extraterrestrial conditions is impossible. Despite these facts, a considerable number of researches have been invested only to find out how these uninhabitable environments can be transformed into a habitable one for humans. The basis of these researches was the view that billions of years ago Earth seemed to be like other planets devoid of life but the great oxidation event (GOE) changed its environment and initiated the evolution of life (Blaustein 2016; Lyons et al. 2014). However, long-term researches proved that creating habitable space for humans in these extraterrestrial environments is hardly achievable at a large scale but can be achieved through a gradual process. For this, we initially have to ensure a few things i.e. continuous oxygen supply, development of a breathable atmosphere and masking the effect of cosmic rays, followed by the establishment and maintenance of human colonies in a sustainable manner using resources available in the target environment. In order to develop oxygen-rich breathable atmosphere in an extraterrestrial environment, photoautotrophic organisms like cyanobacteria, algae and plants are crucial as they perform photosynthesis and liberate oxygen in the atmosphere. Among these, cyanobacteria are supposed to be the most promising as they are known for the GOE and alteration of Earth's atmosphere from uninhabitable to an inhabitable one (Holland 2006). As a result, significant number of studies have been performed using cyanobacteria to investigate their efficiency and mechanisms to survive extraterrestrial conditions. Apart from this, different algae and lichens have also been exploited for the same purpose. Since human colonization is the ultimate target for exploring extraterrestrial habitats, ability of plants to thrive in this condition has also been investigated to unravel possibility of crop production and securing food supply in such completely adverse environments. In the present chapter, we tried to provide a detail account of the responses of the photoautotrophic organisms especially cyanobacteria, algae, plants and lichens in extraterrestrial environments and their adaptation strategies.

12.2 Responses of Photoautotrophs to Simulated Extraterrestrial Environments

12.2.1 Cyanobacterial Responses to Simulated Extraterrestrial Environments

Since the postulation of GOE theory, cyanobacteria have become a subject of astrobiological researches. Many scientific groups around the world invested enormous efforts to unveil the ability of cyanobacteria to withstand extraterrestrial conditions so as to apply them for terraforming the conditions of other planets especially Mars and make them habitable for life forms. For this, cyanobacteria have been exposed to many types of radiation which are common in space and other planets and their responses were recorded. The resistance of cyanobacteria to photon radiations (X-rays and Gamma rays), one of the prominent physical factors in extraterrestrial environments, was first investigated in the early 1950s by Bonham and Palumbo (1951). However, the description of few cyanobacterial species having resistance against photon radiations came in the beginning of 1960s (Shields et al. 1961; Godward 1962). After these pioneering researches, various cyanobacterial strains belonging to different genera were tested for their resistance against photon radiations. These works disentangled that resistance against radiations vary significantly with species type. While some species showed low resistance i.e. D10 below 1 kGy (dose of the radiation which inactivated cyanobacterial population by 90%), some exhibited elevated resistance with D10 above 10 kGy (e.g. Asato 1971; Bruce 1964; Godward 1962; Kraus 1969; Kumar 1964; Shields et al. 1961). However, these results were unable to be interpreted or compared because radiation doses along with other parameters such as temperature, growth phase, dose rates, oxygen levels, light intensity, etc. are presently known to affect cyanobacterial survival. This means radiation dose-based experiments were not controlled ones. In order to eradicated the biasness of the results obtained from previous studies, attention was paid toward performing controlled studies with specified conditions. From the results of those studies, it was found that few cyanobacteria belonging to the genera *Synechococcus* and *Synechocystis* spp. were sensitive to gamma rays, as they reflected D10 value only at 0.3 kGy (Agarwal et al. 2008; Domain et al. 2004) and 0.7 kGy (Domain et al. 2004) radiation doses, respectively. However, two strains of *Anabaena* were observed to survive efficiently upon exposure to a dose of 5 kGy gamma-ray. While the GI50 dose (50% growth inhibition) of the cyanobacterium was found to be at 6–11 kGy, survival was noticed up to the 15 kGy dose (Singh et al. 2010, 2013). Considerable resistance to gamma radiation was also noticed in the cyanobacterium *Arthrospira* sp. PCC 8005 which survived 6.4 kGy dose. Moreover, the same species displayed resistance to respectively 1 and 2 kGy of He and Fe particle radiations as well (Badri et al. 2015). The maximum level of resistance to space-related radiations was noticed in *Chroococcidiopsis* spp. as the hydrated cells of this organism survived 15 kGy of gamma and X-ray doses and also endured heavy ion irradiation (Billi et al. 2000; Verseux et al. 2017). In addition, the dried cells of the same species were found to withstand even 24 kGy dose of gamma-rays (Verseux et al. 2017). Cyanobacterial resistance to UVC, the most damaging part of UV spectrum of Mars, was investigated thoroughly which showed that these organisms are well capable to survive UVC. Experiments performed using *Anabaena variabilis* ATCC 29413 and *Anabaena* sp. PCC 7120 reflected that these two strains can withstand 750 J•m-2 (Levine and Thiel 1987). Similarly, elevated resistance to UVC was also observed in the case of *Synechocystis* sp. PCC6803 and *Synechococcus* sp. PCC7942 as they showed survival at the radiation doses of 500 J•m-2 and 400 J•m-2, respectively (Domain et al. 2004). Like photonic radiation tolerance, considerable level of UVC tolerance was demonstrated for *Chroococcidiopsis* sp. The suspension culture of strain *Chroococcidiopsis* sp. CCMEE 029 showed survival up to 300 J•m-2 of UVC radiation. This tolerance of *Chroococcidiopsis* sp. CCMEE 029 was attributed to their multicellular aggregates on the cell envelope which only allowed the passage of attenuated UVC radiation to the cell interior (Baqué et al. 2013a). Besides, the dried cell of *Chroococcidiopsis* sp. CCMEE 029 and CCMEE 057 were also found to survive 10 kJ•m-2 dose of UVC (Baqué et al. 2013b, 2014). Furthermore, the survival of *Chroococcidiopsis* spp. upon long-term exposure to vacuum or Mars-like atmosphere i.e. extreme temperatures (1 h at −25 °C followed by 1 h at +60 °C) and extreme temperature cycles (2 h at −10 °C followed by 2 h at +45 °C, 66 times) was also shown by Baqué et al. (2013b, 2016).

12.2.2 Responses of Lichens to Simulated Extraterrestrial Environments

The responses of photoautotrophic lichen to simulated extraterrestrial conditions have also been investigated, wherein the lichen *Pleopsidium chlorophanum* along with endolithic cyanobacteria was exposed to environmental stress factors present in the surface of Mars for 34 days. Simulation of the stresses was performed in a Mars Simulation Chamber (MSC) in which high atmospheric CO_2, low temperatures, low water activity and high UV fluxes, and 6 mbar atmospheric pressure were maintained. In order to understand the responses of lichen and endolithic cyanobacteria to unprotected and semi-protected (micro-caves within rocks, cracks, fissures and permafrost soil) Martian surfaces, the experiments were performed in two sets and the stress factors were arranged accordingly. In the semi-protected simulated Martian environment, the lichen *Pleopsidium chlorophanum* was found to survive and actively perform photosynthesis whereas in the unprotected environment, the lichen sample survived but their photosynthetic activity was not clear. In contrast, the photosynthetic activity of the endolithic cyanobacteria was noticed to be strongly inhibited. The survival of lichen in a simulated Martian environment was assumed to be a result of its increased photosynthetic activity over 34 days.

12.2.3 Algal Responses to Simulated Extraterrestrial Environments

Like cyanobacteria and lichens, algal survival efficiency in simulated extraterrestrial environment has also been studied in detail. The algal strain *Chlorella* belonging to desert environment was tested for its ability to tolerate Mars-like near-space environment by Wang et al. (2022). It was found that although there was significant viability of *Chlorella* cells but they experienced considerable damage at intracellular level as there were reduction in photosynthesis, decrease in cell growth and modulation in the ultra-structures of chloroplast and mitochondria (Fig. 12.2). Another study investigating the growth of five different algal species i.e. *Chloromonas brevispina*, *Kremastochrysopsis austriaca*, *Dunaliella salina*, *Chlorella vulgaris* and *Spirulina plantensis* at low pressure (670 ± 20 mbar, 330 ± 20 mbar, 160 ± 20 mbar and 80 ± 2.5 mbar) analogous to that of Mars reflected that *D. salina*, *C. brevispina* and *C. vulgaris* are the most resistant ones. Among the tested algal species, highest level of tolerance under low-pressure conditions was observed in *D. salina* which survived at 160 mbar with a cell number of 4.6 × 105 30.0 ± per ml. The second most tolerant species was found to be *C. brevispina* which showed a cell number of 0.9 × 105 ± 19.8 per ml at 330 mbar. Moreover, the third most tolerant species was *C. vulgaris* whose cell number was 0.5 × 105 ± 13.0 per mL at 160 mbar. Apart from this, considerable growth of *C. brevispina* (2.5 × 104 ± 43.4cells/mL), *D. salina* (1.3 × 104 ± 15.8 cells/mL) and *C. vulgaris* (4.5 × 104 ± 57.1 cells/mL) was also noticed at the lowest pressure (80 mbar) applied in this study.

12.2.4 Plants Responses to Simulated Extraterrestrial Environments

Although a number of photoautotrophs like cyanobacteria, lichens and algae were examined for their tolerance levels under extraterrestrial conditions, maximum importance was given to the higher plants as they were considered to be the most suitable long-term life support systems in space flights. The responses of plants to extraterrestrial conditions were studied mainly by exposing them to space environments. Only a few reports related to their responses under simulated extraterrestrial conditions are available. One such study was performed by Fengler et al. (2016) wherein the cultured cells of the model plant *Arabidopsis thaliana* were subjected to partial-gravitational forces (0.16 g, 0.38 g and 0.5 g) for varied time intervals followed by the changes in their gene expression were thoroughly studied. Exposure of *A. thaliana* cell cultures to partial-g was done under simulated clinostat experiments (5 min, 10 min and 30 min) and real parabolic flight (24 s, 33 s and 20 s). The results obtained from clinostat experiment reflected very small changes in gene expression in cell cultures after 30 min exposure to 0.5 g. Moreover, Pearson's Correlation Coefficient analysis of results obtained from clinorotation experiment denoted that the alterations in gene expression were more positively correlated with the exposure time as compared to the levels of gravitational forces. The clustering of results of simulated clinorotation and real parabolic flight experiments showed very low similarities in gene expression changes, but the screening performed at single gene level pinpointed the occurrence of many differentially expressed common genes, especially at 0.38 g force. These common genes were identified to be cell wall modifying protein-coding genes, peroxidase enzyme coding genes, genes coding for auxin and phytohormones. Some other genes were found to be those which code for transmembrane proteins, intracellular signalling proteins like kinases and transcription factors. This study clearly showed that 30 min exposure of *Arabidopsis* cell cultures to clinorotation results in the loss of specific responses, reduction in cell susceptibility and induction of gene expression changes as observed in parabolic flight study.

The other study performed by Manzano et al. (2012) demonstrated that the semi-solid callus cultures of *A. thaliana*, upon 200 min exposure to simulated microgravity (mg) and hypergravity (2 g) under diamagnetic levitation, exhibited changes in expression of very few genes. However, the Gene Expression Dynamics Inspector (GEDI)-based clustering of similarly expressed genes showed synergistic effect of gravitational and magnetic fields on the gene expression profile. It was found that the application of magnetic field gradient increased the number of altered genes and the GO groups. This gradient of magnetic field affected the expression of genes associated with secondary metabolism, structural organization and abiotic stress management.

A recent study performed by Kamal et al. (2019) investigated the responses of immobilized culture of *Arabidopsis* to simulated microgravity in terms of transcriptomic alterations in cell cycle regulation and progression. For this, two subpopulations of *Arabidopsis* cells at G1 and G2/M phases were subjected to respectively 14 h and 7 h under microgravity, followed by their gene expression changes through microarray analysis. The result was as such that a very small number of genes were found to be differentially expressed in the tested subpopulations. A total of 73 genes, mostly related to NADH/oxidative activity and mitochondrial functioning, were recorded to be up-accumulated whereas 83 genes, mainly associated with ribosomal activity and photosynthesis, were deregulated. Apart from this, differential effects of microgravity on the expression of genes required for cell cycle progression were also noticed. While the deregulation of genes associated with cell cycle regulation, cell organization, ribosome biogenesis and developmental processes was recorded in G2/M population, up-accumulation of transcripts of the same genes was identified in the G1 population. Moreover, G2/M population showed up-accumulation of genes related to abiotic stress responses whereas G1 population displayed significant deregulation of the same genes. Abiotic stress response genes that were observed to be up-accumulated in G2/M and deregulated in G1 populations upon 14 h exposure to microgravity included UV-B, heat, osmotic shock and wounding-related ones. These differential expression patterns of abiotic stress-related genes indicated that different cell cycle phases respond differentially to simulated microgravity.

12.3 Responses of Photoautotrophs to Spaceflight Environments

12.3.1 Cyanobacterial Responses to Spaceflight Environments

In 1994, cyanobacteria were exposed to the real space environment for the first time. For this, a halophilic *Synechococcus* sp. thriving in gypsum-halite crystals was selected and the space exposure was done for 2 weeks onboard ESA's BIOPAN-I facility. The aim was to investigate the ability of UV-attenuating gypsum–halite crystals to help survive the cyanobacterium under space radiation and vacuum (Mancinelli et al. 1998). This halophilic strain was further exposed to space for nearly 2 years on EXPOSE-R (Rabbow et al. 2015) and the strain showed high survival rate (90%) under dark and very low survival (1%) under attenuated UV. However, the strain failed to survive under unattenuated UV (Mancinelli 2015). Furthermore, the resistance of cyanobacterial akinetes to space environments was also assessed by drying them onto limestone rocks followed by sending them into Low Earth Orbit (LEO) for 10 days during BIOPAN VI mission. In the absence of UV radiation, a substantial proportion of the akinetes showed survival upon 10 days exposure to LEO and 28 days exposure to simulated Martian environment (−27 °C, 0.8 kPa, CO (2)). In addition, the akinetes were also found to exhibit high survival rate upon 28 days exposure to desiccation and low temperature (−80 °C). In contrast, long-term exposure to vacuum (0.7×10^{-3} kPa) and high temperature (80 °C) was observed to be detrimental to the akinetes. From this study, it was evident that akinetes possess elevated resistance against extreme space environments and may be useful to study the survival of microbial resting cells in space (Olsson-Francis and Cockell 2010a, b). Apart from this, cells of *Anabaena* mixed with their natural rock substrates were also exposed to space for 10 days during BIOPAN VI mission. Results showed that cyanobacterial cells not receiving UV exposure survived the space condition (de La Torre et al. 2010). Space exposure for 548 days of the *Gloeocapsa* sp. biofilm containing *Anabaena cylindrica* akinetes and *Nostoc commune* and *Chroococcidiopsis* sp. CCMEE 029 vegetative cells was performed during another space mission EXPOSE-E (Rabbow et al. 2012). In this mission, except for *Nostoc commune*, all cyanobacteria survived under attenuated UV. However, under unattenuated UV, only the survival of *Chroococcidiopsis* sp. was noticed (Cockell 2011). The survival of *Chroococcidiopsis* sp. CCMEE 029 in space for 22 months was also recorded during EXPOSE-R mission (Bryce et al. 2014). In a recent report by Kimura et al. (2016), the cyanobacterium *Nostoc* sp. HK-01 was shown to possess significant tolerance against space environments. This strain, as a part of the Japanese Tanpopo space mission, is now being used for space exploration studies. Responses of cyanobacteria to space environments have recently been investigated by the Chinese Scientific Experimental System (Ye et al. 2021). In this experiment, the dried cells of *Nostoc* sp. were exposed to the stratosphere environment (32,508 m altitude). The exposure times were 3 h and 22 min during HH-19-2 flight mission. The reason behind exposing the cyanobacterium to stratosphere was the similarity between stratospheric environmental factors (such as atmospheric pressure, temperature, relative humidity and ionizing and non-ionizing radiation levels) and the environmental factors of the Martian surface. This study reflected that there was decreased photosynthetic activity of the tested cyanobacterium and morphological damages, but the death rate was observed to be slow. RNA-sequencing of the flight sample revealed that a number of metabolic pathways of the cyanobacterium such as were affected by the stratospheric environmental factors. In brief, the studies performed so far advocated that most of the tested cyanobacterial strains are able to survive under real space environments upto a certain extent and the survival rate differs with the species type. Cyanobacterial responses to extraterrestrial environments have been given in Fig. 12.1.

12.3.2 Algal Responses to Spaceflight Environments

Algae are one of the first organisms to be sent for space travel. The first algal strain to be flown in space was *Chlorella pyrenoidosa*. This alga was sent to space in 1960 with the spacecraft "Korabl-Sputnik 2" for only 1 day (Hall and David 2001). Therefore, this flight was only able to assess short-term effects of space exposure on algae. This short-term visit to space was found to exert very least or no effect on algae as the growth, reproduction and development processes remained unaltered. The only hazardous impact of space environment on the algae was an increased population of dead cells and slightly lower photosynthetic efficiency (Semenenko and Vladimirova 1961). Studies performed before mid-60 s clearly demonstrated that algae or any other organisms belonging to Earth can easily survive in space for a short period of time. However, the impact of long-term exposure to space on algae

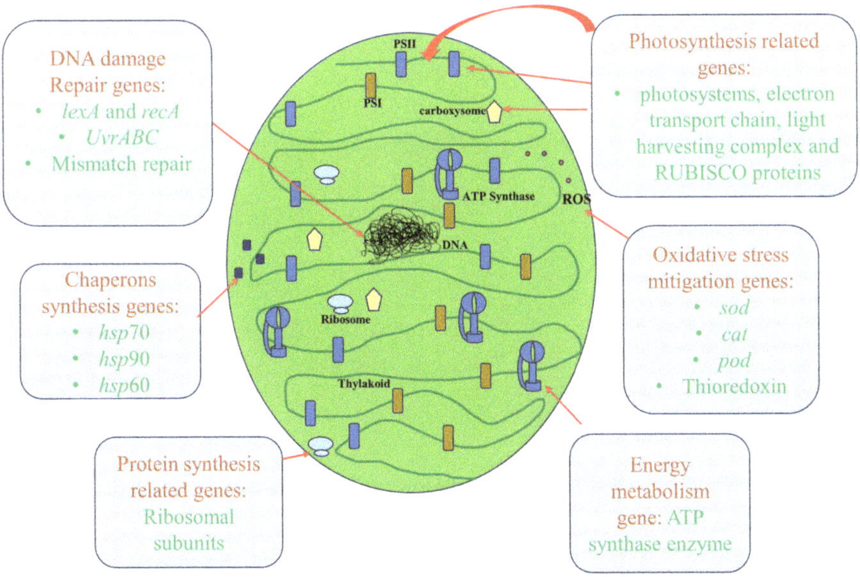

Fig. 12.1 Responses of cyanobacteria to extraterrestrial conditions: upregulated genes

or other organisms was not known then. In order to know the effect of long-term space exposure on organisms, the Soviet spacecraft Cosmos 110 was launched in 1966 with many organisms including dogs and the alga *Chlorella* (LARG-1, LARG-3). This space mission was 22 days long wherein mutations in the *Chlorella* cultures were monitored. However, no significant difference was noticed in the culture when compared with the controls in Earth. The LARG-3 strain of Chlorella only showed reduced survival and the data was found to be statistically significant (Antipov et al. 1969). The survival efficiency of algae in space after 22 days exposure was also deciphered by this study. Moreover, algae not only travelled space before humans but also, visited moon prior to humans. Three spacecrafts, i.e. Zond 5, Zond 6 and Zond 7, were launched on 15th September 1968, tenth November 1968 and seventh August 1969, respectively. These spacecrafts flew to the moon with *Chlorella* cultures and came back to Earth. After this space visit, it was observed that the survival rate of *Chlorella* on Zond 5 and Zond 6 was low whereas the cells on Zond 7 remained unaffected by the space environment (Vaulina et al. 1971). Experiments on the survival of *Chlorella* in space were performed by the Soviet Union between 1977 and 1982 wherein the Soviet space stations Salyut 6 was used. From this experiment, it was proved that low gravitational force has no adverse impact on the growing cells of *Chlorella* (Sychev 2000). The above reports altogether suggested that space exposure does not generally exert an inhibitory effect on the growth and development of algae, except for a few cases. Schematic presentation of algal responses to extraterrestrial conditions has been given in Fig. 12.2.

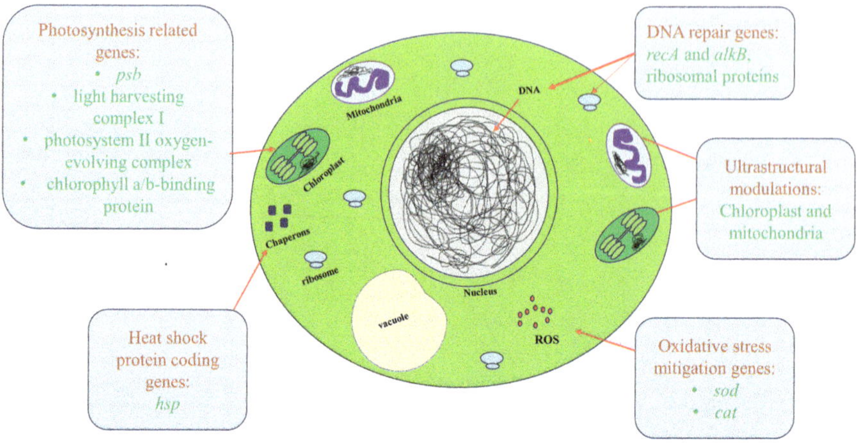

Fig. 12.2 Responses of algae to extraterrestrial conditions: upregulated genes and ultrastructural modulations

12.3.3 Responses of Lichens to Spaceflight Environments

Apart from cyanobacteria and algae, lichen species have also been subjected to space environments and monitored for their survival rate. For instance, one Russian spacecraft named FOTON-M was successfully launched in 2005, which flew lichens *Xanthoria elegans* and *Rhizocarpon geographicum* to space. This spacecraft containing lichens spent 16 days in space and then returned to Earth. The moderately long period exposure of lichens to the conditions prevailing in space such as UV irradiation, vacuum, extreme temperatures and microgravity, showed initial reduction in metabolic activity of lichens, but a rapid recovery to their maximum metabolic activities only after 24 h (García Sancho et al. 2007). Long-term space exposure experiment (1.5 years) in International Space Station was also performed using lichens. Results of this study reflected that *Aspicilia fruticulosa* and *X. elegans* were able to withstand space conditions for a year and sustain their physiological and photosynthetic activities whereas the same conditions caused damage to the lichen *R. geographicum* (De Vera 2012).

12.3.4 Plant Responses to Spaceflight Environments

Among the photoautotrophic organisms, plants were exploited maximally in space research. Even at present times, various plant species are being studied for their survival strategies under space conditions. As a result, a large number of reports related to the responses of plants to space have been accumulated. The major reason underlying such copious number of researches on plants was to investigate their resistance to space environments followed by using them as life support systems in space flights. For this, effects of major physical factors generally encountered in space such as microgravity, cosmic radiation and magnetic field, on the plants were evaluated. In this section, the responses of plants to abovementioned space physical factors will be discussed one by one.

The ability of plants to survive and reproduce under the microgravity of space was demonstrated by many researchers (Ivanova et al. 1993; Link et al. 2003; Musgrave et al. 1997; Sychev et al. 2001, 2007; Merkys et al. 1984; Wolff et al. 2013, 2014; Musgrave 2008; Wolverton and Kisss 2009). During the studies performed in plant-based CRLSS, it was observed that the microgravity in space does not inhibit plant reproduction but their reproductive fitness is somehow affected by this factor (De Micco et al. 2014; Kordyum 2014). This reduced reproductive fitness was thought to be a barrier to the resource utilization process in spaceflights equipped with plant-based life support system. It was reported that the microgravity-induced alterations in plant physiology are mainly due to modulations in the physical environments of the plant. Therefore, physiological changes in plants are secondary effects. This result not only necessitated a thorough examination of the space impacts on the physiologically relevant transport as well as exchange processes in plants but also indicated the importance of maintaining environmental conditions to grow plants in spaceflights. The influences of microgravity on plant's physiological transport and gas exchange were elaborated by Porterfield (2002). As mentioned above, microgravity primarily alters the physical environment of plants which in turn exerts adverse impacts on plant's physiological transport and gas exchange processes. Since the gravitational changes in space do not directly influence plant transport and exchange, these responses are considered indirect effects. The absence of buoyancy-driven thermal convection (BDTC) under reduced gravity of space thickens the boundary layer which is the main reason for hindrance in the gas exchange process of plants (Porterfield 2002). While forced movement of air or proper ventilation successfully alleviates these problems in the aerial parts of plants (Kitaya et al. 2003; Musgrave et al. 1997), diminishing the same problem from the root zones of plants is a challenging task as microgravity alters the distribution of gas and fluid in this zone and causes hypoxia (Liao et al. 2004; Briarty and Maher 2004; Porterfield et al. 2000) The resultant root hypoxia further decreases nutrient uptake and transport in plants (Porterfield 2002). The link between reduced nutrient uptake and stunted growth of plants under microgravity and space environment-induced alterations in plant's nutrient transport process were proposed by many other studies as well (Nechitailo and Gordeev 2001; Aliyev et al. 1987; Belyavskaya 1996; Kordyum et al. 1984). However, interpretation of these results is tedious because these studies did not provide any information about the experimental conditions, especially ventilation, environmental control strategy and the hardware used. Impact of space exposure on the nutrient acquisition capacity of plants were also evaluated by NASA using a plant growth unit in spaceflight (Levine and Krikorian 2008). In this study, a spaceflight experiment was conducted and the ability of a plant to acquire nutrients from its growth medium was assessed. The researchers looked for the disappearance of nutrients from the growth medium and found a very low concentration of potassium. Based on this observation, it was suggested that space exposure induces potassium uptake in plants either because of increased mass of root or microgravity-induced phenomenon. A similar study performed using spaceflight and ground-exposed plants disclosed no deviation in nutrient uptake (Heyenga et al. 2000). Therefore, studies related to the assessment of space impacts on plant nutrient uptake failed to provide any direct conclusion. Although the studies conducted under controlled environmental conditions reflected no effect of microgravity on the photosynthetic, metabolic and gas exchange efficiencies of plants (Stutte et al. 2005; Musgrave 2008), few reports are available in the literature that demonstrated microgravity-induced reduction in the photosynthetic ability of plants (Stutte et al. 2005; Tripathy et al. 1996). Hence, further detailed studies assessing the impact of microgravity on photosynthesis are required in order to reach the conclusion. Moreover, no morphological variations in higher plants were observed after both long- and short-term space exposure. The best-characterized response of plants to microgravity was gravitropism or gravity-directed growth (Kiss et al. 1997). The review of results obtained from gravitropism-related studies performed in space environments can easily be found in many articles (Kiss 2000; Wyatt and Kiss 2013; Blancaflor and Masson 2003; Perbal and Driss-Ecole 2002; Morita 2010; Toyota and Gilroy 2013). While one study unveiled that microgravity leads to automorphogenesis and autotropism in lentil roots (Driss-Ecole et al. 2008), other studies demonstrated plant growth in the direction of light and water availability under space conditions (Cassab 2008; Millar et al. 2010; Takahashi et al. 1999). However, such responses of plants are not always same as observed in ground experiments (Millar et al. 2010). In contrast to the above reports, microgravity has large impacts on the cell organelles of plants as it alters the ultrastructural organization of chloroplasts, causes random distribution of amyloplasts and decreases the content of cell wall polysaccharides (Hoson et al. 2001; Soga et al. 2002; Nedukha 1997).

The most prominent effects of cosmic radiation on plants were found to be damage to DNA, mutation in chromosomes and changes in gene expression (Nevzgodina 1999). Albeit, these changes at chromosome level were not lethal to the growth and development of plants (Karoliussen et al. 2013). Although plants were grown for many generations in low Earth orbit (Sychev et al. 2007), complete evidence on the stability of plant genome under space environments is still missing. Although very few studies related to the exposure of plants to chronic radiation have been performed so far (Real et al. 2004), it is generally thought that long-term exposure of plants to low doses of chronic radiation than high doses of acute radiation is more effective to assess the plant responses to cosmic radiation. The impact of low chronic radiation on the plant genome was observed to be much stronger than that of acute doses (Kovalchuk et al. 2000). It was reported that exposure to Chronic ionizing radiation causes long-term genetic alterations in plants and leads to reduced genetic variability which is possibly an adaptive strategy of plants against chronic radiation (Esnault et al. 2010). Till date, different response mechanisms of plants were noticed after their exposure to chronic and acute radiation doses (Kovalchuk et al. 2007). Exposure to acute radiation was found to affect mainly oxidative stress-related genes whereas general stress and nucleic acid metabolism-related genes mediated responses to chronic radiation. Genes associated with carbohydrate metabolism and photosynthesis were also found to be induced by chronic radiation. In *A. thaliana*, exposure to low doses of chronic gamma radiation decreased fresh weight of stem, root and leaves but no alteration was noticed in the case of oxidative stress-related parameters (Vandenhove et al. 2010; Kovalchuk et al. 2000, 2007). Therefore, the studies on the responses of plants to cosmic radiation suggested that response mechanisms and resistance levels of plants to cosmic radiation vary with species (Bhaskaran and Swaminathan 1961).

Another major space factor i.e. magnetic field (lower as well as higher than geomagnetic field), exerts significant effect on plant growth, metabolism and development (Karoliussen et al. 2013; Galland and Pazur 2005; Belyavskaya 2004; Travkin 1973; Ahmad et al. 2007; Aladjadjiyan and Zahariev 2009; Brettel and Sétif 1987; Hakala-Yatkin et al. 2011; Jovanic and Jevtovic 2002; Yano et al. 2004; Solov'yov and Schulten 2012; Xu et al. 2012). Assessment of effects imposed by magnetic fields on plants is important because there is no global magnetic field in Mars and Moon. Since the magnetic field exhibits a shielding effect against radiation, therefore, understanding the survival rate of plants in such environments receiving chronic cosmic radiation is crucial for future exploration of Mars and Moon. Studies reflected that low magnetic field affects plant processes both positively and negatively. One study showed that plants respond to changes in magnetic field strength through cryptochrome-mediated magneto receptor mechanism (Occhipinti et al. 2014). Despite the importance of these studies (Yamashita et al. 2004), experiments related to the effects of magnetic field on plants were only performed in simulated Earth environment, not in space. The genetic responses of plants to spaceflight conditions have been illustrated in Fig. 12.3.

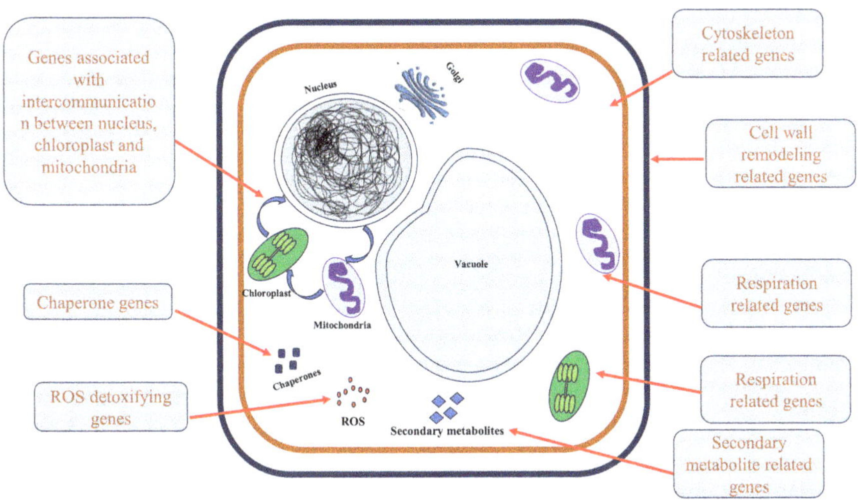

Fig. 12.3 Responses of plants to extraterrestrial conditions: upregulated genes

12.4 Adaptation Strategies of Photoautotrophs under Extraterrestrial Conditions

The adaptative strategies of photoautotrophs under extraterrestrial conditions were studied by many research groups throughout the world. Among the photoautotrophs, higher plants constitute the most exploited group as their survival strategies were rigorously investigated in spaceflights under varied conditions like shade, light, low Earth orbit (LEO), etc. In contrast, such detailed studies are missing in case of other photoautotrophs such as cyanobacteria, algae and lichens. Although these organisms were flown to space and observed for their responses, direct space exposure experiment-based detail analysis of their adaptive strategies has not been executed yet. Instead, the reports available on the survival strategies of cyanobacteria, algae and lichens are very less and mainly based on the experiments performed in ground-based simulated extraterrestrial conditions. In this section, the survival strategies of cyanobacteria, algae, lichens and higher plants will be discussed.

12.4.1 Cyanobacteria

The studies related to space exposure of cyanobacteria and observing their responses were initiated long ago, but their exact survival mechanisms under space environments are still not known as no such studies have yet been performed. However, one recent study managed to decipher the mechanisms of cyanobacteria to survive in space conditions (Ye et al. 2021). In this study, cyanobacterial samples (*Nostoc* sp.) were exposed to the extreme conditions of stratosphere followed by the survival strategies of the organism evaluated through gene expression analyses and compared with the ground-exposed samples. Since the stratospheric environment is quite similar to that of Mars (in terms of desiccation, vacuum, high UV radiation and low temperatures), cyanobacterial exposure to this environment and thorough analysis of their adaptive responses may help understand the possible survival mechanisms of this group of organisms in space. During this experiment, cyanobacterial samples were divided into different groups and all of them were air-dried to eliminate the impact of air drying on the gene expressions of the samples. The major impact of stratospheric UV radiation included damage to D1 protein and consequent reduction in photosynthetic efficiency. Alterations in expressions of many other genes were also observed in cyanobacteria upon their exposure to stratosphere. Expression of genes coding for the proteins belonging to electron transport chain, light-harvesting complex and ATP synthase were enhanced. When this result was compared with that of ground-exposed samples, expressions of the above genes were found to be reduced. These responses indicated that low doses of UV radiation in ground-based experiment caused downregulation of these genes whereas the strong UV dose and association of other factors resulted in the upregulation of these genes as a strategy to repair damaged parts and cope with the extreme environment of stratosphere. Moreover, the expression of *rbcR* gene coding for a Rubisco subunit was also found to be upregulated in this study. The expression of this gene was suggested to be a compensatory mechanism to replace the damaged part of Rubisco. The occurrence of oxidative stress condition was observed in flight-exposed samples of *Nostoc* sp. as there were enhanced expressions of genes coding for antioxidative enzymes like catalase (CAT), peroxidase (POD), superoxide dismutase (SOD) and increased synthesis of non-enzymatic antioxidants such as β-carotene, C-phycocyanin, glutathione (GSH) and thioredoxin. Apart from this, hem gene was also found to be upregulated in the tested cyanobacterium. Based on these results, the authors suggested that stratospheric UV radiation possibly resulted in the generation of free oxygen radicals in the cyanobacterial cells and in order to neutralize the free radicals, the cyanobacterium enhanced the expression of these genes. The upregulation of hem gene was explained by the fact that POD enzyme uses hem as a cofactor, therefore, to support the activity of POD, expression of hem synthesizing gene was also increased. Alongside, DnaJ/DnaK, HtpG and GroESL encoding chaperon genes i.e. *hsp70*, *hsp90* and *hsp60*, were also upregulated under stratospheric conditions. The flight-exposed cyanobacterial samples also showed the upregulation of genes coding for double (DBS) and single-strand break (SSB) repairing proteins, LexA and RecA proteins. Among the SSB genes, abundant expression of genes involved in nucleotide excision repair (NER) was noticed. This included genes coding for UvrC protein belonging to UvrABC system and MutS, MutL, MutT proteins belonging to mismatch repair (MMR) system. The upregulated expression of these DNA-repairing genes indicated their active participation in repairing damaged cyanobacterial DNA resulting from exposure to stratospheric environment and increasing the cyanobacterial resistance to this environment. Moreover, lipid metabolism-related fab gene and genes coding for 50S and 30S ribosomal subunits were also found to be upregulated. This was explained by the fact that Mars-like low temperature of stratosphere might have caused saturation of membrane lipids as well as protein denaturation and the cyanobacterium, in response, upregulated the expression of these related genes for survival purpose. In brief, survival of the cyanobacterium *Nostoc* sp. under Mars-like extreme conditions of stratosphere was mainly determined by the induced expression of genes encoding antioxidants, DNA repair proteins, lipid metabolism-related proteins and proteins involved in folding of denatured proteins (Fig. 12.4).

12.4.2 Algae

Like cyanobacteria, the survival strategies of algae under extraterrestrial environments are less explored in spite of being one of the first organisms to be flown in space. Only one report is available in the literature wherein the green alga *Chlorella* sp. was exposed to Mars-like near-space environment using high-altitude scientific balloons. Like other studies, one ground-based experiment was also set as control (Wang et al. 2021). The near-space exposure was found to cause minute reductions in the chlorophyll fluorescence and Fv/Fm of *Chlorella* as compared to the ground-exposed algal sample. While the reduced Fv/Fm denoted the decreased intrinsic ability of PS II reaction centre to convert light energy, altered structural and functional properties of photosystem were unveiled by minimized chlorophyll a

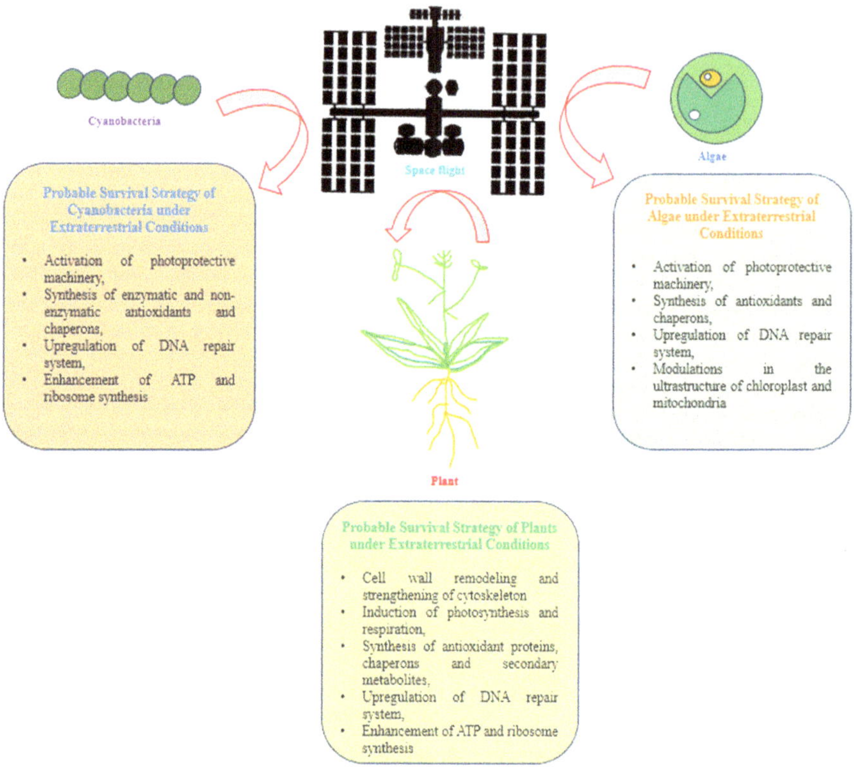

Fig. 12.4 Probable strategies of cyanobacteria, algae and plants to survive in extraterrestrial conditions

fluorescence. These effects on algal PS II were deciphered by the authors as a result of UV exposure since other near-space environmental factors had very little effect on the PSII. The results obtained from transcriptomic analysis revealed that near-space environment upregulated the expression of genes encoding Psb proteins, light-harvesting complex I protein, photosystem II oxygen-evolving complex protein and chlorophyll a/b-binding proteins in *Chlorella*. Increased expression of these genes was explained as a strategy of *Chlorella* to withstand radiation by repairing damaged parts of photosynthetic apparatus. The other notable impacts of near-space environment included ultrastructural modifications in mitochondria and chloroplasts. Exactly like the cyanobacterium *Nostoc* sp., increased expression of genes encoding antioxidant enzymes was found in *Chlorella* upon flight exposure. Additionally, genes of DNA repair proteins (RecA and AlkB), ribosomal proteins and heat shock proteins (HSPs) were also found to be significantly upregulated. The above responses of *Chlorella* were suggested to be the results of UVB exposure in near-space environment. Altogether, the study concluded that survival of *Chlorella* in the extreme conditions of near-space environment was achieved through the activation of photoprotective machinery, PSII and DNA-repairing systems and DNA and protein damage controlling apparatus (Fig. 12.4).

12.4.3 Plants

A large number of studies have been invested in the last few decades to assess the adaptive strategies of higher plants in space (Paul et al. 2012; Kwon et al. 2015; Johnson et al. 2017; Choi et al. 2019; Basu et al. 2017; Johnson et al. 2017; Kruse et al. 2020; Angelos et al. 2021; Beisel et al. 2019; Ferl and Paul 2016; Califar et al. 2020). The majority of these studies were performed using *Arabidopsis thaliana* based on transcriptomic analyses. The interpretation of the results obtained so far is tedious due to various experimental constraints, insights into the adaptive strategies of this plant under spaceflight conditions have largely been advanced. A number of experiments on *A. thaliana* were carried out in International Space Station (ISS) and responses of the test plant were analysed through RNA sequencing and/or microarray. Results obtained from most of the experiments showed the existence of many common response pathways. The major biological processes that were observed to be affected in space-exposed *Arabidopsis* included photosynthesis, remodelling of cell wall, activation of antioxidative defence system, expression of heat shock proteins (HSPs) and other defence processes. The adaptive strategies of *Arabidopsis* against microgravity were disclosed based on gene expression profiles. It was described that microgravity acts as a novel stress factor to the plant and leads to the generation of reactive oxygen species. The stress signal of microgravity is perceived by the plant at the cell membrane followed by transmitted to the nucleus through MAPKs, Ca^{2+} and auxin. In the nucleus, this signal triggers the expression of transcription factors which in turn activate genes capable of regulating different physiological and molecular adaptation processes. Some of these processes were found to be associated with the maintenance of cell integrity (e.g. remodelling of cell wall), and some were identified to be involved in protecting metabolic activities (photosynthesis, mitochondrial electron transportation and inter-organellar communications). Apart from this, the induction of antioxidative machinery and synthesis of molecular chaperones like HSPs were also among the prominent responses of *Arabidopsis* in ISS. However, the connection between these responses of *Arabidopsis* is still not disclosed and it is tough to draw the exact mechanisms of its survival in space. Therefore, it can only be said from these findings that each of these responses is crucial for assuring survival of *Arabidopsis* in space conditions. The probable space survival strategies of *Arabidopsis* have been shown in Fig. 12.4. Furthermore, a recent study performed by NASA provided new insights into the survival strategies of *Arabidopsis* in space (Olanrewaju et al. 2023). In this investigation, germination and growth of *Arabidopsis* seedling were carried out in the International Space Station (ISS) using Light Emitting Diode (LED) hardware, and ground control on Earth was also established. After 10 days exposure to space, the frozen plant samples were brought to Earth and subjected to transcriptomics and proteomic analysis. Results obtained from these analyses reflected spaceflight-induced alteration in photosynthetic machinery of the test plant. It was apparent that the plant tried to cope with the space environment by triggering organ-specific expression of photosystem proteins. Apart from this, genes involved in protease activity, mRNA binding and nuclease activities showed increased mRNA and protein levels in spaceflight-exposed *Arabidopsis*. However, the genes associated with exoribonuclease activity, tRNA binding and RNA helicase activity exhibited suppressed transcript and protein levels. Downregulation of proteins responsible for cellulose deposition in cell wall i.e. cellulose syntheses and cellulose-like proteins, was also evident in spaceflight-exposed *Arabidopsis*. In contrast to this, *Arabidopsis* showed an elevated level of UDP-arabinopyranose mutases, involved in the synthesis of non-cellulosic polysaccharides of cell wall, upon spaceflight exposure. Considerable changes in redistribution of polar auxins, lipid content and intracellular transportation of ions were recorded too in spaceflight conditions. This study also suggested that plants in space require more metabolic energy than that on Earth and as a result, activate different shunt metabolic pathways.

Recently, the roles of telomeres and telomerase activity in the survival of plants in space were assessed using the model plant of *Arabidopsis thaliana*. Unlike astronauts, no alteration in the length of telomeres was observed in space-exposed *Arabidopsis*, but the telomerase activity was found to be increased dramatically. From these results and ground-based studies, it was concluded that the telomerase activity is trigged by different environmental factors but this activity is not related to the length of telomere. Therefore, the telomerase activity, which is inversely correlated with genome oxidation, was proposed as a redox protective strategy for *Arabidopsis thaliana* to survive in space or adverse environmental conditions (Barcenilla et al. 2023).

Apart from *Arabidopsis*, space survival strategies were also evaluated in other plants like *Brassica rapa var. nipposinica*, *Ceratopteris richardii* and *Oryza sativa* after long-term space exposure. The mRNA sequencing-based transcriptome analysis of *Brassica rapa var. nipposinica* grown in ISS was performed and more than two-fold upregulation of 20 oxidative markers including glutaredoxin and thioredoxin was recorded. Besides, elevated transcript levels of NADPH oxidase genes i.e. *rbohF* and *rbohD*, in *Brassica rapa* were also detected. From these results, it was evident that the *Brassica rapa var. nipposinica* plant activates its ROS-producing genes followed by using the ROS signals to induce ROS scavenging machineries as defence system to withstand long-term space exposure (Sugimoto et al. 2014). When the spores of the fern *Ceratopteris richardii* were exposed to space conditions followed by being subjected to microarray-based gene expression analysis, upregulation of chaperon and protein phosphatase 2C coding genes and downregulation of peroxidase coding genes was noticed in the first 20 h of their germination. Furthermore, increased transcript levels of genes associated with reserve lipid mobilization were detected in spaceflight, and lowered mRNA levels of genes responsible for lipid synthesis/hydrolysis/transfer were recorded in microgravity. These observations indicated that the fern *Ceratopteris richardii* exhibits alterations in stress response and utilization of lipid. Both of these features may be common for other plants grown in space conditions. This study not only unveiled early physiologically responses of plants to microgravity but also disclosed genetic basis of single-cell responses to gravity (Salmi and Roux 2008). In order to study the survival strategies of *Oryza sativa* in space, the calli of the plant was flown to space as a part of Shenzhou spaceship mission for 13.5 days and then subjected to transcript analysis. It was found that the genes involved in various metabolic and signalling pathways were influenced by microgravity. The metabolic pathways whose genes showed altered expression included starch, protein degradation, glycolysis and cell wall metabolism. Additionally, microgravity was also found to alter the expression of genes encoding components of electron transport chains and transporters. The most interesting observations included the downregulation of cell wall biosynthesis-related genes and upregulation of genes associated with the modification of primary cell wall. Apart from this, spaceflight-induced upregulation of aquaporine coding genes was also identified in rice calli. Upregulation of cell wall modifying genes indicated cell wall extension as a strategy of rice to survive in space. (Jin et al. 2018). Therefore, it is clear that different plant species exhibit distinct strategies to cope with spaceflight conditions, especially gravity.

12.5 Conclusions

The present chapter describes the responses and survival strategies of photoautotrophs in extraterrestrial environments based on available reports. Although a considerable number of studies have been performed over the decades on different groups of photoautotrophs to understand their exact mechanisms of survival in space conditions, complete knowledge on this issue is still lacking. However, remarkable progress has been made in this field with time and probable mechanisms of different photoautotrophs have been untangled either through direct space exposure-based studies or through simulated extraterrestrial environment-based experiments. From the studies performed so far, it is apparent that space environment acts as a novel stressor to all of the tested photoautotrophs and interestingly, many of these organisms were found to successfully survive in this harsh environment. Nonetheless, the survival strategies were observed to be distinct for each of the photoautotrophs. While cyanobacteria survive in space-like conditions by inducing the expression of genes encoding antioxidants, DNA repair proteins, lipid metabolism-related proteins and proteins involved in folding of denatured proteins, algae ensured their survival in near-space environments by activating photoprotective machinery, PSII and DNA-repairing systems and DNA and protein damage controlling apparatus. In contrast, the space survival mechanisms in plants are rather complex and mainly include maintenance of cell integrity, protection of crucial metabolic activities (photosynthesis, mitochondrial electron transportation and inter-organellar communications), induction of antioxidative machinery, synthesis of molecular chaperones like HSPs, minimizing genome oxidation by increasing telomerase activity and utilization of stored lipid.

12.6 Existing Lacunae and Future Direction of Research

From the information provided in this chapter, one thing is clear that more than 60 years of researches on different groups of photoautotrophs could not completely explore their exact mechanisms of survival in space. However, the knowledge on plants' survival strategies in extraterrestrial environments has greatly been advanced in recent times. In contrast, very less is known about the space-survival strategies of cyanobacteria, algae and lichens due to extremely small number of studies. Since cyanobacteria are considered as one of the earliest colonizers of Earth and known to be responsible for the transformation of Earth's environment from reducing to oxidizing one, significant number of studies on cyanobacteria were invested to evaluate their survivability in space and many genera were identified to possess commendable efficiency to withstand space conditions. Similar observations were also made in case of few algal and lichen species. Despite this fact, studies on the space survival strategies of these organisms are still lacking except for few. Therefore, future researches must be directed towards the exploration of space survival strategies of cyanobacteria, algae and lichens. For this, advanced techniques (RNAseq and microarray) based studies are needed to be employed. As these organisms are small sized, studying the molecular mechanisms of space survival will be easy as compared to higher plants.

Acknowledgments The Head of the Department of Botany, Banaras Hindu University, Varanasi, India is gratefully acknowledged for providing necessary laboratory facilities. We sincerely acknowledge the support from Institute of Eminence (IOE-6031), Banaras Hindu University. Sindhunath Chakraborty is thankful to Institute of Eminence, Banaras Hindu University for Raja Jwala Prasad Post-doctoral Fellowship. Ranjan Kumar Tiwari is thankful to the Council of Scientific and Industrial Research for research fellowship in terms of Junior Research Fellowship.

Conflict of Interest The authors declare no conflict of interest.

References

Abbass K, Qasim MZ, Song H, Murshed M, Mahmood H, Younis I (2022) A review of the global climate change impacts, adaptation, and sustainable mitigation measures. Environ Sci Pollut Res 29(28):42539–42559

Agarwal R, Rane S, Sainis JK (2008) Effects of 60Co γ radiation on thylakoid membrane functions in *Anacystis nidulans*. J Photochem Photobiol B Biol 91(1):9–19

Ahmad M, Galland P, Ritz T, Wiltschko R, Wiltschko W (2007) Magnetic intensity affects cryptochrome-dependent responses in *Arabidopsis thaliana*. Planta 225:615–624

Aladjadjiyan A, Zahariev A (2009) Influence of stationary magnetic field on the absorption spectra of some energy plants. J Environ Prot Ecol 10:1032–1036

Aliyev AA, Abilov ZK, Mashinskiy AL, Ganiyeva RA, Ragimova GK (1987) The ultrastructure and physiological characteristics of the photosynthesis system of shoots of garden pea grown for 29 days on the "salyut-7"space station. USSR Space Life Sci Dig 10:6

Angelos E, Ko DK, Zemelis-Durfee S, Brandizzi F (2021) Relevance of the unfolded protein response to spaceflight-induced transcriptional reprogramming in Arabidopsis. Astrobiology 21(3):367–380

Antipov V, Delone N, Nikitin M, Parfyonov G, Saxonov P (1969) Some results of radiobiological studies performed on Cosmos-110 biosatellite. Life Sci Space Res 7:207–208

Asato Y (1971) Photorecovery of gamma irradiated cultures of blue-green alga, Anacystis nidulans. Radiat Bot 11(4):313–316

Badri H, Monsieurs P, Coninx I, Wattiez R, Leys N (2015) Molecular investigation of the radiation resistance of edible cyanobacterium *Arthrospira* sp PCC 8005. Microbiology 4(2):187–207

Baqué M, de Vera J-P, Rettberg P, Billi D (2013a) The BOSS and BIOMEX space experiments on the EXPOSE-R2 mission: endurance of the desert cyanobacterium *Chroococcidiopsis* under

simulated space vacuum, Martian atmosphere, UVC radiation and temperature extremes. Acta Astronaut 91:180–186
Baqué M, Viaggiu E, Scalzi G, Billi D (2013b) Endurance of the endolithic desert cyanobacterium *Chroococcidiopsis* under UVC radiation. Extremophiles 17:161–169
Baqué M, Verseux C, Rabbow E, de Vera J-PP, Billi D (2014) Detection of macromolecules in desert cyanobacteria mixed with a lunar mineral analogue after space simulations. Orig Life Evol Biosph 44:209–221
Baqué M, Verseux C, Böttger U, Rabbow E, de Vera J-PP, Billi D (2016) Preservation of biomarkers from cyanobacteria mixed with Marslike regolith under simulated Martian atmosphere and UV flux. Orig Life Evol Biosph 46:289–310
Barcenilla BB, Meyers AD, Castillo-González C, Young P, Min J-H, Song J, Phadke C, Land E, Canaday E, Perera IY (2023) Arabidopsis telomerase takes off by uncoupling enzyme activity from telomere length maintenance in space. Nat Commun 14(1):7854
Basu P, Kruse CP, Luesse DR, Wyatt SE (2017) Growth in spaceflight hardware results in alterations to the transcriptome and proteome. Life Sci Space Res 15:88–96
Beisel NS, Noble J, Barbazuk WB, Paul AL, Ferl RJ (2019) Spaceflight-induced alternative splicing during seedling development in Arabidopsis thaliana. NPJ Microgravity 5(1):9
Belyavskaya N (1996) Free and membrane-bound calcium in microgravity and microgravity effects at the membrane level. Adv Space Res 17(6–7):169–177
Belyavskaya N (2004) Biological effects due to weak magnetic field on plants. Adv Space Res 34(7):1566–1574
Bhaskaran S, Swaminathan M (1961) Chromosome aberrations, changes in DNA content and frequency and spectrum of mutations induced by X-rays and neutrons in polyploids. Radiat Bot 1:166–181
Billi D, Friedmann EI, Hofer KG, Caiola MG, Ocampo-Friedmann R (2000) Ionizing-radiation resistance in the desiccation-tolerant cyanobacterium *Chroococcidiopsis*. Appl Environ Microbiol 66(4):1489–1492
Blancaflor EB, Masson PH (2003) Plant gravitropism. Unraveling the ups and downs of a complex process. Plant Physiol 133(4):1677–1690
Blaustein R (2016) The great oxidation event: evolving understandings of how oxygenic life on earth began. Bioscience 66(3):189–195
Brettel K, Sétif P (1987) Magnetic-field effects on primary reactions in photosystem I. Biochim Biophys Acta (BBA) Bioenergetics 893(2):109–114
Briarty L, Maher E (2004) Reserve utilization in seeds of *Arabidopsis thaliana* germinating in microgravity. Int J Plant Sci 165(4):545–551
Bruce A (1964) Extraction of the radioresistant factor of *micrococcus radiodurans*. Radiat Res 22(1):155–164
Califar B, Sng NJ, Zupanska A, Paul A-L, Ferl RJ (2020) Root skewing-associated genes impact the spaceflight response of Arabidopsis thaliana. Front Plant Sci 11:239
Cassab GI (2008) Other tropisms and their relationship to gravitropism. In: Plant tropisms. Wiley, New York, pp 123–139
Choi WG, Barker RJ, Kim SH, Swanson SJ, Gilroy S (2019) Variation in the transcriptome of different ecotypes of Arabidopsis thaliana reveals signatures of oxidative stress in plant responses to spaceflight. Am J Bot 106(1):123–136
Cleland CE (2019) Moving beyond definitions in the search for extraterrestrial life. Astrobiology 19(6):722–729
Cockell CS (2011) Synthetic geomicrobiology: engineering microbe–mineral interactions for space exploration and settlement. Int J Astrobiol 10(4):315–324
de La Torre R, Sancho LG, Horneck G, de los Ríos A, Wierzchos J, Olsson-Francis K, Cockell CS, Rettberg P, Berger T, de Vera J-PP (2010) Survival of lichens and bacteria exposed to outer space conditions–results of the Lithopanspermia experiments. Icarus 208(2):735–748
De Micco V, De Pascale S, Paradiso R, Aronne G (2014) Microgravity effects on different stages of higher plant life cycle and completion of the seed-to-seed cycle. Plant Biol 16:31–38
De Vera J-P (2012) Lichens as survivors in space and on Mars. Fungal Ecol 5(4):472–479

Domain F, Houot L, Chauvat F, Cassier-Chauvat C (2004) Function and regulation of the cyanobacterial genes lexA, recA and ruvB: LexA is critical to the survival of cells facing inorganic carbon starvation. Mol Microbiol 53(1):65–80

Driss-Ecole D, Legué V, Carnero-Diaz E, Perbal G (2008) Gravisensitivity and automorphogenesis of lentil seedling roots grown on board the International Space Station. Physiol Plant 134(1):191–201

Esnault M-A, Legue F, Chenal C (2010) Ionizing radiation: advances in plant response. Environ Exp Bot 68(3):231–237

Fengler S, Spirer I, Neef M, Ecke M, Hauslage J, Hampp R (2016) Changes in gene expression of *Arabidopsis thaliana* cell cultures upon exposure to real and simulated partial-g forces. Microgravity Sci Technol 28:319–329

Ferl RJ, Paul A-L (2016) The effect of spaceflight on the gravity-sensing auxin gradient of roots: GFP reporter gene microscopy on orbit. NPJ Microgravity 2(1):1–9

Galland P, Pazur A (2005) Magnetoreception in plants. J Plant Res 118:371–389

García Sancho L, De la Torre Noetzel R, Horneck G, Ascaso C, Ríos A, Pintado A, Wierzchos J, Schuster M (2007) Lichens survive in space: results from the 2005 lichens experiment. Astrobiology 7(3):443–454

Godward MBE (1962) Invisible radiations. In: Lewin RA (ed) Physiology and biochemistry of algae. Academic Press, New York, pp 551–566

Hakala-Yatkin M, Sarvikas P, Paturi P, Mäntysaari M, Mattila H, Tyystjärvi T, Nedbal L, Tyystjärvi E (2011) Magnetic field protects plants against high light by slowing down production of singlet oxygen. Physiol Plant 142(1):26–34

Hall R, David S (2001) The rocket men: Vostok & Voskhod. The first soviet manned spaceflights. Springer, Cham

Heyenga A, Forsman A, Stodieck L, Hoehn A, Kliss M (2000) Approaches in the determination of plant nutrient uptake and distribution in space flight conditions. Adv Space Res 26(2):299–302

Hoson T, Saiki M, Kamisaka S, Yamashita M (2001) Automorphogenesis and gravitropism of plant seedlings grown under microgravity conditions. Adv Space Res 27(5):933–940

Ivanova T, Bercovich YA, Mashinskiy A, Meleshko G (1993) The first "space" vegetables have been grown in the "SVET" greenhouse using controlled environmental conditions. Acta Astronaut 29(8):639–644

Jin J, Chen H, Cai W (2018) Transcriptomic analysis reveals the effects of microgravity on Rice calli on board the Chinese spaceship Shenzhou 8. Microgravity Sci Technol 30:807–816

Johnson CM, Subramanian A, Pattathil S, Correll MJ, Kiss JZ (2017) Comparative transcriptomics indicate changes in cell wall organization and stress response in seedlings during spaceflight. Am J Bot 104(8):1219–1231

Jovanic B, Jevtovic R (2002) Effect of a permanent magnetic field on the optical and physiological properties on green plant leaves. Int J Environ Stud 59(5):599–606

Kamal KY, van Loon JJ, Medina FJ, Herranz R (2019) Differential transcriptional profile through cell cycle progression in *Arabidopsis* cultures under simulated microgravity. Genomics 111(6):1956–1965

Karoliussen I, Brinckmann E, Kittang A-I (2013) Will plants grow on moon or Mars? Curr Biotechnol 2(3):235–243

Kimura S, Tomita-Yokotani K, Inoue K, Sato S, Arai M, and Katoh H (2016) Space environmental tolerance of a terrestrial cyanobacterium, Nostoc sp. HK-01. International conference on environmental systems

Kiss JZ (2000) Mechanisms of the early phases of plant gravitropism. Crit Rev Plant Sci 19(6):551–573

Kiss JZ, Guisinger MM, Miller AJ, Stackhouse KS (1997) Reduced gravitropism in hypocotyls of starch-deficient mutants of *Arabidopsis*. Plant Cell Physiol 38(5):518–525

Kitaya Y, Kawai M, Tsuruyama J, Takahashi H, Tani A, Goto E, Saito T, Kiyota M (2003) The effect of gravity on surface temperatures of plant leaves. Plant Cell Environ 26(4):497–503

Kordyum E (2014) Plant cell gravisensitivity and adaptation to microgravity. Plant Biol 16:79–90

Kordyum E, Belyavskaya N, Nedukha E, Palladina T, Tarasenko V (1984) The role of calcium ions in cytological effects of hypogravity. Adv Space Res 4(12):23–26

Kovalchuk O, Arkhipov A, Barylyak I, Karachov I, Titov V, Hohn B, Kovalchuk I (2000) Plants experiencing chronic internal exposure to ionizing radiation exhibit higher frequency of homologous recombination than acutely irradiated plants. Mutat Res Fund Mol M Mutagenesis 449(1–2):47–56

Kovalchuk I, Molinier J, Yao Y, Arkhipov A, Kovalchuk O (2007) Transcriptome analysis reveals fundamental differences in plant response to acute and chronic exposure to ionizing radiation. Mutat Res Fund Mol M Mutagenesis 624(1–2):101–113

Kraus MP (1969) Resistance of blue-green algae to 60Co gamma radiation. Radiat Bot 9(6):481–489

Kruse CP, Meyers AD, Basu P, Hutchinson S, Luesse DR, Wyatt SE (2020) Spaceflight induces novel regulatory responses in Arabidopsis seedling as revealed by combined proteomic and transcriptomic analyses. BMC Plant Biol 20:1–16

Kumar H (1964) Effects of radiations on blue-green algae: II. Effects in growth. Ann Bot 28(4):555–564

Kwon T, Sparks JA, Nakashima J, Allen SN, Tang Y, Blancaflor EB (2015) Transcriptional response of *Arabidopsis* seedlings during spaceflight reveals peroxidase and cell wall remodeling genes associated with root hair development. Am J Bot 102(1):21–35

Levine HG, Krikorian A (2008) Changes in plant medium composition after a spaceflight experiment: potassium levels are of special interest. Adv Space Res 42(6):1060–1065

Levine E, Thiel T (1987) UV-inducible DNA repair in the cyanobacteria *anabaena* spp. J Bacteriol 169(9):3988–3993

Liao J, Liu G, Monje O, Stutte G, Porterfield D (2004) Induction of hypoxic root metabolism results from physical limitations in O_2 bioavailability in microgravity. Adv Space Res 34(7):1579–1584

Link B, Durst S, Zhou W, Stankovic B (2003) Seed-to-seed growth of *Arabidopsis thaliana* on the international Space Station. Adv Space Res 31(10):2237–2243

Lyons TW, Reinhard CT, Planavsky NJ (2014) The rise of oxygen in Earth's early ocean and atmosphere. Nature 506(7488):307–315

Mancinelli R (2015) The affect of the space environment on the survival of *Halorubrum chaoviator* and *Synechococcus* (Nägeli): data from the space experiment OSMO on EXPOSE-R. Int J Astrobiol 14(1):123–128

Mancinelli R, White M, Rothschild LJ (1998) Biopan-survival I: exposure of the osmophiles *Synechococcus* sp.(Nageli) and *Haloarcula* sp. to the space environment. Adv Space Res 22(3):327–334

Manzano AI, van Loon JJ, Christianen PC, Gonzalez-Rubio JM, Medina FJ, Herranz R (2012) Gravitational and magnetic field variations synergize to cause subtle variations in the global transcriptional state of *Arabidopsis* in vitro callus cultures. BMC Genomics 13:1–13

Merkys A, Laurinavičius R, Švegždiene D (1984) Plant growth, development and embryogenesis during Salyut-7 flight. Adv Space Res 4(10):55–63

Millar KD, Kumar P, Correll MJ, Mullen JL, Hangarter RP, Edelmann RE, Kiss JZ (2010) A novel phototropic response to red light is revealed in microgravity. New Phytol 186(3):648–656

Morita MT (2010) Directional gravity sensing in gravitropism. Annu Rev Plant Biol 61:705–720

Musgrave ME (2008) Growing plants in space, vol 2007. CABI Reviews, Mumbai, p 9

Musgrave ME, Kuang A, Matthews SW (1997) Plant reproduction during spaceflight: importance of the gaseous environment. Planta 203:S177–S184

Nechitailo G, Gordeev A (2001) Effect of artificial electric fields on plants grown under microgravity conditions. Adv Space Res 28(4):629–631

Nedukha EM (1997) Effects of microgravity on the structure and function of plant cell walls. Int Rev Cytol 170:39–77

Nevzgodina L (1999) Chromosomal aberrations as a biomarker for cosmic radiation. In: Fundamentals for the assessment of risks from environmental radiation. Springer, Cham, pp 203–208

Occhipinti A, De Santis A, Maffei ME (2014) Magnetoreception: an unavoidable step for plant evolution? Trends Plant Sci 19(1):1–4

Olanrewaju GO, Kruse CP, Wyatt SE (2023) Functional meta-analysis of the proteomic responses of arabidopsis seedlings to the spaceflight environment reveals multi-dimensional sources of variability across spaceflight experiments. Int J Mol Sci 24(19):14425

Olsson-Francis K, Cockell CS (2010a) Experimental methods for studying microbial survival in extraterrestrial environments. J Microbiol Methods 80(1):1–13

Olsson-Francis K, Cockell CS (2010b) Use of cyanobacteria for in-situ resource use in space applications. Planet Space Sci 58(10):1279–1285

Paul A-L, Zupanska AK, Ostrow DT, Zhang Y, Sun Y, Li J-L, Shanker S, Farmerie WG, Amalfitano CE, Ferl RJ (2012) Spaceflight transcriptomes: unique responses to a novel environment. Astrobiology 12(1):40–56

Perbal G, Driss-Ecole D (2002) Contributions of space experiments to the study of gravitropism. J Plant Growth Regul 21(2):156

Porterfield DM (2002) The biophysical limitations in physiological transport and exchange in plants grown in microgravity. J Plant Growth Regul 21(2):177

Porterfield D, Barta D, Ming D, Morrow R, Musgrave M (2000) ASTROCULTURE™ root metabolism and cytochemical analysis. Adv Space Res 26(2):315–318

Rabbow E, Rettberg P, Barczyk S, Bohmeier M, Parpart A, Panitz C, Horneck G, von Heise-Rotenburg R, Hoppenbrouwers T, Willnecker R (2012) EXPOSE-E: an ESA astrobiology mission 1.5 years in space. Astrobiology 12(5):374–386

Rabbow E, Rettberg P, Barczyk S, Bohmeier M, Parpart A, Panitz C, Horneck G, Burfeindt J, Molter F, Jaramillo E (2015) The astrobiological mission EXPOSE-R on board of the international Space Station. Int J Astrobiol 14(1):3–16

Real A, Sundell-Bergman S, Knowles J, Woodhead D, Zinger I (2004) Effects of ionising radiation exposure on plants, fish and mammals: relevant data for environmental radiation protection. J Radiol Prot 24(4A):A123

Salmi ML, Roux SJ (2008) Gene expression changes induced by space flight in single-cells of the fern *Ceratopteris richardii*. Planta 229:151–159

Semenenko V, Vladimirova M (1961) Effect of cosmic flight conditions in the Sputnikship on the viability of *chlorella*. Physiol Plants 8:743–749

Schröder K-P, Connon Smith R (2008) Distant future of the Sun and Earth revisited. Mon Not R Astron Soc 386(1):155–163

Shields LM, Durrell LW, Sparrow AH (1961) Preliminary observations on radiosensitivity of algae and fungi from soils of the Nevada test site. Ecology 42(2):440–441

Singh H, Fernandes T, Apte SK (2010) Unusual radioresistance of nitrogen-fixing cultures of anabaena strains. J Biosci 35:427–434

Singh H, Anurag K, Apte SK (2013) High radiation and desiccation tolerance of nitrogen-fixing cultures of the cyanobacterium *anabaena* sp. strain PCC 7120 emanates from genome/proteome repair capabilities. Photosynth Res 118:71–81

Soga K, Wakabayashi K, Kamisaka S, Hoson T (2002) Stimulation of elongation growth and xyloglucan breakdown in *Arabidopsis* hypocotyls under microgravity conditions in space. Planta 215:1040–1046

Solov'yov IA, Schulten K (2012) Reaction kinetics and mechanism of magnetic field effects in cryptochrome. J Phys Chem B 116(3):1089–1099

Stutte G, Monje O, Goins GD, Tripathy B (2005) Microgravity effects on thylakoid, single leaf, and whole canopy photosynthesis of dwarf wheat. Planta 223:46–56

Sugimoto M, Oono Y, Gusev O, Matsumoto T, Yazawa T, Levinskikh MA, Sychev VN, Bingham GE, Wheeler R, Hummerick M (2014) Genome-wide expression analysis of reactive oxygen species gene network in Mizuna plants grown in long-term spaceflight. BMC Plant Biol 14(1):1–11

Sychev V (2000) Investigations of the effects of microgravity on biological objects-components of closed ecological systems of life support and development of cultivation technologies. IMBP-RAN, Moscow

Sychev V, Shepelev EY, Meleshko G, Gurieva T, Levinskikh M, Podolsky I, Dadasheva O, Popov V (2001) Main characteristics of biological components of developing life support system observed during the experiments aboard orbital complex MIR. Adv Space Res 27(9):1529–1534
Sychev VN, Levinskikh MA, Gostimsky SA, Bingham GE, Podolsky IG (2007) Spaceflight effects on consecutive generations of peas grown onboard the Russian segment of the international Space Station. Acta Astronaut 60(4–7):426–432
Takahashi H, Mizuno H, Kamada M, Fujii N, Higashitani A, Kamigaichi S, Aizawa S, Mukai C, Shimazu T, Fukui K (1999) A spaceflight experiment for the study of gravimorphogenesis and hydrotropism in cucumber seedlings. J Plant Res 112:497–505
Toyota M, Gilroy S (2013) Gravitropism and mechanical signaling in plants. Am J Bot 100(1):111–125
Travkin M (1973) Change in the bioelectric activity of *Setcreasea purpurea* due to the effects of constant and pulsating magnetic fields. Biofizika 18(1):172–174
Tripathy BC, Brown CS, Levine HG, Krikorian AD (1996) Growth and photosynthetic responses of wheat plants grown in space. Plant Physiol 110(3):801–806
Vandenhove H, Vanhoudt N, Cuypers A, van Hees M, Wannijn J, Horemans N (2010) Life-cycle chronic gamma exposure of *Arabidopsis thaliana* induces growth effects but no discernable effects on oxidative stress pathways. Plant Physiol Biochem 48(9):778–786
Vaulina E, Anikeeva I, Gubareva I, Shtraukh G (1971) Influence of space-flight factors aboard "Zond" automatic stations on survival and mutability of *chlorella* cells. Cosm Res 9:865
Verseux C, Baqué M, Cifariello R, Fagliarone C, Raguse M, Moeller R, Billi D (2017) Evaluation of the resistance of *Chroococcidiopsis* spp. to sparsely and densely ionizing irradiation. Astrobiology 17(2):118–125
Wang B, Ye T, Li X, Bian P, Liu Y, Wang G (2021) Survival of desert algae Chlorella exposed to Mars-like near space environment. Life Sci Space Res 29:22–29
Wang L, Xie J, Mou C, Jiao Y, Dou Y, Zheng H (2022) Transcriptomic analysis of the interaction between FLOWERING LOCUS T induction and photoperiodic signaling in response to spaceflight. Front Cell Dev Biol 9:813246
Wolff S, Coelho L, Zabrodina M, Brinckmann E, Kittang A-I (2013) Plant mineral nutrition, gas exchange and photosynthesis in space: A review. Adv Space Res 51(3):465–475
Wolff SA, Coelho LH, Karoliussen I, Jost AIK (2014) Effects of the extraterrestrial environment on plants: recommendations for future space experiments for the MELiSSA higher plant compartment. Life 4(2):189–204
Wolverton C, Kisss JZ (2009) An update on plant space biology. Gravitat Space Biol 22(2):13–21
Wyatt SE, Kiss JZ (2013) Plant tropisms: from Darwin to the International Space Station, vol 100. Wiley, New York, pp 1–3
Xu C, Yin X, Lv Y, Wu C, Zhang Y, Song T (2012) A near-null magnetic field affects cryptochrome-related hypocotyl growth and flowering in *Arabidopsis*. Adv Space Res 49(5):834–840
Yamashita M, Tomita-Yokotani K, Hashimoto H, Takai M, Tsushima M, Nakamura T (2004) Experimental concept for examination of biological effects of magnetic field concealed by gravity. Adv Space Res 34(7):1575–1578
Yano A, Ohashi Y, Hirasaki T, Fujiwara K (2004) Effects of a 60 Hz magnetic field on photosynthetic CO_2 uptake and early growth of radish seedlings. Bioelectromagnetics *25*(8):572–581
Ye T, Wang B, Li C, Bian P, Chen L, Wang G (2021) Exposure of cyanobacterium Nostoc sp. to the Mars-like stratosphere environment. J Photochem Photobiol B Biol 224:112307

Stress Granules: Synthesis and Significance

13

Ashutosh Singh, Ravinsh Kumar, and Amrita Srivastava

Abstract

Stress granules as indicated by name, are transient cytoplasmic structures formed in response to different kinds of stresses. These are found in various eukaryotic organisms with an aim to regulate expression of mRNAs as and when required. Being sessile in nature, plants have given more emphasis on post-transcriptional and translational regulation of gene expression so as to respond quickly to the stressful environment. Also, by virtue of their existence in the form of ribonucleoprotein complexes that possess all the potential to initiate translation, they help the cells to recover from stressful conditions rapidly. The basic constituents of stress granules are mRNA and proteins. These granules are formed by liquid–liquid phase separation and do not possess membranes. Nevertheless, their organized structure is a brief illustration of their functional significance. They comprise a shell consisting of untranslated mRNAs, transcription factors, and enzymes; and a core consisting of translation-related proteins with RNA-binding domains. During stress, when the cellular dynamics is disturbed, they act as important centers for storage, translational stalling, and degradation of mRNAs. This chapter discusses structural attributes of stress granules, dynamics of their assembly and disassembly, and their role in diverse plant stresses.

Keywords

Plant stress · Untranslated mRNA · Liquid–liquid phase separation · Post-transcriptional regulation · Cellular dynamics

A. Singh · R. Kumar · A. Srivastava (✉)
Department of Life Science, Central University of South Bihar, Gaya, Bihar, India
e-mail: amritasrivastava@cub.ac.in

A. K. Mishra (ed.), *Stress Biology in Photosynthetic Organisms*,
https://doi.org/10.1007/978-981-97-1883-2_13

13.1 Introduction

Formation of stress granules is a mark of a cell's strategy to alter its transcriptome and proteome in order to survive a stressful condition. Former is a regular phenomenon owing to numerous stresses to which the cells are subjected on a continuous basis. Deleterious effects of the stress as well as adaptive responses of cells, both are manifested through an altered proteome generated through a differential range of mRNA. Thus, mRNAs are extremely essential for normal growth and also for the generation of stress responses. Functionally, mRNAs take part in all cellular processes by forming complexes known as ribonucleoproteins. Most common of such ribonucleoprotein complexes are polysomes, found under all physiological conditions, and stress granules and processing bodies formed under adverse conditions (Fig. 13.1; Chantarachot and Bailey-Serres 2018).

Stress granules were first observed in plant cells. The discovery of stress granules was made in the context of heat stress in tomato cell cultures (Nover et al. 1983). The granules, which mainly comprised untranslated mRNAs and heat shock proteins, were accordingly termed as heat shock granules. The first reference of stress granules in the animal system was made by Collier and Schlesinger (1986) in chick embryo fibroblasts followed by Arrigo et al. (1988) in HeLA cells. Both consisted mainly of heat shock proteins. Mammalian stress granules were reported much later by Kedersha et al. (1999). They implied the stress conditions leading to the formation of stress granules and illustrated the presence of core proteins and mRNA in them. Stress granules in incidences of stresses other than heat were observed in yeast under oxidative stress and glucose starvation (Hoyle et al. 2007; Buchan et al. 2008; Grousl et al. 2009). Based on composition, stress-induced

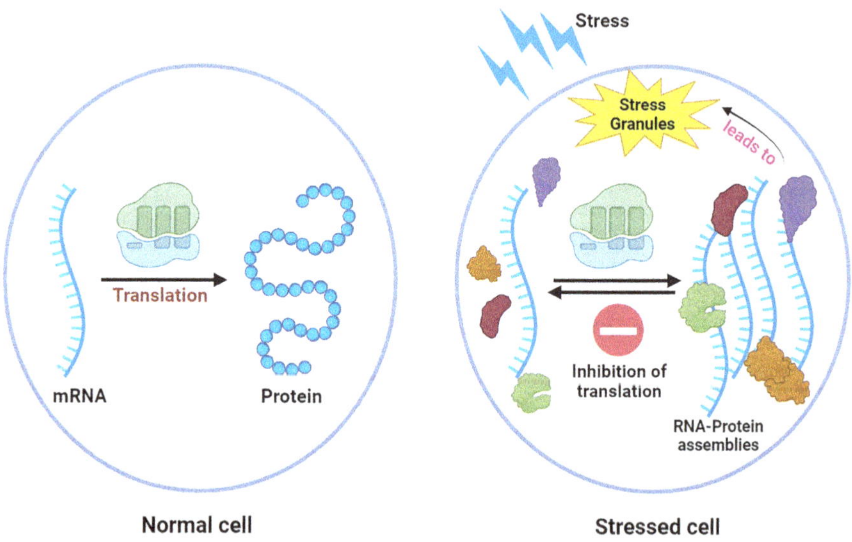

Fig. 13.1 Functional Transformation of Ribonucleoprotein Complexes During Cellular Stress

cytoplasmic granules of plants were classified into stress granules, heat shock granules, and processing bodies. This demarcation was made by Weber et al. (2008) in immunofluorescence studies. Weber's analysis refuted the earlier notion put forward by Nover et al. (1983). Heat shock granules that aggregate in the cytoplasm upon extended duration of heat stress differ from stress granules in the absence of polyA and mRNA. Stress granules are the ones that actually contain mRNA and are formed in response to short-term heat exposure. Processing bodies on the other hand comprise mainly untranslated mRNAs along with several mRNA degradation factors and thereby indicate their role in repression of translation (Chantarachot and Bailey-Serres 2018). Since then numerous studies have been done on structural and functional characterization of stress granules, factors, and mechanisms governing their formation, and their significance in the maintenance of cellular homeostasis.

13.2 Structural Attributes of Stress Granules

Beginning with the initial observation of presence of stress granules, several chromatographic studies have provided insights into the molecular composition of stress granules (Bhasin and Hülskamp 2017; Kosmacz et al. 2019; Gutierrez-Beltran et al. 2021). Further investigation regarding the composition of stress granules was done using techniques like mass spectrometry, proximity labeling, and fluorescence in situ hybridization (Jain et al. 2016; Youn et al. 2018; Markmiller et al. 2018).

Stress granules are characterized as non-membranous structures found in cytoplasm under various stresses leading to translational repression (Buchan et al. 2008; Buchan and Parker 2009; McCormick and Khaperskyy 2017). Their physical attributes indicate their appearance as liquid droplets rather than simple precipitates. Their formation is a multistep phenomenon involving the process of liquid–liquid phase separation (Banani et al. 2017; Courchaine et al. 2016). Microscopic observation coupled with immunofluorescence analyses revealed that stress granules possess several dynamic shells surrounding a much stable core region (Jain et al. 2016). Being composed of RNAs and proteins, stress granules exhibit a dynamic interrelation between the components. Instead of being simple aggregates of their components, stress granules exhibit extensive protein modifications and RNA remodeling. These modifications have implications in conferring specific biophysical characteristics and accordingly stress-specific physiological roles (Siwach and Kaganovich 2017; Mateju et al. 2017).

13.2.1 The Protein Component

Most of the work regarding stress granules has been conducted on the yeast or mammalian system and there is an entire database dedicated to mammalian stress granule proteome. Plant and microbial stress granules are comparatively less explored and most of them have been probed as orthologs of yeast or human stress granules (Kosmacz et al. 2019). The proteins found in stress granules are otherwise located either in the cytoplasm or nucleus. Knowledge about plant stress granules is mainly centered around their presence and function in *Arabidopsis thaliana*. On initiation of stress, these proteins get modified (in most of the cases) and get shuttled into the focus of stress granules. Some prominent SG proteins include TSN1, TSN2, HSP101 found in heat and/or salinity stress that are normally located in cytoplasm (Yan et al. 2014; Gutierrez-Beltran et al. 2015; McLoughlin et al. 2019); Ubp1B found in osmotic, salinity and heat stress and is normally located in nucleus (Nguyen et al. 2016, 2017); Rbp47b in heat stress, G3BP-2 in heat stress and pathogen infection, Ubp1C, RH6, RH8, and RH12 under hypoxic conditions, all of which are normally located in cytoplasm and nucleus (Sorenson and Bailey-Serres 2014; Krapp et al. 2017; Chantarachot et al. 2020; Reuper et al. 2021a, b); and ANGUSTIFOLIA with critical roles in hypoxia, salinity, osmotic, and heat stress that is otherwise localized to cytosol and trans-Golgi networks (Bhasin and Hülskamp 2017). These proteins exhibit diverse roles such as mRNA catabolism, mRNA binding, poly(A) binding, membrane trafficking, transcriptional regulation, protein folding and stabilization, ATP hydrolysis, and helicase activity. Two of the main components of plant stress granules that act as foci, Rbp47b (RNA-binding protein 47b) and UBP1B (Oligouridylate-binding protein 1B), are homologous to mammalian proteins TIA-1 that also play an essential role in the assembly of stress granules (Kedersha et al. 1999; Gilks et al. 2004). Presence of proteins with similar structure and function in organisms as diverse as humans and plants indicates that the mechanism of stress granules to combat various stresses is highly conserved in nature (Jain et al. 2016). This fact is also justified as organisms as small as unicellular microbes to higher vertebrates and plants are constantly exposed to a range of abiotic and biotic stresses and struggle to maintain homeostasis. Different core stress granule proteins also interact actively among themselves. This is exemplified by the association of an important SG constituent of *Arabidopsis*-ANGUSTIFOLIA with several other SG proteins namely poly(A)-binding protein 2 (PAB2), Rbp45 and Rbp47, tandem zinc finger 3 (TZF3), and eukaryotic initiation factor (eIF4E1). Also, ANGUSTIFOLIA is seen to exhibit such interactions even under normal physiological conditions.

Apart from RNA-binding and stabilizing proteins, a whole lot of proteins are dedicated to posttranslational modification activities (Kosmacz et al. 2019). This class of stress granule proteins comprises SNF1-related protein kinase 2.1 (SnRK), mitogen-activated MKK5 and MPK3 kinases, and cyclin-dependent kinase A;1 (CDKA;1). Another group of proteins are those which have roles in ethylene biosynthesis such as ACC oxidases 2 and 4; and in sugar metabolism, e.g., UER1, rhamnose RHM1, and RHM2 enzymes. Enzymatic components of cellular antioxidative machinery such as ascorbate peroxidase, glutathione peroxidase, and glutathione-S-transferase are also some of the prevalent stress granule proteins. Tudor staphylococcal nuclease (TSN) as mentioned earlier is one of the core components of stress granules and as compared to non-stressed conditions, TSN2 interactome was found to have nearly 1.25 times more proteins under stress conditions (Gutierrez-Beltran et al. 2015). Similarly, an investigation conducted by Dubiel et al. (2020) in *Arabidopsis thaliana* revealed presence of heat shock proteins, ribosomal proteins, and translational elongation factors.

Apart from cytoplasmic localization, presence of stress granules like—aggregational centers has been reported in chloroplasts of *Chlamydomonas reinhardtii*, a unicellular green alga (Uniacke and Zerges 2008). These chloroplast stress granules consist of proteins that are similar to cytoplasmic stress granules, including dismantled polysomes, small subunit ribosome proteins. and poly(A)-binding proteins. *A. thaliana* chloroplasts also possess similar stress granules-like structures under heat stress (Chodasiewicz et al. 2020). An elongation factor used in translation by plastids, snowy cotyledon 1 protein (SCO1) was found to be of universal occurrence in heat stress-induced chloroplast stress granules in *A. thaliana*. Proteomic studies using this protein as a marker SG protein helped in identification of other protein counterparts of chloroplast stress granules namely HSP90–5, other chaperons, DEAD-box RNA helicase, translation elongation factor Tu (RABE1b), CP29A and IDRs (intrinsically disordered regions). Interestingly, chloroplast stress granules also consisted of functional enzymes of photosynthesis, RuBisCo, and RuBisCo activase.

Thus, it is evident that proteins related to cell cycle regulation, especially the ones responsible for replication and transcription, RNA stabilization, protein synthesis as well as enzymes involved in physiological processes for energy generation and stress management are integral components of stress granules.

13.2.2 The mRNA Component

For a long time, heat shock granules in plants were considered to be what is now considered as stress granules. Later on, physiological and biochemical studies coupled with the advent of techniques with better resolution techniques revealed stress granules as distinct structures. Initial studies suggested presence of transitionally stored translationally inhibited mRNAs in the stress granules (Kedersha et al. 2002; Anderson and Kedersha 2006). All the cellular mRNAs are not destined to enter stress granules. Similarly, some mRNA population is retained within the nucleus even under extended periods of stress. Several techniques were employed to investigate and identify the RNA components of stress granules. The former included use of fluorescently labeled oligo-dT probes, along with immunogold antibody labeling or use of simple RNA-binding proteins (Scharf et al. 1998; Kedersha et al. 1999; Sheinberger and Shav-Tal 2013). Although several different kinds of proteins are present in the stress granules, mRNA is predominantly present as a complex with RNA-binding proteins. This fact also serves as a tool for the search of stress granule mRNA.

Low-abundance mRNA transcripts, signal-influenced mRNA transcripts, and mRNA transcripts that are localized to certain specific cellular locations are comparatively much less abundant as compared to mRNA transcribed from constitutively expressed genes required for cellular maintenance. Similar distribution biases also occur in stress granules with respect to these mRNA populations. MS2 tagging of mRNPs revealed that the mRNA-protein complexes belonging to stress granules were mobile in nature (Mollet et al. 2008). The inclusion of mRNPs existing in the cytoplasm into stress granules is a very rapid process. Genes for several stress granule proteins are found to be regulated as they serve to form complexes with stalled mRNAs and aid in their protection.

Quantification studies regarding mRNA components of stress granules have mainly been conducted in mammalian systems. According to analyses based on single-molecule RNA FISH technique and use of beta-Gal-MS2 reporter, it was deciphered that stress granule-specific mRNA comprised 7–9% of total cytoplasmic mRNAs (Mollet et al. 2008; Yunger et al. 2013).

13.3 Assembly and Disassembly of Stress Granule

After sensing the trigger for the need for stress granules, a sequentially operated phenomenon is initiated leading to formation of these structures. With an ultimate aim to confer tolerance and cellular protection, this process is highly coordinated and orchestrated by several proteins. Once formed, SGs do not remain in the cell as stationary stagnant bodies. Rather, they are extremely dynamic in nature as discussed below. The way in which the assembly part of the procedure is important, similarly disassembly is equally significant for proper resumption of cellular activities post-stress.

13.3.1 Mechanism of Assembly

In mammals, the assembly of SG is initiated through the phosphorylation of eIF2a under stressful conditions by stress-sensing kinases (Sidrauski et al. 2015). This phosphorylation event results in the depletion of the eIF2/tRNAi Met/GTP ternary complex which is essential for the initiation of translation. RNA-binding proteins (RBPs), including T-cell intracellular antigen 1 (TIA-1), Ras-GAP SH3 domain-binding protein (G3BP) and TRISTETRAPROLIN (TTP) play a role in promoting the formation of stress granules (Kang et al. 2021). Similar processes with related counterparts have been discovered in other organisms as well.

As already stated, being a non-membranous organelle the formation of stress granules is driven by liquid–liquid phase separation (LLPS). It occurs through a phenomenon in which liquid and nucleic acids aggregate from the aqueous environment and form membraneless organelle (MLO) that have the tendency to concentrate more molecules than the surrounding environment. In the stress condition, eukaryotic cells completely shut down the translation by polysome disassembly leading to mRNA release resulting in the establishment of mRNP complexes. Analysis of the mechanism of assembly of SGs in mammals, yeasts, and plants suggests that it is conserved in eukaryotic cells. Initiation of SG assembly is propelled by LLPS of mRNA complexes and it proceeds in following steps—nucleation, core growth, and shell assembly. Out of the above-mentioned steps nucleation is a primary step followed by core formation and shell condensation. Together these steps lead to creation of a physical state that is biphasic in nature. Disassembly of SGs follows backward steps as those involved in assembly. The process begins with diffusion of the shell and then dissipation of core.

Nucleation step is triggered when the concentration of mRNP complexes is increased locally (Protter and Parker 2016). Formation of SGs takes place by cross-linking of mRNAs that were not further used for translation and used as a stage for mRNA-binding proteins. Thus, allowing multivalent interactions among mRNPs and leading to the formation of SGs. Modulation of the nucleation process is affected by the quantity of available interacting domains between proteins and RNA molecules. In addition to this, post-translational modifications like addition of phosphate and methyl groups to the proteins as well as similar modifications in nucleic acid components, glycosylation, and hypusination also positively or negatively regulate the nucleation events. The assembly of SGs is primarily driven by hydrophobic, weak electrostatic, and homo- and heterotypic protein–protein interactions between RBPs that contain intrinsically disordered domains. Protein–protein interactions play a crucial role in SGs assembly. RNPs having IDRs play important roles in SG assembly by providing access to Short Linear Motifs (SLiMs), enhancing the stability of the granules by forming cross-strand beta zipper through hetero- and homotypic interactions, weak interactions between IDRs promote LLPS by providing region for the RNAs to bind. Initially in the nucleation process, mRNPs that are not undergoing the process of translation form oligomeric assemblies which further grow by joining of additional mRNPs to form SGs. A study by Maharana et al. (2018) deciphered that RNA–RNA interactions are induced in stress conditions. Thus, their interactions stimulate LLPS and trigger SG assembly (Van Treeck and Parker 2018).

There were various models proposed regarding SG assembly. According to the "Core First Model" (Jain et al. 2016), core formation is initiated by nucleation of untranslated mRNPs into oligomeric assembly that grows into SG cores. The SG cores so formed, merge, and are overlaid by shells, and are thus transformed into a large granule that could be easily visualized in microscope. Another proposed model is "LLPS First" model (Jain et al. 2016). In this model, aggregation of repressed RNPs occurs in initial phase-separated droplets which further grow due to addition of untranslated RNPs in the following phase. Third phase is marked by formation of the core. The difference in both models is that core formation takes place in the initial stage in core first model while its formation takes place in the third phase in LLPS first model.

13.3.2 Mechanism of Disassembly

SGs are dynamic in nature and exhibit liquid-like behavior. Their disassembly is as important as their assembly. This process proceeds in reverse order in eukaryotes and yeast in which dissipation of shell occurs first followed by dissolution of core. A study on plants by Chodasiewicz et al. (2020) observed that 1,6-hexanediol interrupts the weak hydrophobic protein–protein interactions and protein–RNA interactions in heat-induced stress granule-like structures. Thus, it becomes evident that weak IDR interactions result in shell dissipation in plants and non-plant models. Ubiquitination is found to be involved in SG disassembly in humans. UBQNL2, an adaptor capable of identifying ubiquitin, degrades them, is found to co-localize with SGs in vitro and drives LLPS in vivo (Dao et al. 2018). In the presence of ubiquitin or poly-ubiquitin chain, stress granules get recognized by UBQNL2 leading to their dissolution by removing the shuttle proteins out of SGs.

SGs are dynamic in nature as ATP-driven remodeling events keep the RNP assemblies in a dynamic state. This can be understood with an example of nucleoli that maintain their liquid-like behaviour by using the ATP and its dynamicity decreases by ten-fold when the ATP is depleted (Brangwynne et al. 2011). Similarly, the stable interactions between mRNA and proteins can be disrupted by ATPases. In stress conditions, granule components like non-translating RNAs accumulated. Disruption of interactions by ATPases in the recovery phase leads to its disassembly. The ATPases include chaperones and RNA helicases-like complexes that directly affect the interactions in the SGs (Protter and Parker 2016).

RNA/DNA helicases that use ATP hydrolysis also play significant roles in assembly and disassembly of SGs. Proteins such as small heat shock proteins (sHSPs), RNA helicases, cytoskeletal proteins, and disassembly engaged proteins (DEPs) are involved in disassembly of SGs in mammalian cells (Marmor-Kollet et al. 2020).

Disassembly of large SGs proceeds in two steps. In the first step, large SGs are converted to small foci and then disassembly of the smaller foci takes place wherein RNA is moved out of the stress granule and involved in translation. Further, the structural instability caused by disassembly of a stress granule complex is cleared by autophagy (Wheeler et al. 2016).

13.4 Stress Granule Formation in Plants

Plants being immobile organisms, necessitate adaptation to environmental fluctuations for their survival. One pivotal environmental factor is temperature which undergoes rapid changes. Notably, exposure of leaves to direct sunlight can lead to substantial temperature increases. Typically, plants dissipate heat through transpiration (Campbell 1977). Nevertheless, certain circumstances particularly in arid and high-temperature settings may impede plant's ability to regulate their internal temperature. To endure such harsh conditions, plants have evolved diverse survival strategies, the prominent one being synthesis of of heat shock proteins (Wahid et al. 2007). Various translational repression responses take place in plants when they are subjected to elevated temperatures (Nover et al. 1989). In response to the preceding alterations in cellular mechanisms, mRNAs facing translation repression and their associated proteins initiate the assembly of stress granules. As observed in other organisms, stress granules in plants are non-membranous cytoplasmic RNA granules that comprise translationally arrested messenger RNAs (mRNAs), translation initiation factors, and numerous RNA-binding proteins (RBPs). Stress granules are highly conserved across evolution and have been documented in various cell types including plant cells, animal cells, and yeast cells. Their primary function is to regulate the entire plethora of components involved in transcription and translation by selectively sequestering and safeguarding mRNAs and proteins respectively. These granules are generated in reaction to diverse stress stimuli and play a role in reshaping cellular metabolism to support cell survival. Furthermore, specific components unique to plants have been identified within stress granules, shedding light on the molecular processes governing their formation (Kearly et al. 2024). In the study conducted by Hamada et al. (2018), the authors investigated the impact of elevated temperatures on stress granule formation in *Arabidopsis thaliana*. The findings revealed that stress granule formation occurred when temperatures rose beyond 34 °C. In plants, a comprehensive characterization of the precise composition of stress granules in plants and other systems under diverse stress conditions remains to be fully elucidated.

SGs represent dynamic structures found in plants, emerging at a critical temperature of 34 °C in the absence of transpiration. The quantity and size of these granules exhibit responsiveness to both the intensity and duration of heat shock. In *Arabidopsis* cells, the movement of stress granules over long distances is facilitated by actin filaments (Hamada et al. 2018). The formation of stress granules relies on a delicate equilibrium between microtubule polymerization and depolymerization. During periods of stress, RNA degradation gives rise to 2′,3′-cAMP, a molecule conserved across various contexts (Lee et al. 2020). The polyadenylate-binding protein Rbp47b, recognized as a pivotal component of stress granules engages with 2′,3′-cAMP thereby promoting the assembly of stress granules.

The roles of SGs in addressing heat and hypoxia stresses have been extensively described and are succinctly outlined here.

13.4.1 Relevance in Heat Stress

High-temperature stress induces cellular protein misfolding, leading to the formation of aggregates. Within plant cells, heat shock proteins (HSPs), such as HSP70 and HSP100, work collaboratively to prevent protein aggregation and facilitate refolding. The onset of heat stress triggers the development of stress granules (SGs) that contain mRNAs and eIF4E. These SGs serve as temporary storage for mRNAs, helping alleviate the influx of proteins. As the cell undergoes stress recovery, SGs disassemble, a process facilitated by HSPs. In higher plants, HSP101, a homolog of yeast HSP104, plays a crucial role in aiding SG disassembly, thereby promoting the resumption of translation. Furthermore, during heat stress, LARP1 functions as a heat-specific cofactor for XRN4, guiding mRNA degradation (Merret et al. 2013). Heat stress granules (HSGs), distinct from SGs, are composed of small heat shock proteins and associate with specific mRNAs and ribosomes, particularly during the recovery phase. SGs and HSGs fulfill unique roles, with mRNAs predominantly associated with SGs.

13.4.2 Relevance in Hypoxia Stress

Hypoxia stress in *Arabidopsis* seedlings significantly reduces translation initiation. In adverse conditions, plant UBP1C akin to mammalian TIA-1 and TIAR proteins forms stress granules, crucial for *Arabidopsis* growth and survival under hypoxia (Maruri-López et al. 2021). With three RNA-recognition motifs, UBP1C efficiently binds and sequesters mRNAs into stress granules induced by oxygen/ATP deprivation, leading to an overall reduction in protein synthesis.

13.4.3 Stress Granules as Cellular Stress Response Mechanisms in Plants

In general, the growth and development of plants depend on the cell shape and cell division/expansion regulation. The process of cellular growth is intricately governed by factors such as cell metabolism, nutrient availability and cellular energy and stress levels. One extensively studied gene in *Arabidopsis thaliana* is ANGUSTIFOLIA (Gachomo et al. 2013). Homolog of C-terminal-binding protein/brefeldin A-ADP ribosylated substrate is encoded by AN gene. These BARS proteins are situated in the cytosol and play a crucial role in membrane trafficking and Golgi fission. According to the findings of Bhasin and Hülskamp (2017) AN interacts with various stress granule components and localizes to SGs.

Cells exhibit a swift and comprehensive reprogramming of gene expression in response to diverse stress conditions acting at the transcriptional, post-transcriptional, and translational levels. Posttranscriptional regulation is commonly overseen by RBPs that frequently associate with mRNA-ribonucleoprotein (mRNP) complexes such as SGs and Processing Bodies (P-bodies). These RBPs engage dynamically with both single and double-stranded RNAs playing essential roles in mRNA stability, localization, maturation, decay, and translation. Cellular stress triggers the formation of SG which is serving as sequestration sites for untranslated messenger RNAs (mRNAs). Scientific evidence indicates their crucial role in promoting cell survival by inhibiting crowding of proteins that have lost their functional conformation and abating certain apoptosis-regulating factors. These granules exhibit high dynamism and disassemble when cells return to their standard growth conditions. Several recent works have analyzed the roles of stress granules in conferring protection to plants under diverse stress conditions (Table 13.1).

In the realm of plant biology, the regulation of stress granule formation is influenced by a limited number of identified factors. Notable among these are OLIGOURIDYLATE-BINDING PROTEIN 1 (UBP1), as well as RNA-BINDING PROTEIN 45 and 47 (RBP45/47) along with the PAB (poly-A binding) protein families (Yan et al. 2022), which have been acknowledged as plant RBPs exhibiting a close relationship to their animal counterpart, TIA-1 (Xu et al. 2020). Despite these recognized factors the process and composition of mRNP formation in plants remain predominantly unexplored.

Table 13.1 Relevant works on role of stress granules in stress management in plants

Serial no	Work title	Key findings	Reference
1.	Cauliflower mosaic virus protein P6 is a multivalent node for RNA granule proteins and interferes with stress granule responses during plant infection	They recognized G3BP7 and UBP1 family members as essential factors for SG nucleation and identified them as integral components of viral factories (VFs)	Hoffmann et al. (2023)
2.	*Arabidopsis* translation factor eEF1Bγ impacts plant development and is associated with heat-induced cytoplasmic foci	By creating *Arabidopsis* mutants characterized by significantly decreased eEF1Bγ levels the researchers elucidated the pivotal role of eEF1Bγ in plant growth and development. They systematically examined its influence on translation processes. Through protein–protein interaction studies and co-expression analyses it was revealed that eEF1Bβ actively engaged with both other subunits, facilitating their recruitment to cytoplasmic foci	Lohmann et al. (2023)
3.	The RasGAP-associated endoribonuclease G3BP mediates stress granule assembly	G3BP, an endoribonuclease activated through phosphorylation, forms associations with RasGAP and is mobilized to SGs in cells subjected to arsenite exposure. Consequently, G3BP plays a pivotal role in determining the destiny of mRNAs during cellular stress. This specific region which binds to RasGAP encompasses serine 149. The dephosphorylation of serine 149 is stimulated by arsenite treatment. G3BP functions as an effector in SG assembly with Ras signaling influencing this process by modulating the dephosphorylation of G3BP	Tourrière et al. (2023)
4.	Genomic identification of CCCH-type zinc finger protein genes reveals the role of *HuTZF3* in tolerance of heat and salt stress of pitaya (*Hylocereus polyrhizus*)	Identified 81 CCCH-type zinc finger protein genes within the pitaya genome HuTZF3 emerges as responsive to both salt and heat stress, with expression observed across various organs, particularly pronounced in roots and stems. The research affirms that the expression of HuTZF3 is triggered by both salt and heat stress Subcellular localization reveals HuTZF3's presence in PBs and SGs Introducing HuTZF3 through heterologous expression in *Arabidopsis* enhances tolerance to salt and heat, diminishes oxidative stress and improves catalase and peroxidase activity	Xu et al. (2023)

(continued)

Table 13.1 (continued)

Serial no	Work title	Key findings	Reference
5.	Stress granule-associated TaMBF1c confers thermotolerance through regulating specific mRNA translation in wheat (*Triticum aestivum*)	The research demonstrated the pivotal role of *TaMBF1c* in the wheat heat stress response by overseeing the translation of specific genes The gene exhibited evolutionary conservation across diverse species of wheat. Experiments with knockdown and knockout lines revealed heightened heat sensitivity TaMBF1c colocalizes with the stress granule complex and interacts with TaG3BP These findings underscore the importance of TaMBF1c in regulation of heat stress through modifications in protein synthesis suggesting it as a potential approach towards generation of heat-tolerant wheat breeds	Tian et al. (2022)
6.	RNA-binding protein ostzf7 traffics between the nucleus and processing bodies/stress granules and positively regulates drought stress in rice	OsTZF7, recently identified tandem zinc finger protein in rice, belongs to the tandem CCCH zinc finger (TZF) protein category and plays a crucial role in the formation of processing bodies (PBs) and stress granules (SGs). These structures are vital for growth development and stress management through facilitating the posttranscriptional management of mRNA in both animals and plants OsTZF7 demonstrates a positive impact on drought response in rice through the abscisic acid signaling pathway and appears to be implicated in the turnover of mRNA	Guo et al. (2022)
7.	ALBA proteins confer thermotolerance through stabilizing *HSF* messenger rnas in cytoplasmic granules	The research identifies three proteins, namely ALBA4, ALBA5, and ALBA6 that play a crucial role in the formation of SGs and PBs under conditions of heat stress These proteins establish direct interactions with specific messenger RNAs, including those associated with heat shock factors (HSFs) and facilitate their recruitment to SGs and PBs This process serves to shield the messenger RNAs from degradation in *Arabidopsis* The study underscores an innovative regulatory mechanism governing HSFs where ALBA proteins function to stabilize HSF messenger RNAs within SGs and PBs, thereby contributing to the enhancement of plant thermotolerance	Tong et al. (2022)

13.5 Conclusion

mRNAs are extremely essential for normal growth as they take part in execution of all cellular processes by forming complexes known as ribonucleoproteins. Under normal circumstances, mRNA protein complexes are formed for the purpose of translation. In adverse conditions, most common polysomes are stress granules and processing bodies. SGs that appear as liquid droplets are formed from pools of untranslating mRNAs. It is found to be conserved in eukaryotes such as yeast, plants, and mammals. SGs are dynamic in nature and their assembly and disassembly are regulated by post-translational modifications, RNP, and protein modelling complexes. SGs in plants get assembled during multiple stresses such as heat, hypoxia, high salt, and darkness. Mobility of SGs depends on microtubules. In plants, both microtubule and actin filaments are involved in the assembly and transport of SGs. It is evident that SGs are dynamic cellular entities and are formed under stress conditions to improve chances of survival.

Acknowledgement Ashutosh Singh and Ravinsh Kumar are thankful to UGC, New Delhi, for the financial assistance in the form of fellowship.

Conflict of Interest Authors declare that there is no conflict of interest.

References

Anderson P, Kedersha N (2006) RNA granules. J Cell Biol 172(6):803–808. https://doi.org/10.1083/jcb.200512082

Arrigo AP, Suhan JP, Welch WJ (1988) Dynamic changes in the structure and intracellular locale of the mammalian low-molecular-weight heat shock protein. Mol Cell Biol 8(12):5059–5071. https://doi.org/10.1128/mcb.8.12.5059-5071.1988

Banani SF, Lee HO, Hyman AA, Rosen MK (2017) Biomolecular condensates: organizers of cellular biochemistry. Nat Rev Mol Cell Biol 18(5):285–298. https://doi.org/10.1038/nrm.2017.7

Bhasin H, Hülskamp M (2017) ANGUSTIFOLIA, a plant homolog of CtBP/BARS localizes to stress granules and regulates their formation. Front Plant Sci 8:1004. https://doi.org/10.3389/fpls.2017.01004

Brangwynne CP, Mitchison TJ, Hyman AA (2011, March 15) Active liquid-like behavior of nucleoli determines their size and shape in Xenopus laevis oocytes. Proc Natl Acad Sci USA 108(11):4334–4339. https://doi.org/10.1073/pnas.1017150108

Buchan JR, Parker R (2009) Eukaryotic stress granules: the ins and outs of translation. Mol Cell 36(6):932–941. https://doi.org/10.1016/j.molcel.2009.11.020

Buchan JR, Muhlrad D, Parker R (2008) P bodies promote stress granule assembly in Saccharomyces cerevisiae. J Cell Biol 183(3):441–455. https://doi.org/10.1083/jcb.200807043

Campbell GS (1977) Plants and their environment. In: An introduction to environmental biophysics. Heidelberg science library. Springer, Cham, pp 115–126. https://doi.org/10.1007/978-1-4684-9917-9_9

Chantarachot T, Bailey-Serres J (2018) Polysomes, stress granules, and processing bodies: a dynamic triumvirate controlling cytoplasmic mRNA fate and function. Plant Physiol 176(1):254–269. https://doi.org/10.1104/pp.17.01468

Chantarachot T, Sorenson RS, Hummel M, Ke H, Kettenburg AT, Chen D, Aiyetiwa K, Dehesh K, Eulgem T, Sieburth LE, Bailey-Serres J (2020) DHH1/DDX6-like RNA helicases maintain ephemeral half-lives of stress-response mRNAs. Nat Plants 6(6):675–685. https://doi.org/10.1038/s41477-020-0681-8

Chodasiewicz M, Sokolowska EM, Nelson-Dittrich AC, Masiuk A, Beltran JCM, Nelson ADL, Skirycz A (2020) Identification and characterization of the heat-induced plastidial stress granules reveal new insight into *Arabidopsis* stress response. Front Plant Sci 11:595792. https://doi.org/10.3389/fpls.2020.595792

Collier NC, Schlesinger MJ (1986) The dynamic state of heat shock proteins in chicken embryo fibroblasts. J Cell Biol 103(4):1495–1507. https://doi.org/10.1083/jcb.103.4.1495

Courchaine EM, Lu A, Neugebauer KM (2016) Droplet organelles? EMBO J 35(15):1603–1612. https://doi.org/10.15252/embj.201593517

Dao TP, Kolaitis RM, Kim HJ, O'Donovan K, Martyniak B, Colicino E, Hehnly H, Taylor JP, Castañeda CA (2018) Ubiquitin modulates liquid–liquid phase separation of UBQLN2 via disruption of multivalent interactions. Mol Cell 69(6):965–978.e6. https://doi.org/10.1016/j.molcel.2018.02.004

Dubiel M, De Coninck T, Osterne VJS, Verbeke I, Van Damme D, Smagghe G, Van Damme EJM (2020) The ArathEULS3 lectin ends up in stress granules and can follow an unconventional route for secretion. Int J Mol Sci 21(5):1659. https://doi.org/10.3390/ijms21051659

Gachomo EW, Jimenez-Lopez JC, Smith SR, Cooksey AB, Oghoghomeh OM, Johnson N, Baba-Moussa L, Kotchoni SO (2013) The cell morphogenesis ANGUSTIFOLIA (AN) gene, a plant homolog of CtBP/BARS, is involved in abiotic and biotic stress response in higher plants. BMC Plant Biol 13:79. https://doi.org/10.1186/1471-2229-13-79

Gilks N, Kedersha N, Ayodele M, Shen L, Stoecklin G, Dember LM, Anderson P (2004) Stress granule assembly is mediated by prion-like aggregation of TIA-1. Mol Biol Cell 15(12):5383–5398. https://doi.org/10.1091/mbc.e04-08-0715

Grousl T, Ivanov P, Frydlová I, Vašicová P, Janda F, Vojtová J, Malínská K, Malcová I, Nováková L, Janosková D, Valásek L, Hasek J (2009) Robust heat shock induces eIF2α-phosphorylation-independent assembly of stress granules containing eIF3 and 40S ribosomal subunits in budding yeast, Saccharomyces cerevisiae. J Cell Sci 122(12):2078–2088. https://doi.org/10.1242/jcs.045104

Guo C, Chen L, Cui Y, Tang M, Guo Y, Yi Y, Li Y, Liu L, Chen L (2022) RNA binding protein OsTZF7 traffics between the nucleus and processing bodies/stress granules and positively regulates drought stress in rice. Front Plant Sci 13:802337. https://doi.org/10.3389/fpls.2022.802337

Gutierrez-Beltran E, Moschou PN, Smertenko AP, Bozhkov PV (2015) Tudor staphylococcal nuclease links formation of stress granules and processing bodies with mRNA catabolism in Arabidopsis. Plant Cell 27(3):926–943. https://doi.org/10.1105/tpc.114.134494

Gutierrez-Beltran E, Elander PH, Dalman K, Dayhoff GW, Moschou PN, Uversky VN, Crespo JL, Bozhkov PV (2021) Tudor staphylococcal nuclease is a docking platform for stress granule components and is essential for SnRK1 activation in Arabidopsis. EMBO J 40(17):e105043. https://doi.org/10.15252/embj.2020105043

Hamada T, Yako M, Minegishi M, Sato M, Kamei Y, Yanagawa Y, Toyooka K, Watanabe Y, Hara-Nishimura I (2018) Stress granule formation is induced by a threshold temperature rather than a temperature difference in *Arabidopsis*. J Cell Sci 131(16):jcs216051. https://doi.org/10.1242/jcs.216051

Hoffmann G, López-González S, Mahboubi A, Hanson J, Hafrén A (2023) Cauliflower mosaic virus protein P6 is a multivalent node for RNA granule proteins and interferes with stress granule responses during plant infection. Plant Cell 35(9):3363–3382. https://doi.org/10.1093/plcell/koad101

Hoyle NP, Castelli LM, Campbell SG, Holmes LE, Ashe MP (2007) Stress-dependent relocalization of translationally primed mRNPs to cytoplasmic granules that are kinetically and spatially distinct from P-bodies. J Cell Biol 179(1):65–74. https://doi.org/10.1083/jcb.200707010

Jain S, Wheeler JR, Walters RW, Agrawal A, Barsic A, Parker R (2016) ATPase-modulated stress granules contain a diverse proteome and substructure. Cell 164(3):487–498. https://doi.org/10.1016/j.cell.2015.12.038

Kang W, Wang Y, Yang W, Zhang J, Zheng H, Li D (2021) Research progress on the structure and function of G3BP. Front Immunol 12:718548. https://doi.org/10.3389/fimmu.2021.718548
Kearly A, Nelson ADL, Skirycz A, Chodasiewicz M (2024) Composition and function of stress granules and P-bodies in plants. In: Seminars in cell and developmental biology, vol 156. Academic, New York, pp 167–175. https://doi.org/10.1016/j.semcdb.2022.11.008
Kedersha NL, Gupta M, Li W, Miller I, Anderson P (1999) RNA-binding proteins TIA-1 and TIAR link the phosphorylation of eIF-2 alpha to the assembly of mammalian stress granules. J Cell Biol 147(7):1431–1442. https://doi.org/10.1083/jcb.147.7.1431
Kedersha N, Chen S, Gilks N, Li W, Miller IJ, Stahl J, Anderson P (2002) Evidence that ternary complex (eIF2-GTP-tRNA(i)(Met))-deficient preinitiation complexes are core constituents of mammalian stress granules. Mol Biol Cell 13(1):195–210. https://doi.org/10.1091/mbc.01-05-0221
Kosmacz M, Gorka M, Schmidt S, Luzarowski M, Moreno JC, Szlachetko J, Leniak E, Sokolowska EM, Sofroni K, Schnittger A, Skirycz A (2019) Protein and metabolite composition of *Arabidopsis* stress granules. New Phytol 222(3):1420–1433. https://doi.org/10.1111/nph.15690
Krapp S, Greiner E, Amin B, Sonnewald U, Krenz B (2017) The stress granule component G3BP is a novel interaction partner for the nuclear shuttle proteins of the Nanovirus pea necrotic yellow dwarf virus and geminivirus abutilon mosaic virus. Virus Res 227:6–14. https://doi.org/10.1016/j.virusres.2016.09.021
Lee Y, Choe J, Park OH, Kim YK (2020) Molecular mechanisms driving mRNA degradation by m6A modification. Trends Genet 36(3):177–188. https://doi.org/10.1016/j.tig.2019.12.007
Lohmann J, de Luxán-Hernández C, Gao Y, Zoschke R, Weingartner M (2023) *Arabidopsis* translation factor eEF1Bγ impacts plant development and is associated with heat-induced cytoplasmic foci. J Exp Bot 74(8):2585–2602. https://doi.org/10.1093/jxb/erad050
Maharana S, Wang J, Papadopoulos DK, Richter D, Pozniakovsky A, Poser I, Bickle M, Rizk S, Guillén-Boixet J, Franzmann TM, Jahnel M, Marrone L, Chang YT, Sterneckert J, Tomancak P, Hyman AA, Alberti S (2018, May 25) RNA buffers the phase separation behavior of prion-like RNA binding proteins. Science 360(6391):918–921. https://doi.org/10.1126/science.aar7366
Markmiller S, Soltanieh S, Server KL, Mak R, Jin W, Fang MY, Luo EC, Krach F, Yang D, Sen A, Fulzele A, Wozniak JM, Gonzalez DJ, Kankel MW, Gao FB, Bennett EJ, Lécuyer E, Yeo GW (2018) Context-dependent and disease-specific diversity in protein interactions within stress granules. Cell 172(3):590–604.e13. https://doi.org/10.1016/j.cell.2017.12.032
Marmor-Kollet H, Siany A, Kedersha N, Knafo N, Rivkin N, Danino YM, Moens TG, Olender T, Sheban D, Cohen N, Dadosh T, Addadi Y, Ravid R, Eitan C, Toth Cohen B, Hofmann S, Riggs CL, Advani VM, Higginbottom A et al (2020) Spatiotemporal proteomic analysis of stress granule disassembly using APEX reveals regulation by SUMOylation and links to ALS pathogenesis. Mol Cell 80(5):876–891.e6. https://doi.org/10.1016/j.molcel.2020.10.032
Maruri-López I, Figueroa NE, Hernández-Sánchez IE, Chodasiewicz M (2021) Plant stress granules: trends and beyond. Front Plant Sci 12:722643. https://doi.org/10.3389/fpls.2021.722643
Mateju D, Franzmann TM, Patel A, Kopach A, Boczek EE, Maharana S, Lee HO, Carra S, Hyman AA, Alberti S (2017) An aberrant phase transition of stress granules triggered by misfolded protein and prevented by chaperone function. EMBO J 36(12):1669–1687. https://doi.org/10.15252/embj.201695957
McCormick C, Khaperskyy DA (2017) Translation inhibition and stress granules in the antiviral immune response. Nat Rev Immunol 17(10):647–660. https://doi.org/10.1038/nri.2017.63
McLoughlin F, Kim M, Marshall RS, Vierstra RD, Vierling E (2019, August) HSP101 interacts with the proteasome and promotes the clearance of ubiquitylated protein aggregates. Plant Physiol 180(4):1829–1847. https://doi.org/10.1104/pp.19.00263
Merret R, Descombin J, Juan YT, Favory JJ, Carpentier MC, Chaparro C, Charng YY, Deragon JM, Bousquet-Antonelli C (2013) XRN4 and LARP1 are required for a heat-triggered mRNA decay pathway involved in plant acclimation and survival during thermal stress. Cell Rep 5(5):1279–1293. https://doi.org/10.1016/j.celrep.2013.11.019

Mollet S, Cougot N, Wilczynska A, Dautry F, Kress M, Bertrand E, Weil D (2008) Translationally repressed mRNA transiently cycles through stress granules during stress. Mol Biol Cell 19(10):4469–4479. https://doi.org/10.1091/mbc.e08-05-0499
Nguyen CC, Nakaminami K, Matsui A, Kobayashi S, Kurihara Y, Toyooka K, Tanaka M, Seki M (2016) Oligouridylate binding protein 1b plays an integral role in plant heat stress tolerance. Front Plant Sci 7:853. https://doi.org/10.3389/fpls.2016.00853
Nguyen CC, Nakaminami K, Matsui A, Watanabe S, Kanno Y, Seo M, Seki M (2017) Overexpression of oligouridylate binding protein 1b results in ABA hypersensitivity. Plant Signal Behav 12(2):e1282591. https://doi.org/10.1080/15592324.2017.1282591
Nover L, Scharf KD, Neumann D (1983) Formation of cytoplasmic heat shock granules in tomato cell cultures and leaves. Mol Cell Biol 3(9):1648–1655. https://doi.org/10.1128/mcb.3.9.1648-1655.1983
Nover L, Scharf KD, Neumann D (1989) Cytoplasmic heat shock granules are formed from precursor particles and are associated with a specific set of mRNAs. Mol Cell Biol 9(3):1298–1308. https://doi.org/10.1128/mcb.9.3.1298-1308.1989
Protter DSW, Parker R (2016) Principles and properties of stress granules. Trends Cell Biol 26(9):668–679. https://doi.org/10.1016/j.tcb.2016.05.004
Reuper H, Amari K, Krenz B (2021a) Analyzing the G3BP-like gene family of Arabidopsis thaliana in early turnip mosaic virus infection. Sci Rep 11(1):2187. https://doi.org/10.1038/s41598-021-81276-7
Reuper H, Götte B, Williams L, Tan TJC, McInerney GM, Panas MD, Krenz B (2021b) Arabidopsis thaliana G3BP ortholog rescues mammalian stress granule phenotype across kingdoms. Int J Mol Sci 22(12):6287. https://doi.org/10.3390/ijms22126287
Scharf KD, Heider H, Höhfeld I, Lyck R, Schmidt E, Nover L (1998) The tomato Hsf system: HsfA2 needs interaction with HsfA1 for efficient nuclear import and may be localized in cytoplasmic heat stress granules. Mol Cell Biol 18(4):2240–2251. https://doi.org/10.1128/MCB.18.4.2240
Sheinberger J, Shav-Tal Y (2013) The dynamic pathway of nuclear RNA in eukaryotes. Nucleus 4(3):195–205. https://doi.org/10.4161/nucl.24434
Sidrauski C, McGeachy AM, Ingolia NT, Walter P (2015) The small molecule ISRIB reverses the effects of eIF2α phosphorylation on translation and stress granule assembly. eLife 4:e05033. https://doi.org/10.7554/eLife.05033
Siwach P, Kaganovich D (2017) Getting stress out of stressed-out stress granules. EMBO J 36(12):1647–1649. https://doi.org/10.15252/embj.201797176
Sorenson R, Bailey-Serres J (2014) Selective mRNA sequestration by OLIGOURIDYLATE-BINDING PROTEIN 1 contributes to translational control during hypoxia in Arabidopsis. Proc Natl Acad Sci USA 111(6):2373–2378. https://doi.org/10.1073/pnas.1314851111
Tian X, Qin Z, Zhao Y, Wen J, Lan T, Zhang L, Wang F, Qin D, Yu K, Zhao A, Hu Z, Yao Y, Ni Z, Sun Q, De Smet I, Peng H, Xin M (2022) Stress granule-associated TaMBF1c confers thermotolerance through regulating specific mRNA translation in wheat (Triticum aestivum). New Phytol 233(4):1719–1731. https://doi.org/10.1111/nph.17865
Tong J, Ren Z, Sun L, Zhou S, Yuan W, Hui Y, Ci D, Wang W, Fan LM, Wu Z, Qian W (2022) ALBA proteins confer thermotolerance through stabilizing HSF messenger RNAs in cytoplasmic granules. Nat Plants 8(7):778–791. https://doi.org/10.1038/s41477-022-01175-1
Tourrière H, Chebli K, Zekri L, Courselaud B, Blanchard JM, Bertrand E, Tazi J (2023) The rasgap-associated endoribonuclease G3BP mediates stress granule assembly. J Cell Biol 222(11):e200212128072023new. https://doi.org/10.1083/jcb.200212128072023new
Uniacke J, Zerges W (2008) Stress induces the assembly of RNA granules in the chloroplast of Chlamydomonas reinhardtii. J Cell Biol 182(4):641–646. https://doi.org/10.1083/jcb.200805125
Van Treeck B, Parker R (2018, August 9) Emerging roles for intermolecular RNA-RNA interactions in RNP assemblies. Cell 174(4):791–802. https://doi.org/10.1016/j.cell.2018.07.023
Wahid A, Gelani S, Ashraf M, Foolad MR (2007) Heat tolerance in plants: an overview. Environ Exp Bot 61(3):199–223. https://doi.org/10.1016/j.envexpbot.2007.05.011

Weber C, Nover L, Fauth M (2008) Plant stress granules and mRNA processing bodies are distinct from heat stress granules. Plant J 56(4):517–530. https://doi.org/10.1111/j.1365-313X.2008.03623.x

Wheeler JR, Matheny T, Jain S, Abrisch R, Parker R (2016) Distinct stages in stress granule assembly and disassembly. eLife 5:e18413. https://doi.org/10.7554/eLife.18413

Xu M, Mazur MJ, Tao X, Kormelink R (2020) Cellular RNA hubs: friends and foes of plant viruses. Mol Plant Microbe Interact 33(1):40–54. https://doi.org/10.1094/MPMI-06-19-0161-FI

Xu W, Jian S, Li J, Wang Y, Zhang M, Xia K (2023) Genomic identification of CCCH-type zinc finger protein genes reveals the role of HuTZF3 in tolerance of heat and salt stress of pitaya (Hylocereus polyrhizus). Int J Mol Sci 24(7):6359. https://doi.org/10.3390/ijms24076359

Yan Y, Gan J, Tao Y, Okita TW, Tian L (2022) RNA-Binding Proteins: The Key Modulator in Stress Granule Formation and Abiotic Stress Response. Front Plant Sci13:882596.. https://doi.org/10.3389/fpls.2022.882596

Yan C, Yan Z, Wang Y, Yan X, Han Y (2014) Tudor-SN, a component of stress granules, regulates growth under salt stress by modulating GA20ox3 mRNA levels in *Arabidopsis*. J Exp Bot 65(20):5933–5944. https://doi.org/10.1093/jxb/eru334

Youn JY, Dunham WH, Hong SJ, Knight JDR, Bashkurov M, Chen GI, Bagci H, Rathod B, MacLeod G, Eng SWM, Angers S, Morris Q, Fabian M, Côté JF, Gingras AC (2018) High-density proximity mapping reveals the subcellular organization of mRNA-associated granules and bodies. Mol Cell 69(3):517–532.e11. https://doi.org/10.1016/j.molcel.2017.12.020

Yunger S, Rosenfeld L, Garini Y, Shav-Tal Y (2013) Quantifying the transcriptional output of single alleles in single living mammalian cells. Nat Protoc 8(2):393–408. https://doi.org/10.1038/nprot.2013.008

Nanobionics in Bioenergy and Crop Production

14

Deepanshu Pant, Balkrishna Tiwari, Taruna,
and Himani Sharma

Abstract

The use of nanoparticles (NPs) to augment or regulate specific processes that may significantly boost crop output is known as "nanobionics" in agriculture. NPs penetrate and interact with the plant system at the cellular and subcellular levels via a variety of pathways, mostly from roots and leaves. Metal-based NPs have been found to be effective in enhancing crop production by accelerating photosynthesis and protecting crops from a number of biotic and abiotic stresses. Recently, nanotechnology has emerged as a promising tool in biofuel production. The use of NPs as nanocarriers enhances the efficiency of enzymes used in biofuel production from biowaste. Therefore, the yield of biofuels such as biogas, biodiesel, biohydrogen and bioethanol can be enhanced by the use of NPs in the production processes. Considering the importance of Nanobionics, various aspects of its utilisation in agriculture as well as bioenergy production, and the mechanisms of NPs action in enhancing crop and biofuel production will be discussed in the chapter.

Keywords

Nanobionics · Agriculture · Bioenergy · Crop productivity · Nanoparticles

D. Pant · B. Tiwari (✉) · Taruna · H. Sharma
Genetics and Tree Improvement Division, ICFRE-Himalayan Forest Research Institute, Conifer Campus, Shimla, India
e-mail: tiwarib@icfre.org

A. K. Mishra (ed.), *Stress Biology in Photosynthetic Organisms*,
https://doi.org/10.1007/978-981-97-1883-2_14

14.1 Introduction

Food security and energy demand are the most concerning universal challenges for the rapidly rising population of the world, especially in developing countries (Duro et al. 2020; Neme et al. 2021; Sherwood 2020). Intensive cropping and use of chemical fertilisers and pesticides after the Green Revolution undoubtedly contributed significantly to alleviating hunger throughout the world, but the adverse effects like soil quality deterioration, reduced nutrient use efficiency, environmental pollution and detrimental effects on human health and other organisms have emerged as a major problem. A major part of chemical fertilisers (i.e. around 70–90%) is leached out or lost after their application in the field. Their indiscriminate uses are causing environmental damage through contamination of groundwater, eutrophication of water bodies, emissions of greenhouse gases and affecting the biogeochemical cycles. Their effects on soil microbiome and nutrient imbalance have also impacted plant growth, yield and quality of crops (Neme et al. 2021; Bhardwaj et al. 2022). On the other hand, the limited stocks of fossil fuels and their rapid depletion due to high energy demand throughout the world may lead to an energy crisis in the near future. Emissions of CO_2 and other greenhouse gases due to fossil fuels burning have also significantly contributed to global warming. Hence, enormous attention has been given to the development of alternate renewable energy resources for their sustainable use. To address the petroleum crisis, biofuels derived from materials of biological origin (i.e. food grains and oils, bio-waste, lignocellulosic agricultural wastes, microalgae, etc.) can be a potential alternative. However, low yields and high production costs of biofuels are regarded as major problems in this sector (Antunes et al. 2017; Bidir et al. 2021).

Owing to its wide applicability across various sectors, the field of nanotechnology is growing rapidly, and there is a lot of possibility for further advancements in the near future (Fig. 14.1).

The distinctive physicochemical characteristics of nanoparticles (NPs) in comparison to their bulk material enable them to have the potential for applications in agriculture and biofuel production to ensure food and energy security. Nanomaterials (NMs), used in nanotechnology, are a class of material characterised by at least one dimension within the range of 1–100 nm (Bhardwaj et al. 2022; Garg et al. 2023). With the advancement of nanotechnology, several smart nanofertilisers are being developed with the signal-controlled release of minerals to reduce loss and enhance use efficiency. Such nanofertilisers are required in a very small amount to achieve the desired growth and yield of crop plants. Several nanopesticides have also been developed to improve their stability and efficiency after field application. In agriculture, the term nanobionics is simply referred to as the development and use of nanomaterials to engineer or modulate plant structure and function for enhancing photosynthetic efficiency, nutrient use efficiency, stress tolerance and detection of pathogens and pesticide residues (Giraldo et al. 2014; Sharma and Kar 2019; Ranjan et al. 2022). It ensures biomass accumulation in plants and biosafety as well. It is an emerging field, and a lot of scope is available to develop new methods

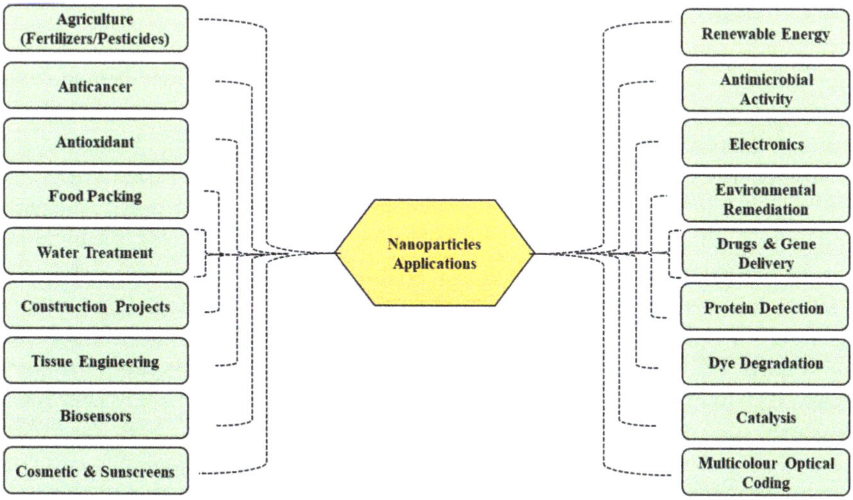

Fig. 14.1 Various applications of nanoparticles in different sectors

and techniques for engineered NPs synthesised to solve various problems in the agricultural sector.

Uses of NMs in the current biofuel production technology have shown promising results in terms of yield improvement. Nano-catalysts developed by enzyme embolisation on nanomaterials significantly improve stability, recovery and reusability of biocatalysts which ultimately leads to improved biofuel production. Nanotechnology can also be applied in feedstock modifications (Antunes et al. 2017), enhancement in microalga growth and induce lipid production, fatty acid harvesting from microalgae, transesterification of lipids, etc. (Akubude et al. 2019; Bin Rashid 2023). Therefore, an overview of recent developments and various aspects of the application of nanobionics in agriculture to improve crop productivity and bioenergy sectors have been given in this chapter.

14.2 Nanobionics in Agriculture

The field of nanobionics in agriculture explores the application of NPs and their interplay with a plant system to regulate particular functions, leading to a significant enhancement in crop yield (Giraldo et al. 2014). Nanoparticles are designed to minimise the dispersion of chemical substances, decrease nutrient loss during fertiliser application and enhance both quality and yield (Sangeetha et al. 2021). The potential application of nanoscale agrochemicals, including nanofertilisers, nanopesticides, nanosensors and other nanoformulations, has revolutionised conventional agricultural methods, enhancing their sustainability and efficiency (Neme et al. 2021).

14.2.1 Nano-Fertilisers

Agricultural crops require both macro and micronutrients for their growth. Traditionally, nitrogen fertilisers and NPKs are mostly used by farmers to obtain better crop growth and yield. Fertilisers are commonly administered to the soil and then taken up by the roots for transport to the different parts of the plants. Generally, artificial chemical fertilisers are water-soluble and facilitate their easy application and nutrient uptake by plant roots. However, due to this characteristic, the nutrients can be washed away with rainwater or leached down into groundwater, eventually causing pollution in lakes, rivers and aquifers (Bhardwaj et al. 2022). It leads to the reduced use efficiency for the majority of synthetic fertilisers viz., 18–20% for phosphorus (P), 30–35% for nitrogen (N) and 35–40% for potassium (K) (Haris et al. 2023). The production cost of fertilisers is increasing day by day, which has also increased the burden on farmers and government agencies. Now, the recent trend is towards nanofertilisers, which are highly efficient and required in very low quantities in comparison to traditional fertilisers. Furthermore, smart nanofertilisers are also being developed for the regulated release of minerals under different environmental conditions. Many metal oxide NPs, such as CeO_2, ZnO, TiO_2, Al_2O_3 and FeO, have signified their potential application as nanofertilisers (Sabir et al. 2014). The application of ZnO nanofertiliser to the barley crop elevates the rate of photosynthesis, root length and yield by 60% (Kale and Kane 2017). Enhanced nutrient absorption can be achieved through the utilisation of encapsulated fertilisers, leading to decreased nutrient wastage, enhanced crop quality and yield, and reduced environmental risks (Mittal et al. 2020). Of all the essential mineral nutrients, nitrogen stands out as the primary requirement for crop plants. It serves as a vital component of chlorophyll, numerous proteins and enzymes, playing a crucial role in the vegetative growth phase of crops (Kumar et al. 2021). Nevertheless, conventional urea fertilisers exhibit inherent drawbacks, including ineffective nutrient absorption, potential losses environmental contamination and economic inefficiencies (Yadav et al. 2023). Nanourea comprises nitrogen particles at the nanoscale, featuring increased surface area and a higher particle count, enhancing its overall effectiveness (Kumar et al. 2021). It can be easily administered either through foliar application (the direct spraying of nanourea onto plant leaves) or through roots (introduction into the irrigation systems or soil) (Lohar et al. 2023). Similarly, Rathnayaka et al. (2018) found nanourea to be a better substitute for conventional urea for improving rice yield. Manikandan and Subramanian (2016) assessed the impact of nanourea and traditional urea on maize growth, yield and quality. The nanourea was found to be more effective in terms of growth, yield and nutrient uptake compared to conventional urea.

Nanobionics approaches, such as smart delivery system allowing controlled release of nutrients under the influence of different environmental conditions, have significantly improved nutrient use efficiency (NUE). In this technique, small amounts of fertilisers are generally incorporated or encapsulated in the nanocarriers that widen their distribution. Further, the release of nutrients depends on the choice of fertiliser, the type of materials used as carriers and the specific design to sense the stimuli. As per the requirement, nanomaterials are designed to sense stimuli like light, temperature, pH, humidity, ultrasonic waves and specific molecules. Many engineered metal oxides (MOx) including ZnO and SiO_2 have been synthesised for controlled and sustained nutrient release (Taurozzi et al. 2011; Yoon et al. 2020; Martins et al. 2020; Mushtaq et al. 2018).

Globally, Zinc (Zn) deficiency is a prevalent problem that severely affects agricultural productivity. However, the Zn salts used in conventional fertilisers are leached rapidly from the soil, which leads to their reduced bioavailability and eco-toxicity. A bio-polymeric ZnO NP designed to work at specific organic matter content and pH displayed prolonged and slow release of Zn according to the requirements of maize plants. Thus, this system can be effectively used to reduce the bioavailability of Zn and environmental toxicity (Martins et al. 2020). developed an engineered composite nanomaterial carrying both SiO_2 and NPK NPs inside the core of beads that consists of chitosan as the first coating, followed by coatings of sodium alginate and kaolin. At room temperature, it showed the slow release of nutrients continuously for a period of 6 months. In addition, this formulation also possesses water-holding capacity, which helps in improving the salinity and drought stress tolerance of plants.

Polyacrylate emulsion is widely used for the coating of urea granules owing to its film-forming properties, appropriate viscosity, low price and environmentally friendly nature. Shen et al. (2019) observed that coating urea with a hybrid emulsion prepared after the mixing of polyacrylate emulsion and aqueous silica nanoparticles (1.0 wt %) effectively delayed the release of urea. The addition of aqueous silica nanoparticles to the polyacrylate emulsion dramatically increases the tensile stress, hardness and glass-transition temperature of the coating. In addition, it also decreases the rate of water absorption and the water vapour permeability of the coating. In order to develop potential cost-effective and eco-friendly nanocarriers for controlled release of fertilisers, several materials are also being explored as well (Ranjan et al. 2022).

14.2.2 Nanomaterials and Chloroplast Interaction to Augment Photosynthesis

Plant nanobionics involves enhancing the capabilities of plant organelles, tissues and entire organisms by incorporating NMs, as proposed by Giraldo et al. (2014) and Wong et al. (2017). NMs are becoming increasingly recognised as carriers for transporting biomolecules within plant systems (Torney et al. 2007; Wang et al. 2019; Demirer et al. 2019; Kwak et al. 2017, 2019). Interacting engineered nanomaterials (ENMs) with plants has resulted in notable progress in tackling essential issues related to the delivery of plant genetic elements (Demirer et al. 2019), biochemical sensing (Giraldo et al. 2014, 2015; Wong et al. 2017) and the delivery of nutrients and pesticides (Borgatta et al. 2018; White and Gardea-Torresdey 2018).

Chloroplasts are crucial cellular structures present in the cells of green plants and algae, which play a vital role in photosynthesis and thus provide the necessary fuel for plant growth (Weise et al. 2004). Additionally, it also serves as a signalling organelle, contributing to plant stress responses (Van Aken et al. 2016; Su et al. 2019). Therefore, it is considered an important target for improving crop growth, stress tolerance, facilitating biopharmaceutical production and advancing the creation of tools in synthetic biology (Jin and Daniell 2015). Photosynthesis is an essential process in which atmospheric CO_2 is captured and transformed into three primary types of sugars, viz., maltose, triose phosphate and glucose, using light energy. During this complex biochemical reaction, glucose and oxygen are produced, a vital for sustaining life on Earth. The chloroplasts have the capability to absorb the visible spectrum, which constitutes 50% of the incoming solar radiation. Furthermore, plants typically utilise only 10% of the sunlight available to them (Zhu et al. 2010). As a result, scientists are actively engaged in endeavours to improve the efficiency of photosynthetic processes by expanding the range of sunlight absorption (Blankenship et al. 2011). NPs have proven to be effective in modulating plant photosynthesis. They act as an artificial photosynthetic system, benefiting fast-growing plants and improving overall processes (Noji et al. 2011; Singh and Bell 2016). Enhancing the efficiency of photosynthesis in crop plants is recognised as a promising approach to boost overall crop productivity (Makino 2011; Ort et al. 2015). NPs application could lead to an enhancement in electron transfer within the photosystems, pigment content, the absorption of light throughout the UV-visible spectrum and ultimately the photosynthetic capacity (Fig. 14.2) (Ranjan et al. 2022).

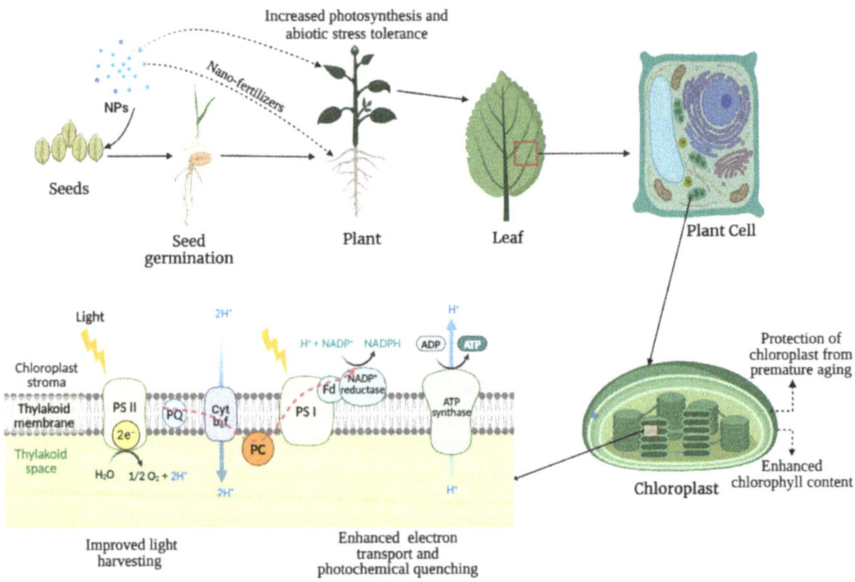

Fig. 14.2 Role of NPs in enhancing seed germination, photosynthesis efficiency and plant growth. (The image was generated using BioRender, Website: https://app.biorender.com)

Several metal and metal oxide nanoparticles (MO_X NPs) have been explored with the goal of impacting the photosynthesis process and improving the growth and overall traits of plants. Au, ZnO and Ag-based NPs have demonstrated their capacity to improve photosynthesis by enhancing chlorophyll content, photochemical quenching, electron transport rate and aiding photosystem functions (Torres et al. 2018; Li et al. 2020; Shoyhet et al. 2021). According to Ze et al. (2011), the TiO_2 NPs increase light-harvesting complex II (LHCII) expression and speed up light energy conversion to electronic energy, water photolysis and oxygen evolution. Another study by Yang et al. (2006) states that utilisation of TiO_2 NPs safeguards chloroplasts from premature ageing and extends the photosynthetic duration of chloroplasts. It safeguards the chloroplast by enhancing the activity of antioxidant enzymes such as catalase, peroxidase and superoxide dismutase (Hong et al. 2005). Ribulose bisphosphate carboxylase oxygenase (RuBisCO) is the key enzyme in the CO_2 fixation process of photosynthesis. Therefore, researchers have shown a significant interest in this enzyme to enhance the photosynthetic capacity using a genetic engineering approach (Lin et al. 2014; Von Caemmerer 2020). It has been reported that the activity of RuBisCO can be effectively modulated by the use of nanotechnology. In addition, targeting various processes of photosynthesis, TiO_2 NPs enhance RuBisCO activity as well (Farahi et al. 2023). The cerium oxide nanoparticles with Ce^{3+} and Ce^{4+} ions (nanoceria) enhance the carbon reactions of photosynthesis by facilitating higher rates of RuBisCO carboxylation and overall photosynthetic capacity, even under light and temperature stresses (Wu et al. 2017).

It has also been documented by Zand et al. (2020) that the application of chemically synthesised TiO_2 NPs at 250 mg kg^{-1}, with Plant Growth Promoting Rhizobacteria (PGPR), notably augments the overall chlorophyll content. Nano-SiO_2 positively influences plant growth and development by augmenting various gas exchange and chlorophyll fluorescence parameters. It leads to enhancement in the rate of net photosynthesis and transpiration, stomatal conductance, PSII potential activity, electron transport rate and photochemical quenching (Siddiqui et al. 2014). SiO_2 NPs enhance the photosynthetic rate through the augmentation of carbonic anhydrase activity and photosynthetic pigment contents, as reported by Siddiqui et al. (2014) and Xie et al. (2012). The increased activity of carbonic anhydrase facilitates the supply of CO_2 to Rubisco, potentially leading to an enhancement in the CO_2 fixation rate (Siddiqui et al. 2012). Lei et al. (2007) reported that nano-anatase TiO_2 significantly enhances the overall electron transport throughout the entire chain as well as the photoreduction activity of photosystem II. Moreover, it exhibits increased O_2-evolving and photophosphorylation activities of chlorophyll under both visible and ultraviolet light conditions.

Carbon-based NMs are very useful in diverse fields, and their significant effects on plant functions have also been reported. Carbon nanotubes (CNTs) are carbon allotropes (different forms of carbon) with a cylindrical nanostructure (Zhao et al. 2017). By introducing carbon nanotubes into chloroplasts, researchers have improved the capacity of plants to absorb more light energy. CNTs act as artificial antennae that enable chloroplasts to absorb light wavelengths outside of their standard range, such as ultraviolet, green and near-infrared (Cossins 2014; Giraldo et al. 2014). The integration of single-walled carbon nanotubes (SWNTs) into chloroplasts holds promise for enhancing the essential light reactions crucial for photosynthesis. This potential arises from the unique optical characteristics exhibited by SWNTs. Unlike chloroplast pigments with limited absorption rates, SWNTs demonstrate the capability to capture light across the visible and near-infrared spectra. Therefore, SWNTs are suggested as a prospect for optimising photosynthetic processes (Dufil et al. 2021). After entry into the cell, SWNTs interact with the chloroplast, which leads to improved light sensitivity and photosynthesis (Bhaskar et al. 2024). Giraldo et al. (2014) implanted SWCNTs in the isolated chloroplast, which increased electron transport rates and photosynthetic activity by three times compared to controls. Additionally, SWCNTs allowed the plants to detect nitric oxide, a signalling molecule. They proposed that the use of nanobionics in plants would allow photosynthetic organelles to have new and improved functional features. Further, more thorough investigation is required to determine how carbon nanotubes (CNTs) affect the final products of photosynthesis, such as glucose and other sugars.

14.2.3 Nano-Pesticides

The use of synthetic pesticides and fertilisers in modern agriculture practices has substantially contributed to nourishing the world. However, their use has also created several types of health and environmental issues (Hassaan and El Nemr 2020; Carvalho 2017). Pesticides are used to increase crop yield by preventing or controlling potentially harmful pathogens and insect pests (Jampílek and Kráová 2017). However, overuse of pesticides in agricultural practices raises a number of issues, such as resistance to diseases and insect pests, effects on non-target organisms, loss of soil biodiversity, pesticide buildup in the surroundings and adverse health effects on humans (Mahmood et al. 2016; Adisa et al. 2019).

In the year 2019, IUPAC identified 10 chemical technologies with the greatest potential to influence human society in the coming years. Among those, nanopesticides claimed the top position due to their anticipated minimal effects on both human and environmental health (Ten 2019). Nano-pesticides exhibit remarkable thermal stability, improved solubility of active ingredients, undergo biodegradation upon successful administration, and demonstrate heightened affinity for specific targets (Fincheira et al. 2023). Consequently, various nano-formulations of pesticides have been innovatively designed, including polymeric NPs, nano-emulsions, solid lipid NPs, nano-fibres, nano-gels and nano-micelles (Fincheira et al. 2023). The goal of synthesising and utilising nanopesticides is to improve the effectiveness and longevity of pesticides while simultaneously decreasing the quantity of active ingredients and mitigating or eliminating potential risks (Bratovcic et al. 2021).

The utilisation of nanoscale carriers becomes instrumental for the effective delivery of antibiotics, insecticides, herbicides and plant growth regulators. Encapsulation of pesticide-active ingredients in nanoscale carriers offers several advantages in terms of enhancing efficiency, delivery and adverse environmental impact (Dasgupta et al. 2015). Diverse nanocarriers have been specifically crafted, featuring controlled or gradual release characteristics. These innovations hold the potential to release pesticide-active ingredients (AIs) based on crop requirements and environmental circumstances (Camara et al. 2019). Numerous release mechanisms associated with nanoencapsulation, including diffusion, dissolution, biodegradation and osmotic pressure influenced by specific pH levels (Vidhyalakshmi et al. 2009; Ding and Shah 2009), ensure prolonged and effective action against pests. This controlled release not only improves the overall efficiency of the pesticide but also minimises the frequency of application and operational costs for farmers.

Xiang et al. (2017) developed a pH-responsive controlled-release pesticide (PRCRP) system with magnetic collectability for enhanced pesticide utilisation efficiency. Herbicide (Glyphosate) and insecticide (Cypermethrin) were loaded into micro-nano pores of diatomite/Fe_3O_4, a magnetic nanocarrier. Further, the developed pesticide/diatomite/Fe_3O_4 composites were coated with chitosan, allowing for controlled pesticide release in acidic conditions. PRCRP demonstrated strong adhesion to weed surfaces and pest epidermis, along with notable controlled-release capabilities, high utilisation efficiency and remarkable efficacy in controlling both weeds and pests. Owing to their magnetic nature, these nanocarriers can be easily separated and recycled from water and soil. Therefore, this technology may be promising in reducing pesticide residue and environmental risk. Liang et al. (2018) introduced lanthanum-modified chitosan oligosaccharide nanoparticles (Cos-La) as a novel formulation to load the pesticide avermectin. The water-based Cos-La formulation demonstrated a high avermectin loading capacity of 46.3%, enhanced avermectin persistence by over 25%, and reduced the photolysis rate by around 20%. Application of different Cos-La concentrations to rice cultivation resulted in growth promotion, with increased plant height and fresh weight. Higher Cos-La concentrations also reduced rice wettability, potentially lowering the risk of plant diseases and pests. Moreover, Cos-La-treated rice exhibited a significant defense response against rice blast, surpassing the effects of the equivalent Cos and $LaCl_3 \cdot 7H_2O$ mixture solution two times. Overall, Cos-La nanoparticles prove to be a promising and eco-friendly nanocarrier for pesticides in agriculture, improving stability, persistence and crop growth while enhancing disease resistance.

Shan et al. (2019) developed a mesoporous silica nanoparticle (MSN) pesticide carrier ornamented with sulfonic acid groups (MSN-SO_3) bearing negative charge. A positively charged herbicide, diquat dibromide (DQ), was effectively loaded onto the negatively charged MSN-SO_3 by electrostatic interaction, with an increased loading rate of 12.73%. The electrostatic nature of pesticide and nanomaterial interaction ensured the pH-dependent release of DQ. This composite nanoherbicide was also proposed to be highly effective in controlling jimsonweed (*Datura stramonium* L.).

Chi et al. (2021) developed a magnetic-responsive controlled-release herbicide using a composite nanomaterial derived from ferroferric oxide (Fe_3O_4), palygorskite, amino silicon oil (ASO) and glyphosate. Palygorskite, with its micro/nanonetwork structure, binds to a large quantity of Fe_3O_4 and glyphosate. In this system, ASO was used as a coating agent to slow down the release of the herbicide

Glyphosate. It also protected the herbicide from direct environmental exposure, which reduced loss. However, the Fe_3O_4 ensured the controlled magnetic field-driven release of herbicide. Thus, the system was developed in such a way that the herbicide release could be accelerated by applying the magnetic field. Similarly, other nanomaterials coated with pyraclostrobin have also been developed to reduce pesticide toxicity by preventing the direct interaction of AIs and non-target organism (An et al. 2022).

14.2.4 Nanotechnology in Abiotic Stress Tolerance

Various environmental factors, including biotic and abiotic, can cause stress in plants under certain circumstances and significantly affect the plant's metabolism and physiology (Jan and Ahmad 2019). Abiotic stress is considered one of the main factors limiting agricultural output globally, which lowers the average yields of the majority of important crops by over 50% (Bray 2000; Wang et al. 2003). Drought, salt, flooding, mineral shortages and a variety of contaminants, including radiation, are thought to serve as the key abiotic stressors that have an impact on crop growth and yield (Pereira 2016; Husen et al. 2014, 2016). Generally, in most crops, including groundnuts, wheat and rice, abiotic stress leads to an excess of reactive oxygen species (ROS) production, responsible for severe cellular damage (Abhinandan et al. 2018; Trinh et al. 2014). Various kinds of nanomaterials have been designed with the potential to improve plant growth and development of plants under abiotic stress by modulating photosynthesis, soil quality and antioxidant systems (Fiol et al. 2021). According to Hu and Xianyu (2021), functionalised NPs distribute macro-and micronutrients more actively throughout the plant than nano-fertilisers and boost plant growth. In addition, it also shields the plant from the harmful impacts of abiotic and biotic stress through the induction of stress-tolerant genes (Van Aken 2015). Metallic nanoparticles interact with plants and exert their impact on various aspects such as growth, crop productivity and nutritional quality (Zulfiqar et al. 2019; Giraldo et al. 2019). In a recent study, CuNPs were administered to millet crops under heat stress after being synthesised using a green synthesis procedure. Results showed that the growth, photosynthesis and efficiency of the antioxidant defense systems of the crop were all markedly enhanced (Pelegrino et al. 2021). Application of CuO NPs to millet crops exposed to salinity stress greatly enhances crop growth, yield and physiological parameters by strengthening antioxidant defense systems and minimising oxidative stress caused by salinity (Großmann et al. 2015; Wrigley 2017). Tarafdar et al. (2014) have reported that ZnO NPs greatly enhanced the growth, photosynthetic efficiency, water usage efficiency and antioxidant defense mechanisms of pearl millet when applied under drought stress (Mohanasundar et al. 2021). The adverse effects of abiotic stress on seed germination may also be alleviated by priming the seeds with nanoparticles (Fig. 14.2). It has been reported that priming *Festuca ovina* seed with silver NPs (AgNPs) seed germination even under drought (Abbasi Khalaki et al. 2019). Similarly, *Brassica napus* L. seeds primed with zinc oxide (ZnO) NPs show considerable enhancement in percent seed germination as compared to unprimed and hydro-primed seeds under salt stress (El-Badri et al. 2021). Moreover, under drought stress, supplementation of multi-walled CNTS (MWCNTs) with seeds of *Dodonaea viscosa* L. enhanced the percent seed germination and also reduced the mean germination time (Yousefi et al. 2017). Silicon oxide (SiO_2) NPs also assist in alleviating the impact of drought stress on germination of wheat seeds by boosting water uptake and amylase activity (Rai-Kalal et al. 2021). *Zea mays* L. seed is also less likely to germinate under salt stress; however, when treated with NPs made up of mango peel (nMPs), seed germination is induced (Elkhatib et al. 2019). Along with the role in seed germination, MWCNTs functionalised with carboxylic group significantly triggers plant growth under stress condition (Bhaskar et al. 2024).

Treatment of cerium oxide nanoparticles in plants under abiotic stress exhibits increased rates of carbon assimilation, higher quantum yields of photosystem II and enhanced efficiency in CO_2 utilisation compared to control plants (Wu et al. 2017, 2018). To increase ROS scavenging potential and photosynthetic performance in *Arabidopsis thaliana* Wu et al. (2017) treated its chloroplast with anionic and cationic cerium oxide nanoparticles (nanoceria). The study indicated that anionic nanoceria with a low Ce^{3+}/Ce^{4+} ratio act as ROS scavengers and contribute to safeguarding the chloroplast photosynthetic system from abiotic stresses.

14.2.5 Nanobionics in DNA Delivery and Crop Improvement

The common techniques employed for DNA delivery encompass agrobacterium-mediated transformation (AMT) (Gelvin 2003), particle bombardment (Altpeter et al. 2005) and polyethylene glycol (PEG)-mediated protoplast transfection (Hayashimoto et al. 1990). The biolistic system for plant genetic transformation stands out as effective tool for the biotechnologists. Its application extends to delivering DNA into the nuclear or plastid genomes of various plant species (Warzecha and Hennig 2010). However, in numerous species, it is still difficult to achieve reliable genetic transformation using the established technique of DNA delivery. The exploration of NP engagement in DNA or cargo delivery for biotechnological purposes has been carried out using isolated chloroplasts (Wong et al. 2016), plant protoplasts (Lew et al. 2018) and within the leaves of terrestrial plants (Hu et al. 2020). Martin-Ortigosa et al. (2012) introduced a novel approach by involving the gold plating of mesoporous silica nanoparticles (MSNs), aiming to enhance density and efficacy in biolistic-mediated delivery. The researchers demonstrated that optimising DNA-coating protocols and adjusting the size of gold nanoparticles for capping could further improve MSN-mediated DNA delivery. The utilisation of nanoparticles for facilitating transformation holds significant potential in plant genetic engineering. In another study by Chang et al. (2013), a unique DNA delivery approach was explored using functionalised MSNs as carriers. MSNs were utilised to transport foreign DNA into intact *Arabidopsis thaliana* roots without the need for mechanical force, facilitating gene expression in diverse root tissues. The investigators also examined MSN uptake and the mechanism of subcellular distribution of these particles. The results indicated effective internalisation of MSNs by *Arabidopsis* roots, showcasing a strong uptake capability and efficient delivery at elevated levels.

The manipulation of chloroplast function in plants, achieved through the delivery of biomolecules or chemicals is essential for gaining insights into the role of these plastids in plant biology and strategies development for chloroplast bioengineering (Santana et al. 2020). SWNCTs integrated into chloroplasts possess the capability to augment the light-dependent reactions in the process of photosynthesis (Giraldo et al. 2014). Kwak et al. (2019) designed chitosan-complexed SWNTs using the lipid exchange envelope penetration mechanism. Their experiment successfully demonstrated chloroplast-targeted transgene delivery and transient expression in mature plants such as *Eruca sativa*, *Nasturtium officinale*, *Nicotiana tabacum* and *Spinacia oleracea*, as well as in mesophyll protoplasts of isolated *Arabidopsis thaliana*. Results revealed that the created platform has the potential for additional optimisation which could facilitate consistent chloroplast transformation in crop species, offering advantages for both crop enhancement and agronomic applications.

14.3 Nanobionics in Bioenergy Production

The extensive use of fossil fuels is regarded as a primary source of greenhouse gases that contribute to climate change (Masson-Delmotte et al. 2019). Among the energy-consuming industries, the transport sector consumes a major part of petroleum oil, accounting for 92% of fuel demand (World Energy Balances 2020). Environmental pollution policies such as the Paris Agreement and the European Green Deal consider that the widespread use of biofuels could significantly contribute to reaching reduction targets of 80–95% for greenhouse gas emissions by 2050 (Biofuels: Energy 2021; UNFCCC 2015; European Commission 2021). Due to high octane numbers and high heat of vaporisation, bioethanol and biodiesel can both be utilised in internal combustion engines (Erdiwansyah et al. 2019). The primary source of biofuel production at the moment is food crops, which are associated with environmental and socioeconomic issues. These issues include the introduction of invasive species, direct land-use change, soil degradation and the risk of food security (Correa et al. 2019). Biofuels extracted from edible feedstocks like wheat, rice, potatoes, sugarcane, barley and vegetable oils are referred to as first-generation biofuels. Oils are removed from the biomass, fermented and transesterified to produce bioethanol or biodiesel (Bhargavi et al. 2018; Rodionova et al. 2017). Non-edible plant oils like Mahua, Jatropha, Neem, Karanja and lignocellulosic materials from agricultural and plant-based industrial wastes can also be used as common sources of feedstock to produce second-generation biofuel (Shahabuddin et al. 2012; Ghorbani et al. 2011; Liaquat et al. 2012; Atadashi et al. 2012; Qiu et al. 2010). Lignocellulosic materials are initially processed by physical and chemical methods, into the cellulose which is further transformed into glucose through cellulase-mediated enzymatic hydrolysis. However, due to the low stability and lack of reusability of cellulase, the traditional production system requires a high amount of this enzyme, which leads to a rise in the price per unit of biofuels (Jaroenkhasemmeesuk and Tippayawong 2015; Lu et al. 2012).

Microalgae and cyanobacteria possess promising potential for biofuel production, and they are considered third-generation biofuels. These organisms grow faster than plants and do not require a large or arable land area for their biomass production. Thus, for biomass production, there will be no issue with the farming sector or food security (Chew et al. 2018). According to Shah et al. (2018), the considerable volatility of microalgal biofuel makes it less stable than that of the first and second generations. Furthermore, the extraction of lipids and the drying of biomass demand a lot of energy (Bennion et al. 2015; Dasan et al. 2019). The genetic engineering of microalgae is the main focus of the fourth-generation biofuel. The goal is to use cutting-edge technologies to generate microalgae capable of absorbing high CO_2, boost their output as biofuel and make them more adaptable in wastewater (Zhu et al. 2017). Unfortunately, owning to a lack of genetic and biological data, this application cannot be used with all microalgae species (Abdullah et al. 2019).

Various unique characteristics of nanomaterials, such as high surface areas, catalytic activity, high crystallinity, stability, durability, adsorption capacity, potential for recovery and reusability, make them exceptional candidates for their use in various systems of biofuel production (Garcia-Martinez 2010). One important prerequisite for the manufacturing of biofuels, especially biodiesel and bioethanol, is the development of highly efficient biocatalysts. Nanocatalysts fabrication and their potential applications in biofuel production have been the focus of the last few decades (Verma 2017; Zhang et al., 2010). It is feasible to convert biomass into biofuels by immobilising the enzymes in nanomaterials, which creates nanocatalysts (Abraham et al. 2014; Verma et al. 2013a,b,c,d; Puri et al. 2013; Guo et al. 2012). By using nanoparticles as biocatalyst carriers, nanotechnology opens the door to overcoming the low efficiency of biocatalysts which are under current use. The resulting nanobiocatalysts are used as a tool to convert a variety of biomass-related compounds into biofuels. Novel nanobiocatalysts have the potential to outperform traditional biocatalysts in terms of benefits such as mass transfer, poisoning, catalyst deactivation and shortened reaction times (Arya et al. 2021). Biohydrogen, biodiesel, bioethanol and biogas are just a few of the biofuels being produced using nanotechnology.

14.3.1 Different Nanomaterials Used in Bioenergy Generation

In order to boost the productivity of biofuels, high-quality nanocatalyst systems, including MO_X NPs, magnetic nanoparticles (MNPs) and CNTs, have been recently developed by utilising contemporary techniques for their synthesis and characterisation (Fig. 14.3). There are several types of MOx nanocatalysts: alkaline earth MOx (CaO and MgO), transition Mox (NiO, CuO, TiO_2, etc.), mixed MOx (Iron-Cadmium and Iron-Tin) and supported MOx nanocatalysts (KF/CaO-MgO, CsNanoMgO-C, etc.) (Zhang et al. 2017).

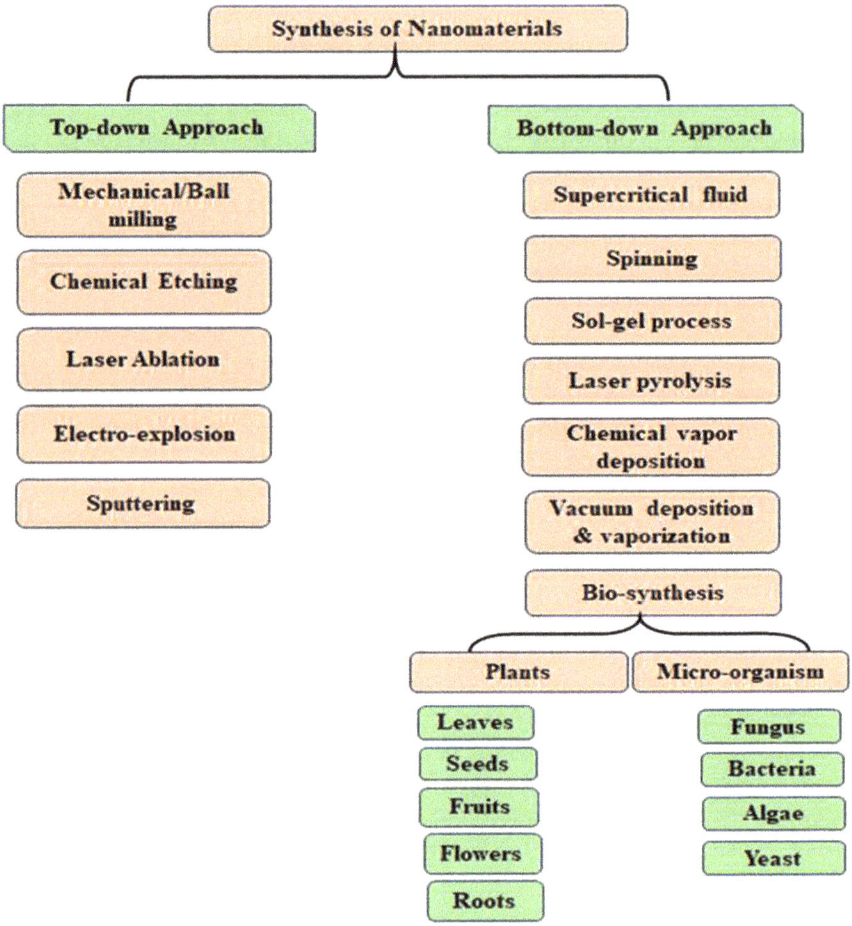

Fig. 14.3 Top-down and bottom-up approaches for NP synthesis. Generally, two basic approaches viz., top-down and bottom-up is used in the synthesis of nanoparticles. In top-down approach, the size of the solid materials is broken down or reduced into tiny particles. Mechanical milling, laser ablation, sputtering and thermal decomposition etc. are some techniques which are involved in top-down approach. Changes in the surface chemistry and physicochemical characteristics of nanoparticles are the main issues associated with the top-down approach (Nadagouda et al. 2011). In bottom-up approach, atoms cluster themselves to become nanoscale particles. Sol-gel, spinning, chemical vapours deposition (CVD), laser pyrolysis, biosynthesis, etc. are various techniques that are employed in bottom-up approach for nanoparticles formation

The biomass conversion efficiency of the NPs is determined by various characteristics such as the types of metal or material used in the synthesis, porous nature and adjustable acid–base property (Zhang et al. 2017). Furthermore, the characteristics of the nanocatalysts can be influenced by factors like size, shape and surface chemistry as well (Somorjai et al. 2009). A reduction in particle size enhances the available catalytic active sites due to increased surface-to-volume ratio. Changes in particle shape, such as spherical, tetrahedral and cubic forms, may also lead to altered catalytic properties (Zhang et al. 2017). Therefore, different types of nanomaterials have been developed and employed to enhance the cost-effective production of biofuel using various types of feedstocks.

14.3.1.1 Magnetic Nanoparticles (MNPs)

Because of the enhanced surface-to-volume ratio, immobilising impact, controllable particle size, modifiable surface and quantum properties, MNPs are in high demand (Ahmed and Douek 2013). Most importantly, the MNPs can be easily separated from the reaction solutions by applying an external magnetic field. Recovery and repeated use of immobilised enzymes reduce biocatalyst requirements by roughly 30% and enhance biofuel production (Zhang et al. 2013). The surface of MNPs can also be modified by coating them with a range of complex materials, viz., polymers, carbon-based materials, silica and metallic-organic frameworks (MOFs). Coating allows improvements in enzyme loading, stability and potential industrial applicability of the MNPs. Such NPs of size below the crucial value of 10–20 nm display the best function as they behave as a single magnetic domain under the applied magnetic field and act like enormous paramagnetic atoms (Antunes et al. 2017; Singh et al. 2020). Platinum, nickel, iron and cobalt are a few examples of commonly used materials to create MNPs (Lai et al. 2018; Zanuso et al. 2022). The key constraint of using MNPs is their instability and inherent ability to aggregate over extended periods of time. MNPs can also readily lose their magnetic characteristics through oxidation and dispersion due to their high chemical activity. To prevent the MNPs from being destroyed chemically both during and after the production steps, techniques like coating or modification with surfactants or inorganic layers of silica and carbon may be crucial (Zhong and Sels 2018). It makes them very stable (i.e. highly dispersible), reactive and easily separable at a later stage.

14.3.1.2 Carbon Nanotubes (CNTs)

Nanometrically sized cylindrical graphite sheets are wrapped into CNTs, which display high biocompatibility and mechanical strengths as well. The large surface area of CNTs allows for low diffusion resistance and a high enzyme loading capacity (Bianco et al. 2011). However, the volume and length of CNTs, as well as the blended substance influence the biocompatibility of CNT composites (Aoki and Saito 2020). Due to their remarkable physicochemical properties and enzyme-binding ability, carbon nanotubes (CNTs) have received significant attention in the field of biofuel production since their discovery in 1991. However, the recovery of CNT-immobilised enzymes from the reaction mixture in the biofuel synthesis process for repeated uses was initially encountered as a bottleneck. Further, the development of CNTs endowed with a magnetic property was regarded as a potential solution to this problem (Fan et al. 2017). Magnetic CNTs can be developed by either surface loading or filling iron oxide NPs inside the nanotubes. Immobilised enzymes like lipase and amyloglucosidase on magnetic CNTs developed using both methods display increased activity (Tan et al. 2012; Goh et al. 2012). Enzyme loading and activity can also be enhanced by the coating of synthetic dendrimer on magnetic CNTs. Fan et al. (2017) developed superparamagnetic multi-walled carbon nanotubes (mMWCNTs) by filing iron oxide in multi-walled carbon nanotubes (MWCNTs) and surface-linking polyamidoamine (PAMAM) dendrimers. Immobilised bacterial lipase on the developed mMWCNTs displayed 17 fold higher activities than the free enzyme. It enhanced the pH resistance and thermostability of the enzyme, and better recycling performance improved biodiesel production with a 92.8% conversion rate. Even after 20 cycles of use, this enzyme-immobilised system retained about 90% of its original activity.

14.3.1.3 Metal Oxides (MOx)

MOx NPs of various structural forms are generally developed by the reaction of many metallic elements with oxygen under different conditions. MOx NPs with sizes in the nano range showed various important applications, viz., gas, optical and fluorescence sensors, catalysts, bioenergy, biomedicine, photovoltaic, etc. Variations in shape and size of the MOx mainly affect their optical, mechanical, electrical, magnetic and catalytic properties (Alhalili 2023). Usually, porous substrates are used to incorporate and stabilise nanoparticles. Several mesoporous MOx NPs are used in nano-biocatalyst preparation and their efficient applications in biofuel production (Genna et al. 2013; Jin et al. 2012a, b). The incorporation of porous substrates not only prevents the aggregation of nanoparticles but also induces synergistic effects useful in many applications (Yang et al. 2017). Generally, mesoporous MOx is synthesised using either soft or hard templates. The numerous openings in the substrate facilitate the incorporation or encapsulation of metal nanoparticles and also provide access to gases and various other molecules. The physicochemical attributes of the mesoporous supports and their interactions with biocatalyst determine the stability, catalytic activity and selectivity of mesoporous MO_X NPs (Bastakoti et al. 2021). A variety of templates, including block copolymers (Jagtap et al. 2011), amines (Chandra et al. 2008) and surfactants (Parida et al. 2010), can be used to create mesoporosity in MOx. Soft templates like aspartic acid and salicylic acid are generally used to induce mesopores on MOx particles. A self-assembled mesoporous TiO_2 nanoparticle catalyst developed using sodium salicylate as a template displayed promising potential in the conversion of carbohydrates to 5-hydroxymethylfurfural (HMF), an intermediate in the process of biofuel production (Dutta et al. 2011). According to a report, Lewis's acidity and the large surface area of the mesoporous TiO_2 NPs allow considerable catalytic activity, and enhance production of HMF 34.3% and 54.1% in DMSO and water, respectively. The sodium alginate-coated porous TiO_2 nanocatalyst with large pores in nanoscale dimension at the material surface also contributed to the high catalytic activity (De et al. 2012). In hydrothermal conditions, this material was able to catalyse the transformation of unutilised sugar derivatives (i.e. D-mannose, D-galactose and lactose) to HMF under microwave-assisted irradiation at 140 °C with a 42% yield. Various MOx, such as highly ordered mesoporous sulfated zirconium (Gopinath et al. 2017), mesoporous Al_2O_3 (Liu et al. 2015) and reduced copper porous MOx (McClelland et al. 2019), have been developed and employed in biofuel production using various substrates.

14.3.1.4 Silica Nanoparticles

Numerous enzymes can be effectively immobilised to the mesoporous silica owing to its wide surface area and pore volume, simple surface functionalisation, hydrothermal stability and biocompatibility (Tang et al. 2012; Trisunaryanti and Fatmawati 2020). High concentrations of silanol groups found in silica-based materials play a crucial role in the covalent immobilisation of active sites, which permits the addition of organic functionality to regulate surface hydrophobicity. These are highly desirable properties for improving catalytic performance in acid-catalysed chemical reactions, as they increase durability, activity and selectivity (Bhanja and Bhaumik 2016). It has been observed in numerous studies that functionalised mesoporous silica's active sites are desirable for converting biomass. For instance, mesoporous silica (SBA-15) with active sites like sulfonated acid employing organic–inorganic hybrid silane improves catalytic activity and selectivity (Dufaud and Davis 2003). Poorakbar et al. (2018) used magnetic gold MSNs core shells as a support for immobilising cellulase via covalent bonding which leads to retention of 58% of initial catalytic activity even after 9 h. Furthermore, it demonstrated improved thermal stability of enzymes and was found to be suitable for a wider temperature and pH range. Lee et al. (2014) used enzyme-immobilised Fe_3O_4-encapsulated MSNs for cellulosic biomass conversion, and obtained maximum amount of fructose (i.e. about 51%) under optimum reaction temperature (50 °C), time (24 h) and pH 4.8. It has been reported that molybdenum-incorporated KIT-5 mesoporous silica catalysts can be used to produce biofuels and value-added chemicals (furans and phenols) through catalytic fast pyrolysis of pine biomass (Budhi et al. 2015), while organically modified biogenic MSNs (RKIT-6) can be used to hydrolyse lignocellulosic biomass by immobilising xylanase (Ariaeenejad et al. 2020). Moreover, most of these NPs can be stored for longer time period and can be easily separated using an external magnet.

14.3.1.5 Metal Organic Frameworks

Metal Organic Frameworks (MOFs) are a class of crystalline porous materials consisting of multidentate organic molecules holding a three-dimensional (3D) metal ion network (Leus et al. 2010; Baumann et al. 2019). Owing to their high surface area and changeable MOFs, they are particularly helpful in the development of heterogeneous catalysts. Since MOFs have several advantages, researchers have used them to create MOF-derived solid acid with high porosity and an ordered structure than other heterogeneous catalysts. It allows greater substrate transfer within the MOF catalyst during the biotransformation of carbohydrates to HMF. Zhang et al. (2011) reported the first instance of MOFs application in carbohydrate dehydration utilising MIL-101, a chromium-based MOFs that is well-known for its wide surface area, large size of pores and good stability. Additionally, under aqueous hydrogenation conditions, cellulose and cellobiose were preferentially converted to sorbitol over phosphotungstic acid (PTA)-MOF hybrid-supported ruthenium catalysts Ru-PTA/MIL-100 (Cr) under ideal circumstances (Chen et al. 2013). With this work, new insights into the rational design of acid-metal bifunctional catalysts for effective biomass conversion were opened up.

14.3.1.6 Zeolites

Zeolites are frequently employed as a primary heterogeneous catalyst in petrochemical and fine industrial facilities because of their strong Bronsted and Lewis acid sites, well-arranged micropores and exceptional temperature stability owning their crystalline inorganic structure. Stocker (2008) has reported that the active acidic sites of zeolite (i.e. H-ZSM), which act as carbonium ions, can be useful in the pyrolysis of wood biomass. By adjusting the Si or Al content of the zeolite, ZSM-5-based catalysts can adjust their acidity and porosity. For instance, Srinivasan et al. (2014) examined the SiO_2/Al_2O_3 ratios of the zeolite (ZSM-5) catalyst in the catalytic pyrolysis of cellulose to produce aromatic hydrocarbons. The results demonstrated that the catalyst acidity is crucial in removing anhydro sugars and other oxygenated compounds to produce more aromatic compounds. The strength and density of acidic sites in zeolites can also be effectively controlled by doping ZSM-5-based catalysts with additional metal components like Ni, Mn, Co, Pt, Ga and Ce. Vichaphund et al. (2015) used Ni-ZSM-5 catalyst for catalytic pyrolysis of jatropha residues (cellulose: 59.2%, hemicellulose: 18%, lignin: 22.8%) under the perfect conditions for the dehydrogenation, cyclisation and oligomerisation of small olefins. Similarly, Zhou et al. (2018) utilised silicon carbide foam based on a ZSM-5 catalyst for microwave-assisted pyrolysis of biomass. They reported that with an increase in catalyst-to-biomass ratio (0–12), the quality of bio-oil was enhanced, while a decrease in yield was recorded from 40.2% to 33.5%. A group of novel catalysts made of magnetically recoverable zeolites exchanged with transition metals demonstrated considerable promise for the catalytic valorisation of hemicellulosic biomass (Jorge et al. 2017).

14.3.2 Enzyme Immobilisation

Enzyme characteristics, such as high activity at most pH ranges, reusability, selectivity and reduced inhibition can be improved by immobilising enzymes on support materials. For instance, immobilisation of the cellulase enzyme increases its resistance to structural changes induced by a rise in temperature (Baskar et al. 2016). Immobilisation on solid substrates leads to a uniform distribution of catalysts and a reduction in the chance of agglomeration. Three main strategies are used to immobilise enzymes: covalent binding, entrapment and physical adsorption (Fig. 14.4). Of the three, the covalent binding method is recommended since it minimises protein desorption (Abraham et al. 2014). A variety of enzymes are immobilised, including cellulases (Cherian et al. 2015), lipases (Xie and Wang 2014) and beta-glucosidases (Lee et al. 2010). Most notably, MNPs' huge surface-area-to volume ratio makes it very beneficial to load a lot of enzymes onto their surface during enzyme immobilisation, which raises the enzymes' catalytic activity. Another advantage of enzyme-immobilised MNPs and other derived nanomaterials is that they can be recovered and reused. Enzymes can be immobilised on a variety of nanomaterials, like polymeric magnetic nanofibres, silica-functionalised MNPs,

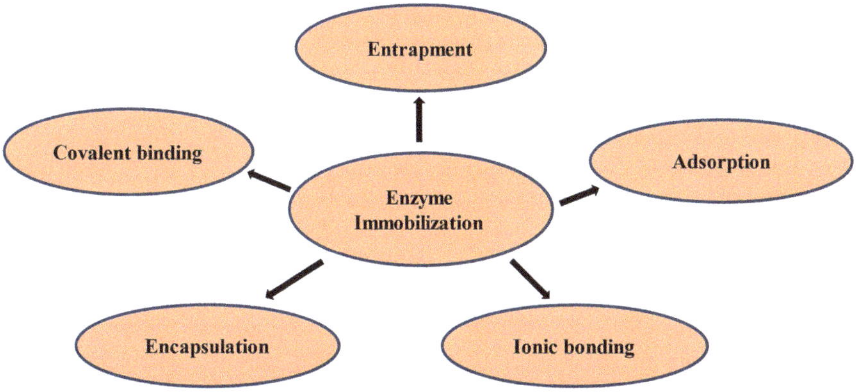

Fig. 14.4 Different methods for immobilisation of enzymes

magnetic chitosan microspheres, amino-functionalised MNPs and composite-functionalised MNPs (Antunes et al. 2017).

Ren et al. (2011) have reported that lipase immobilisation on polydopamine-coated magnetic nanoparticles improved the pH and heat enzyme stability during the biodiesel production process. It resulted in to increase binding efficiency of the enzyme up to 73.9%, retention of more than 70% activity even after repeated use for 21 cycles, and simple recovery from the reaction mixture. The impact of nanotechnology on immobilising commercially expensive enzymes and enabling multiple enzyme reusability is now a defining feature of bioenergy research. It is possible to improve the biocatalytic property and produce more biofuels at a profit by examining and researching the fundamental mechanisms underlying nano-developed enzymes (Srivastava et al. 2019).

14.3.3 Nanotechnology in Biofuel Production

14.3.3.1 Biogas Production

Anaerobic digestion is among the most preferable methods for the conversion of organic matter to energy, specifically biogas (Hao et al. 2019; Feng et al. 2021). In this method, organic matter is decomposed by the anaerobic bacteria into methane gas and other by-products in the absence of O_2 (Seadi et al. 2008). This method has wide applicability in the conversion of a broad range of organic waste materials, including municipal waste, organic food waste, agricultural residues, animal manure, sludge and wastewater into biogas (Vasco-Correa et al. 2018; Holm-Nielsen et al. 2009; Feng et al. 2014). There are four main phases in biogas production: hydrolysis, acidogenesis, acetogenesis and methanogens. The extent of microbial interactions during each of the four phases determines the yield of biogas.

Recent research has revealed that biogas production can be enhanced by the implications of NPs in production plants. NMs offer more exposed sites for anaerobic bacteria (Rahman et al. 2016) and contribute to organic matter solubilisation as well as their chemical modification (Nyberg et al. 2008). In addition, NPs interaction with the bacterial cells in the sludge, makes their membrane permeable through structural changes. It increases the bacterial interaction with the sludge, which also leads to enhanced biogas production (Faisal et al. 2018). Nanoscaled Zero-Valent Iron (NZVI) has been extensively used in various kinds of waste treatment, where it releases electrons required for methanogenesis in anaerobic digestion processes. Its high surface-to-volume ratio increases the sites for the chemical reaction and positively influences biogas production. Increased methane production by the application of Ni and Co NPs in the process of anaerobic decomposition of cattle manure and green microalgae has also been reported (Khan et al. 2022). Impacts of various MOx, nanoscaled transition metal carbides, and carbon-based nanomaterials on biogas production using various waste materials have been investigated, and many of them have demonstrated their promising abilities for biogas augmentation (Nguyen et al. 2015; Zhang and Lu 2016; Li et al. 2018). It has been found that carbon-based NPs, are beneficial in each of the four stages of biogas production. For instance, carbon-based NMs in the hydrolysis stage act as electron donors and diminish the aqueous environment potential for oxygen reduction. It is considered as advantageous for the hydrolysis reaction. Various NPs viz., ZVI, Fe_2O_3, Ag and MgO, also boost the population of methanogeneic bacteria in the anaerobic digester, and increased bacterial activity that leads to enhanced biogas production (Wang et al. 2016; Yang et al. 2013).

14.3.3.2 Biohydrogen Production

Biohydrogen is mainly produced through biological processes like photolysis, fermentation and electrolysis (Chai et al. 2021; Varma 2019; Rai et al. 2022). It requires specialised microorganisms and a diverse kind of organic substrate, including lignocellulosic materials and bio-waste. The lower yield and high cost of the feed and enzymes are the major limitations of the production of biohydrogen on a large scale. Production efficiency and effective utilisation of various enzymes in the H_2 production process are also considered to be very critical bottlenecks. In this regard, several strategies including process optimisation, use of low-cost feeds like bio-waste and identification of potential H_2 producers have been embarked on to promote biohydrogen production (Arora et al. 2020; Rai et al. 2019; Aoki and Saito 2020). NPs can be employed to improve bio-hydrogen production in various processes of photolysis and fermentation owing to their novel properties. In bio-hydrogen production, hydrogenase enzymes play a key role, and their activities are influenced by the Ni^{2+} and Fe^{2+} metal ion cofactors. A pronounced effect of various NPs derived from Ni^{2+} and Fe^{2+} metal ions, along with activated carbon, CNTs, Au, Si, Ti, Ag and Pd, on bio-hydrogen production and an increase in yield up to 6.7-fold has been reported (Lai et al. 2018; Baskar et al. 2016; McClelland et al. 2019). H_2 yield can be enhanced up to 200% by adding a mixture of Ni and Fe NPs in anaerobic sludge augmented with cultured microbes (Chen et al. 2014). However, many NPs, such as CuNPs, exert negative effects on bio-hydrogen production that ultimately result in a reduction in yield. Thus, the production of H_2 mainly rely on the properties concentrations of NPs. They mostly augment H_2 production by positively affecting microbial growth, the efficiency of feed degradation and the profile of intermediate metabolites. Additions of NPs change the microbial community composition in the sludge and enrich the H_2 producers during the fermentation process (Patel et al. 2018). They have been found to be effective in pre-treatment of lignocellulosic biomass in H_2 production process through whole-cell or enzyme immobilisation (Bosu and Rajamohan 2022).

14.3.3.3 Bioethanol Production

Bioethanol is largely being used as an alternative of fossil fuels, mainly in the transportation sector, due to its high-octane number (108), high evaporation enthalpy and flammability in a wide temperature range. At present, most of the bioethanol is being produced from edible crops through a bacteria- or yeast-mediated fermentation process. The use of edible crops as feedstock for bioethanol production is considered a threat to food security (Singhvi and Kim 2020; Sankaran et al. 2021). Therefore, non-edible feedstocks like lignocellulosic biomass are proposed as alternative resources for bioethanol production. To overcome the drawbacks of conventional enzyme-mediated decomposition processes of lignocellulosic material, nano-catalysts have been suggested as an alternative approach. In this regard, new kinds of nano-catalysts are being developed and employed to increase the bioethanol yield from lignocellulosic agricultural waste. Three main steps are typically involved in the transformation of lignocellulosic materials into bioethanol: (1) pre-treatment for delignification, which releases the cellulose and hemicellulose; (2) enzymatic hydrolysis that yields glucose, xylose, arabinose, galactose or mannose; and (3) fermentation of reducing sugars. Saeed et al. (2022) showed increased bioethanol generation from potato waste by using laser irradiation and graphitic carbon nitride (g-C_3N_4) nanomaterials. ZnO NPs have also shown their potential implications productivity in bioethanol using rice straw as feedstock (Gupta and Chundawat 2020) (Fig. 14.5).

Similarly, alginate MNPs ensnared in yeast cells were found to improve bioethanol production (Ivanova et al. 2011). The fact that magnetic particles containing fixed yeast cells remained stable at 4 °C in saline conditions for almost a month. In a similar vein, Kim et al. (2014) used methyl-functionalised silica NPs for increased bioethanol yield in syngas fermentation. It is anticipated that NMs will impact the enzymatic activity or mass transfer rate during the biochemical conversion process in fermentation-based liquid fuel production. A cellulase enzyme complex trapped in MNPs with properties like stability at higher pH and temperature, higher storage stability, higher substrate affinity and reusability in reactions could be used to produce ethanol from lignocellulose (Sudheer et al. 2022). Lupoi and Smith (2011) employed an entrapment technique to immobilise the glucosidase enzyme on polymer magnetic nanofibres in order to synthesise cellulosic ethanol. The enzyme glucosidase transforms cellobiose into glucose, which is further fermented into ethanol by yeast or bacterial activity. In order to develop a cost-effective novel for nanotechnologies for bioethanol production, various physical, chemical and biological aspects need to be maximised and taken into account. Still, the nanosystems based

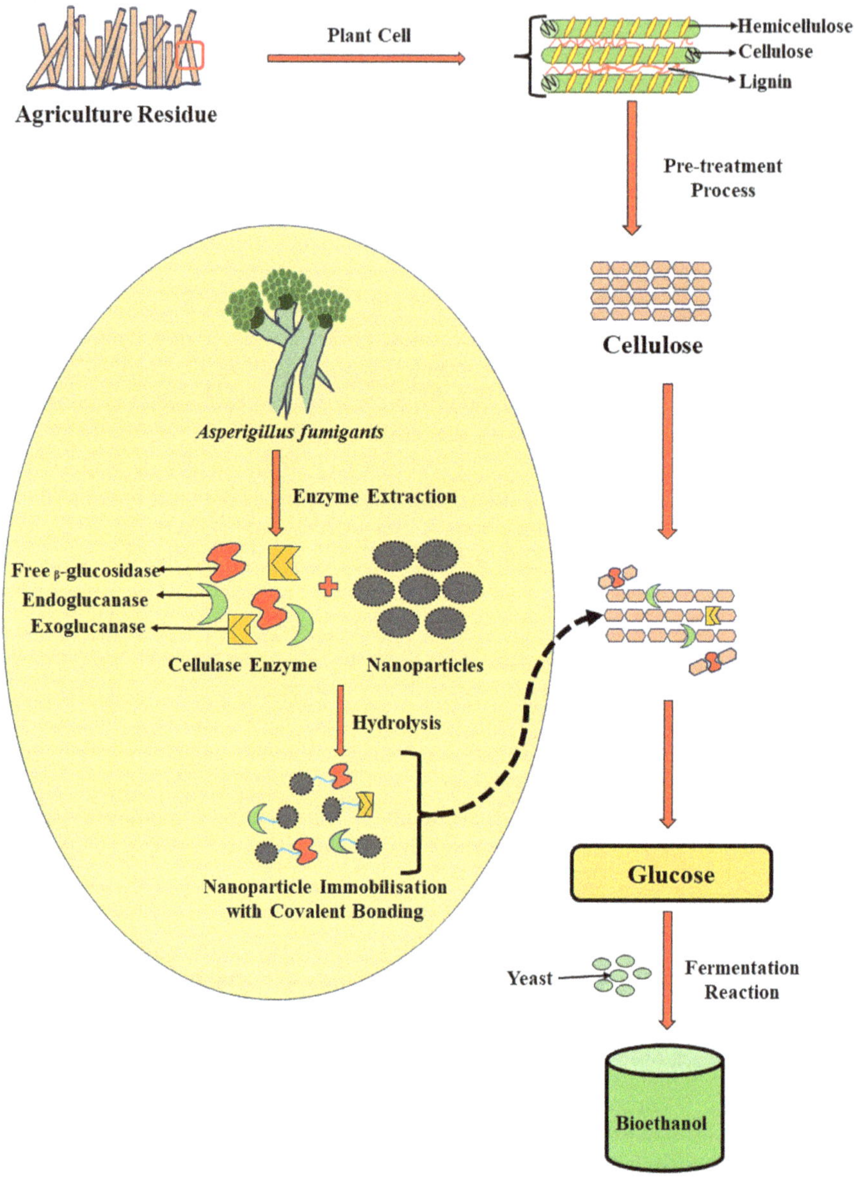

Fig. 14.5 Implications of nanobiocatalyst in bioethanol production using lignocellulosic materials as a feedstock

on lignocellulosic biomass are the early stages of development. Ethanol production from non-food substitutes presents several benefits, such as reduced conversion expenses and a negligible ecological footprint (Rodríguez-Couto 2019). Thus, the biofuel sector needs innovative biocatalysts to meet the growing demand for ethanol.

14.3.3.4 Biodiesel

Edible oils, vegetable oils, animal fats, soaps and greases are utilised as low-cost feedstocks in biodiesel production process. The use of enzymes immobilised on nanomaterials enhances biodiesel production from soybean and other vegetable oils due to improved stability, recovery and reusability of the catalyst. Tahvildari et al. (2015) studied the catalytic properties of magnesium oxide (MgO) and calcium oxide (CaO) nanoparticles for transesterification reactions in the biodiesel formation process using recycled cooking oil as feedstock. Nano MgO and CaO are the basic heterogenic catalysts, and, owing to their weaker basic affinity, Nano MgO was not able to catalyse the transesterification reaction. However, it showed catalytic properties in the presence of nano-CaO, as its surface structure helped increase the basic properties. Thus, the increased transesterification reaction leads to a significant improvement in the yield of biodiesel.

Intense use of vegetable oils in biodiesel production may lead to a price rise that will ultimately affect the kitchen expenses of common people. Consequently, the non-food raw resource known as microalgae is becoming more significant in the modern economy of biofuel production. Quick growth, high biomass production and high energy yield make microalgae useful for producing biofuel (Sekoai et al. 2019). Algae oil extraction without damaging their cells using nanocatalysts has appeared as a novel approach in biodiesel production (Gordon and Seckbach 2012). Lin et al. (2016) developed an approach for fatty acid harvesting from microalgae using mesoporous NPs. These sponges, like NPs, possess the tendency to absorb hydrophobic lipid or fatty acid molecules and thereby draw oils from the cells without killing them. Furthermore, nanocatalysts such as strontium oxides and CaO can be incorporated into the pores of these nanomaterials for the subsequent transesterification of lipids entrapped in the pores. To improve the efficiency of this novel approach, work is being carried out to develop nonporous carbons and many other derivatives as adsorbents for the separation of biofuel (Akubude et al. 2019).

14.4 Conclusions

Nanobionics offer enormous ability to enhance agricultural productivity as well as biofuel production. NMs are not only used as an efficient source of nutrients but also interact with the various metabolic processes in the cells and affect the plant's functional traits, including photosynthesis and defense mechanisms to tolerate various abiotic stresses. Recent developments in the synthesis of functionalised composite NMs have enabled the regulated release of fertilisers and pesticides and the delivery of cargo in various organelles, including chloroplasts. It has created a new hope for nanomaterial-mediated delivery and expression of genes in the nucleus and chloroplast for improvement in crop productivity. However, additional research is mandatory for safe assurance and acceptable utilisation of this genetic engineering tool in agriculture. The NM-plant interaction is very complex, and various factors like type of material, size, duration of exposure and plant species determine the fate of the interaction. For instance, many NMs positively affect photosynthesis, whereas others exert a negative impact and cause growth reductions in plants. It is still an emerging field, and in order to understand the mechanisms of NMs-plant interaction and optimise the delivery process, dose and duration to obtain a positive response from plants in terms of their growth and yield, more intensive efforts are required. The application of nanotechnology in the current biofuel production system has shown its potential to improve catalytic efficiency and reduce biofuel production costs. NMs improve catalytic efficiency and production mainly due to the easy recovery of catalysts, high selectivity, reusability, easy operating procedures and stability. The procedure of preparation and composition of nanocatalysts significantly impact biofuel production using different feedstocks. Furthermore, there is a lot of scope to develop novel, cost-effective nanobiocatalysts with different compositions and improve the catalytic efficiency. Meanwhile, concerns about the negative environmental impacts of the NMs also need to be addressed prior to their commercial exploitation.

Acknowledgments We are all thankful to the Director General, ICFRE, Dehradun, India, and the Director, ICFRE-HFRI, Shimla, India, for providing the necessary facilities. DP, T and HS are thankful to the Compensatory Afforestation Fund Management and Planning Authority (CAMPA), MoEFCC, New Delhi, India, for the fellowship.

Conflict of Interest The authors declare no conflict of interest.

References

Abbasi Khalaki M, Ghorbani A, Dadjou F (2019) Influence of nano-priming on *Festuca ovina* seed germination and early seedling traits under drought stress, in laboratory condition. Ecopersia 7(3):133–139

Abdullah B, Muhammad SAFAS, Shokravi Z, Ismail S, Kassim KA, Mahmood AN, Aziz MMA (2019) Fourth generation biofuel: a review on risks and mitigation strategies. Renew Sust Energ Rev 107:37–50

Abhinandan K, Skori L, Stanic M, Hickerson NM, Jamshed M, Samuel MA (2018) Abiotic stress signaling in wheat–an inclusive overview of hormonal interactions during abiotic stress responses in wheat. Front Plant Sci 9:734

Abraham RE, Verma ML, Barrow CJ, Puri M (2014) Suitability of magnetic nanoparticle immobilised cellulases in enhancing enzymatic saccharification of pretreated hemp biomass. Biotechnol Biofuels 7:1–12

Adisa IO, Pullagurala VLR, Peralta-Videa JR, Dimkpa CO, Elmer WH, Gardea-Torresdey JL, White JC (2019) Recent advances in nano-enabled fertilizers and pesticides: a critical review of mechanisms of action. Environ Sci Nano 6(7):2002–2030

Ahmed M, Douek M (2013) The role of magnetic nanoparticles in the localization and treatment of breast cancer. BioMed Res Int 2013:281230

Akubude VC, Nwaigwe KN, Dintwa E (2019) Production of biodiesel from microalgae via nano-catalyzed transesterification process: a review. Mater Sci Energy Technol 2(2):216–225

Alhalili Z (2023) Metal oxides nanoparticles: general structural description, chemical, physical, and biological synthesis methods, role in pesticides and heavy metal removal through wastewater treatment. Molecules 28(7):3086

Altpeter F, Baisakh N, Beachy R, Bock R, Capell T, Christou P et al (2005) Particle bombardment and the genetic enhancement of crops: myths and realities. Mol Breed 15:305–327

An C, Sun C, Li N, Huang B, Jiang J, Shen Y, Wang Y (2022) Nanomaterials and nanotechnology for the delivery of agrochemicals: strategies towards sustainable agriculture. J Nanobiotechnol 20(1):1–19

Antunes FAF, Gaikwad S, Ingle AP, Pandit R, dos Santos JC, Rai M, da Silva SS (2017) Bioenergy and biofuels: nanotechnological solutions for sustainable production. In: Nanotechnology for bioenergy and biofuel production. Springer, Cham, pp 3–18

Aoki K, Saito N (2020) Biocompatibility and carcinogenicity of carbon nanotubes as biomaterials. Nano 10(2):264

Ariaeenejad S, Jokar F, Hadian P, Ma'mani L, Gharaghani S, Fereidoonnezhad M, Salekdeh GH (2020) An efficient nano-biocatalyst for lignocellulosic biomass hydrolysis: xylanase immobilization on organically modified biogenic mesoporous silica nanoparticles. Int J Biol Macromol 164:3462–3473

Arora A, Nandal P, Singh J, Verma ML (2020) Nanobiotechnological advancements in lignocellulosic biomass pretreatment. Mater Sci Energy Technol 3:308–318

Arya I, Poona A, Dikshit PK, Pandit S, Kumar J, Singh HN et al (2021) Current trends and future prospects of nanotechnology in biofuel production. Catalysts 11(11):1308

Atadashi IM, Aroua MK, Aziz AA, Sulaiman NMN (2012) Production of biodiesel using high free fatty acid feedstocks. Renew Sust Energ Rev 16(5):3275–3285

Baskar G, Kumar RN, Melvin XH, Aiswarya R, Soumya S (2016) Sesbania aculeate biomass hydrolysis using magnetic nanobiocomposite of cellulase for bioethanol production. Renew Energy 98:23–28

Bastakoti BP, Kuila D, Salomon C, Konarova M, Eguchi M, Na J, Yamauchi Y (2021) Metal-incorporated mesoporous oxides: synthesis and applications. J Hazard Mater 401:123348

Baumann AE, Burns DA, Liu B, Thoi VS (2019) Metal-organic framework functionalization and design strategies for advanced electrochemical energy storage devices. Commun Chem 2(1):86

Bennion EP, Ginosar DM, Moses J, Agblevor F, Quinn JC (2015) Lifecycle assessment of microalgae to biofuel: comparison of thermochemical processing pathways. Appl Energy 154:1062–1071

Bhanja P, Bhaumik A (2016) Porous nanomaterials as green catalyst for the conversion of biomass to bioenergy. Fuel 185:432–441

Bhardwaj AK, Arya G, Kumar R, Hamed L, Pirasteh-Anosheh H, Jasrotia P, Singh GP (2022) Switching to nanonutrients for sustaining agroecosystems and environment: the challenges and benefits in moving up from ionic to particle feeding. J Nanobiotechnol 20(1):19

Bhargavi G, Nageswara Rao P, Renganathan S (2018, March) Review on the extraction methods of crude oil from all generation biofuels in last few decades. In: IOP conference series: materials science and engineering, vol 330. IOP Publishing, London, p 012024

Bhaskar R, Pandey SP, Kumar U, Kim H, Jayakodi SK, Gupta MK, Han SS (2024) Nanobionics for sustainable crop production: recent development to regulate plant growth and protection strategies from pests. OpenNano 15:100198

Bianco A, Kostarelos K, Prato M (2011) Making carbon nanotubes biocompatible and biodegradable. Chem Commun 47(37):10182–10188

Bidir MG, Millerjothi NK, Adaramola MS, Hagos FY (2021) The role of nanoparticles on biofuel production and as an additive in ternary blend fuelled diesel engine: a review. Energy Rep 7:3614–3627

Bin Rashid A (2023) Utilization of nanotechnology and nanomaterials in biodiesel production and property enhancement. J Nanomater 2023:1–14

BiofuelslEnergy (2021). https://ec.europa.eu/energy/topics/renewable-energy/biofuels/overview_en. Accessed 7 Jan 2021

Blankenship RE, Tiede DM, Barber J, Brudvig GW, Fleming G, Ghirardi M, Sayre RT (2011) Comparing photosynthetic and photovoltaic efficiencies and recognizing the potential for improvement. Science 332(6031):805–809

Borgatta J, Ma C, Hudson-Smith N, Elmer W, Plaza Perez CD, De La Torre-Roche R, Hamers RJ (2018) Copper based nanomaterials suppress root fungal disease in watermelon (*Citrullus lanatus*): role of particle morphology, composition and dissolution behavior. ACS Sustain Chem Eng 6(11):14847–14856

Bosu S, Rajamohan N (2022) Nanotechnology approach for enhancement in biohydrogen production-review on applications of nanocatalyst and life cycle assessment. Fuel 323:124351

Bratovcic A, Hikal WM, Said-Al Ahl HA, Tkachenko KG, Baeshen RS, Sabra AS, Sany H (2021) Nanopesticides and nanofertilizers and agricultural development: scopes, advances and applications. Open J Ecol 11(4):301–316

Bray EA (2000) Responses to abiotic stresses. Biochemistry and molecular biology of plants. American Society of Plant Phsiologists, Rockville, pp 1158–1203

Budhi S, Mukarakate C, Iisa K, Pylypenko S, Ciesielski PN, Yung MM et al (2015) Molybdenum incorporated mesoporous silica catalyst for production of biofuels and value-added chemicals via catalytic fast pyrolysis. Green Chem 17(5):3035–3046

Camara MC, Campos EVR, Monteiro RA, do Espirito Santo Pereira A, de Freitas Proença PL, Fraceto LF (2019) Development of stimuli-responsive nano-based pesticides: emerging opportunities for agriculture. J Nanobiotechnol 17(1):1–19

Carvalho FP (2017) Pesticides, environment, and food safety. Food Energy Secur 6(2):48–60

Chai WS, Bao Y, Jin P, Tang G, Zhou L (2021) A review on ammonia, ammonia-hydrogen and ammonia-methane fuels. Renew Sust Energ Rev 147:111254

Chandra D, Mukherjee N, Mondal A, Bhaumik A (2008) Design and synthesis of nanostructured porous SnO_2 with high surface areas and their optical and dielectric properties. J Phys Chem C 112(23):8668–8674

Chang FP, Kuang LY, Huang CA, Jane WN, Hung Y, Yue-ie CH, Mou CY (2013) A simple plant gene delivery system using mesoporous silica nanoparticles as carriers. J Mater Chem B 1(39):5279–5287

Chen J, Wang S, Huang J, Chen L, Ma L, Huang X (2013) Conversion of cellulose and cellobiose into sorbitol catalyzed by ruthenium supported on a polyoxometalate/metal–organic framework hybrid. ChemSusChem 6(8):1545–1555

Chen J, Li K, Chen L, Liu R, Huang X, Ye D (2014) Conversion of fructose into 5-hydroxymethylfurfural catalyzed by recyclable sulfonic acid-functionalized metal–organic frameworks. Green Chem 16(5):2490–2499

Cherian E, Dharmendirakumar M, Baskar G (2015) Immobilization of cellulase onto MnO2 nanoparticles for bioethanol production by enhanced hydrolysis of agricultural waste. Chin J Catal 36(8):1223–1229

Chew KW, Chia SR, Show PL, Ling TC, Arya SS, Chang JS (2018) Food waste compost as an organic nutrient source for the cultivation of Chlorella vulgaris. Bioresour Technol 267:356–362

Chi Y, Chen C, Zhang G, Ye Z, Su X, Ren X, Wu Z (2021) Fabrication of magnetic-responsive controlled-release herbicide by a palygorskite-based nanocomposite. Colloids Surf B: Biointerfaces 208:112115

Correa DF, Beyer HL, Fargione JE, Hill JD, Possingham HP, Thomas-Hall SR, Schenk PM (2019) Towards the implementation of sustainable biofuel production systems. Renew Sust Energ Rev 107:250–263

Cossins D (2014) Next generation: nanoparticles augment plant functions. The incorporation of synthetic nanoparticles into plants can enhance photosynthesis and transform leaves into biochemical sensors. The scientist, news & opinion

Dasan YK, Lam MK, Yusup S, Lim JW, Lee KT (2019) Life cycle evaluation of microalgae biofuels production: effect of cultivation system on energy, carbon emission and cost balance analysis. Sci Total Environ 688:112–128

Dasgupta N, Ranjan S, Mundekkad D, Ramalingam C, Shanker R, Kumar A (2015) Nanotechnology in agro-food: from field to plate. Food Res Int 69:381–400

De S, Dutta S, Patra AK, Rana BS, Sinha AK, Saha B, Bhaumik A (2012) Biopolymer templated porous TiO_2: An efficient catalyst for the conversion of unutilized sugars derived from hemicellulose. Appl Catal A Gen 435:197–203

Demirer GS, Zhang H, Matos JL, Goh NS, Cunningham FJ, Sung Y, Landry MP (2019) High aspect ratio nanomaterials enable delivery of functional genetic material without DNA integration in mature plants. Nat Nanotechnol 14(5):456–464

Ding WK, Shah NP (2009) Effect of various encapsulating materials on the stability of probiotic bacteria. J Food Sci 74(2):M100–M107

Dufaud V, Davis ME (2003) Design of heterogeneous catalysts via multiple active site positioning in organic-inorganic hybrid materials. J Am Chem Soc 125(31):9403–9413

Dufil G, Bernacka-Wojcik I, Armada-Moreira A, Stavrinidou E (2021) Plant bioelectronics and biohybrids: the growing contribution of organic electronic and carbon-based materials. Chem Rev 122(4):4847–4883

Duro JA, Lauk C, Kastner T, Erb KH, Haberl H (2020) Global inequalities in food consumption, cropland demand and land-use efficiency: a decomposition analysis. Glob Environ Chang 64:102124

Dutta S, De S, Patra AK, Sasidharan M, Bhaumik A, Saha B (2011) Microwave assisted rapid conversion of carbohydrates into 5-hydroxymethylfurfural catalyzed by mesoporous TiO_2 nanoparticles. Appl Catal A Gen 409:133–139

El-Badri AM, Batool M, Mohamed IA, Khatab A, Sherif A, Wang Z, Zhou G (2021) Modulation of salinity impact on early seedling stage via nano-priming application of zinc oxide on rapeseed (Brassica napus L.). Plant Physiol Biochem 166:376–392

Elkhatib E, Attia MG, Mahdy AM, Mostafa RA (2019) Priming with mango peels nanoparticles enhances seed germination of maize (*Zea mays* L.) under salt stress. Alex Sci Exch J 40:767–780

Erdiwansyah MR, Sani MSM, Sudhakar K, Kadarohman A, Sardjono RE (2019) An overview of higher alcohol and biodiesel as alternative fuels in engines. Energy Rep 5:467–479

European Commission (2021) Sustainable transport|mobility and transport. https://ec.europa.eu/transport/themes/sustainable_en. Accessed 11 Feb 2021

Faisal S, Yusuf Hafeez F, Zafar Y, Majeed S, Leng X, Zhao S et al (2018) A review on nanoparticles as boon for biogas producers—nano fuels and biosensing monitoring. Appl Sci 9(1):59

Fan Y, Su F, Li K, Ke C, Yan Y (2017) Carbon nanotube filled with magnetic iron oxide and modified with polyamidoamine dendrimers for immobilizing lipase toward application in biodiesel production. Sci Rep 7(1):45643

Farahi SMM, Yazdi MET, Einafshar E, Akhondi M, Ebadi M, Azimipour S, Iranbakhsh A (2023) The effects of titanium dioxide (TiO_2) nanoparticles on physiological, biochemical, and antioxidant properties of vitex plant (Vitex agnus-Castus L). Heliyon 9(11):e22144

Feng Y, Zhang Y, Quan X, Chen S (2014) Enhanced anaerobic digestion of waste activated sludge digestion by the addition of zero valent iron. Water Res 52:242–250

Feng R, Li Q, Zaidi AA, Peng H, Shi Y (2021) Effect of autoclave pretreatment on biogas production through anaerobic digestion of green algae. Period Polytechn Chem Eng 65(4):483–492

Fincheira P, Hoffmann N, Tortella G, Ruiz A, Cornejo P, Diez MC, Rubilar O (2023) Eco-efficient systems based on nanocarriers for the controlled release of fertilizers and pesticides: toward smart agriculture. Nano 13(13):1978

Fiol DF, Terrile MC, Frik J, Mesas FA, Álvarez VA, Casalongué CA (2021) Nanotechnology in plants: recent advances and challenges. J Chem Technol Biotechnol 96(8):2095–2108

Garcia-Martinez J (2010) Foreword by: Ernest J. Moniz. In: Nanotechnology for the energy challenge. Wiley, New York

Garg D, Sridhar K, Stephen Inbaraj B, Chawla P, Tripathi M, Sharma M (2023) Nano-biofertilizer formulations for agriculture: A systematic review on recent advances and prospective applications. Bioengineering 10(9):1010

Gelvin SB (2003) Agrobacterium-mediated plant transformation: the biology behind the "gene-jockeying" tool. Microbiol Mol Biol Rev 67(1):16–37
Genna DT, Wong-Foy AG, Matzger AJ, Sanford MS (2013) Heterogenization of homogeneous catalysts in metal–organic frameworks via cation exchange. J Am Chem Soc 135(29):10586–10589
Ghorbani A, Bazooyar B, Shariati A, Jokar SM, Ajami H, Naderi A (2011) A comparative study of combustion performance and emission of biodiesel blends and diesel in an experimental boiler. Appl Energy 88(12):4725–4732
Giraldo JP, Landry MP, Faltermeier SM, McNicholas TP, Iverson NM, Boghossian AA, Strano MS (2014) Plant nanobionics approach to augment photosynthesis and biochemical sensing. Nat Mater 13(4):400–408
Giraldo JP, Landry MP, Kwak SY, Jain RM, Wong MH, Iverson NM, Strano MS (2015) A ratiometric sensor using single chirality near-infrared fluorescent carbon nanotubes: application to in vivo monitoring. Small 11(32):3973–3984
Giraldo JP, Wu H, Newkirk GM, Kruss S (2019) Nanobiotechnology approaches for engineering smart plant sensors. Nat Nanotechnol 14(6):541–553
Goh WJ, Makam VS, Hu J, Kang L, Zheng M, Yoong SL, Pastorin G (2012) Iron oxide filled magnetic carbon nanotube–enzyme conjugates for recycling of amyloglucosidase: toward useful applications in biofuel production process. Langmuir 28(49):16864–16873
Gopinath S, Kumar PSM, Arafath KY, Thiruvengadaravi KV, Sivanesan S, Baskaralingam P (2017) Efficient mesoporous SO42−/Zr-KIT-6 solid acid catalyst for green diesel production from esterification of oleic acid. Fuel 203:488–500
Gordon R, Seckbach J (eds) (2012) The science of algal fuels: phycology, geology, biophotonics, genomics and nanotechnology, vol 25. Springer, Cham
Großmann D, Dreier A, Lehmann CW, Grünert W (2015) Encapsulation of copper and zinc oxide nanoparticles inside small diameter carbon nanotubes. Microporous Mesoporous Mater 202:189–197
Guo F, Fang Z, Xu CC, Smith RL Jr (2012) Solid acid mediated hydrolysis of biomass for producing biofuels. Prog Energy Combust Sci 38(5):672–690
Gupta K, Chundawat TS (2020) Zinc oxide nanoparticles synthesized using fusarium oxysporum to enhance bioethanol production from rice-straw. Biomass Bioenergy 143:105840
Hao Y, Wang Y, Ma C, White JC, Zhao Z, Duan C, Xing B (2019) Carbon nanomaterials induce residue degradation and increase methane production from livestock manure in an anaerobic digestion system. J Clean Prod 240:118257
Haris M, Hussain T, Mohamed HI, Khan A, Ansari MS, Tauseef A, Akhtar N (2023) Nanotechnology–a new frontier of nano-farming in agricultural and food production and its development. Sci Total Environ 857:159639
Hassaan MA, El Nemr A (2020) Pesticides pollution: classifications, human health impact, extraction and treatment techniques. Egypt J Aquat Res 46(3):207–220
Hayashimoto A, Li Z, Murai N (1990) A polyethylene glycol-mediated protoplast transformation system for production of fertile transgenic rice plants. Plant Physiol 93(3):857–863
Holm-Nielsen JB, Al Seadi T, Oleskowicz-Popiel P (2009) The future of anaerobic digestion and biogas utilization. Bioresour Technol 100(22):5478–5484
Hong F, Zhou J, Liu C, Yang F, Wu C, Zheng L, Yang P (2005) Effect of nano-TiO_2 on photochemical reaction of chloroplasts of spinach. Biol Trace Elem Res 105:269–279
Hu J, Xianyu Y (2021) When nano meets plants: A review on the interplay between nanoparticles and plants. Nano Today 38:101143
Hu P, An J, Faulkner MM, Wu H, Li Z, Tian X, Giraldo JP (2020) Nanoparticle charge and size control foliar delivery efficiency to plant cells and organelles. ACS Nano 14(7):7970–7986
Husen A, Iqbal M, Aref IM (2014) Growth, water status, and leaf characteristics of Brassica carinata under drought and rehydration conditions. Rev Bras Bot 37:217–227
Husen A, Iqbal M, Aref IM (2016) IAA-induced alteration in growth and photosynthesis of pea (*Pisum sativum* L.) plants grown under salt stress. J Environ Biol 37(3):421
Ivanova V, Petrova P, and Hristov J (2011) Application in the ethanol fermentation of immobilized yeast cells in matrix of alginate/magnetic nanoparticles, on chitosan-magnetite microparticles and cellulose-coated magnetic nanoparticles. arXiv preprint arXiv:1105.0619

Jagtap S, Yenkie MKN, Labhsetwar N, Rayalu S (2011) Defluoridation of drinking water using chitosan based mesoporous alumina. Microporous Mesoporous Mater 142(2–3):454–463

Jampílek J, Kráová K (2017) Nanopesticides: preparation, targeting, and controlled release. In: New pesticides and soil sensors. Academic, Boca Raton, pp 81–127

Jan S, Ahmad P (2019) Ecometabolomics: metabolic fluxes versus environmental stoichiometry. Academic, Boca Raton

Jaroenkhasemmeesuk C, Tippayawong N (2015) Technical and economic analysis of a biomass pyrolysis plant. Energy Procedia 79:950–955

Jin S, Daniell H (2015) The engineered chloroplast genome just got smarter. Trends Plant Sci 20(10):622–640

Jin M, Park JN, Shon JK, Kim JH, Li Z, Park YK, Kim JM (2012a) Low temperature CO oxidation over pd catalysts supported on highly ordered mesoporous metal oxides. Catal Today 185(1):183–190

Jin Z, Xiao M, Bao Z, Wang P, Wang J (2012b) A general approach to mesoporous metal oxide microspheres loaded with noble metal nanoparticles. Angew Chem Int Ed 51(26):6406–6410

Jorge EY, Lima TDM, Lima CG, Marchini L, Castelblanco WN, Rivera DG et al (2017) Metal-exchanged magnetic β-zeolites: valorization of lignocellulosic biomass-derived compounds to platform chemicals. Green Chem 19(16):3856–3868

Kale RD, Kane PB (2017) Colour removal using nanoparticles. Text Cloth Sustain 2(1):1–7

Khan SZ, Zaidi AA, Naseer MN, AlMohamadi H (2022) Nanomaterials for biogas augmentation towards renewable and sustainable energy production: a critical review. Front Bioeng Biotechnol 10:868454

Kim YK, Park SE, Lee H, Yun JY (2014) Enhancement of bioethanol production in syngas fermentation with clostridium ljungdahlii using nanoparticles. Bioresour Technol 159:446–450

Kumar R, Singh RK, Panda A, Singh SK (2021) Nano urea: an efficient tool for precision agriculture and sustainability. Vigyan Varta 2(9):72–74

Kwak SY, Giraldo JP, Wong MH, Koman VB, Lew TTS, Ell J, Strano MS (2017) A nanobionic light-emitting plant. Nano Lett 17(12):7951–7961

Kwak SY, Lew TTS, Sweeney CJ, Koman VB, Wong MH, Bohmert-Tatarev K, Strano MS (2019) Chloroplast-selective gene delivery and expression in planta using chitosan-complexed single-walled carbon nanotube carriers. Nat Nanotechnol 14(5):447–455

Lai HK, Chou YZ, Lee MH, Lin KYA (2018) Coordination polymer-derived cobalt nanoparticle-embedded carbon nanocomposite as a magnetic multi-functional catalyst for energy generation and biomass conversion. Chem Eng J 332:717–726

Lee SM, Jin LH, Kim JH, Han SO, Na HB, Hyeon T et al (2010) β-Glucosidase coating on polymer nanofibers for improved cellulosic ethanol production. Bioprocess Biosyst Eng 33:141–147

Lee YC, Dutta S, Wu KCW (2014) Integrated, cascading enzyme–/chemocatalytic cellulose conversion using catalysts based on mesoporous silica nanoparticles. ChemSusChem 7(12):3241–3246

Lei Z, Mingyu S, Chao L, Liang C, Hao H, Xiao W, Fashui H (2007) Effects of nanoanatase TiO_2 on photosynthesis of spinach chloroplasts under different light illumination. Biol Trace Elem Res 119:68–76

Leus K, Muylaert I, Van Speybroeck V, Marin GB, Van Der Voort P (2010) A coordinative saturated vanadium containing metal organic framework that shows a remarkable catalytic activity. In: Studies in surface science and catalysis, vol 175. Elsevier, Amsterdam, pp 329–332

Lew TTS, Wong MH, Kwak SY, Sinclair R, Koman VB, Strano MS (2018) Rational design principles for the transport and subcellular distribution of nanomaterials into plant protoplasts. Small 14(44):1802086

Li X, Yun S, Zhang C, Fang W, Huang X, Du T (2018) Application of nano-scale transition metal carbides as accelerants in anaerobic digestion. Int J Hydrog Energy 43(3):1926–1936

Li X, Sun H, Mao X, Lao Y, Chen F (2020) Enhanced photosynthesis of carotenoids in microalgae driven by light-harvesting gold nanoparticles. ACS Sustain Chem Eng 8(20):7600–7608

Liang W, Yu A, Wang G, Zheng F, Hu P, Jia J, Xu H (2018) A novel water-based chitosan-La pesticide nanocarrier enhancing defense responses in rice (*Oryza sativa* L) growth. Carbohydr Polym 199:437–444
Liaquat AM, Masjuki HH, Kalam MA, Varman M, Hazrat MA, Shahabuddin M, Mofijur M (2012) Application of blend fuels in a diesel engine. Energy Procedia 14:1124–1133
Lin MT, Occhialini A, Andralojc PJ, Parry MA, Hanson MR (2014) A faster Rubisco with potential to increase photosynthesis in crops. Nature 513(7519):547–550
Lin R, Cheng J, Ding L, Song W, Liu M, Zhou J, Cen K (2016) Enhanced dark hydrogen fermentation by addition of ferric oxide nanoparticles using Enterobacter aerogenes. Bioresour Technol 207:213–219
Liu J, Li H, Liu YC, Lu YM, He J, Liu XF, Yang S (2015) Catalytic conversion of glucose to 5-hydroxymethylfurfural over nano-sized mesoporous Al_2O_3–B_2O_3 solid acids. Catal Commun 62:19–23
Lohar D, Pathak MK, Srivastava RP (2023) Nano urea's role in agricultural innovation: a comprehensive review of efficacy and environmental impact. Int J Res Pub Rev 4(8):3291–3298
Lu Q, Zhang ZB, Zhang CJ, Su SH, Li WY, Dong CQ (2012) Overview of chemical characterization of biomass fast pyrolysis oils. Appl Mech Mater 130:422–425
Lupoi JS, Smith EA (2011) Evaluation of nanoparticle-immobilized cellulase for improved ethanol yield in simultaneous saccharification and fermentation reactions. Biotechnol Bioeng 108(12):2835–2843
Mahmood I, Imadi SR, Shazadi K, Gul A, Hakeem KR (2016) Effects of pesticides on environment. In: Plant, soil and microbes: volume 1: implications in crop science. Springer, Cham, pp 253–269
Makino A (2011) Photosynthesis, grain yield, and nitrogen utilization in rice and wheat. Plant Physiol 155:125–129
Manikandan A, Subramanian K (2016) Evaluation of zeolite-based nitrogen nano-fertilizers on maize growth, yield and quality on inceptisols and alfisols. Int J Plant Soil Sci 9(4):1–9
Martin-Ortigosa S, Valenstein JS, Sun W, Moeller L, Fang N, Trewyn BG et al (2012) Parameters affecting the efficient delivery of mesoporous silica nanoparticle materials and gold nanorods into plant tissues by the biolistic method. Small 8(3):413–422
Martins NC, Avellan A, Rodrigues S, Salvador D, Rodrigues SM, Trindade T (2020) Composites of biopolymers and ZnO NPs for controlled release of zinc in agricultural soils and timed delivery for maize. ACS Appl Nano Mat 3(3):2134–2148
Masson-Delmotte V, Pörtner HO, Skea J, Zhai P, Roberts D, Shukla PR, Buendía EC (2019) Climate change and land: an IPCC special report on climate change, desertification, land degradation, sustainable land management, food security, and greenhouse gas fluxes in terrestrial ecosystems. Intergovernmental Panel on Climate Change, Geneva
McClelland DJ, Galebach PH, Motagamwala AH, Wittrig AM, Karlen SD, Buchanan JS, Huber GW (2019) Supercritical methanol depolymerization and hydrodeoxygenation of lignin and biomass over reduced copper porous metal oxides. Green Chem 21(11):2988–3005
Mittal D, Kaur G, Singh P, Yadav K, Ali SA (2020) Nanoparticle-based sustainable agriculture and food science: recent advances and future outlook. Front Nanotechnol 2:579954
Mohanasundar C, Ramamoorthy K, Latha KR, Santhy P, Chandrasekhar CN, Lakshmanan A (2021) Effect of bio-seed priming and nano zinc oxide foliar application on quality and productivity of finger millet + green gram intercropping system. J Appl Nat Sci 13(4):1552–1562
Mushtaq A, Jamil N, Rizwan S, Mandokhel F, Riaz M, Hornyak GL, Naeem Shahwani M (2018, September) Engineered silica nanoparticles and silica nanoparticles containing controlled release fertilizer for drought and saline areas. In: IOP conference series: materials science and engineering, vol 414. IOP Publishing, London, p 012029
Nadagouda MN, Speth TF, Varma RS (2011) Microwave-assisted green synthesis of silver nanostructures. Acc Chem Res 44(7):469–478
Neme K, Nafady A, Uddin S, Tola YB (2021) Application of nanotechnology in agriculture, postharvest loss reduction and food processing: food security implication and challenges. Heliyon 7(12):e08539

Nguyen D, Visvanathan C, Jacob P, Jegatheesan V (2015) Effects of nano cerium (IV) oxide and zinc oxide particles on biogas production. Int Biodeterior Biodegradation 102:165–171
Noji T, Kamidaki C, Kawakami K, Shen JR, Kajino T, Fukushima Y, Itoh S (2011) Photosynthetic oxygen evolution in mesoporous silica material: adsorption of photosystem II reaction center complex into 23nm nanopores in SBA. Langmuir 27(2):705–713
Nyberg L, Turco RF, Nies L (2008) Assessing the impact of nanomaterials on anaerobic microbial communities. Environ Sci Technol 42(6):1938–1943
Ort DR, Merchant SS, Alric J, Barkan A, Blankenship RE, Bock R, Croce R, Hanson MR, Hibberd JM, Long SP et al (2015) Redesigning photosynthesis to sustainably meet global food and bioenergy demand. Proc Natl Acad Sci USA 112:8529–8536
Parida KM, Sahu N, Mohapatra P, Scurrell MS (2010) Low temperature CO oxidation over gold supported mesoporous Fe–TiO2. J Mol Catal A Chem 319(1-2):92–97
Patel SK, Lee JK, Kalia VC (2018) Nanoparticles in biological hydrogen production: an overview. Indian J Microbiol 58:8–18
Pelegrino MT, Pieretti JC, Lange CN, Kohatsu MY, Freire BM, Batista BL, Seabra AB (2021) Foliar spray application of CuO nanoparticles (NPs) and S-nitrosoglutathione enhances productivity, physiological and biochemical parameters of lettuce plants. J Chem Technol Biotechnol 96(8):2185–2196
Pereira A (2016) Plant abiotic stress challenges from the changing environment. Front Plant Sci 7:1123
Poorakbar E, Shafiee A, Saboury AA, Rad BL, Khoshnevisan K, Ma'mani L, Hosseini M (2018) Synthesis of magnetic gold mesoporous silica nanoparticles core shell for cellulase enzyme immobilization: improvement of enzymatic activity and thermal stability. Process Biochem 71:92–100
Puri M, Barrow CJ, Verma ML (2013) Enzyme immobilization on nanomaterials for biofuel production. Trends Biotechnol 31(4):215–216
Qiu H, Huang J, Yang J, Rozelle S, Zhang Y, Zhang Y, Zhang Y (2010) Bioethanol development in China and the potential impacts on its agricultural economy. Appl Energy 87(1):76–83
Rahman KM, Melville L, Huq SI, Khoda SK (2016) Understanding bioenergy production and optimisation at the nanoscale–a review. J Exp Nanosci 11(10):762–775
Rai M, Ingle AP, Pandit R, Paralikar P, Biswas JK, da Silva SS (2019) Emerging role of nanobiocatalysts in hydrolysis of lignocellulosic biomass leading to sustainable bioethanol production. Catal Rev 61(1):1–26
Rai AK, Al Makishah NH, Wen Z, Gupta G, Pandit S, Prasad R (2022) Recent developments in lignocellulosic biofuels, a renewable source of bioenergy. Fermentation 8(4):161
Rai-Kalal P, Tomar RS, Jajoo A (2021) Seed nanopriming by silicon oxide improves drought stress alleviation potential in wheat plants. Funct Plant Biol 48(9):905–915
Ranjan A, Rajput VD, Kumari A, Mandzhieva SS, Sushkova S, Prazdnova EV et al (2022) Nanobionics in crop production: an emerging approach to modulate plant functionalities. Plan Theory 11(5):692
Rathnayaka RMNN, Mahendran S, Iqbal YB, Rifnas LM (2018) Influence of urea and nano-nitrogen fertilizers on the growth and yield of rice (Oryza sativa L.) cultivar Bg 250. Int J of Res Pub 5:1
Ren Y, Rivera JG, He L, Kulkarni H, Lee DK, Messersmith PB (2011). Facile, high efficiency immobilization of lipase enzyme on magnetic iron oxide nanoparticles via a biomimetic coating. BMC biotechnology 11:1–8.
Rodionova MV, Poudyal RS, Tiwari I, Voloshin RA, Zharmukhamedov SK, Nam HG, Allakhverdiev SI (2017) Biofuel production: challenges and opportunities. Int J Hydrog Energy 42(12):8450–8461
Rodríguez-Couto S (2019) Green nanotechnology for biofuel production. In: Sustainable approaches for biofuels production technologies: from current status to practical implementation. Springer, Cham, pp 73–82
Sabir S, Arshad M, Chaudhari SK (2014) Zinc oxide nanoparticles for revolutionizing agriculture: synthesis and applications. Sci World J 2014:925494

Saeed S, Samer M, Mohamed MS, Abdelsalam E, Mohamed YM, Abdel-Hafez SH, Attia YA (2022) Implementation of graphitic carbon nitride nanomaterials and laser irradiation for increasing bioethanol production from potato processing wastes. Environ Sci Pollut Res 29(23):34887–34897

Sangeetha J, Hospet R, Thangadurai D, Adetunji CO, Islam S, Pujari N, Al-Tawaha ARMS (2021) Nanopesticides, nanoherbicides, and nanofertilizers: the greener aspects of agrochemical synthesis using nanotools and nanoprocesses toward sustainable agriculture. In: Handbook of nanomaterials and nanocomposites for energy and environmental applications. Springer, Cham, pp 1663–1677

Sankaran R, Markandan K, Khoo KS, Cheng CK, Ashokkumar V, Deepanraj B, Show PL (2021) The expansion of lignocellulose biomass conversion into bioenergy via nanobiotechnology. Front Nanotechnol 3:96

Santana I, Wu H, Hu P, Giraldo JP (2020) Targeted delivery of nanomaterials with chemical cargoes in plants enabled by a biorecognition motif. Nat Commun 11(1):2045

Seadi TA, Rutz D, Prassl H, Köttner M, Finsterwalder T, Volk S, Janssen R (2008) Biogas handbook. Igarss, Brussels

Sekoai PT, Ouma CNM, Du Preez SP, Modisha P, Engelbrecht N, Bessarabov DG, Ghimire A (2019) Application of nanoparticles in biofuels: an overview. Fuel 237:380–397

Shah SH, Raja IA, Rizwan M, Rashid N, Mahmood Q, Shah FA, Pervez A (2018) Potential of microalgal biodiesel production and its sustainability perspectives in Pakistan. Renew Sust Energ Rev 81:76–92

Shahabuddin M, Masjuki HH, Kalam MA, Mofijur M, Hazrat MA, Liaquat AM (2012) Effect of additive on performance of CI engine fuelled with bio diesel. Energy Procedia 14:1624–1629

Shan Y, Cao L, Xu C, Zhao P, Cao C, Li F et al (2019) Sulfonate-functionalized mesoporous silica nanoparticles as carriers for controlled herbicide diquat dibromide release through electrostatic interaction. Int J Mol Sci 20(6):1330

Sharma SR, Kar D (2019) An insight into plant nanobionics and its applications. In: Plant nanobionics: volume 1, advances in the understanding of nanomaterials research and applications. Springer, Cham, pp 65–82

Shen Y, Zhou J, Du C (2019) Development of a polyacrylate/silica nanoparticle hybrid emulsion for delaying nutrient release in coated controlled-release urea. Coatings 9(2):88

Sherwood J (2020) The significance of biomass in a circular economy. Bioresour Technol 300:122755

Shoyhet H, Pavlopoulos NG, Amirav L, Adir N (2021) Functionalized gold-nanoparticles enhance photosystem II driven photocurrent in a hybrid nano-bio-photoelectrochemical cell. J Mater Chem A 9(32):17231–17241

Siddiqui MH, Mohammad F, Khan MMA, Al-Whaibi MH (2012) Cumulative effect of nitrogen and Sulphur on Brassica juncea L. genotypes under NaCl stress. Protoplasma 249:139–153

Siddiqui MH, Al-Whaibi MH, Faisal M, Al Sahli AA (2014) Nano-silicon dioxide mitigates the adverse effects of salt stress on Cucurbita pepo L. Environ Toxicol Chem 33(11):2429–2437

Singh MR, Bell AT (2016) Design of an artificial photosynthetic system for production of alcohols in high concentration from CO_2. Energy Environ Sci 9(1):193–199

Singh N, Dhanya BS, Verma ML (2020) Nano-immobilized biocatalysts and their potential biotechnological applications in bioenergy production. Mater Sci Energy Technol 3:808–824

Singhvi M, Kim BS (2020) Current developments in lignocellulosic biomass conversion into biofuels using nanobiotechology approach. Energies 13(20):5300

Somorjai GA, Frei H, Park JY (2009) Advancing the frontiers in nanocatalysis, biointerfaces, and renewable energy conversion by innovations of surface techniques. J Am Chem Soc 131(46):16589–16605

Srinivasan V, Adhikari S, Chattanathan SA, Tu M, Park S (2014) Catalytic pyrolysis of raw and thermally treated cellulose using different acidic zeolites. Bioenergy Res 7:867–875

Srivastava N, Srivastava M, Malhotra BD, Gupta VK, Ramteke PW, Silva RN et al (2019) Nanoengineered cellulosic biohydrogen production via dark fermentation: a novel approach. Biotechnol Adv 37(6):107384

Stocker M (2008) Biofuels and biomass-to-liquid fuels in the biorefinery: catalytic conversion of lignocellulosic biomass using porous materials. Angew Chem Int Ed 47(48):9200–9211
Su T, Li W, Wang P, Ma C (2019) Dynamics of peroxisome homeostasis and its role in stress response and signaling in plants. Front Plant Sci 10:705
Sudheer S, Bai RG, Muthoosamy K, Tuvikene R, Gupta VK, Manickam S (2022) Biosustainable production of nanoparticles via mycogenesis for biotechnological applications: a critical review. Environ Res 204:111963
Tahvildari K, Anaraki YN, Fazaeli R, Mirpanji S, Delrish E (2015) The study of CaO and MgO heterogenic nano-catalyst coupling on transesterification reaction efficacy in the production of biodiesel from recycled cooking oil. J Environ Health Sci Eng 13:1–9
Tan H, Feng W, Ji P (2012) Lipase immobilized on magnetic multi-walled carbon nanotubes. Bioresour Technol 115:172–176
Tang F, Li L, Chen D (2012) Mesoporous silica nanoparticles: synthesis, biocompatibility and drug delivery. Adv Mater 24(12):1504–1534
Tarafdar JC, Raliya R, Mahawar H, Rathore I (2014) Development of zinc nano fertilizer to enhance crop production in pearl millet (Pennisetum americanum). Agric Res 3:257–262
Taurozzi JS, Hackley VA, Wiesner MR (2011) Ultrasonic dispersion of nanoparticles for environmental, health and safety assessment–issues and recommendations. Nanotoxicology 5(4):711–729
Ten GB (2019) Chemical innovations that will change our world: IUPAC identifies emerging technologies in chemistry with potential to make our planet more sustainable. Chem Int 41(2):12–17
Torney F, Trewyn BG, Lin VSY, Wang K (2007) Mesoporous silica nanoparticles deliver DNA and chemicals into plants. Nat Nanotechnol 2(5):295–300
Torres R, Diz VE, Lagorio MG (2018) Effects of gold nanoparticles on the photophysical and photosynthetic parameters of leaves and chloroplasts. Photochem Photobiol Sci 17(4):505–516
Trinh NN, Huang TL, Chi WC, Fu SF, Chen CC, Huang HJ (2014) Chromium stress response effect on signal transduction and expression of signaling genes in rice. Physiol Plant 150(2):205–224
Trisunaryanti W, Fatmawati DA (2020) Synthesis of co-nh 2/mesoporous silica bifunctional catalyst using sidoarjo mud and bovine bone gelatin template for conversion of used cooking oil into biofuel. Rasayan J Chem 13(1):1
UNFCCC (2015) Adoption of the Paris Agreement
Van Aken B (2015) Gene expression changes in plants and microorganisms exposed to nanomaterials. Curr Opin Biotechnol 33:206–219
Van Aken O, De Clercq I, Ivanova A, Law SR, Van Breusegem F, Millar AH, Whelan J (2016) Mitochondrial and chloroplast stress responses are modulated in distinct touch and chemical inhibition phases. Plant Physiol 171(3):2150–2165
Varma RS (2019) Biomass-derived renewable carbonaceous materials for sustainable chemical and environmental applications. ACS Sustain Chem Eng 7(7):6458–6470
Vasco-Correa J, Khanal S, Manandhar A, Shah A (2018) Anaerobic digestion for bioenergy production: global status, environmental and techno-economic implications, and government policies. Bioresour Technol 247:1015–1026
Verma ML (2017) Nanobiotechnology advances in enzymatic biosensors for the Agri-food industry. Environ Chem Lett 15(4):555–560
Verma ML, Chaudhary R, Tsuzuki T, Barrow CJ, Puri M (2013a) Immobilization of β-glucosidase on a magnetic nanoparticle improves thermostability: application in cellobiose hydrolysis. Bioresour Technol 135:2–6
Verma ML, Naebe M, Barrow CJ, Puri M (2013b) Enzyme immobilisation on amino-functionalised multi-walled carbon nanotubes: structural and biocatalytic characterisation. PLoS One 8(9):e73642
Verma ML, Rajkhowa R, Wang X, Barrow CJ, Puri M (2013c) Exploring novel ultrafine Eri silk bioscaffold for enzyme stabilisation in cellobiose hydrolysis. Bioresour Technol 145:302–306
Verma ML, Barrow CJ, Puri M (2013d) Nanobiotechnology as a novel paradigm for enzyme immobilisation and stabilisation with potential applications in biodiesel production. Appl Microbiol Biotechnol 97:23–39

Vichaphund S, Aht-ong D, Sricharoenchaikul V, Atong D (2015) Production of aromatic compounds from catalytic fast pyrolysis of jatropha residues using metal/HZSM-5 prepared by ion-exchange and impregnation methods. Renew Energy 79:28–37
Vidhyalakshmi R, Bhakyaraj R, Subhasree RS (2009) Encapsulation "the future of probiotics"-a review. Adv Biol Res 3(3–4):96–103
Von Caemmerer S (2020) Rubisco carboxylase/oxygenase: from the enzyme to the globe: a gas exchange perspective. J Plant Physiol 252:153240
Wang W, Vinocur B, Altman A (2003) Plant responses to drought, salinity and extreme temperatures: towards genetic engineering for stress tolerance. Planta 218:1–14
Wang T, Zhang D, Dai L, Chen Y, Dai X (2016) Effects of metal nanoparticles on methane production from waste-activated sludge and microorganism community shift in anaerobic granular sludge. Sci Rep 6(1):25857
Wang JW, Grandio EG, Newkirk GM, Demirer GS, Butrus S, Giraldo JP, Landry MP (2019) Nanoparticle-mediated genetic engineering of plants. Mol Plant 12(8):1037–1040
Warzecha H, Hennig A (2010) Plastid transformation. Genetic Modification of plants: agriculture, horticulture and forestry. Springer, Cham, pp 23–37
Weise SE, Weber AP, Sharkey TD (2004) Maltose is the major form of carbon exported from the chloroplast at night. Planta 218:474–482
White JC, Gardea-Torresdey J (2018) Achieving food security through the very small. Nat Nanotechnol 13(8):627–629
Wong MH, Misra RP, Giraldo JP, Kwak SY, Son Y, Landry MP, Strano MS (2016) Lipid exchange envelope penetration (LEEP) of nanoparticles for plant engineering: a universal localization mechanism. Nano Lett 16(2):1161–1172
Wong MH, Giraldo JP, Kwak SY, Koman VB, Sinclair R, Lew TTS, Strano MS (2017) Nitroaromatic detection and infrared communication from wild-type plants using plant nanobionics. Nat Mater 16(2):264–272
World Energy Balances (2020)—Analysis—IEA. Accessed 19 Dec 2020
Wrigley C (2017) The cereal grains: providing our food, feed and fuel needs. In: Cereal grains. Woodhead Publishing, Sawston, pp 27–40
Wu H, Tito N, Giraldo JP (2017) Anionic cerium oxide nanoparticles protect plant photosynthesis from abiotic stress by scavenging reactive oxygen species. ACS Nano 11(11):11283–11297
Wu H, Shabala L, Shabala S, Giraldo JP (2018) Hydroxyl radical scavenging by cerium oxide nanoparticles improves arabidopsis salinity tolerance by enhancing leaf mesophyll potassium retention. Environ Sci Nano 5(7):1567–1583
Xiang Y, Zhang G, Chi Y, Cai D, Wu Z (2017) Fabrication of a controllable nanopesticide system with magnetic collectability. Chem Eng J 328:320–330
Xie W, Wang J (2014) Enzymatic production of biodiesel from soybean oil by using immobilized lipase on Fe3O4/poly (styrene-methacrylic acid) magnetic microsphere as a biocatalyst. Energy Fuel 28(4):2624–2631
Xie Y, Li B, Tao G, Zhang Q, Zhang C (2012) Effects of nano-silicon dioxide on photosynthetic fluorescence characteristics ofIndocalamus barbatus McClure. J Nanjing For Univ 55(02):59
Yadav A, Upadhyay A, Kumar R, Prajapati J, Pal S (2023) Nanotechnology based nano urea to increase agricultural sustainability. Just Agric 3(12):266–274
Yang F, Hong F, You W, Liu C, Gao F, Wu C, Yang P (2006) Influence of nano-anatase TiO_2 on the nitrogen metabolism of growing spinach. Biol Trace Elem Res 110:179–190
Yang Y, Guo J, Hu Z (2013) Impact of nano zero valent iron (NZVI) on methanogenic activity and population dynamics in anaerobic digestion. Water Res 47(17):6790–6800
Yang Q, Xu Q, Jiang HL (2017) Metal–organic frameworks meet metal nanoparticles: synergistic effect for enhanced catalysis. Chem Soc Rev 46(15):4774–4808
Yoon HY, Lee JG, Esposti LD, Iafisco M, Kim PJ, Shin SG et al (2020) Synergistic release of crop nutrients and stimulants from hydroxyapatite nanoparticles functionalized with humic substances: toward a multifunctional nanofertilizer. ACS Omega 5(12):6598–6610

Yousefi S, Kartoolinejad D, Naghdi R (2017) Effects of priming with multi-walled carbon nanotubes on seed physiological characteristics of hopbush (*Dodonaeaviscosa* L.) under drought stress. Int J Environ Stud 74(4):528–539

Zand AD, Mikaeili Tabrizi A, Vaezi Heir A (2020) Application of titanium dioxide nanoparticles to promote phytoremediation of cd-polluted soil: contribution of PGPR inoculation. Biorem J 24(2-3):171–189

Zanuso E, Ruiz HA, Domingues L, Teixeira JA (2022) Magnetic nanoparticles as support for cellulase immobilization strategy for enzymatic hydrolysis using hydrothermally pretreated corn cob biomass. Bioenergy Res 15(4):1946–1957

Ze Y, Liu C, Wang L, Hong M, Hong F (2011) The regulation of TiO_2 nanoparticles on the expression of light-harvesting complex II and photosynthesis of chloroplasts of Arabidopsis thaliana. Biol Trace Elem Res 143:1131–1141

Zhang L, Xu CC, Champagne P (2010) Overview of recent advances in thermo-chemical conversion of biomass. Energy Convers Manag 51(5):969–982

Zhang J, Lu Y (2016) Conductive Fe_3O_4 nanoparticles accelerate syntrophic methane production from butyrate oxidation in two different lake sediments. Front Microbiol 7:1316

Zhang Y, Degirmenci V, Li C, Hensen EJ (2011) Phosphotungstic acid encapsulated in metal–organic framework as catalysts for carbohydrate dehydration to 5-hydroxymethylfurfural. ChemSusChem 4(1):59–64

Zhang XL, Yan S, Tyagi RD, Surampalli RY (2013) Biodiesel production from heterotrophic microalgae through transesterification and nanotechnology application in the production. Renew Sust Energ Rev 26:216–223

Zhang L, Zheng Y, Zhang Y, Lin T, Echeverria AS (2017) Nanotechnology for bioenergy and biofuel production. Ind Crop Prod 1:46–97

Zhao Q, Ma C, White JC, Dhankher OP, Zhang X, Zhang S, Xing B (2017) Quantitative evaluation of multi-wall carbon nanotube uptake by terrestrial plants. Carbon 114:661–670

Zhong R, Sels BF (2018) Sulfonated mesoporous carbon and silica-carbon nanocomposites for biomass conversion. Appl Catal B Environ 236:518–545

Zhou N, Liu S, Zhang Y, Fan L, Cheng Y, Wang Y et al (2018) Silicon carbide foam supported ZSM-5 composite catalyst for microwave-assisted pyrolysis of biomass. Bioresour Technol 267:257–264

Zhu XG, Long SP, Ort DR (2010) Improving photosynthetic efficiency for greater yield. Annu Rev Plant Biol 61:235–261

Zhu B, Chen G, Cao X, Wei D (2017) Molecular characterization of CO_2 sequestration and assimilation in microalgae and its biotechnological applications. Bioresour Technol 244:1207–1215

Zulfiqar F, Navarro M, Ashraf M, Akram NA, Munné-Bosch S (2019) Nanofertilizer use for sustainable agriculture: advantages and limitations. Plant Sci 289:110270

Nitric Oxide: A Double-Edged Sword in Photosynthetic Stress Responses

15

Neha Gupta, Ankit Srivasatava, Anirbana Parida, and Arun Kumar Mishra

Abstract

Nitric oxide (NO), a small gaseous molecule with redox activityserves as a central regulator of growth, development, immunity. This remarkable signaling molecule orchestrates a complex symphony of biological responses, navigating through intricate pathways that influence essential aspects of plant life. The absence of canonical Nitric Oxide Synthases (NOS) in higher plants has led to the discovery of alternative NO production pathways, featuring key contributors like nitrate reductase (NR) in mitochondria, peroxisomes, and chloroplasts. In the realm of plant physiology, NO emerges as a multifaceted player, significantly impacting reproduction, symbiotic interactions, senescence, and defense mechanisms. In symbiotic relationships, NO plays a pivotal role in interactions with rhizobia, mycorrhizae, and lichens, influencing nodule development and senescence. Furthermore, NO governs plant senescence, acting both as a signaling molecule and a modulator of hormone-induced processes. In defense responses, NO collaborates with reactive oxygen species, influencing hypersensitive cell death. Beyond growth, NO showcases its potential in postharvest applications, preserving sensory attributes, enhancing nutritional quality, and serving as an eco-friendly alternative for pest and disease control in horticultural crops. As research delves deeper, unveiling additional dimensions of NO's contributions, it holds promise for innovative applications in agriculture, environmental management, and sustainable horticulture.

N. Gupta · A. Srivasatava · A. Parida · A. K. Mishra (✉)
Laboratory of Microbial Genetics, Department of Botany, Institute of Science, Banaras Hindu University, Varanasi, India

A. K. Mishra (ed.), *Stress Biology in Photosynthetic Organisms*,
https://doi.org/10.1007/978-981-97-1883-2_15

Keywords

Nitric oxide synthase · S-nitrosoglutathione · Nitrite · Nitrate Reductase · Nitrite · Agriculture

15.1 Introduction

In the mid-eighties, the revelation of animal cells possessing the capacity to produce nitric oxide (NO), initially identified as endothelium-derived relaxing factor (EDRF), marked inception of a transformative research domain (Ignarro et al. 1987). This discovery later led to the identification of nitric oxide synthases (NOSs) derived NO generation from L-arginine, unveiling extensive involvement of NO in diverse cellular processes (Alderton et al. 2001). While research on NO in animal cells progressed rapidly, its exploration in green cells, specifically plant cells, lagged. Nevertheless, the increasing volume of articles on plant NO signifies its growing importance (Fig. 15.1). Recent advancements, particularly in understanding plant lipids and the emergence of nitrofatty acids (NO_2-FA), highlight the expanding significance of NO in diverse signaling pathways.

Moreover, NO operates as a crucial signaling molecule, influencing protein functionality (Corpas et al. 2015; Khan et al. 2023). The chapter not only delineates the

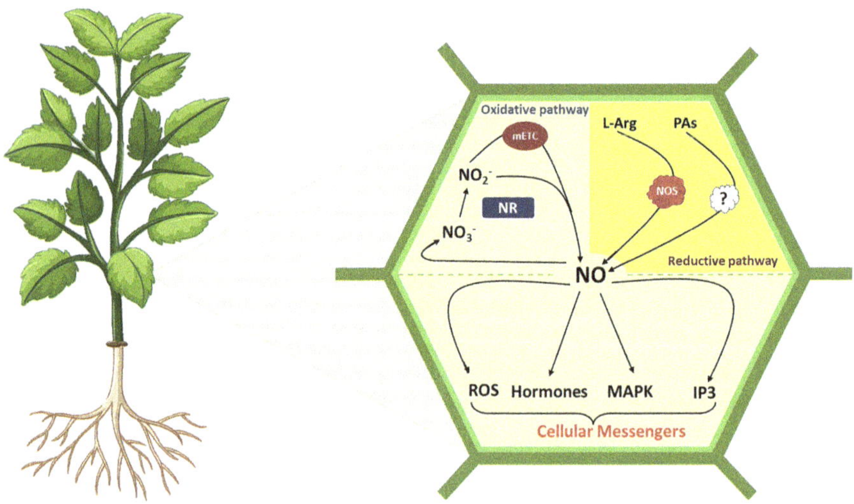

Fig. 15.1 Schemetic diagram depicting the primary pathways of nitric oxide (NO) biosynthesis NO synthesis occurs through either reductive or oxidative mechanisms. Reductive pathways predominantly take place in various cellular compartments, including the plasma membrane, chloroplasts, apoplast, peroxisomes, cytoplasm, and mitochondria, involving the reduction of nitrate (NO_3^-) or nitrite (NO_2^-) by nitrate reductase (NR). Conversely, oxidative mechanisms are initiated by the oxidation of L-arginine, leading to NO production, with this process occurring notably in chloroplasts and leaf peroxisomes

strides made in comprehending the role of NO in plant cells but also delves into the contentious identifications of endogenous enzymatic sources of NO. Beyond its physiological implications, the chapter explores the versatile and promising nature of NO. In summary, this comprehensive exploration underscores the dual significance of NO as a key player in plant cellular processes and as a potential tool for innovative applications in biotechnology.

15.2 Nitrate Reductase (NR) Orchestrating the NO Symphony in Plants

An extensively investigated enzyme associated with NO synthesis in plants is nitrate reductase (NR) (Planchet et al. 2005). NR relies on molybdopterin, heme, and FAD as cofactors (Hageman 1990) (Fig. 15.2). *Arabidopsis* nitrate reductase (NR), from NIA1 and NIA2 genes, was initially linked to nitric oxide (NO) production, but its limited nitrite affinity and low tissue concentrations suggest a minor role under standard conditions (Yamasaki et al. 1999).

An increasing body of research has documented the role of NR-dependent nitric oxide in stress responses in plants (Rockel et al. 2002). Under increased fermentation, nitrite reductase (NiR) faces inhibition, fostering elevated nitrite levels and concurrent

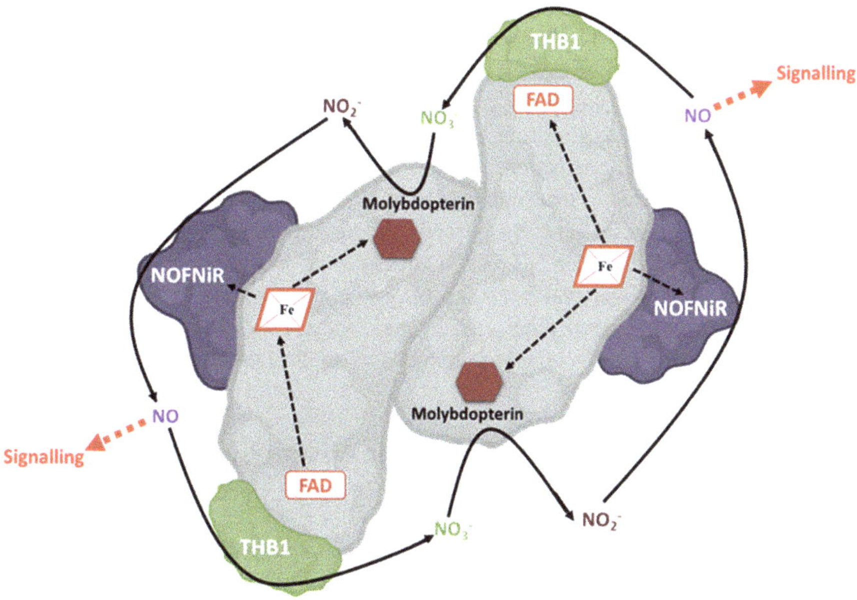

Fig. 15.2 Schematic representation of the nitrate reductase enzyme structure: FAD (flavine adenine dinucleotide), Mo (molybdenum), Mpt (molybdopterin), NAD(P)H (nicotinamide adenine dinucleotide phosphate) and THB1 (truncated hemoglobin). These structural elements play integral roles in the catalytic functions of nitrate reductase during the conversion of nitrate (NO_3^-) to nitrite nitrite (NO_2^-) in biological processes

NO production. Experiments with antisense construct of NiR demonstrated constitutive NO generation in tobacco, indicating a potential regulatory mechanism. Moreover, nitrite, generated through this process, emerged as a beneficial factor in mitigating acidification of cytoplasm (Allegre et al. 2004; Libourel et al. 2006). Genetically modified tobacco exhibits diminished NR activity in roots showcased increased sensitivity to oxygen deficiency in root, revealing the significance of NR in plant responses to low oxygen conditions. Similarly, under hypoxic scenarios, the provision of ammonium as a nitrogen source led to NR activity inhibition, reduced NO production and lowered ATP levels. This observation hints at the involvement of NO produced through NR activity in conferring tolerance to reduced (Wany et al. 2019).

Mutations in the site of phosphorylation in NR led to elevated accumulation of nitrite and release of NO, emphasizing the regulatory intricacies governing NO production (Lillo et al. (2004). Recent revelations indicate that NO synthesis is not solely confined to NR's direct activity. Instead, NR serves as a conduit for electron transfer facilitating the reduction of nitrite (Fig. 15.2).

Recent revelations by Vicente et al. (2017) paint a dynamic picture, linking NO produced through NR in maintaining the stability of ERF-VII amid various abiotic stresses.

Beyond this symphony of transcriptional modulation, NR exhibits a dynamic crescendo in plant resilience. Sang and Blecha (2008) showcase NR's involvement in inducing antioxidant metabolism, fortifying plants against the strains of stress conditions. NR's versatility extends further into the realm of defense orchestration, playing a key role in NO production triggered by pathogens and elicitors.

Moreover, recent findings suggest that the participation of NR in the synthesis of NO seems to be indirect. Concurrently, exploration of NO synthesis has extended at the subcellular scale. The mitochondrion, peroxisome, and chloroplast have emerged as key players in relation to NO production. This growing understanding underscores the intricate subcellular mechanisms that contribute to NO synthesis, providing valuable insights into the broader regulatory networks governing plant responses to various stimuli. The multifaceted interplay between NR and NO has become a focal point in contemporary plant biology, shaping our understanding of how plants respond to various stresses. These findings not only deepen our insights into fundamental physiological processes but also hold promise for potential biotechnological applications in agriculture.

15.3 Lost Symphony: The Enigmatic Absence of NOS in Plants

Though nitric oxide (NO) assumes a substantial role in plant physiology, the absence of Nitric Oxide Synthases (NOS) in higher plants is still a botanical mystery. It is speculated that evolutionary processes, involving the intricate loss of a specific gene (NOS), may have resulted in its disappearance among plants. This absence is compensated by alternative pathways, as suggested by Mur et al. (2013). However, the exact timing and mechanism of this loss, as well as how these alternative pathways collaborate to create the complex network of involving NO in plants, remain uncertain.

In higher plants, various pathways work together to produce nitric oxide, resembling a symphony of alternative processes. Understanding how these pathways collaborate under specific cellular conditions poses a challenging task. The quest for plant NOS persists, especially as Arg-dependent NO formation continues to be measured. While the elusive plant NOS remains hidden, the focus shifts to unraveling the characteristics of the diverse NO producers already identified.

The alga *Ostreococcus*, the sole "plant" possessing a NOS homolog, notably generates substantial NO amounts, particularly during the exponential growth phase (Foresi et al. 2022). In contrast to animals and *Ostreococcus*, land plants under non-stressed conditions maintain a notably low basal level of NO. The absence of identified NOS enzymes in plants suggests potentially distinct way of NO pathway in comparison to animals (Fig. 15.3).

Unlike conventional second messengers, such as reactive oxygen species (ROS), NO exhibits high reactivity and has the ability to alter diverse variety of proteins. Its ability to diffuse through cellular membranes poses a challenge (Sami et al. 2018). However, plants employ different NO-producing systems, raising questions about whether known plant NO producers can substitute for NOS or if an undiscovered NO-producing system is required for specificity in signaling (Luiking et al. 2012). Despite their demonstrated physiological roles, their integration into a putative NO

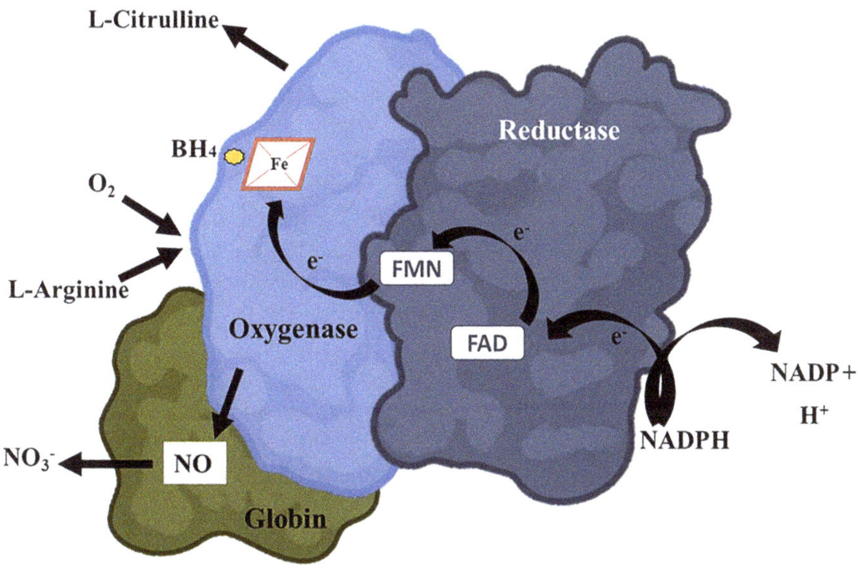

Fig. 15.3 Structure of Nitric oxide synthase (NOS) in cyanobacteria: Diagram depicting the simplified structure and catalytic process of *Synechococcus* NOS. The enzyme comprises distinct oxygenase, globin domain, and reductase domains, featuring heme and BH_4 cofactors. These domains are connected by a calmodulin (CaM)-binding motif. NADPH donates electrons, which traverse through the redox cofactors, FAD and FMN, in the reductase domain. At the catalytic site, electrons oxidize L-arginine, generating L-citrulline and nitric oxide (NO). Additionally, the globin domain aids in further oxidation of NO to nitrate

signaling network in plants remains unclear (Fig. 15.1). It is plausible that land plants utilize the enzyme responsible for NO production for specific, specific functions based on physiology (Kolbert et al. 2019). For instance, Arg-dependent mechanisms predominantly come into play during defense against pathogens, exposure to salt stress, and deficiency in phosphorus.

There's also the possibility of NO production rely on complex peptides encoded separately or proteins with enzymatic activity alike to animal NOS. Regulation of such a complex could involve signal-dependent assembly or disassembly, creating localized pockets of concentrated NOS activity (Fröhlich and Durner 2011; Santana et al. 2017). Moreover, substrate availability appears to be a crucial factor in the generation of NO in land plants (Yarkoni 2012). There are seven possible contenders for NO generation in land plants, and in specific circumstances, these localized NO producers may work in synergy (Fröhlich and Durner 2011). NO can readily move through membranes, potentially leading to NO gradients or hotspots within cells or specific tissues (Trewavas and Malhó 1997). Signaling may rely more on manipulating reservoir of NO than in the synthesis of NO directly. Notably, GSNO (S-nitrosoglutathione) serves as an essential molecular messenger in all biological organism, acting serving as a stable reservoir and donor of NO (Broniowska and Hogg 2012).

15.4 Unveiling NO inside Peroxisomes, Chloroplasts, and Mitochondria

Peroxisomes in plants were the first organelles to potentially exhibit activity resembling NOS, characterized by inhibitor sensitivity similar to NOS in animals. Although not observed in plants, subsequent studies following this initial report confirmed the existence of inducible NOS (iNOS) in peroxisomes. Notably, the synthesis of iNOS protein increased in the presence of under sepsis Loughran et al. 2005). These findings lend support to the idea that peroxisomes may indeed harbor this enzymatic activity. The existence of nitric oxide (NO) was validated through additional techniques, including Electron Paramagnetic Resonance (EPR) spectroscopy employing the use of the spin trap $Fe(MGD)_2$ and fluorometry using probe DAF-2 DA. These results offered additional compelling confirmation of L-arginine-dependent NO generation through NOS in plant peroxisomes (Corpas et al. 2004).

Further, chloroplast is specific to photosynthetic plant tissues and exhibits an active ROS metabolism due to photosynthetic activity. Cooney et al. (1994) initially provided evidence of NO production in chloroplasts, highlighting the non-enzymatic light-induced non-enzymatic conversion of NO_2 to NO facilitated by pigmented compounds. Subsequent studies on purified chloroplasts from soybean leaves, utilizing EPR with the spin trap, further substantiated NO production in chloroplasts (Jasid et al. 2006).

Further, mitochondria serve as significant contributors to NO production, a phenomenon first identified in mitochondria of mammalian cells (Nohl et al. 1999). Mitochondria participation in the NO dynamics has been illustrated by studies on

Chlorella sorokiniana. Under anoxic conditions, *Chlorella sorokiniana* has been shown to generate NO when supplied with nitrite. Furthermore, *Chlorella sorokiniana* cultured in ammonium-rich condition was found to generate NO, a process responsive to the inhibitor of alternative oxidase (AOX), indicating the role of AOX in the generation of NO through nitrite under low oxygen condition. The mitochondria from root cells sourced from pea and tobacco were demonstrated to convert nitrite into NO with myxothiazol and SHAM inhibiting this production (Gupta et al. 2017). Notably, oxygen was found to inhibit nitrite-dependent NO production, with the inhibitory constant (KiO2) at 0.05% and the nitrite saturation constant (Km nitrite) at 175 μM. In root nodules of *Medicago truncatula* under low oxygen conditions, nitrite addition increased NO generation, inhibited by myxothiazol and antimycin A, highlighting the participation of nodule mitochondria in NO generation at the expense of nitrite. The participation of Complex III in NO generation was highlighted by Alber et al. (2017), who observed increased NO production upon Complex III inhibition under normoxia. Electron pressure in Complex III was identified as a NO generator, while AOX served to remove elevated NO levels under normal oxygen conditions (Cvetkovska et al. 2014). Cytochrome oxidase (COX) also participates in generation of NO, as evidenced by the inhibition of nitrite-dependent NO with KCN in isolated mitochondria. While the process by which COX generates NO in plants remains to be demonstrated, in animal systems, it has been associated with the oxidation of iron following its binding to nitrite at the completely reduced Fea3CuB center (Castello et al. 2006). Recent findings demonstrated the protective role of conversion of nitrite into NO in preserving the structure and function of mitochondria (Gupta et al. 2017). The introduction of nitrite to anoxic mitochondria elevated NO levels, reduced ROS content resulting in lipid peroxidation and an elevation in ATP levels.

15.5 Multifaceted Role of NO in Plant Physiology

15.5.1 NO in Plant Reproduction

NO plays a crucial role in various aspects of reproduction in plants, spanning from the development of flowers to germination of seed (Fig. 15.4). The use of NO-induced germination in lettuce (Šírová et al. 2011) while sodium nitroprusside break seed dormancy. Concerns about SNP effects being potentially facilitated by cyanide, a derivative of sodium nitroprusside, prompted additional studies, ultimately confirming that NO itself played a pivotal role in the process of seed dormancy release (Bethke et al. 2006). Subsequent validation involved the application of NO immediately to seeds, bypassing the use of a donor molecule. NO, facilitated by two donors, SNP or SNAP, was demonstrated to alleviate embryonic dormancy in apples, as evidenced by research conducted by Gniazdowska et al. (2007). The study revealed a correlation between the action of NO and the subsequent production of ethylene. Moreover, NO donor triggered a rise in the activity of β-amylase without influencing α-amylase. Since similar effects were observed in other species,

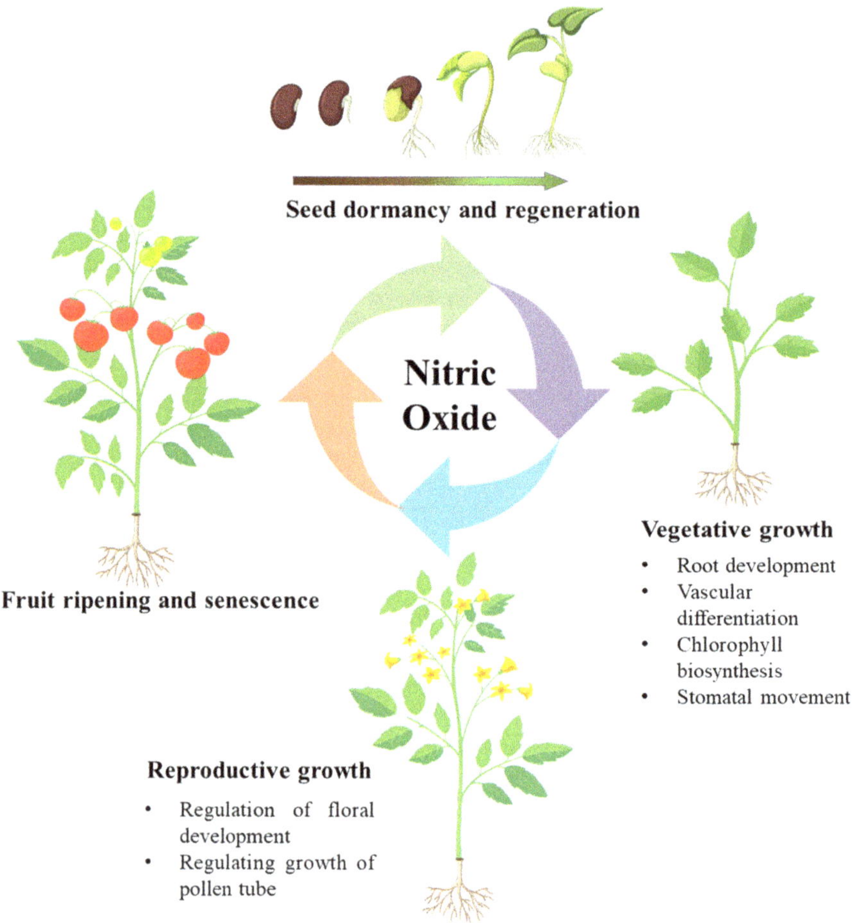

Fig. 15.4 Nitric Oxide's vital role in plant growth regulation and adaptation to changing environments: Nitric oxide (NO) intricately influences various aspects of plant morphogenesis and development. It breaks seed dormancy and promotes germination. NO regulates root formation, primary growth, adventitious roots, and root hair development. Additionally, it impacts chlorophyll levels, vegetative growth, stomatal movement, and iron homeostasis. NO participates in hormonal responses, coordinates Ca_2^+ and reactive oxygen species (ROS) signaling in plant immunity, and modulates reproductive development. In fruit ripening and senescence, NO delays ethylene-induced processes, suppressing respiration and ethylene production

Zhang et al. (2005) proposed that this NO-induced impact on amylase activity might be a universal phenomenon. Seed germination experiments were conducted across various species to substantiate this observation.

In 2004, Prado et al. identified NO as a key regulator of pollen tube growth. Their study showed that NO exposure influenced pollen tube orientation through cGMP signaling. Feijó et al. (2001) also centered their review paper on this noteworthy research in the same year. Further exploration into the alignment of pollen tubes revealed the NO's definite involvement, steering expansion towards the micropyle in ovule. Intriguingly, extracellular ATP was identified as a signal hindering both the initiation and extension of pollen (Reichler et al. 2009; Bright et al. 2009). Subsequently, Zafra et al. (2010) investigated the presence and location of NO in the reproductive tissues of olive plants.

Simpson (2005) demonstrated that the modulation of NO levels led to changes in gene expression, specifically affecting a flowering mediator, FLC, which acts as a repressor of flowering. Moreover, the NR double mutant (nia1nia2) study brought to light the crucial role of this enzyme in the development of flowers, particularly evident in *Arabidopsis* (Seligman et al. 2008). Further, NO's involvement in flower senescence was elucidated through experiments showing that the introduction of NO led to a decrease in both xanthine oxidase activity and superoxide dismutase (SOD) activity. These alterations, coupled with changes in antioxidant capacity, contributed to a reduction in superoxide and hydrogen peroxide levels. The comprehensive findings concluded that NO plays a pivotal role in governing flower senescence, impacting multiple redox pairs and the cellular non-protein thiol status (Tewari et al. 2009).

Recently, a substantial amount of supporting facts has shown the involvement of NO in the response to self-incompatibility (SI), leading to programmed cell death in pollen capable of self-fertilization. The study reveals that pollen *of Papaver*, when germinated in vitro, exhibits transient accumulation of ROS/NO, with variations based on whether the pollen grain is self-compatible or self-incompatible. Soon after the initiation of SI, there is an elevation in cytosolic ROS, followed by a later, transient increase in NO. Recent findings indicate that ROS and NO exert their influence on the actin cytoskeleton through PTM, specifically via S-nitrosylation. These modifications disrupt actin polymerization, leading to significant disruptions in the function of actin cytoskeleton. Uncovering plant proteins directly targeted by NO and ROS in the context of self-incompatibility in understanding the role of post-translational modifications such as S-nitrosylation and carbonylation could provide crucial insights for characterizing the signaling network that governs this process. This approach holds significance in unraveling both the functional and biochemical aspects of the intricate signaling pathways involved in SI (Rodríguez-Serrano et al. 2014).

Moreover, studies involving tubular structure from pollen grains have provided intricate insights into the interplay Integration of NO pathways with other signaling components. For example, research has explored the coordination between NO and key signaling elements such as calcium ions, reactive oxygen species, MAP kinases, and ascorbate. These investigations aim to decipher the complex web of interactions that contribute to the finely tuned orchestration of reproductive processes in plants (Wang et al. 2016).

In essence, the ongoing scientific exploration in this field not only highlights the diverse roles played by NO in plant reproduction but also underscores the interconnectedness of NO signaling with various molecular players, shedding light on the regulatory networks governing these essential biological processes.

15.5.2 NO in Symbiotic Interactions

The generation of NO in the initial stages of symbiotic interactions has been linked to NOS-like activity and Nitrate Reductase in plants (Boscari et al. 2013). Additionally, NOS-like activity, both from plants and bacteria, along with NR and respiratory chains, have been considered potential sources of NO in nodules of nitrogen-fixing organisms (Horchani et al. 2011). An intriguing aspect is the involvement of MtNOA1, which influences the initiation and operation of symbiotic relationships between *M. truncatula and S. meliloti.* NO was identified at various sites throughout the infection process of *M. truncatula–S. meliloti* interactions, encompassing the formation of nodule primordia. Depleting NO resulted in a significant delay in the appearance of nodules, highlighting its importance in the process (Del Giudice et al. 2011).

Microarray examination of genes in *M. truncatula* responsive to NO in roots further supported the idea that NO might play a regulatory role in the establishment of symbiosis or development of nodules (Ferrarini et al. 2008). However, despite its impact on symbiosis, MtNOA1 does not seem to influence NO production within the nodules, as noted in the investigation by Pauly et al. (2011). NO's involvement in symbiotic interactions extends beyond nodules and includes various associations, as extensively discussed in reviews (Hichri et al. 2016; Martínez-Medina et al. 2019). A noteworthy discovery was the response of symbiotic rhizobia to NO, resulting in the upregulation of over 100 genes. Nagata et al. (2008) made crucial contribution to our understanding of the role of nitric oxide (NO) in the distinct plant reactions to both symbiotic and pathogenic microbes. In their study, they observed that the expression of Phytogb1 and the generation of NO in roots of *Lotus* remained unaffected by interactions that are not symbiotic. However, in the presence of symbiotic rhizobia, there was only a transient increase. By contrast, inoculation with plant pathogens induced continuous NO production but did not trigger Phytogb1 expression. Additionally, it was noted that NO itself induced the gene expression of Phytogb1 in *Lotus japonicus.*

Moreover, studies in transcriptomic exploring NO-responsive genes during symbiotic interactions have unveiled intriguing insights. In a study on *M. truncatula* roots exposed to NO donors such as sodium nitroprussiate and S-nitroso-glutathione. 999 possible responsive genes of NO were identified, with 290 also controlled during nodule formation. These genes included those related to development of root hairs, carbon homeostasis, and proteolysis dependent on the proteasome, as well as genes controlling cellular redox states, such as those involved in glutathione and H_2O_2 metabolism. A subsequent RNA-Seq analysis of roots of *M. truncatula* NO depleted NO showcased 2030 differentially expressed genes (Chakraborty et al. 2021). Notably, NO reduction hindered the increased expression of genes crucial for nodule development, including transcription factors and proteins essential for cortical cell dedifferentiation and stimulation of cell division in the process of nodule formation. Strikingly, NO scavenger treatment reversed the reduction of defense-related genes usually initiated by the inoculation of rhizobia, indicating that NO might contribute to suppressing plant defense responses, thereby promoting the formation of advantageous plant–microbe associations. It contrasts with NO's role in pathogenic interactions, where it collaborates with reactive oxygen species (ROS) to trigger hypersensitive cell death (Delledonne et al. 1998), highlighting the diverse functions of NO in different symbiotic contexts. The collective body of literature underscores NO as a versatile orchestrator, intricately participating in various aspects of the symbiotic procedure. Its roles span from symbiont recognition and the modulation of plant defense reactions to steering the development of organs during the establishment of symbiosis. As the symbiotic relationship matures, NO continues to have a pivotal role in governing nitrogen/carbon metabolism within mature nodules, while also contributing to the initiation of senescence.

Compellingly, emerging evidence suggests that these mechanisms are not confined to specific symbiotic interactions but are shared across various symbiotic relationships. Furthermore, NO reacts intricately in conjunction with other key regulators of plant development, including hormones. Unraveling the intricate web of these interplay and understanding the ways in which NO is sensed signaling emerges as a formidable challenge for the next decade, holding the promise of unlocking new dimensions in our comprehension of symbiotic processes in plants.

15.5.3 NO in Plant Senescence

Senescence, a vital mechanism for plant viability, is formally described as the programmed and synchronized deterioration of leaf cells regulated by precise genetic mechanisms. This process significantly contributes to the comprehensive growth of plants by enabling the efficient relocation of crucial nutrients, ensuring successful reproduction, and providing the necessary phenotypic adaptability to adapt to fluctuating environmental cues (Guo et al. 2021). As posited by Schippers et al. (2015), plant senescence embodies a paradoxical nature—living with the purpose of eventual demise and undergoing the aging process as a fundamental aspect of survival.

Several studies have briefly outlined immediate involvement of NO during the aging process in plants (Fig. 15.4) (Corpas and Palma 2018; Bruand and Meilhoc 2019).

Over the past two decades, NO has gained prominence in leaf senescence regulation, notably influencing ABA, ET, and JA phytohormones. NO's intricate interplay with these hormones results in complex signaling cascades, impacting transcriptome and proteome levels across diverse plant species (Hussain et al. 2016). Exogenous NO application effectively extends shelf life, retards flower maturation, and counteracts leaf ageing induced by ABA in rice plant (Ya'Acov and Haramaty, 1996).

In a 2013 study, the physio-biochemical effects of decreased NO on the breakdown of chlorophyll in the course of leaf senescence increased by darkness.

In a broader context, NO's involvement in gene expression is evident. The induction of death-like phenotype through the increased expression of a NO degrading dioxygenase precedes a substantial shift in transcript expression in *Arabidopsis*. This shift involves a decrease in expression of photosynthetic genes and an increase in numerous senescence-associated (SAGs), as highlighted by Mishina et al. (2007). Moreover, aging within seed structure is marked by cellular-level loss of stability in membrane, degradation of nucleic acids, increased oxidative damage attributable to the accumulation of super oxide radicals, and a simultaneous decrease in antioxidant potential (El-Maarouf-Bouteau et al. 2011). While specific function of NO in seed development was not fully elucidated, but recognized for breaking and inducing dormancy of seed promoting the production of ethylene (ET). Moreover, the introduction application of NO donors and arginine to apple (*Malus domestica*) seeds has been found to eliminate embryo dormancy and promote the process of germination (Krasuska et al. 2016).

The rhizobium–legume symbiosis exemplifies another instance where NO is involved in plant organ senescence. However, the molecular mechanisms governing nodule senescence remain largely unknown. Similar to leaf senescence, observations indicate pivotal roles of plant hormones—in legume root nodule senescence (Chen et al. 2014). Overall decline in antioxidant defenses during the ageing of nodules. and manipulation of the microbial symbiont's antioxidant response hint at the potential role of redox balance in controlling nodule ageing (Puppo et al. 2005; Loscos Aranda et al. 2008; Jamet et al. 2003; Harrison et al. 2005; Redondo et al. 2009; Coba de la Peña et al. 2010).

Several reports underscore NO serving as a significant contributor to the ageing of nodules. NO is identified in various stages in the symbiotic relationship between legumes and rhizobia, evidence suggests higher concentrations during late symbiotic stages. Notably, NO is more prevalent in the older, root-proximal region of *M. truncatula* nodules where the ageing initiation occurs, and its presence is indicated by nitrated leghemoglobins in aging soybean and NO accumulation in old *Lotus japonicus* nodules (Cam et al. 2012; Chungopast et al. 2017).

Alternatively, NO may directly influence senescence through transcriptional regulation, enzymatic activity, or structural protein modifications. Rhizobia possess the NO-responsive transcriptional regulator NnrR, while in plants, ethylene response factors group-VII are implicated in NO signaling (Gibbs et al. 2014). In mature *M. truncatula* nodules, 80 proteins, including glutamine synthetase (GS), undergo S-nitrosylation, correlates accompanied by alterations in enzymatic activity during NO-modulated nodule ageing (Puppo et al. 2013; Melo et al. 2011; Blanquet et al. 2015). In soybean and bean nodules, leghemoglobin is tyrosine-nitrated, but its decrease in senescent nodules suggests proteolysis targeting the modified protein (Sainz et al. 2015).

Understanding how NO triggers senescence is a compelling question. Unraveling NO mysteries influence on plant organ senescence holds the promise of a green revolution in plant biotechnology. In this landscape, NO emerges as a secret weapon, fortifying plants against the trials of nature.

15.5.4 NO in Hypersensitive Cell Death

The defense response in plants following pathogen recognition unveils a fascinating phenomenon named hypersensitive response (HR), marked by programmed cell execution in the area of known infection (Garcion et al. 2014). Recent studies suggest that the orchestration of cell death during HR involves the dynamic interplay of reactive oxygen intermediates (ROIs) produced by NADPH oxidases plus NO liberation in the nitrosative stress (Yu et al. 2014). Notably, mutations in the AtGSNOR1 gene emphasize its role in regulating global S-nitrosothiol (SNO) concentrations during HR cell death. Surprisingly, heightened SNO concentrations facilitate cell death with a lack of SA, a well-known activator of cell death (Hashem 2019). This unexpected finding challenges traditional assumptions, revealing that the occurrence of cell death, with increased kinetics and magnitude, can confer resistance against pathogens, even in conditions of low SA. Moreover, changes in SNO levels regulate apoplastic reaction oxygen intermediates accumulation (ROI), with S-nitrosylation of NADPH oxidases at Cys 890 playing a key role in modulating their activity. Interestingly, such redox modification, observed with relatively high SNO concentrations, reduces NADPH activity, thereby curbing ROI accumulation during later stages of HR. The conservation of this S-nitrosylation mechanism in NADPH oxidase within humans and flies suggests similar regulatory pathway in responses of immune across species (Yu et al. 2014). Additionally, NO's impact on peroxynitrite detoxification, particularly through S-nitrosylation of peroxiredoxin II E (PrxII E), further underscores the intricate control of NO over its radicals during HR cell death, potentially influencing downstream tyrosine kinase signaling (Romero-Puertas et al. 2007). As our understanding deepens, uncovering additional reactive cysteine residues subject to oxidative modifications promises to reveal more about the molecular intricacies of cell death development in plant immunity.

15.5.5 NO Interactions with Phytohormones

Auxins (AUXs) are pivotal in governing diverse developmental processes in plants, particularly in root system development, where optimal indole-3-acetic acid (IAA) levels play a crucial role (Casanova-Sáez et al. 2021). AUX and NO collaborate in orchestrating various plant actions, encompassing root organogenesis, gravitropic, nodule formation, response against iron deficiency, cell division activation, and embryogenic cell development (Sun et al. 2017). The interaction between AUX and NO is complex and context-dependent. Exogenous AUX administration or stress conditions often lead to increased NO production, and NO is also produced during the conversion of IBA into IAA (Rinaldi et al. 2012). The interplay between AUX and NO under metal stress conditions yields conflicting outcomes, impacting root meristem growth, auxin levels, and transport. Recent research reveals bidirectional interactions, with NO influencing AUX metabolism, transport, and signaling (Hu et al. 2017). Under cadmium stress, NO increases root IAA concentrations by reducing IAA breakdown, mitigating stress effects. Conversely, excessive NO inhibits acropetal transport of AUX in *Arabidopsis* modulating the abundance of the AUX efflux protein PIN-FORMED 1 (PIN1) in roots leading to reduced meristem size of root and activity (Martínez-Lorente et al. 2022). Additionally, NO directly impacts AUX perception and signal transduction by S-nitrosylating the AUX receptor protein TIR1, promoting

AUX-dependent gene expression by promoting AUX/IAA degradation (Chauhan et al. 2022). In essence, the intricate interplay between AUX and NO governs fundamental aspects of plant development and responses to environmental stimuli, highlighting their interconnected regulatory roles.

Moreover, NO exhibits synergistic as well as antagonistic effects on cytokinin (CK) pathway, influencing CK responses depending on the plant cell's condition. Surprisingly, synergistic effects often arise when CKs combine with molecules that decrease NO content, such as NO-deficient mutants (Khan et al. 2023). For example, the NO-deficient mutant "noa1" exhibits reduced sensitivity towards suppressing root growth when exposed to CKs. Additionally, NO can initiate CYCD3;1 by CKs in the course of cell division, correcting abnormalities related to the meristems in the mutant of noa1. However, upon combining CKs with increased NO content, achieved through NO contributors or mutants overexpressing NO, demonstrates antagonistic effects. In mutants with elevated endogenous NO levels, CKs reverse shoot growth phenotypes (Khan et al. 2023). Conversely, NO treatment reduces CK activity, preventing primary root formation. Elevated endogenous NO levels inhibit CK signaling, as observed in mutants like gsnor1 and nox1, leading to reduced sensitivity of CK's towards root growth restriction (Khan et al. 2019). Additional researches are required to enhance our understanding regarding the intricate interactions among NO and CK.

Further, seed dormancy and germination processes are intricately regulated by NO and abscisic acid (ABA). ABA induces dormancy and inhibits germination however, NO has the opposite effect, breaking seed dormancy and promoting germination. Seeds treated with NO donors exhibit reduced ABA content. However, NO-deficient mutants demonstrate increased seed dormancy and sensitivity towards ABA suppression of seed germination, highlighting important role of NO in breaking seed dormancy (León et al. 2014). S-nitrosylation at Cys-153 leads to the degradation of ABA insensitive 5 (ABI5), promoting seed germination and seedling growth. In stomata regulation, NO and ABA play crucial roles (Prakash et al. 2019). ABA, known for controlling stomatal closure, triggers increased NO synthesis in guard cells, leading to cGMP and cADPR activity and subsequently leads to closing of stomata. Interaction between NO and ABA is significant in leaf senescence and fruit ripening (Prakash et al. 2019). Exogenous NO application counterbalances leaf senescence by increasing ABA levels, and it delays fruit ripening by activating antioxidant enzymes. Under abiotic stress conditions, NO serves as a chemical/molecular messenger, enhancing the generation of phytohormones, antioxidative enzymes, chlorophyll synthesis, gene expression, and closure of stomata (Kumar and Ohri 2023). NO and ABA exhibit close interaction in response to stress, modulating ABA pathways, cGMP production, and MAPK activity, crucial for ABA-induced closure of stomata (Neill et al. 2008). Additionally, NO seems to play a role in ABA-triggered CAM expression and interacts with type 2C protein phosphatases (PP2Cs) in intricate networks regulating ABA-induced stomatal closure (Chen et al. 2021). The intricate relationship between NO and ABA extends to seed dormancy, stomatal regulation, and stress response, emphasizing their integrative function in the adaptive mechanisms of plants.

Gibberellic acid (GAs) and NO are key players in plant developmental processes, with a particularly well-explored interaction involved in controlling the germination of seeds and hypocotyl elongation during de-etiolation of seedling (Falak et al. 2021). NO acts upstream of GAs, influencing their production, perception, and transduction. Evidence suggests an antagonistic relationship between NO and GAs in various physiological processes, and DELLA proteins, important regulators in hormonal signaling, play a crucial role in their crosstalk (Simontacchi et al. 2013). Research indicates that NO has negative impact on the gibberellic acid (GAs) signaling network leading to the accumulation of GA-regulated DELLA proteins, thereby reducing tissue sensitivity to GA. The antagonistic relationship extends to their endogenous levels, where exogenous GA administration reduces NO production, and vice versa. However, in some instances, NO can stimulate GA biosynthetic machinery, leading to a beneficial relationship, such as in the transcription of GA3oxidase genes during seed dormancy breaking in *Arabidopsis* seeds (Castro-Camba et al. 2022). Despite their intricate interactions, both NO and GA stimulation of seed germination can be hampered by high concentrations of ABA, emphasizing the complexity of their regulatory networks (Parwez et al. 2022). While it is established that both NO and GA encourage germination in various species, the detailed mechanisms and interactions during seed dormancy breaking remain subjects for further exploration in future studies.

Numerous researches have consistently demonstrated the antagonistic relationship between NO and ethylene (ETH). This antagonism is particularly evident in regulating fruit ripening as well as leaf and floral senescence (Xiang et al. 2021). NO acts as an antagonist to ETH, delaying senescence in both vegetative and reproductive organs. Exogenous NO administration, either through fumigation or the addition of NO-liberating compounds, has been shown to inhibit factors that promote ETH generation, leading to a delay in senescence (Liu et al. 2023).

The measurements of ETH and NO levels during fruit ripening and plant senescence support these findings, revealing opposing trends where ETH production increases while levels of NO decrease during the process of ripening and senescence (Manjunatha et al. 2012). It has been proposed that NO may limit formation of ETH by influencing the transcription and/or activity of key ETH biosynthesis enzymes. In climacteric fruits, exogenous NO has been observed to alter the activity and transcription of ACC oxidase (ACO) and ACC synthase (ACS), enzymes crucial for ETH production (Singh et al. 2013). The inhibitory impact of NO on ACC, ACS, and ACO likely contributes to the reduction in ETH synthesis in plant tissues (Vanderstraeten et al. 2019).

Further, NO and salicylic acid (SA) play pivotal roles as signaling compounds in plant growth, development, and defence responses to both biotic and abiotic stress conditions (Prakash et al. 2021). Application of both NO and SA has been shown to significantly enhance photosynthesis, regulate plant water balance, and ultimately improve crop yield under normal conditions (Tahjib-Ul-Arif et al. 2018). Additionally, these compounds contribute significantly to plant defence mechanisms against various stresses. Studies have demonstrated that both NO and SA can mitigate the adverse effects of stressors such as cadmium (Cd) on plant growth. For example, the application of NO and SA was found to be more effective than their individual application in alleviating Cd stress in rice (Khan et al. 2023). Numerous investigations have reported the positive impact of exogenously supplied NO and SA in reducing the harmful effects of extreme temperature, drought, salinity, and heavy metal stress on various plant species. Furthermore, NO and SA are key players in promoting different modes of plant immunity, including effector-triggered immunity (ETI), pathogen-triggered immunity (PTI), and systemic acquired resistance (SAR) against virulent and avirulent bacteria (Zhang et al. 2018). In *Arabidopsis thaliana*, genes such as IAA-leucine resistant (AtILL6) and salicylic acid induction deficient 2 (AtSID2), induced by NO, positively regulate PTI, ETI, and SAR when plants are exposed to pathogenic bacteria (Khan et al. 2021). The induction of these genes by NO is considered crucial for promoting plant immunity against pathogens.

15.5.6 NO Action in Root Hair Development and Iron Homeostasis Regulation

In the intricate world beneath the soil, NO takes center stage, orchestrating a mesmerizing performance that influences root hair development and the dynamic interplay of iron nutrition dynamics (Lombardo et al. 2006). Root hairs, the prima donnas of the plant world, delicately sway between water absorption and anchorage, guided by the rhythmic cues of NO. As trichoblasts, these specialized cells undergo a captivating transformation, marked by vacuolation, all under the watchful influence of NO, highlighting its pivotal role in this botanical spectacle.

The NO saga doesn't end with initiation; it gracefully permeates the cytoplasm of nature cells and the vacuoles of actively growing root hairs. Depleting NO in *Arabidopsis* root hairs disrupts the harmony of essential processes—endocytosis, vesicle formation, and trafficking—affecting the very rhythm of root hair development. Mutations in NO-related genes introduce a new twist, unraveling distinct phenotypes and underscoring NO's intricate role in regulating the vesicle dynamics of this root hair ballet (Lombardo et al. 2006).

But NO's influence extends beyond the roots, seamlessly transitioning to a central role of iron metabolism, and homeostasis. Guided by NO, the formation of iron–nitrosyl complexes takes center stage, enhancing iron availability, a virtuoso performance under scarcity of iron. Here, NO takes the lead, regulating in regulating the iron reductase as well as transporting iron which is gracefully choreographed by the transcription maestro FIT (FER-LIKE IRON DEFICIENCY INDUCED TRANSCRIPTION FACTOR) (Ramirez et al. 2011). This regulatory interplay extends the oxidative stress in leaves during any pathogen infection, where nitric oxide (NO) dynamically modulates NADPH oxidase activity through S-nitrosylation. These revelations cast NO as the lead dancer, captivating scientists, and farmers alike with the promise of resilient crops and increased yields. The ongoing crescendo of advancements, from manipulating NO function to biotechnological marvels addressing iron fluctuations, aligns with the potential benefits of understanding and harnessing NO's versatile roles. The harmonious interplay between NO, root hair development, and iron nutrition paints a panoramic portrait of how this signaling virtuoso contributes to the overall vitality and productivity of plants.

15.5.7 NO Function in Plant Nutrition and Horticulture Innovations

Nitric oxide is recognized as a pivotal regulator influencing the uptake as well as homeostasis of crucial macronutrients, like nitrogen (N), phosphorus (P), and potassium (K) (de Bang et al. 2021). These nutrients play essential roles in the formation of key macromolecules like nucleic acid, amino acids, and proteins. Notably, nitrate (NO3−), a major nitrogen source in agricultural systems, is metabolically linked to NO through the enzyme NR, impacting nitrogen acquisition and NO generation. This connection manifests as rapid NO production in response to increased NO3− supplies, and in turn, NO participates in root architecture adjustments, highlighting its involvement in nitrogen perception and uptake (Sun et al. 2021).

Beyond nitrogen dynamics, NO exerts multifaceted regulatory roles in response to deficiencies in other nutrients, such as phosphorus (P) and potassium (K). Under Phosphorus deficient conditions, NO contributes to root system remodelling and enhances element transport. Similarly, in K restriction scenarios, NO negatively regulates K+ channel, indicating its role in cellular-level potassium homeostasis. While interactions between NO and micronutrient deficiencies are less explored, the existing studies emphasize the positive effects of NO under increased copper (Cu) levels, mitigating toxicity and enhancing plant resilience (Sun et al. 2021).

Additionally, NO play a vital role in plant responses against abiotic stresses, including salinity, drought, and extreme temperatures. Under salt stress, NO has shown to improved seed germination and plant growth by mitigating oxidative stress in various horticultural crops. In drought conditions, NO aids in preserving cell structure and promoting root development. Moreover, NO contributes to enhanced antioxidant systems and osmolyte metabolism, conferring drought tolerance in several horticultural crops. Both low and high temperatures induce NO accumulation, with exogenous NO application triggering cold acclimation and tolerance, as well as mitigating heat-induced oxidative stress (Nabi et al. 2019).

Considering the intricate interplay of NO with nutrient dynamics and stress responses, its role in horticulture holds significant promise for improving crop performance under challenging conditions. Numerous studies highlight the potential of NO in enhancing nutritional quality, and extending the shelf life of fruits like strawberry, apple, and plum, as well as vegetables such as mushrooms, lettuce, and broccoli. Previous reports suggest that NO acts as a fumigant to control postharvest diseases and pests, presenting a safe option for disease and pest control in fresh produce (Liu et al. 2023).

Notably, NO under ultralow oxygen conditions demonstrates inhibitory effects on surface browning of apple slices and lettuce, while also extending the postharvest life of various flowers. Additionally, NO diminishes pesticide toxicity, reducing residues in horticultural crops and mitigating potential health risks. Its role in mediating antioxidant defence systems contributes to pesticide tolerance in plants, protecting against cellular damage induced by herbicides. Moreover, NO enhances specific enzymes like hydrolytic enzymes, glutathione, cytochrome P450 oxidases, and peroxidases involved in pesticide detoxification. This underscores the strong potential of NO in protecting horticultural crops and reducing pesticide residues in food crops (Huque et al. 2013). NO plays diverse physiological and biological role in development, environmental interactions, and postharvest storage of horticultural crops. Despite significant studies in model plant species, the practical application of NO in horticulture is still in its early stages, hindered by the instability of NO donors. A promising avenue involves NO donor-entrapped nanomaterials, a technique successfully applicable in biomedicine. This innovation, exemplified by chitosan nanoparticles releasing NO protecting maize from salt stress, holds significant economic potential for agriculture and horticulture (Mukherjee, 2019).

The significant role of S-nitrosation in transducing NO bioactivity influencing plant growth and stress responses underscores its importance. Future research should delve into the downstream signaling cascades activated by S-nitrosation, especially focused on the identification and characterization of major transcription factors subject to S-nitrosation in crop plants (Aranda-Caño et al. 2019).

Advancements in sequencing technology have enabled the sequencing of genomes in some horticultural crops. The room for genetic manipulations and molecular tools available may facilitate the transfer of desirable NO-related features identified in model plant species into horticultural species for improved production, however, challenges in genome editing persist.

15.6 NO Symphony with Reactive Signaling Partners in Plant Cells

The production of NO enhances under stress conditions, leading to the generation of NO-derived molecules like, nitrogen dioxide (NO_2), dinitrogen trioxide (N_2O_3), peroxynitrite ($ONOO^-$), S-nitrosoglutathione (GSNO), and S-nitrosothiols (SNOs). This collective group of compounds based on nitrogen is often referred to as RNS (Corpas et al. 2023).

Under conditions of over-accumulation, these reactive molecules can induce a nitrosative stress, a term introduced in plant sciences in 2003. Nitrosative stress, in addition to oxidative stress, may contribute to cellular damage. While oxidative stress markers include processes like lipid peroxidation and protein carbonylation, RNS primarily use post-translational modifications (PTMs) such nitroalkylation, S-nitrosation, and tyrosine nitration to transfer their bioactivity. These PTMs are essential for both oxidative and nitrosative signaling pathways and control the activity of proteins.

The potential for $ONOO^-$ generation arises when superoxide anions and NO are produced simultaneously in plant cells. $ONOO^-$ possesses signaling properties and causes nitrosylation of tyrosine residues in target proteins, influencing the functions (Pacher et al. 2007). Additionally, GSNO, produced when the Glutathione reacts with the NO, serves as a reservoir of NO bioactivity and contributes to NO transport within organisms.

The redox nature of NO has long been recognized, with interactions documented with various reactive signals, including ROS, H_2S, and H_2. NO engages in protein interactions through S-nitrosation, thereby exerting influence over cellular redox status. The scrutiny of NO's impact on proteins has garnered attention, facilitated by the development of methods for its assay. These techniques enable the elucidation of roles played by S-nitrosated proteins, exemplified by RBOHD, a protein implicated in the generation of reactive signaling molecules. Furthermore, interaction of NO and fatty acids, resulting in the synthesis of nitro-fatty acids (NO_2-FAs) or nitrolipids. NO_2-FAs libarate NO and regulate gene expression linked with antioxidant responses, illustrating the diverse interactions of NO within cellular processes.

Overall, NO does not function in isolation, as evidenced by extensive investigations into its interactions with ROS, H_2S, H_2, proteins, fatty acids, and redox potential in recent years (Aranda-Caño, 2019). The complex interplay within this network of interactions plays a crucial role in modulating signaling pathways and cellular responses.

15.7 NO-Dependent Post-Translational Modifications in Plants

In the animal kingdom, NO, mediated by NOS, engages with soluble guanylate cyclase, initiating a cascade that culminates in the formation of cyclic 3′,5′-guanosine monophosphate (cGMP). Despite the detection of reduced amounts of cGMP in plants, bioinformatic analyses of more than 1000 plant species cast doubt on the presence of signaling pathways of NO-cGMP in plants. In the absence of a discernible NO receptor, it appears that NO communicates its bioactivity in plants coordinating post-translational modifications (PTMs) that rely on NO through complex chemical interactions with specific residues of target proteins (Denninger and Marletta, 1999).

In the intricate realm of post-translational modifications (PTMs), one standout player is S-nitrosation—a process where NO chemically attaches and detaches from reactive thiol groups, forming S-nitrosothiols. Pioneering work using biotin switch approaches uncovered a treasure trove of S-nitrosated proteins in *Arabidopsis*, spanning metabolic pathways, cytoskeleton dynamics, cellular signaling, redox balance, and stress responses. AtGSNOR1, a gene similar to S-nitroglutathione reductase, dictating the ebb and flow of S-nitrosothiols in *Arabidopsis* and holding sway over global S-nitrosylation extent. NO, employing its regulatory precision, strategically employs S-nitrosation to mitigate cellular apoptosis during hypersensitive responses. This adept modulation involves the inhibition of ROS production through S-nitrosation of NADPH oxidase at the cysteine residue.

In plants, a parade of reports highlights NO's symphony in orchestrating S-nitrosation's influence over enzymes and proteins governing hormone transport and signal transduction, thereby shaping development and defense responses (Anand and Stamler, 2012). A stellar example unfolds in the auxin domain, where NO influences an auxin receptor, with S-nitrosation at specific sites namely, Cys 140 and Cys 480. This kind of PTM enhances the interaction of TIR1 with AUX/IAA proteins, setting the stage for AUX/IAA degradation and orchestrating the process of auxin signal transduction. NO, the maestro of negative regulation, orchestrates a similar symphony in ABA and cytokinin signaling (Wang et al. 2016).

As NO's redox ballet unfolds through S-nitrosation, the spotlight turns to the pivotal question: how does the show end? Here enter GSNOR, the conductor orchestrating the denitrosylation finale. This evolutionarily conserved NADH-dependent reductase, is a maestro in its own right, which fine-tunes GSNO level and, consequently, global S-nitrosylation. The GSNOR-led denitrosylation takes center stage in regulating growth and development of plant, tolerance to disease, temperature and salt, iron toxicity, and responses to low oxygen—a narrative awaiting exploration in the diverse tapestry of horticultural crop species (Choi, 2018).

Moreover, NO manages detailed post-translational modifications (PTMs) in plants, influencing various metalloproteins occurs through the influence of metal S-nitrosylation reactions, thereby modulating their functions (Astier and Lindermayr, 2012). Another facet of NO-mediated PTMs is tyrosine nitration, yielding 3-nitrotyrosine, linked to nitro-oxidative damage in plants. Under unfavorable conditions, like elevated temperature, salinity stress, and arsenic stress, plants accumulate tyrosine-nitrated proteins, raising questions about the potential role of tyrosine nitration in the transmission of NO signals through proteasomal degradation (Corpas and Barroso 2013). Further, recent revelations unveil the formation of various nitro-fatty acids, including nitro-oleic acid and nitro-linolenic acid, as products of NO-related species in plants. Accumulation of nitro-fatty acids in response to stress suggests their role as novel mediators in NO-mediated defense mechanisms against various biotic and abiotic stressors (Aranda-Caño et al. 2019).

A multilayered molecular framework, encompassing the time and space dimension of PTMs-based NO, governs signaling specificity in distinct cellular microenvironments. Various redox based modifications of Cysteine thiols, including SNO, S–OH, S–S, S-SG, S–O_2H, and S-O_3H, each leading to diverse conformational changes and cellular outcomes. While proximity-based mechanisms in mammals maintain NO signaling specificity, this aspect remains unexplored in various plants (Chung et al. 2013; Foster et al. 2009). Intriguingly, NO also influences other fundamental PTMs like SUMOylation, phosphorylation, persulfidation, and acetylation, providing a versatile means of tailoring cellular responses to stimuli. Distinct differences between ROS and RNS further contribute to the specificity of redox signaling outputs (Mendel and Schwarz, 2023).

15.8 Future Perspectives

The versatility of NO in the detailed weave of environmental interactions, growth, and development, and responses to immune system is increasingly evident. While strides have been made in understanding plant NO biology, numerous challenges persist. Beyond the synthesis of NO, a crucial frontier for exploration lies in unraveling how NO signaling is switched off or downgraded—a particularly intricate matter given the transient nature of biological cues. While AtGSNOR1 has been identified, its role in indirectly controlling S-nitrosylation lacks the specificity required for fine-tuning NO signaling. Exploring additional enzymes that may revert specific S-nitrosylated cysteine residues to thiol, while maintaining the neighboring S-nitroso groups could be beneficial for comprehending the complexities of redox signal transmission. Moreover, the detection and characterization of denitrosylases could provide valuable insights into the molecular machinery integral to transduction of NO bioactivity. Achieving specificity in S-nitrosylation poses another critical challenge. The existence of analogous enzymes orchestrating the transmission of NO-related biological activities in plants remains a question.

Moreover, the connection between nitrate assimilation and NO generation via NR adds complexity to understanding the interplay between NO assimilation, biosynthesis, and turnover—a promising area for future exploration. As our knowledge of NO biology in plants deepens, translating these insights into practical applications becomes imperative. Unveiling the redox-based molecular machinery governing the function of NO holds the potential to shape future breeding or design strategies for key plant traits, marking a transformative journey in agricultural innovation.

Acknowledgement The Head, Department of Botany, Banaras Hindu University, Varanasi, India is gratefully acknowledged for providing laboratory facilities. Arun kumar Mishra sincerely acknowledge the support from Institute of Eminence. (IoE-6031), Banaras Hindu University. Neha Gupta, Ankit Srivasatava is thankful to the CSIR for research fellowship in terms of JRF and SRF. Anirbana Parida is thankful to UGC-Non-NET.

Conflict of Interest The authors affirm that they have no competing interests.

References

Alber NA, Sivanesan H, Vanlerberghe GC (2017) The occurrence and control of nitric oxide generation by the plant mitochondrial electron transport chain. Plant Cell Environ 40(7):1074–1085

Alderton WK, Cooper CE, Knowles RG (2001) Nitric oxide synthases: structure, function and inhibition. Biochem J 357(3):593–615

Allegre A, Silvestre J, Morard P, Kallerhoff J, Pinelli E (2004) Nitrate reductase regulation in tomato roots by exogenous nitrate: a possible role in tolerance to long-term root anoxia. J Exp Bot 55(408):2625–2634

Anand P, Stamler JS (2012) Enzymatic mechanisms regulating protein S-nitrosylation: implications in health and disease. Journal of molecular medicine 90:233–244

Aranda-Caño L, Sánchez-Calvo B, Begara-Morales JC, Chaki M, Mata-Pérez C, Padilla MN et al (2019) Post-translational modification of proteins mediated by nitro-fatty acids in plants: nitroalkylation. Plants 8(4):82

Astier J, Lindermayr C (2012) Nitric oxide-dependent posttranslational modification in plants: an update. International journal of molecular sciences 13(11):15193–15208

Bethke PC, Libourel IG, Jones RL (2006) Nitric oxide reduces seed dormancy in Arabidopsis. Journal of experimental botany 57(3):517–526

Blanquet P, Silva L, Catrice O, Bruand C, Carvalho H, Meilhoc E (2015) Sinorhizobium meliloti controls nitric oxide–mediated post-translational modification of a Medicago truncatula nodule protein. Molecular Plant-Microbe Interactions 28(12):1353–1363

Boscari A, Meilhoc E, Castella C, Bruand C, Puppo A, Brouquisse R (2013) Which role for nitric oxide in symbiotic N2-fixing nodules: toxic by-product or useful signaling/metabolic intermediate? Front Plant Sci 4:384

Bright J, Hiscock SJ, James PE, Hancock JT (2009) Pollen generates nitric oxide and nitrite: a possible link to pollen-induced allergic responses. Plant Physiol Biochem 47(1):49–55

Broniowska KA, Hogg N (2012) The chemical biology of S-nitrosothiols. Antioxid Redox Signal 17(7):969–980

Bruand C, Meilhoc E (2019) Nitric oxide in plants: pro-or anti-senescence. J Exp Bot 70(17):4419–4427

Cam Y, Pierre O, Boncompagni E, Hérouart D, Meilhoc E, Bruand C (2012) Nitric oxide (NO): a key player in the senescence of Medicago truncatula root nodules. New Phytologist 196(2):548–560

Casanova-Sáez R, Mateo-Bonmatí E, Ljung K (2021) Auxin metabolism in plants. Cold Spring Harb Perspect Biol 13(3):a039867

Castello PR, David PS, McClure T, Crook Z, Poyton RO (2006) Mitochondrial cytochrome oxidase produces nitric oxide under hypoxic conditions: implications for oxygen sensing and hypoxic signaling in eukaryotes. Cell Metab 3(4):277–287

Castro-Camba R, Sánchez C, Vidal N, Vielba JM (2022) Plant development and crop yield: the role of gibberellins. Plan Theory 11(19):2650

Chakraborty N (2021) Salicylic acid and nitric oxide cross-talks to improve innate immunity and plant vigor in tomato against Fusarium oxysporum stress. Plant Cell Reports 40(8): 1415–1427

Chauhan P, Mir RA, Khah MA (2022) Ascorbate–glutathione cycle: nitric oxide and phytohormone interactions for plant stress tolerance. Nitric Oxide in Plants: A Molecule with Dual Roles, 148–178

Chen Q, Bai L, Wang W, Shi H, Ramón Botella J, Zhan Q et al (2021) COP1 promotes ABA-induced stomatal closure by modulating the abundance of ABI/HAB and AHG3 phosphatases. New Phytol 229(4):2035–2049

Chen Y, Chen W, Li X, Jiang H, Wu P, Xia K et al (2014) Knockdown of LjIPT3 influences nodule development in Lotus japonicus. Plant Cell Physiol 55(1):183–193

Choi MS (2018) Pathophysiological role of s-nitrosylation and transnitrosylation depending on s-nitrosoglutathione levels regulated by s-nitrosoglutathione reductase. Biomolecules & Therapeutics 26(6):533

Chung KW, Jeong HO, Jang EJ, Choi YJ, Kim DH, Kim SR et al (2013) Characterization of a small molecule inhibitor of melanogenesis that inhibits tyrosinase activity and scavenges nitric oxide (NO). Biochimica et Biophysica Acta (BBA)-General Subjects 1830(10):4752–4761

Chungopast S, Duangkhet M, Tajima S, Ma JF, Nomura M (2017) Iron-induced nitric oxide leads to an increase in the expression of ferritin during the senescence of Lotus japonicus nodules. J Plant Physiol 208:40–46

Coba de la Pena T, Redondo FJ, Manrique E, Lucas MM, Pueyo JJ (2010) Nitrogen fixation persists under conditions of salt stress in transgenic Medicago truncatula plants expressing a cyanobacterial flavodoxin. Plant Biotechnology Journal 8(9):954–965

Cooney RV, Harwood PJ, Custer LJ, Franke AA (1994) Light-mediated conversion of nitrogen dioxide to nitric oxide by carotenoids. Environ Health Perspect 102(5):460–462

Corpas FJ, Barroso JB, Carreras A, Quirós M, León AM, Romero-Puertas MC et al (2004) Cellular and subcellular localization of endogenous nitric oxide in young and senescent pea plants. Plant Physiol 136(1):2722–2733

Corpas FJ, Barroso JB (2013) Nitro-oxidative stress vs oxidative or nitrosative stress in higher plants. New Phytologist 199(3):633–635

Corpas FJ, Begara-Morales JC, Sánchez-Calvo B, Chaki M, Barroso JB (2015) Nitration and S-nitrosylation: two post-translational modifications (PTMs) mediated by reactive nitrogen species (RNS) and their role in signalling processes of plant cells. Reactive oxygen and nitrogen species signaling and communication in plants, 267–281

Corpas FJ, Palma JM (2018) Assessing nitric oxide (NO) in higher plants: an outline. Nitrogen 1(1):12–20

Corpas FJ, & Palma JM (2023). Functions of NO and H2S signal molecules against plant abiotic stress. In Plant Abiotic Stress Signaling (pp. 97-109). New York, NY: Springer US.

Cvetkovska M, Dahal K, Alber NA, Jin C, Cheung M, Vanlerberghe GC (2014) Knockdown of mitochondrial alternative oxidase induces the 'stress state'of signaling molecule pools in Nicotiana tabacum, with implications for stomatal function. New Phytol 203(2):449–461

de Bang TC, Husted S, Laursen KH, Persson DP, Schjoerring JK (2021) The molecular–physiological functions of mineral macronutrients and their consequences for deficiency symptoms in plants. New Phytologist 229(5):2446–2469

Del Giudice J, Cam Y, Damiani I, Fung-Chat F, Meilhoc E, Bruand C et al (2011) Nitric oxide is required for an optimal establishment of the Medicago truncatula–Sinorhizobium meliloti symbiosis. New Phytol 191(2):405–417

Delledonne M, Xia Y, Dixon RA, Lamb C (1998) Nitric oxide functions as a signal in plant disease resistance. Nature 394(6693):585–588

Denninger JW, Marletta MA (1999) Guanylate cyclase and the NO/cGMP signaling pathway. Biochimica et Biophysica Acta (BBA)-Bioenergetics 1411(2-3):334–350

El-Maarouf-Bouteau H, Mazuy C, Corbineau F, Bailly C (2011) DNA alteration and programmed cell death during ageing of sunflower seed. J Exp Bot 62(14):5003–5011

Falak N, Imran QM, Hussain A, Yun BW (2021) Transcription factors as the "Blitzkrieg" of plant defense: a pragmatic view of nitric oxide's role in gene regulation. Int J Mol Sci 22(2):522

Feijó JA, Sainhas J, Holdaway-Clarke T, Cordeiro MS, Kunkel JG, Hepler PK (2001) Cellular oscillations and the regulation of growth: the pollen tube paradigm. BioEssays 23(1):86–94

Ferrarini A, De Stefano M, Baudouin E, Pucciariello C, Polverari A, Puppo A, Delledonne M (2008) Expression of Medicago truncatula genes responsive to nitric oxide in pathogenic and symbiotic conditions. Mol Plant-Microbe Interact 21(6):781–790

Foresi N, Calo G, Del Castello F, Nejamkin A, Salerno G, Lamattina L et al (2022) Arginine as the sole nitrogen source for Ostreococcus tauri growth: insights on nitric oxide synthase enzyme. Front Mar Sci 9:1064077

Foster MW, Hess DT, Stamler JS (2009) Protein S-nitrosylation in health and disease: a current perspective. Trends in molecular medicine 15(9):391–404

Fröhlich A, Durner J (2011) The hunt for plant nitric oxide synthase (NOS): is one really needed? Plant Sci 181(4):401–404

Garcion C, Lamotte O, Cacas JL, Métraux JP (2014) Mechanisms of defence to pathogens: biochemistry and physiology. Induced Resistance for Plant Defense, 106–136

Gibbs DJ, Isa NM, Movahedi M, Lozano-Juste J, Mendiondo GM, Berckhan S et al (2014) Nitric oxide sensing in plants is mediated by proteolytic control of group VII ERF transcription factors. Mol Cell 53(3):369–379

Gniazdowska A, Dobrzyńska U, Babańczyk T, Bogatek R (2007) Breaking the apple embryo dormancy by nitric oxide involves the stimulation of ethylene production. Planta 225:1051–1057

Guo Y, Ren G, Zhang K, Li Z, Miao Y, Guo H (2021) Leaf senescence: progression, regulation, and application. Molecular Horticulture 1:1–25

Gupta KJ, Lee CP, Ratcliffe RG (2017) Nitrite protects mitochondrial structure and function under hypoxia. Plant Cell Physiol 58(1):175–183

Hageman RH (1990) Historical perspectives of the enzymes of nitrate assimilation by crop plants and potential for biotechnological application. In: Inorganic nitrogen in plants and microorganisms: uptake and metabolism. Springer, Berlin Heidelberg, pp 3–11

Harrison J, Jamet A, Muglia CI, Van de Sype G, Aguilar OM, Puppo A, Frendo P (2005) Glutathione plays a fundamental role in growth and symbiotic capacity of Sinorhizobium meliloti. J Bacteriol 187(1):168–174

Hashem HA (2019) Nitric oxide A regulator of plant signaling and defense against abiotic stress. Plant Tolerance to Environmental Stress, 145–158

Hichri I, Meilhoc E, Boscari A, Bruand C, Frendo P, Brouquisse R (2016) Nitric oxide: jack-of-all-trades of the nitrogen-fixing symbiosis? Adv Bot Res 77:193–218

Horchani F, Prévot M, Boscari A, Evangelisti E, Meilhoc E, Bruand C et al (2011) Both plant and bacterial nitrate reductases contribute to nitric oxide production in Medicago truncatula nitrogen-fixing nodules. Plant Physiol 155(2):1023–1036

Hu Y, Vandenbussche F, Van Der Straeten D (2017) Regulation of seedling growth by ethylene and the ethylene–auxin crosstalk. Planta 245:467–489

Hussain S, Khan F, Cao W, Wu L, Geng M (2016) Seed priming alters the production and detoxification of reactive oxygen intermediates in rice seedlings grown under sub-optimal temperature and nutrient supply. Front Plant Sci 7:439

Huque R, Wills RBH, Pristijono P, Golding JB (2013) Effect of nitric oxide (NO) and associated control treatments on the metabolism of fresh-cut apple slices in relation to development of surface browning. Postharvest biology and technology 78:16–23

Ignarro LJ, Buga GM, Wood KS, Byrns RE, Chaudhuri G (1987) Endothelium-derived relaxing factor produced and released from artery and vein is nitric oxide. Proc Natl Acad Sci 84(24):9265–9269

Jamet A, Sigaud S, Van de Sype G, Puppo A, Hérouart D (2003) Expression of the bacterial catalase genes during Sinorhizobium meliloti-Medicago sativa symbiosis and their crucial role during the infection process. Mol Plant-Microbe Interact 16(3):217–225

Jasid S, Simontacchi M, Bartoli CG, Puntarulo S (2006) Chloroplasts as a nitric oxide cellular source. Effect of reactive nitrogen species on chloroplastic lipids and proteins. Plant Physiol 142(3):1246–1255

Khan M, Al Azawi TNI, Pande A, Mun BG, Lee DS, Hussain A et al (2021) The role of nitric oxide-induced ATILL6 in growth and disease resistance in Arabidopsis thaliana. Front Plant Sci 12:685156

Khan M, Ali S, Al Azzawi TNI, Yun BW (2023) Nitric oxide acts as a key signaling molecule in plant development under stressful conditions. Int J Mol Sci 24(5):4782

Khan MA, Khan AL, Imran QM, Asaf S, Lee SU, Yun BW et al (2019) Exogenous application of nitric oxide donors regulates short-term flooding stress in soybean. PeerJ 7:e7741

Kolbert ZS, Barroso JB, Brouquisse R, Corpas FJ, Gupta KJ, Lindermayr C et al (2019) A forty year journey: the generation and roles of NO in plants. Nitric Oxide 93:53–70

Krasuska U, Ciacka K, Orzechowski S, Fettke J, Bogatek R, Gniazdowska A (2016) Modification of the endogenous NO level influences apple embryos dormancy by alterations of nitrated and biotinylated protein patterns. Planta 244:877–891

Kumar D, Ohri P (2023) Say "NO" to plant stresses: unravelling the role of nitric oxide under abiotic and biotic stress. Nitric Oxide 130:36–57

León J, Castillo MC, Coego A, Lozano-Juste J, Mir R (2014) Diverse functional interactions between nitric oxide and abscisic acid in plant development and responses to stress. J Exp Bot 65(4):907–921

Libourel IGL, Van Bodegom PM, Fricker MD, Ratcliffe RG (2006) Nitrite reduces cytoplasmic acidosis under anoxia. Plant Physiol 142(4):1710–1717

Lillo C, Meyer C, Lea US, Provan F, Oltedal S (2004) Mechanism and importance of post-translational regulation of nitrate reductase. J Exp Bot 55(401):1275–1282

Liu Y, Chen T, Tao N, Yan T, Wang Q, Li Q (2023) Nitric oxide is essential to keep the postharvest quality of fruits and vegetables. Horticulturae 9(2):135

Lombardo MC, Graziano M, Polacco JC, Lamattina L (2006) Nitric oxide functions as a positive regulator of root hair development. Plant signaling & behavior 1(1):28–33
Loscos J, Matamoros MA, Becana M (2008) Ascorbate and homoglutathione metabolism in common bean nodules under stress conditions and during natural senescence. Plant Physiology 146(3):1282–1292
Loughran PA, Stolz DB, Vodovotz Y, Watkins SC, Simmons RL, Billiar TR (2005) Monomeric inducible nitric oxide synthase localizes to peroxisomes in hepatocytes. Proc Natl Acad Sci 102(39):13837–13842
Luiking YC, Ten Have GA, Wolfe RR, Deutz NE (2012) Arginine de novo and nitric oxide production in disease states. Am J Physiol Endocrinol Metab 303(10):E1177–E1189
Manjunatha G, Gupta KJ, Lokesh V, Mur LA, Neelwarne B (2012) Nitric oxide counters ethylene effects on ripening fruits. Plant Signal Behav 7(4):476–483
Martínez-Lorente SE, Pardo-Hernández M, Martí-Guillén JM, López-Delacalle M, Rivero RM (2022) Interaction between melatonin and no: action mechanisms, main targets, and putative roles of the emerging molecule nomela. Int J Mol Sci 23(12):6646
Martínez-Medina A, Pescador L, Fernández I, Rodríguez-Serrano M, García JM, Romero-Puertas MC, Pozo MJ (2019) Nitric oxide and phytoglobin PHYTOGB1 are regulatory elements in the Solanum lycopersicum–Rhizophagus irregularis mycorrhizal symbiosis. New Phytol 223(3):1560–1574
Melo PM, Silva LS, Ribeiro I, Seabra AR, Carvalho HG (2011) Glutamine synthetase is a molecular target of nitric oxide in root nodules of Medicago truncatula and is regulated by tyrosine nitration. Plant Physiol 157(3):1505–1517
Mendel RR, Schwarz G (2023) The History of Animal and Plant Sulfite Oxidase—A Personal View. Molecules 28(19):6998
Mishina TE, Lamb C, Zeier J (2007) Expression of a nitric oxide degrading enzyme induces a senescence programme in Arabidopsis. Plant Cell Environ 30(1):39–52
Mukherjee S (2019) Recent advancements in the mechanism of nitric oxide signaling associated with hydrogen sulfide and melatonin crosstalk during ethylene-induced fruit ripening in plants. Nitric Oxide 82:25–34
Mur LA, Mandon J, Persijn S, Cristescu SM, Moshkov IE, Novikova GV et al (2013) Nitric oxide in plants: an assessment of the current state of knowledge. AoB plants 5:pls052
Nabi RBS, Tayade R, Hussain A, Kulkarni KP, Imran QM, Mun BG, Yun BW (2019) Nitric oxide regulates plant responses to drought, salinity, and heavy metal stress. Environmental and Experimental Botany 161:120–133
Nagata M, Murakami EI, Shimoda Y, Shimoda-Sasakura F, Kucho KI, Suzuki A et al (2008) Expression of a class 1 hemoglobin gene and production of nitric oxide in response to symbiotic and pathogenic bacteria in Lotus japonicus. Mol Plant-Microbe Interact 21(9):1175–1183
Neill S, Barros R, Bright J, Desikan R, Hancock J, Harrison J et al (2008) Nitric oxide, stomatal closure, and abiotic stress. J Exp Bot 59(2):165-176
Nohl H, Gille L, Kozlov AV (1999) Critical aspects of the antioxidant function of coenzyme Q in biomembranes. Biofactors 9(2–4):155–161
Pacher P, Beckman JS, Liaudet L (2007) Nitric oxide and peroxynitrite in health and disease. Physiological reviews 87(1):315–424
Parwez R, Aftab T, Gill SS, Naeem M (2022) Abscisic acid signaling and crosstalk with phytohormones in regulation of environmental stress responses. Environ Exp Bot 199:104885
Pauly N, Ferrari C, Andrio E, Marino D, Piardi S, Brouquisse R et al (2011) MtNOA1/RIF1 modulates Medicago truncatula–Sinorhizobium meliloti nodule development without affecting its nitric oxide content. J Exp Bot 62(3):939–948
Planchet E, Jagadis Gupta K, Sonoda M, Kaiser WM (2005) Nitric oxide emission from tobacco leaves and cell suspensions: rate limiting factors and evidence for the involvement of mitochondrial electron transport. Plant J 41(5):732–743
Prado AM, Porterfield DM, Feijó JA (2004) Nitric oxide is involved in growth regulation and re-orientation of pollen tubes. Development 131 (11): 2707–2714

Prakash V, Singh VP, Tripathi DK, Sharma S, Corpas FJ (2019) Crosstalk between nitric oxide (NO) and abscisic acid (ABA) signalling molecules in higher plants. Environ Exp Bot 161:41–49
Prakash V, Singh VP, Tripathi DK, Sharma S, Corpas FJ (2021) Nitric oxide (NO) and salicylic acid (SA): a framework for their relationship in plant development under abiotic stress. Plant Biol 23:39–49
Puppo A, Groten K, Bastian F, Carzaniga R, Soussi M, Lucas MM et al (2005) Legume nodule senescence: roles for redox and hormone signalling in the orchestration of the natural aging process. New Phytol 165(3):683–701
Puppo A, Pauly N, Boscari A, Mandon K, Brouquisse R (2013) Hydrogen peroxide and nitric oxide: key regulators of the legume—rhizobium and mycorrhizal symbioses. Antioxid Redox Signal 18(16):2202–2219
Ramirez L, Simontacchi M, Murgia I, Zabaleta E, Lamattina L (2011) Nitric oxide, nitrosyl iron complexes, ferritin and frataxin: a well equipped team to preserve plant iron homeostasis. Plant Science 181(5):582–592
Redondo FJ, de la Pena TC, Morcillo CN, Lucas MM, Pueyo JJ (2009) Overexpression of flavodoxin in bacteroids induces changes in antioxidant metabolism leading to delayed senescence and starch accumulation in alfalfa root nodules. Plant Physiol 149(2):1166–1178
Reichler SA, Torres J, Rivera AL, Cintolesi VA, Clark G, Roux SJ (2009) Intersection of two signalling pathways: extracellular nucleotides regulate pollen germination and pollen tube growth via nitric oxide. J Exp Bot 60(7):2129–2138
Rinaldi MA, Liu J, Enders TA, Bartel B, Strader LC (2012) A gain-of-function mutation in IAA16 confers reduced responses to auxin and abscisic acid and impedes plant growth and fertility. Plant Mol Biol 79:359–373
Rockel P, Strube F, Rockel A, Wildt J, Kaiser WM (2002) Regulation of nitric oxide (NO) production by plant nitrate reductase in vivo and in vitro. J Exp Bot 53(366):103–110
Rodríguez-Serrano M, Pazmiño DM, Sparkes I, Rochetti A, Hawes C, Romero-Puertas MC, Sandalio LM (2014) 2, 4-Dichlorophenoxyacetic acid promotes S-nitrosylation and oxidation of actin affecting cytoskeleton and peroxisomal dynamics. J Exp Bot 65(17):4783–4793
Romero-Puertas MC, Laxa M, Matte A, Zaninotto F, Finkemeier I, Jones AM et al (2007) S-nitrosylation of peroxiredoxin II E promotes peroxynitrite-mediated tyrosine nitration. Plant Cell 19(12):4120–4130
Sainz M, Calvo-Begueria L, Pérez-Rontomé C, Wienkoop S, Abián J, Staudinger C et al (2015) Leghemoglobin is nitrated in functional legume nodules in a tyrosine residue within the heme cavity by a nitrite/peroxide-dependent mechanism. Plant J 81(5):723–735
Sami F, Faizan M, Faraz A, Siddiqui H, Yusuf M, Hayat S (2018) Nitric oxide-mediated integrative alterations in plant metabolism to confer abiotic stress tolerance, NO crosstalk with phytohormones and NO-mediated post translational modifications in modulating diverse plant stress. Nitric Oxide 73:22–38
Sang Y, Blecha F (2008) Antimicrobial peptides and bacteriocins: alternatives to traditional antibiotics. Anim Health Res Rev 9(2):227–235
Santana MM, Gonzalez JM, Cruz C (2017) Nitric oxide accumulation: the evolutionary trigger for phytopathogenesis. Front Microbiol 8:1947
Schippers JH, Schmidt R, Wagstaff C, Jing HC (2015) Living to die and dying to live: the survival strategy behind leaf senescence. Plant Physiol 169(2):914–930
Seligman K, Saviani EE, Oliveira HC, Pinto-Maglio CAF, Salgado I (2008) Floral transition and nitric oxide emission during flower development in Arabidopsis thaliana is affected in nitrate reductase-deficient plants. Plant Cell Physiol 49(7):1112–1121
Simontacchi M, García-Mata C, Bartoli CG, Santa-María GE, Lamattina L (2013) Nitric oxide as a key component in hormone-regulated processes. Plant Cell Rep 32:853–866
Simpson GG (2005) NO flowering. BioEssays 27(3):239–241
Singh Z, Khan AS, Zhu S, Payne AD (2013) Nitric oxide in the regulation of fruit ripening: challenges and thrusts. Stewart Postharvest Rev 9(4):1–11
Šírová J, Sedlářová M, Piterková J, Luhová L, Petřivalský M (2011) The role of nitric oxide in the germination of plant seeds and pollen. Plant Sci 181(5):560–572

Sun H, Feng F, Liu J, Zhao Q (2017) The interaction between auxin and nitric oxide regulates root growth in response to iron deficiency in rice. Front Plant Sci 8:2169

Sun Y, Ma C, Kang X, Zhang L, Wang J, Zheng S, Zhang T (2021) Hydrogen sulfide and nitric oxide are involved in melatonin-induced salt tolerance in cucumber. Plant Physiology and Biochemistry 167:101–112

Tahjib-Ul-Arif M, Siddiqui MN, Sohag AAM, Sakil MA, Rahman MM, Polash MAS et al (2018) Salicylic acid-mediated enhancement of photosynthesis attributes and antioxidant capacity contributes to yield improvement of maize plants under salt stress. J Plant Growth Regul 37:1318–1330

Tewari RK, Kumar P, Kim S, Hahn EJ, Paek KY (2009) Nitric oxide retards xanthine oxidase-mediated superoxide anion generation in Phalaenopsis flower: an implication of NO in the senescence and oxidative stress regulation. Plant Cell Rep 28:267–279

Trewavas AJ, Malhó R (1997) Signal perception and transduction: the origin of the phenotype. Plant Cell 9(7):1181

Vanderstraeten L, Depaepe T, Bertrand S, Van Der Straeten D (2019) The ethylene precursor ACC affects early vegetative development independently of ethylene signaling. Front Plant Sci 10:1591

Vicente J, Mendiondo GM, Movahedi M, Peirats-Llobet M, Juan YT, Shen YY et al (2017) The Cys-Arg/N-end rule pathway is a general sensor of abiotic stress in flowering plants. Curr Biol 27(20):3183–3190

Wang YJ, Wei XY, Jing XQ, Chang YL, Hu CH, Wang X, Chen KM (2016) The fundamental role of NOX family proteins in plant immunity and their regulation. Int J Mol Sci 17(6):805

Wany A, Gupta AK, Kumari A, Mishra S, Singh N, Pandey S et al (2019) Nitrate nutrition influences multiple factors in order to increase energy efficiency under hypoxia in Arabidopsis. Ann Bot 123(4):691–705

Xiang W, Wang HW, Sun DW (2021) Phytohormones in postharvest storage of fruit and vegetables: mechanisms and applications. Crit Rev Food Sci Nutr 61(18):2969–2983

Ya'Acov YL, Haramaty E (1996) The characterization and contrasting effects of the nitric oxide free radical in vegetative stress and senescence of Pisum sativum Linn. foliage. Journal of Plant Physiology 148(3-4):258–263

Yamasaki H, Sakihama Y, Takahashi S (1999) An alternative pathway for nitric oxide production in plants: new features of an old enzyme. Trends Plant Sci 4(4):128–129

Yarkoni O (2012) Engineering an inducible NO pathway to facilitate cell-electronics communication (Doctoral dissertation, Newcastle University)

Yu M, Lamattina L, Spoel SH, Loake GJ (2014) Nitric oxide function in plant biology: a redox cue in deconvolution. New Phytol 202(4):1142–1156

Zafra A, Rodríguez-García MI, Alché JDD (2010) Cellular localization of ROS and NO in olive reproductive tissues during flower development. BMC Plant Biology 10:1–14

Zhang H, Shen WB, Zhang W, Xu LL (2005) A rapid response of β-amylase to nitric oxide but not gibberellin in wheat seeds during the early stage of germination. Planta 220:708–716

Zhang W, Zhao F, Jiang L, Chen C, Wu L, Liu Z (2018) Different pathogen defense strategies in Arabidopsis: more than pathogen recognition. Cells 7(12):252

Case Studies on Stress Responses in Various Phytoplanktons

16

Savita Singh

Abstract

The ocean accounts for 70% of Earth's surface and serves as a habitat for a wide variety of life forms. The phytoplanktons are mobile microscopic organisms comprising both photosynthetic prokaryotes and eukaryotes, inhabiting marine and fresh-water systems. They play vital roles in marine ecosystems by acting as primary producers, affecting biogeochemical cycles, and regulating productivity outcomes. In the context of marine environments, the phytoplankton's productivity is influenced by a range of natural and human-induced activities. Various stresses that phytoplankton face include temperature, CO_2 concentration, acidification, ozone depletion, nutrient availability, and pollution caused by pollutants such as pesticides, organic pollutants, crude oil spills etc. These factors significantly affect survival, behavior, nutritional procurement, reproduction, and overall population dynamics. In this chapter, our objective is to systematically present all the discoveries related to the various stresses that affect the phytoplankton community by outlining the challenges, exploring their impacts, and analyzing their significance in relation to climate change, ecological shifts, environmental protection, and fisheries oceanography.

Keywords

Phytoplankton · Stress · Temperature · CO_2 concentration · Pollution · Climate change

S. Singh (✉)
Babu Shivnath Agrawal (PG) College, Mathura, India

A. K. Mishra (ed.), *Stress Biology in Photosynthetic Organisms*,
https://doi.org/10.1007/978-981-97-1883-2_16

16.1 Introduction

Earth's oceans are very old and comprise a myriad of habitats that have shaped the evolution of the incredible diversity of marine life we see today. The hypothesis suggests that the genesis of Earth's oceans transpired roughly 3.8–4.4 billion years ago, in close temporal proximity to the planet's formation approximately 4.5 billion years ago. About 98% of marine species are benthic, meaning they live on or within the seafloor, whereas only 2% inhabit the water column—that is, they are pelagic. Organisms that live in the water column are pelagic. Pelagic organisms can either drift with the currents, a life strategy called planktonic or propel themselves by swimming, a life strategy called nektonic. (A planktonic organism is called a plankton; a nektonic organism is called a nekton.) All other marine organisms, comprising a whopping 98 percent of described marine species, can be classified as benthic: that is, they live on or in the seafloor, or benthos.

Phytoplanktons are drifting photosynthetic prokaryotes and eukaryotes inhabiting marine and fresh-water systems. Approximately 4000–5000 legitimate species of marine phytoplanktons have been known (Sournia et al. 1991; Tett and Barton 1995). The size of phytoplankton cells differs significantly among various genera and even among different individuals. It spans from a few micrometers (or less than 1 mm) to a few millimeters. The phytoplankton can be classified into two main evolutionary functional groups: prokaryotes, exemplified by cyanobacteria (formerly known as blue-green algae), and eukaryotic forms, which encompass diatoms, cryptophytes, haptophytes, green algae, chrysophyte algae, and other variations (Yoon et al. 2004; Olson and Blankenship 2004). Phytoplankton assumes a critical role as it constitutes the foundational tier of the marine trophic structure and serves as a pivotal participant in the carbon cycle of Earth (Mouw et al. 2016; Guidi et al. 2015). Microalgae and cyanobacteria are significant primary producers in aquatic ecosystems, contributing to the biological carbon pump that efficiently absorbs CO2 from the atmosphere (Field et al. 1998). The productivity of marine phytoplankton is influenced by a combination of environmental variables and anthropogenic interventions. The first step in this direction would be identifying the type of stress experienced by phytoplanktons, marking out visible changes in physiology, cytology, or molecular aspects as it may be, and then finally outlining the response of the phytoplankton to the stress confronted by them. Various stresses that phytoplankton face include changes in temperature, CO_2 concentration, acidification, ozone depletion, nutrient availability, and pollution caused by pollutants such as pesticides, organic pollutants, crude oil spills, etc. These factors significantly affect the survival, behavior, nutritional procurement, reproduction, and overall population dynamics of phytoplankton species in both freshwater and marine forms. Here, we aim to systematically report findings on various types of stresses the phytoplankton community is exposed to, enlisting and discussing their impacts and analyzing their significance with respect to climate change, fisheries oceanography, ecological variability, and environmental protection.

16.2 Various Stress Factors Confronted by Phytoplanktons and their Responses

Phytoplankton face various stresses, both natural and anthropogenic. These phytoplankton can be either prokaryotic or eukaryotic, leading to diverse and complex responses. Microalgae and cyanobacteria play a crucial role in the biogeochemical cycling of both major and trace elements (Robinson 2017; Sunda 1994; Fig. 16.1). These microorganisms facilitate different pathways for the movement of elements associated with their cells, each pathway having distinct biological, biogeochemical, and geological implications (Fisher et al. 2000; Lobus et al. 2021).

16.2.1 Temperature

Temperature stands as a crucial variable capable of influencing the composition of phytoplankton communities, giving rise to both direct and indirect effects. Indirect effects include changes in water column stability, nutrient recirculation, phosphorus release from sediment, and shifts in grazer populations, influencing the structure of phytoplankton communities (Winder and Sommer 2012; Jeppesen et al. 2010).

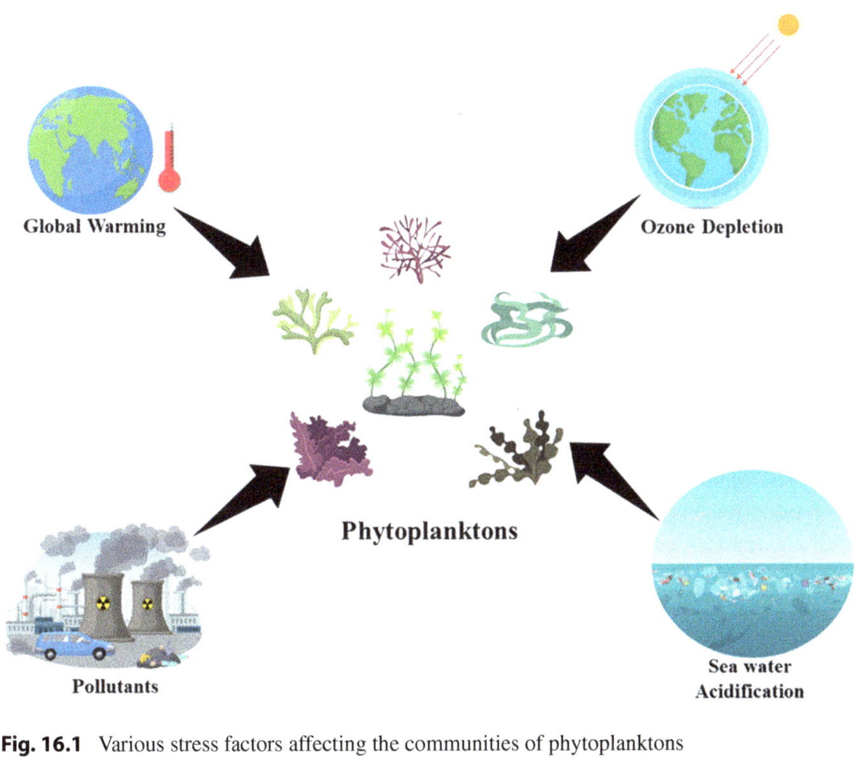

Fig. 16.1 Various stress factors affecting the communities of phytoplanktons

Elevated temperatures can lead to extended and increased thermal stratification within deep lakes. Studies indicate that as temperatures rise, the average cell size of phytoplankton tends to decrease (Schlesinger et al. 1981). Rising temperatures and heatwaves can worsen eutrophication symptoms, leading to increased cyanobacteria blooming and shifts in trophic structure. Flavia et al. proposed that warming could increase phytoplankton biovolume, favoring taxa with higher temperature preferences. Warmer temperatures might intensify thermal stratification, influencing nutrient dynamics and causing changes in phytoplankton community structure.

Marine heatwaves commonly induce a reduction in content of chlorophyll a (Chl a) in mid to low latitudes, while concurrently increasing Chl content in high latitudes, as evidenced by studies conducted by Montie et al. (2020), Le Grix et al. (2021), and Noh et al. (2022). In the investigation by Zhan et al. (2023), an analysis was conducted to investigate the adaptive responses of phytoplankton to an enduring marine heatwave occurrence in the southeastern tropical Indian Ocean throughout the period of 2015–2016. Diverse fluctuations in Chl a concentration were observed during distinct phases of the extended marine heatwave. Furthermore, the study observed the alterations in phytoplankton biomass, physiological adaptations, and size structure in reaction to the heatwave, elucidating associated mechanisms.

16.2.2 Atmospheric CO_2 Concentration and Sea Water Acidification

Increasing atmospheric CO_2 concentrations result in greater CO_2 absorption into surface waters. During the onset of the industrial revolution, the atmospheric concentration of carbon dioxide (CO_2) stood at approximately 270 parts per million (ppm) as reported by Aresta and Dibenedetto 2021. Presently, these levels have surpassed 400 ppm, according to Kaiser (2014), with expectations of further escalation, as anticipated by Meinshausen et al. (2011). Despite the effective buffering capacity of marine waters, elevated CO_2 levels globally contribute to a pH decrease of approximately 0.1 units, attributed to human-induced CO_2 emissions into the atmosphere (Gao et al. 2012; Häder and Gao 2015). Surface waters have undergone a 30% elevation in hydrogen ion (H^+) concentration, and anticipated future trends that indicate potential pH decreases by 0.3–0.4 units, corresponding to a projected 100–150% elevation in H^+ concentration over the ensuing century (Feely et al. 2004). Persistent acidification of the ocean affects the calcification process of calcifying phytoplankton. Coccolithophores, significant contributors to oceanic carbonate production, show contradictory responses to elevated CO_2 levels in experiments, both between species and within the same species (Beaufort et al. 2011). Wu et al. (2008) documented that increased CO_2 levels can down-regulate Carbon Concentrating Mechanisms (CCM), potentially enhancing the growth of marine algae under light-limiting conditions. Adaptation of marine algae to increased concentrations of CO_2 in the ocean's environment suggests that future populations may exhibit genetic differences compared to contemporary ones. For diatoms such *as Skeletonema costatum*, responses to elevated CO2 concentration vary; while laboratory conditions with 800 µatm did not stimulate growth, growth was enhanced in an enclosed ecosystem at 750 µatm CO_2 concentration (Chen and Gao 2004; Kim et al. 2006). Other diatoms, including *Phaeodactylum tricornutum, Navicula pelliculosa, and Attheya* sp., showed enhanced growth under higher CO_2 levels in laboratory conditions (Low-Décarie et al. 2011; King et al. 2011). When subjected to different levels of sunlight and increased CO_2 concentrations (1000 µatm), specific diatoms (*S. costatum, P. tricornutum, and Thallasiora pseudonana*) exhibited enhanced growth in lower light conditions but experienced inhibition under higher light levels. This underscores the interconnected influence of elevated CO_2 and light levels on diatom responses to ocean acidification. However, no analogous interaction was observed in the coccolithophore *Emiliania huxleyi* (Gao et al. 2012).

16.2.3 Nutrient Availability

Phytoplankton assumes an integral function in laying the groundwork for oceanic productivity. Across the majority of global oceans, the productivity of phytoplankton is constrained by the presence of essential nutrients such as nitrogen, iron, and phosphorus within the euphotic zone.(Häder and Gao 2015). Alterations in temperature and nutrient concentrations are recognized factors that influence the composition and diversity of phytoplankton. Warmer temperatures and lower nutrient levels typically favor smaller phytoplankton species (Litchman et al. 2007; Winder and Sommer 2012). Additionally, variations in the ratios of silicon to nitrogen (Si:N) and nitrogen to phosphorus (N:P) influence the relative importance of diatoms compared to other groups like dinoflagellates and cyanobacteria (Makareviciute-Fichtner et al. 2020).

16.2.4 Pollutants

Human activities have resulted in pollution of both freshwater and seawater. Water bodies are contaminated by untreated effluents from domestic and industrial sources. Increased concentrations of pollutants such as ammonium and nutrients, along with high levels of BOD (Biochemical Oxygen Demand), hardness, & conductivity, contribute to the dominance of unialgal species population. Conversely, improved quality of water promotes greater diversity among algae and crustaceans. The prevalence of heavy metal pollution in lakes and rivers within developing nations is a prevalent concern, exerting an influence on the productivity of phytoplankton (Grosse et al. 2017). Nutrient addition experiments conducted by Grosse et al. (2017) showed nutrient-specific effects on phytoplankton metabolism. The growth-limiting nutrient addition led to an increase in amino acid synthesis and a simultaneous decrease in storage compound synthesis. Nitrogen (N)-limited flagellates exhibited a five-fold increase in amino acid synthesis within 24 h upon nutrient limitation relief, while phosphorus (P) limitation had a slower response. The metabolic adaptability of phytoplankton to varying nutrient availability holds ramifications for the nutritional dynamics within the food web and the carrying capacity of coastal ecosystems. Crude oil spills inflict harmful consequences on phytoplankton, with aqueous fractions of heavy oil, particularly pyrene, identified as profoundly toxic constituents. The accumulation of pyrene in sediment is notable, and its toxicity is increased through exposure to solar UV-B radiation and increasing temperatures (Grenvald et al. 2013). The lipophilic nature of polychlorinated biphenyls (PCBs) poses a substantial challenge in marine and freshwater ecosystems. These compounds easily traverse through the phytoplankton cell membranes, resulting in considerable bioconcentration, as evidenced by observations in marine phytoplankton from the Baltic Sea (Gerofke et al. 2005).

16.2.5 Ozone Depletion

The Earth's stratospheric ozone layer, located in the upper atmosphere, plays a pivotal role in shielding life on our planet from harmful ultraviolet (UV) radiation emitted by the sun. However, anthropogenic activities, particularly the release of chlorofluorocarbons (CFCs), have led to the depletion of this vital layer. This environmental concern extends its consequences to phytoplankton, microscopic photosynthetic organisms inhabiting marine ecosystems and forming the foundation of the oceanic food web. The consequence of ozone layer depletion is manifested by the increased penetration of UV radiation into the Earth's atmosphere, posing a distinct threat to phytoplankton, which are inherently sensitive to such radiation. In response, phytoplankton deploy an array of photoprotective mechanisms, including the synthesis of specialized pigments like mycosporine-like amino acids (MAAs) that function as natural sunscreens, absorbing and dissipating UV radiation to shield cellular components (Núñez-Pons et al. 2018). Moreover, these organisms activate intricate DNA repair mechanisms when exposed to elevated UV radiation due to ozone layer depletion, thereby ensuring genetic integrity (Wu et al. 2023) The outcome of ozone layer depletion extends beyond individual phytoplankton, influencing the composition and distribution of entire phytoplankton communities (Williamson et al. 2019). Certain species may exhibit resilience to increased UV radiation, while others may face challenges in adaptation. These shifts through marine ecosystems, affecting the abundance and diversity of higher trophic levels, with potential implications for fisheries and biodiversity. As primary producers, phytoplankton's health and abundance are integral to marine food webs, and disruptions therein may have far-reaching results. A comprehensive understanding of the refined responses of phytoplankton to ozone layer depletion is imperative for predicting ecological outcomes in marine ecosystems.

16.3 Climate Change and Global Warming

The challenge of climate change arises as a crucial global issue underscored by the IPCC in 2007. The substantial augmentation in Greenhouse Gases (GHGs) within Earth's atmosphere stems from increased anthropogenic activities since the industrialization of the mid-eighteenth century. The extensive combustion of fossil fuels, alterations in land use and modifications in forestry practices have markedly increased the concentration of greenhouse gases. For instance, the concentration of CO_2 has ascended from 275–285 ppm during the pre-industrial era to 398 ppm in 2015. Moreover, synthetic Greenhouse Gases (GHGs) such as CFCs, HFCs, and SF6 are progressively accumulating. This accumulation, coupled with global warming, has induced alterations in rainfall patterns, disruptions in hydrological cycles, the retreat of ice caps and glaciers, elevated sea levels, and an increased occurrence and severity of extreme incidents, including intense precipitation and cyclones (Allen and Ingram 2002). Radiatively active gases like CO_2, CH_4, N_2O, and CFCs, can absorb long-wave infrared radiation. The escalating concentration of these gases, attributed to industrialization and urbanization, contributes to Global Climatic Change. Predicted future global warming has the potential to impact weather patterns, climate conditions, sea levels, organism distribution, and timing, as well as food production and oceanic fishery resources (Ullah et al. 2018; Singh 2019). The global average temperature has risen by approximately 0.6 °C in the twentieth century, with estimation suggesting a potential increase of 1.4–5.8 °C by 2100 compared to 1990 levels, with the most significant changes anticipated in middle and higher latitudes. Anticipated impacts of climate change are expected to result in a worldwide reduction in marine net primary production (NPP) and phytoplankton biomass, particularly in regions characterized by middle and low latitudes (Bopp et al. 2013; Lotze et al. 2019; Tittensor et al. 2021).

Elsworth et al. (2023) found that, under diverse regional trends and a high emission scenario (RCP8.5), marked decrease in the inherent variability of global phytoplankton carbon biomass was observed. Specifically, a notable reduction in biomass internal variability occurred in the sub-polar North Atlantic and North Pacific regions, influenced predominantly by bottom-up factors such as temperature and nutrient availability. The study by Elsworth et al. (2023) also suggests that efforts to manage and adapt to the effects of climate change, especially in fisheries and marine carbon cycling, should consider adjustments to the internal variability of phytoplankton caused by human-induced warming, particularly at regional scales. Recent decades have seen a decline in oceanic primary production, linked to increasing sea surface temperatures and reduced iron input. Here, the role of the ocean as a significant carbon sink, a potential further decrease in primary production due to climate change poses significant implications for the carbon cycle, particularly in high latitudes (Gregg et al. 2003). Changes in phytoplankton composition can trigger transformations in the entire food web, necessitating adjustments by predators to new food sources (Käse and Geuer 2018). Environmental fluctuations can create conditions favorable for the invasion of new species. There are speculations that invasive species may even have better adaptive capacity than indigenous species, potentially establishing dominance (Walther et al. 2002; Olenina et al. 2010). As per Lowery et al. 2020, plankton extinction seems to be primarily influenced by significant climatic changes that impact water column stratification, temperature, and potential chemistry studies by Cloern et al. (2023) suggest that phytoplankton ecology is bidirectional, asynchronous, and not associated with global warming. They highlighted the dynamic nature of phytoplankton ecology concerning seasonal patterns, with variability influenced by numerous processes impacting both phytoplankton growth and grazing rates. While the effects of global warming are evident in certain ecosystems, they may be obscured in others due to responses to alternative human disturbances and the multidecadal oscillations of the climate system.

16.4 Conclusion

In the current scenario of global change, the impact of various stresses predominantly anthropogenic in nature are affecting the aquatic ecosystems, especially the essential microorganisms, phytoplankton leading toward a biodiversity crisis. Though at present the effects are subtle however in the coming decade the impact will be seriously affecting the population dynamics, ecological variability, and fisheries oceanography. Sustainability turns out to be the only answer to anthropogenic perturbations affecting the oceans and phytoplankton.

Acknowledgement The Head, Department of Botany, Babu Shivnath Agrawal (PG) College, Mathura, India is gratefully acknowledged for providing necessary facilities.

References

Allen MR, Ingram WJ (2002) Constraints on future changes in climate and the hydrologic cycle. Nature 419:224–232

Aresta M, Dibenedetto A (2021) The CO2 revolution. Eds. Carbon Dioxide Revolution Springer Nature:31–43

Beaufort L, Probert I, de Garidel-Thoron T, Bendif EM, Ruiz-Pino D, Metzl N et al (2011) Sensitivity of coccolithophores to carbonate chemistry and ocean acidification. Nature 476(7358):80–83

Bopp L, Resplandy L, Orr JC, Doney SC, Dunne JP, Gehlen M et al (2013) Multiple stressors of ocean ecosystems in the 21st century: projections with CMIP5 models. Biogeosciences 10(10):6225–6245

Chen X, Gao K (2004) Characterization of diurnal photosynthetic rhythms in the marine diatom Skeletonema costatum grown in synchronous culture under ambient and elevated CO2. Funct Plant Biol 31(4):399–404

Cloern JE, Schraga TS, Nejad E, Eddy T (2023) Phytoplankton as indicators of global warming? Limnol Oceanogr Lett

Elsworth GW, Lovenduski NS, Krumhardt KM, Marchitto TM, Schlunegger S (2023) Anthropogenic climate change drives non-stationary phytoplankton internal variability. Biogeosciences 20(21):4477–4490

Feely RA, Sabine CL, Lee K, Berelson W, Kleypas J, Fabry VJ, Millero FJ (2004) Impact of anthropogenic CO2 on the CaCO3 system in the oceans. Science 305(5682):362–366

Field CB, Behrenfeld MJ, Randerson JT, Falkowski P (1998) Primary production of the biosphere: integrating terrestrial and oceanic components. Science 281(5374):237–240

Fisher NS, Stupakoff I, Sañudo-Wilhelmy S, Wang WX, Teyssié JL, Fowler SW, Crusius J (2000) Trace metals in marine copepods: a field test of a bioaccumulation model coupled to laboratory uptake kinetics data. Mar Ecol Prog Ser 194:211–218

Gao K, Helbling EW, Häder DP, Hutchins DA (2012) Responses of marine primary producers to interactions between ocean acidification, solar radiation, and warming. Mar Ecol Prog Ser 470:167–189

Gerofke A, Kömp P, McLachlan MS (2005) Bioconcentration of persistent organic pollutants in four species of marine phytoplankton. Environ Toxicol Chem 24(11):2908–2917

Gregg WW, Conkright ME, Ginoux P, O'Reilly JE, Casey NW (2003) Ocean primary production and climate: global decadal changes. Geophys Res Lett 30:15
Grenvald JC, Nielsen TG, Hjorth M (2013) Effects of pyrene exposure and temperature on early development of two co-existing Arctic copepods. Ecotoxicology 22:184–198
Grosse J, van Breugel P, Brussaard CP, Boschker HT (2017) A biosynthesis view on nutrient stress in coastal phytoplankton. Limnol Oceanogr 62(2):490–506
Guidi L, Legendre L, Reygondeau G, Uitz J, Stemmann L, Henson SA (2015) A new look at ocean carbon remineralization for estimating deepwater sequestration. Glob Biogeochem Cycles 29(7):1044–1059
Häder DP, Gao K (2015) Interactions of anthropogenic stress factors on marine phytoplankton. Front Environ Sci 3:14
Jeppesen E, Meerhoff M, Holmgren K, González-Bergonzoni I, Teixeira-de Mello F, Declerck SA, Lazzaro X (2010) Impacts of climate warming on lake fish community structure and potential effects on ecosystem function. Hydrobiologia 646:73–90
Kaiser KL (2014) The Carbon Cycle and Royal Society Math. *Principia Scientific International. Keeling, CD, JFS Chin and TP*
Käse L, Geuer JK (2018) Phytoplankton responses to marine climate change–an introduction. In *YOUMARES 8–Oceans Across Boundaries: Learning from each other: Proceedings of the 2017 conference for YOUng MARine RESearchers in Kiel, Germany* (pp. 55–71). Springer International Publishing
Kim JM, Lee K, Shin K, Kang JH, Lee HW, Kim M et al (2006) The effect of seawater CO2 concentration on growth of a natural phytoplankton assemblage in a controlled mesocosm experiment. Limnol Oceanogr 51(4):1629–1636
King AL, Sanudo-Wilhelmy SA, Leblanc K, Hutchins DA, Fu F (2011) CO2 and vitamin B12 interactions determine bioactive trace metal requirements of a subarctic Pacific diatom. ISME J 5(8):1388–1396
Le Grix N, Zscheischler J, Laufkötter C, Rousseaux CS, Frölicher TL (2021) Compound high-temperature and low-chlorophyll extremes in the ocean over the satellite period. Biogeosciences 18(6):2119–2137
Litchman E, Klausmeier CA, Schofield OM, Falkowski PG (2007) The role of functional traits and trade-offs in structuring phytoplankton communities: scaling from cellular to ecosystem level. Ecol Lett 10(12):1170–1181
Lobus NV, Kulikovskiy MS, Maltsev YI (2021) Multi-element composition of diatom Chaetoceros spp. from natural phytoplankton assemblages of the Russian Arctic Seas. Biology 10(10):1009
Lotze HK, Tittensor DP, Bryndum-Buchholz A, Eddy TD, Cheung WW, Galbraith ED et al (2019) Global ensemble projections reveal trophic amplification of ocean biomass declines with climate change. Proc Natl Acad Sci 116(26):12907–12912
Low-Décarie E, Fussmann GF, Bell G (2011) The effect of elevated CO2 on growth and competition in experimental phytoplankton communities. Glob Chang Biol 17(8):2525–2535
Lowery CM, Bown PR, Fraass AJ, Hull PM (2020) Ecological response of plankton to environmental change: thresholds for extinction. Annu Rev Earth Planet Sci 48:403–429
Makareviciute-Fichtner K, Matthiessen B, Lotze HK, Sommer U (2020) Decrease in diatom dominance at lower Si: N ratios alters plankton food webs. J Plankton Res 42(4):411–424
Meinshausen M, Smith SJ, Calvin K, Daniel JS, Kainuma ML, Lamarque JF et al (2011) The RCP greenhouse gas concentrations and their extensions from 1765 to 2300. Clim Chang 109:213–241
Montie S, Thomsen MS, Rack W, Broady PA (2020) Extreme summer marine heatwaves increase chlorophyll a in the Southern Ocean. Antarct Sci 32(6):508–509
Mouw CB, Barnett A, McKinley GA, Gloege L, Pilcher D (2016) Phytoplankton size impact on export flux in the global ocean. Glob Biogeochem Cycles 30(10):1542–1562
Noh KM, Lim HG, Kug JS (2022) Global chlorophyll responses to marine heatwaves in satellite ocean color. Environ Res Lett 17(6):064034
Núñez-Pons L, Avila C, Romano G, Verde C, Giordano D (2018) UV-protective compounds in marine organisms from the Southern Ocean. Mar Drugs 16(9):336

Olenina I, Wasmund N, Hajdu S, Jurgensone I, Gromisz S, Kownacka J, Olenin S (2010) Assessing impacts of invasive phytoplankton: the Baltic Sea case. Mar Pollut Bull 60(10):1691–1700
Olson JM, Blankenship RE (2004) Thinking about the evolution of photosynthesis. Photosynth Res 80:373–386
Robinson C (2017) Phytoplankton biogeochemical cycles. OUP
Schlesinger DA, Molot LA, Shuter BJ (1981) Specific growth rates of freshwater algae in relation to cell size and light intensity. Can J Fish Aquat Sci 38(9):1052–1058
Singh S (2019) Cyanobacteria and their role under elevated CO2 conditions. In: Climate change and agricultural ecosystems. Woodhead Publishing, pp 249–263
Sournia A, Chrdtiennot-Dinet MJ, Ricard M (1991) Marine phytoplankton: how many species in the world ocean? J Plankton Res 13(5):1093–1099
Stocker TF, Qin D, Plattner GK (2013) Intergovernmental Panel on Climate Change 2013. Cambridge University Press
Sunda WG (1994) Trace metal/phytoplankton interactions in the sea. In: Chemistry of aquatic systems: local and global perspectives. Springer Netherlands, Dordrecht, pp 213–247
Tett P, Barton ED (1995) Why are there about 5000 species of phytoplankton in the sea? J Plankton Res 17(8):1693–1704
Tittensor DP, Novaglio C, Harrison CS, Heneghan RF, Barrier N, Bianchi D, Blanchard JL (2021) Next-generation ensemble projections reveal higher climate risks for marine ecosystems. Nat Clim Chang 11(11):973–981
Ullah H, Nagelkerken I, Goldenberg SU, Fordham DA (2018) Climate change could drive marine food web collapse through altered trophic flows and cyanobacterial proliferation. PLoS Biol 16(1):e2003446
Walther GR, Post E, Convey P, Menzel A, Parmesan C, Beebee TJ et al (2002) Ecological responses to recent climate change. Nature 416(6879):389–395
Williamson CE, Neale PJ, Hylander S, Rose KC, Figueroa FL, Robinson SA et al (2019) The interactive effects of stratospheric ozone depletion, UV radiation, and climate change on aquatic ecosystems. Photochem Photobiol Sci 18(3):717–746
Winder M, Sommer U (2012) Phytoplankton response to a changing climate. Hydrobiologia 698:5–16
Wu H, Zou D, Gao K (2008) Impacts of increased atmospheric CO2 concentration on photosynthesis and growth of micro-and macro-algae. Sci China Ser C Life Sci 51(12):1144–1150
Wu T, Lei Y, Jian Z (2023) DNA repair-related genes are the key for stony coral ancestors to survive under elevated levels of UVR. Front Mar Sci 10:1170565
Yoon HS, Hackett JD, Ciniglia C, Pinto G, Bhattacharya D (2004) A molecular timeline for the origin of photosynthetic eukaryotes. Mol Biol Evol 21(5):809–818
Zhan W, Zhang Y, He Q, Zhan H (2023) Shifting responses of phytoplankton to atmospheric and oceanic forcing in a prolonged marine heatwave. Limnol Oceanogr 9999:1–14

Cell Death in Photoautotrophs 17

Samujjal Bhattacharjee, Prashansa Singh, Alka Bhardwaj, and Arun Kumar Mishra

Abstract

Programmed cell death (PCD) is a genetically controlled mechanism regulating cellular demise. Though commonly associated with multicellular organisms, it has also been observed in unicellular organisms such as photoautotrophs, which are organisms proficient in generating their sustenance through photosynthesis. This chapter delves into the fundamental role of programmed cell death in photoautotrophs, elucidating its pivotal contributions to growth, development, and adaptive responses to environmental challenges. The exploration unveils the sophisticated mechanisms these organisms have evolved to ensure survival and reproductive success amid changing conditions. Through an in-depth analysis of distinct mechanisms and regulatory pathways governing PCD in photoautotrophs, this chapter provides valuable insights into the broader understanding of cell death processes. It accentuates unique features, specific pathways, molecular players, and regulatory elements that PCD in photoautotrophs apart from other systems, particularly animals. Key discoveries underscore the significance of PCD in sculpting the life cycle of photoautotrophs, resonating with implications for plant biology, ecological dynamics, and beyond. Comparative analyses with PCD in diverse organisms shed light on the evolutionary dimensions of cell death mechanisms. The presented findings not only propel our comprehension of PCD in photoautotrophs but also pave the way for future research, unraveling the intricate interplay between cellular life and death in these vital organisms.

Samujjal Bhattacharjee and Prashansa Singh contributed equally with all other contributors.

S. Bhattacharjee · P. Singh · A. Bhardwaj · A. K. Mishra (✉)
Laboratory of Microbial Genetics, Department of Botany, Institute of Science, Banaras Hindu University, Varanasi, India

A. K. Mishra (ed.), *Stress Biology in Photosynthetic Organisms*,
https://doi.org/10.1007/978-981-97-1883-2_17

Keywords

Programmed cell death · Regulated cell death · Necrosis · Caspase homologs · Autolytic-PCD · Autophagy

17.1 Introduction

Programmed Cell Death (PCD) has emerged as a paradigm-shifting discovery within the complex framework of photoautotrophic organisms, encompassing cyanobacteria, microalgae, and plants. The phenomenon of PCD goes beyond mere biological occurrences, it encapsulates a rich history of scientific inquiry, profound significance in developmental processes, and a pivotal role in responding to environmental stresses. To understand the refined details of PCD in photoautotrophs, it is essential to trace the historical trajectory of the study, beginning with its early exploration. The journey of understanding PCD in plants and algae traces its roots to the mid-twentieth century when researchers first recognized that these seemingly continuous and dynamic organisms also possessed mechanisms for controlled cellular demise. Initial observations centered around morphological changes in plant tissues, leading to the coining of terms such as "programmed cell death" and "senescence." However, it wasn't until the late twentieth century that the molecular underpinnings of PCD began to be unraveled. The advent of molecular biology techniques in the 1980s and 1990s marked a turning point in PCD research. Scientists started identifying key players in the complex regulatory networks governing cell demise. The identification of PCD-related genes, such as Bcl-2 family members and caspases, provided pivotal insights into the molecular machinery facilitating cell death in photoautotrophs (Bidle 2015). As research progressed into the twenty-first century, technological advancements, including genomics and advanced imaging techniques enabled a more comprehensive understanding of the diverse triggers and pathways involved in PCD. Traditionally associated with metazoans, the revelation that such a mechanism is operative in organisms, fundamental to global ecosystems has reframed the understanding of life at the cellular level. The trajectory of this discovery follows an interesting path from initial skepticism to widespread acceptance, with each experiment, observation, and molecular insight solidifying the existence of PCD in the plant kingdom.

As dissecting the complexities of PCD, its role in the coordination of altruistic adaptation comes to the forefront. The sacrificial demise of selective cells for the benefit of the organism constitutes a phenomenon that underscores the adaptive prowess of photoautotrophs. This selfless cellular behavior, driven by programmed mechanisms, is integral to the survival and ecological success of these organisms, a role that unfolds as a captivating revelation. Consequently, PCD manifests its significance through the formation of specialized structures such as hormogonia and akinetes, essential components of reproductive strategies in cyanobacteria and microalgae (Aguilera et al. 2021). Vascular tissue development in plants, intricately linked to PCD, stands as a testament to the regulatory role this process plays in shaping the structural integrity of higher plants. Additionally, the rapid and localized hypersensitive responses in plants, a form of PCD, exemplify its role as a front-line defense mechanism against pathogenic threats (Hegedus and Rimmer 2005).

The comparative analysis of PCD mechanisms across cyanobacteria, microalgae, and plants unravels evolutionary adaptations and variations. This comparative lens, a vital component of scientific exploration, reveals not only the conserved aspects of PCD regulation but also the detailed divergence that has shaped the diverse strategies employed by photoautotrophs in governing cellular demise. At the molecular level, the discovery of caspase homologs and their significance in PCD adds a layer of intricacy to our understanding. This molecular machinery, akin to the caspases in metazoans, serves as the executioners of cell death, directing a cascade of events that culminate in the demise of specific cells. Unraveling the molecular signatures of PCD not only enhances the comprehension of the regulatory networks governing this process but also opens avenues for targeted interventions and manipulations.

In essence, the significance of PCD in photoautotrophs extends beyond the confines of individual cells, it is deeply ingrained in the fabric of their survival and adaptation. This chapter serves not only as an exposition of the scientific intricacies of PCD but also as an invitation for further inquiry, exploration, and dialogue. As navigating through the molecular pathways, evolutionary landscapes, and ecological implications, the discovery of PCD emerges not merely as a scientific revelation but as a profound cornerstone in our quest to decipher the elaborated mechanisms that govern life in photoautotrophic organisms.

17.2 Types and Mechanisms of PCD across Photosynthetic Organisms

Death is a fundamental aspect of life cycle of the cells and plays an important role in various biological processes, encompassing development, tissue homeostasis, and immune responses. Understanding cell death is essential for comprehending normal physiological processes as well as pathological conditions. Earlier apoptosis was considered synonymous with PCD. However, this perspective was challenged as additional cell death morphotypes was emerged, suggesting that apoptosis was just one modality of PCD. Further, the Nomenclature Committee on Cell Death (NCCD) has put forward the functional classification of cell death subroutines (Galluzzi et al. 2012, 2018). They emphasized the importance of molecular definitions over morphological ones, recognizing the potential for substantial functional, biochemical, and immunological heterogeneity underlying related morphological appearances. The NCCD's suggested shift toward precise molecular definitions is driven by the need to align with evolving knowledge in cell death, particularly in photosynthetic organisms. The committee highlights that standardized nomenclature is crucial for accurately referring to diverse cell death pathways and their functions. Unfortunately, the significance of standardizing these terms has often been overlooked. It is imperative that the terminology evolves in parallel with our understanding of cell execution in photosynthetic organisms. In various groups, notably within photosynthetic organisms, there has been an evolution in the nomenclature for the specific cell death modalities. The terms "regulated cell death" (RCD) and "programmed cell death" (PCD) are now both employed, highlighting the dynamic nature of the field. Specifically, "RCD" is designated to any form of cell demise that encompasses cascades signaling and utilizes intracellular molecular machinery.

The NCCD's classification delineated various cell death subroutines, including apoptosis, autophagy, and mitotic catastrophe, alongside a few more cell death modalities. Extrinsic apoptosis, characterized as a caspase-mediated process, necessitates death receptor signaling, distinguishing it from intrinsic apoptosis, which relies on outer membrane permeabilization of mitochondria. Intrinsic apoptosis is linked to the decrease in the mitochondrial transmembrane potential ($\Delta\psi$ m) and the release of proteins from the intermembrane space of mitochondria Notably, intrinsic apoptosis can be independent of caspases, particularly in the presence of molecules that inhibit caspases. This alternative pathway may be facilitated actively by endonuclease G and apoptosis-stimulating factors, or passively through ATP depletion (David et al. 2006; Kroemer et al. 2007; Galluzzi et al. 2012). Autophagosomes are a hallmark of autophagic cell death, which is defined by caspase-independent demise brought on by an excessive autophagic flux. This represents a general cytoprotective response that paradoxically accelerates the cell death processes (Kroemer et al. 2009). The idea of necrosis as an accidental type of cell death has changed throughout time, giving way to a comprehension of its control and importance in both physiological and pathological reactions. When it comes to cases of receptor-interacting protein kinase (RIP) 1- and 3-dependent necrosis, the name "necroptosis" has been particularly suggested (Vandenabeele et al. 2010; Galluzzi et al. 2012). Recently, a distinct form of necrosis, driven by mitochondrial permeability transition, has been delineated. This kind of necrosis is characterized by the disruption of the transmembrane potential of mitochondria ($\Delta\psi$ m) as a result of the osmotic breakdown and loss of internal permeability of mitochondrial membrane, which eventually results in the necrotic-type morphotype (Galluzzi et al. 2018). Furthermore, the term "mitotic catastrophe," which designates the removal of cells during mitosis or in the interphase that follows mitotic abnormalities, is defined as an onco-suppressive route (Vakifahmetoglu et al. 2008). Moreover, several alternative mechanisms of cell death have been clarified, such as Anoikis, which entails the loss of integrin-dependent attachment and consequent induction of intrinsic apoptosis; Autosis, signifying the death of autophagic cells reliant on the plasma membrane's $Na+/K+$ Ω ATPase; Entosis is the term for a process known as cell-in-cell hemolysis that results in the enveloped cell's breakdown within a lysosome; ferroptosis, which is prevented by iron chelators or antioxidants and is triggered by intracellular oxidative disruptions; Parthanatos causes poly [ADP-ribose] polymerase 1 to get activated early; pyroptosis, which is caspase-1-dependent and linked to pyrogenic mediators such as IL-1β and IL-18; NETosis, the ejection of neutrophil extracellular traps linked to ROS-dependent neutrophil cell death; Cornification is a keratinocyte-specific dying process that involves lipid extrusion and buildup that is reliant on caspase-14 (Galluzzi et al. 2012, 2018). These diverse cell death pathways underscore the complexity and versatility of cellular responses in different physiological and pathological contexts. As our understanding of cell death progresses, the need for accurate terminology becomes increasingly important for effective communication and collaboration in the scientific community.

PCD is associated with both developmental and defense mechanisms in plants, however, the comprehensive biochemical understanding of several PCD morphotypes, their homology to animal systems and classification remains debatable. During certain developmental programs, plant cell death displays autophagic characteristics (Beers 1997), while processes involving DNA damage and protoplast retraction in response to other developmental, stress, or pathological conditions are considered as apoptosis and referred as AL-PCD (apoptotic-like PCD). A functionally similar but distinct mechanism of cell death is implied by the absence of genes expressing caspase in *Arabidopsis* and the occurrence of caspase homologs (3 and 6 genes expressing metacaspases 1 and 2, respectively), even though caspase-like activity is closely associated with apoptosis and is linked to the execution of AL-PCD (Danon et al. 2000; Reape and McCabe 2008). However, the transport of cytochrome c molecules from mitochondria to the cytosol during conditions of stress-induced cell death was not witnessed during the AL-PCD (Balk et al. 1999, 2003; Watanabe and Lam 2004).

All categories of PCDs identified in plants are broadly divided into two major categories: autolytic and non-autolytic PCDs. In the former case, hydrolases are released from the vacuole when they collapse, rapidly clearing the cytoplasm, primarily occurring during normal plant development. On the other hand, non-autolytic PCD is characterized by the absence of rapid cytoplasmic clearance. Although there may be increased permeability or tonoplast rupture, these changes do not result in the rapid release of a large quantity of hydrolases that would clear the residual cytoplasm quickly. Non-autolytic PCD primarily occurs during plant–pathogen interactions. In autolytic PCD, the onset is characterized by the emergence of small vacuoles within the cytoplasm, that subsequently amalgamate. The merging continues, progressing from intermediate-sized vacuoles to a larger single vacuole. Notably, vesicles enclosed in a single membrane are frequently observed within both the smaller and larger vacuoles. As a consequence, the cytoplasm undergoes a transformation, gradually being replaced by vacuolar volume. Throughout this process, a substantial number of cytoplasmic organelles, notably plastids, ribosomes, membranes of endoplasmic reticulum (ER), and peroxisomes disappear (Van Doorn 2011). These alterations bear a striking resemblance to the autophagic processes observed in yeast and animal cells, although the regulatory mechanisms of this process in plants and the specific organelles involved are yet to be fully understood. Processes associated with autolytic PCD encompass various sequential events. These include an elevation in cytoplasmic Ca^{2+} quantity, acidification of the cytosol, induction of MAP-kinase (mitogen-activated protein kinase) signaling, and alterations in the cytoskeletons made of microtubule and actin (Hoeberichts and Woltering 2003; Smertenko et al. 2003; Bosch et al. 2008; Fagerstedt 2010). Moreover, it encompasses organellar swelling, breakdown of chloroplast contents, DNA degradation, chromatin aggregation, and translocation of condensed chromatin to the peripheral space of nucleus (Inada et al. 2000; Gunawardena et al. 2001; Yamada et al. 2006a, b; Bosch et al. 2008). The nucleus undergoes further changes, including condensation to a smaller diameter and breakup into smaller fragments (Yamada et al. 2001, 2003, 2006a, b). Notably, these alterations generally precede tonoplast rupture, although, in some instances, nuclear changes become apparent only after the rupture of tonoplast (Obara et al. 2001; Kladnik et al. 2004). Closing of plasmodesmata occurs as an early event in certain cases, such as petal senescence (Van Doorn et al. 2003). This closure can impede the flow of sugars between cells, potentially leading to ATP depletion if sugars aren't transported from apoplast to the cells. Notably, the PCD of Tulip petal (senescence) was linked to depletion of ATP molecules (Azad et al. 2008).

Autolytic PCDs often involve the presence of cysteine proteases and/or serine proteases. However, the exact functions of these proteases in PCD, beyond their role in bulk protein degradation, remain largely unknown (Pak and Van Doorn 2005). Cysteine proteinases with KDEL tails are situated in organelles derived from the rough endoplasmic reticulum, known as ricinosomes. The proposition is that the ricinosomes burst due to the acidification of the cytoplasm followed by tonoplast rupture, releasing active proteases that aid in the cytoplasm's disintegration (Senatore et al. 2009). Moreover, the activation of *cathepsin B*-encoding gene (cysteine protease) is linked to the autolytic PCD in aleurone cells of barley seeds (Martinez et al., 2003). A similar up-regulation of a cathepsin B-encoding gene is observed in the cell death of *Populus* xylem fiber (Courtois-Moreau et al. 2009). Knocking out 3 cathepsin B-encoding genes in *Arabidopsis* not only delays leaf PCD but also results in a significant seven-fold reduction in the expression of senescence marker *SAG12* gene. Cathepsins, implicated in certain animal PCD types, are normally found in lysosomes, equivalent to plant vacuoles (Kroemer and Jäättelä 2005; McLellan et al. 2009). Cathepsins B and D can be released into the cytoplasm via lysosome permeabilization, which then causes the pro-apoptotic protein Bid to be activated by proteolysis. Bid then induces permeabilization of mitochondrial outer membrane and PCD via activating the caspases. It has been reported that PCDs in xylem fibers are linked to an increase in the transcript level of autophagy-associated genes including *atg8c*, *8d*, and *8f*. Additionally, it also involves a sphingosine-1-phosphate phosphatase-encoding gene and an orthologue of a mammalian Bcl-2-associated athanogene (BAG gene). These autophagy-associated genes likely play a role in remobilization processes. BAG proteins, mainly function as chaperone regulators, thus modulating various processes including PCD (Courtois-Moreau et al. 2009). Plant BAG proteins remarkably resemble their animal counterparts in that they control PCD during developmental processes and in response to pathogen invasions (Doukhanina et al. 2006). Furthermore, this autolytic PCD occurs during the course of general plant development including the senescence of several plant parts like cotyledons, leaves, petals, and roots. Moreover, it also involves the maturation of laticifers and other ducts, the development of dead xylem conduits and bark, as well as pollen and ovary formation. Additionally, this form of PCD contributes to the demise of various parts during the development and germination of seeds.

The non-autolytic PCD is characterized by absence of cytoplasmic clearance. Despite this, there may be an increase in permeability or even tonoplast rupture; however, these alterations do not result in the massive release of hydrolases that can quickly clear the remaining cytoplasm. The non-autolytic PCD predominantly occurs in three contexts—Firstly, it is associated with the hypersensitive response (HR). Secondly, it occurs in PCD induced by necrotrophic plant pathogens. Thirdly, it is observed in the PCD within the cereal seed's endosperm. The HR-related PCD can be further classified into two groups: in the first group, physiologically active proteins are released from the vacuoles while the other group lacks this feature. When a plant encounters a pathogen (bacteria, viruses, oomycetes, and fungi), its defense mechanism operates in two stages. Firstly, conserved regions of pathogen-related molecules are recognized by plant pattern recognition receptors, triggering a basal immune response that hinders pathogen growth. However, pathogens deploy effector molecules to suppress this response. Here, the second level of defense comes into play. Plants have developed resistance (R) genes that encode receptors capable of recognizing pathogen effector molecules, thus initiating a signaling cascade leading to gene expression. Consequently, a layer of plant cells collapses which surrounds the interfering pathogen, and this death morphotype is referred to as HR-related PCD (Mur et al. 2008; Hayward et al. 2009). It is often preceded by the influx of calcium ions, generation of salicylic acid, ROS, and RNS, as well as the activation of a MAPK cascade. Moreover, the continuous overexpression of AtCYS1, a cystatin acting as a natural cysteine protease inhibitor, effectively suppresses HR-related PCD in both cultured *Arabidopsis* cells and tobacco plants. This finding suggests that cysteine proteinases may play a regulatory role in HR-related PCD (Belenghi et al. 2003). The HR-related PCD type II, which does not require the tonoplast rupture and the vacuolar release of physiologically active proteins for cell collapse, occurs in the hypersensitive response in the epidermal cells of cowpea against the rust (Heath et al. 1997). Further, *Cochliobolus victoriae* responsible for causing victoria blight in oats, secretes a host-selective toxin i.e., victorin that induces HR-related PCD in oats. In this case, victorin-induced PCD is defined by the collapse of the mitochondrial transmembrane potential, activation of caspase-like proteases, cytoplasmic shrinkage, and requirement of Ca^{2+} import. Remarkably, this PCD occurs without plasmodesmatal closure, tonoplast rupture, or even plasma membrane permeabilization (Curtis and Wolpert 2004; Williams and Dickman 2008). Necrotrophic plant pathogens utilize toxic molecules to induce cell death in host cells and subsequently consume the dead plant cells. These pathogens, exemplified by fungi like *Botrytis cinerea*, employ botrydial and oxalic acid-like toxic molecules to kill host cells. In response to infection, a significant ROS accumulation occurs. This process may result in lipid peroxidation, antioxidant depletion, and ultimately contributing to PCD in the plant cells (Gechev et al. 2006). Significantly, a mutation in the Vacuolar Processing Enzyme γ (VPEγ) and cysteine protease having caspase-1 activity in plants reduces *B. cinerea*-induced PCD in *Arabidopsis*. However, mutations in other VPEs (a, b, d) do not exhibit the same effect (Rojo et al. 2004). The next necrotrophic fungal pathogen is *Sclerotinia sclerotiorum*, which primarily functions by secreting oxalic acid, which then results in increased production of H_2O_2 that leads to PCD in plant cells. Mutation in the oxalic acid synthesis or inhibition in the oxalate-induced H_2O_2 production can prevent PCD in the host plant cells.

Furthermore, two instances of PCD in plant development have been identified where cells depicted deceased prior to rupturing of tonoplast or expire without evident tonoplast rupturing. One case is observed in cereal, while the other occurs during embryogenesis using somatic cells. Moreover, at the time of seed formation, inner endosperm cells of cereals (Poaceae) undergo PCD. Although the exact cell death mechanism remains unknown, and there is no evidence of tonoplast permeabilization. During this PCD, loss of plasma membrane integrity; a decrease in soluble protein content as well as RNA, and DNA content; an increase in the abundance of cysteine proteases, specifically caspase-like protease; RNase and DNase induction (Young and Gallie 2000; Nguyen et al., 2007). Further, the dead cells of the endosperm remain intact till the germination of seeds. Subsequently, the material within these dead cells undergoes degradation by the secretion of hydrolases from the surrounding aleurone tissue (Borén et al., 2006). This hydrolyzed substance is then consumed by the developing embryo (Young and Gallie 2000). This form of cell death is completely distinct from the process observed in the seeds of various other species, where endosperm cells only undergo death at the onset of seed germination (Sreenivasulu et al. 2004). In this case, the death of endosperm cells is facilitated by the hydrolases released from the vacuole of the cells undergoing death (Greenwood et al. 2005). The only known example of in situ PCD during normal plant growth is the PCD of cereal endosperm (DeBono and Greenwood 2006). Notably, it differs from the autolytic category and falls under the classification of non-autolytic PCD. Another important example of non-autolytic PCD has been observed during the in vitro somatic embryogenesis of *Picea abies* where embryogenesis initiates by selecting some cells from the mother plant, cultivating them in the influence of cytokinin and auxin. Subsequently, the addition of abscisic acid (ABA) induces PCD in some cells of the amorphous cell mass, leading to its differentiation into suspensor cells and embryos (Bozhkov et al. 2002). Notably, the dying cells have intact tonoplasts, but their plasma membranes are no longer semipermeable. It was further confirmed through Evans blue staining, revealing the absence of stain within vacuoles of the dying cells, while it was present in the suspensor cells. The staining with Evans blue affirmed this, as this stain was absent from the vacuoles of the dying cells, although it was present within the suspensor cells (Filonova et al. 2000; Smertenko et al. 2003). Moreover, the cells undergoing death exhibited a gradual reduction in cytoplasmic content and an enlargement of vacuolar size. Nuclear changes included lobing and fragmentation of nuclear DNA (Filonova et al. 2000). In dying embryonal tube cells, the microtubule network exhibited partial disorganization, while in dying suspensor cells, it was completely degraded. Actin appears thin in cells undergoing non-PCD, whereas it exhibited thick strings in the dying suspensor. The use of F-actin depolymerization drugs impeded PCD in the suspensor cells, highlighting the significance of actin network alterations. After the rupturing of tonoplasts the lifeless cytoplasm dissolved, leaving behind only the cell wall (Filonova et al. 2000; Smertenko et al. 2003).

The two categories of plant PCD, namely autolytic and non-autolytic PCDs, do not align with the distinctive morphological characteristics of apoptosis, such as the development of apoptotic bodies, and their subsequent engulfment by another cell.

Autolytic plant PCD is often grouped under animal's autophagic PCD, primarily due to its consistent association with cytoplasmic autophagy-like structures. This classification acknowledges that our understanding of plant autophagic structures is still rudimentary, and may not be able to precisely mirror those in animals. On the other hand, non-autolytic PCD in plants doesn't exhibit the characteristics of animal apoptotic or autophagic PCD. Consequently, it is generally classified under necrosis in animal cells, characterized by the lack of morphological attributes associated with apoptosis and autophagy. However, certain instances of non-autolytic PCD linked to HR have been demonstrated to require *atg* genes, leading to their placement under autophagic PCD in animals (Hofius et al. 2009). In cases where progressive vacuolization is observed in non-autolytic PCD examples (Greenberg and Yao 2004; Hara-Nishimura et al. 2005; Kwon et al. 2010), must also be classified under animal autophagic PCD. Despite some morphological and biochemical similarities, there are fundamental differences in the processes of PCD between plants and animals. Remarkably, the lack of structures resembling autolysosomes or autophagosomes, the absence of apoptosome formation, and the non-existence of archetype caspase genes indicate a diverse origin and evolution of PCD in these two life forms. Nevertheless, the mutual capability of key cell death executioners in plants, i.e., metacaspases and animals, i.e., caspases to cleave a conserved substrate, Tudor staphylococcal nuclease, provides evidence for a shared origin of PCD in both these lifeforms (Sundström et al. 2009). Therefore, to unravel the evolution of PCD in plants, it is crucial to compare the cyanobacterial cell death morphotypes, which serve as the progenitors of oxygenic photoautotrophs.

Research on cyanobacterial cell death mechanisms has unveiled several resemblances to eukaryotic PCD, including phosphatidylserine externalization, cytoplasmic clearance, caspase-like activity and DNA damage. Nevertheless, these attributes are not categorized into distinct PCD morphotypes, instead, they are grouped under a common mechanism of cell demise. Extensive research on the *Azolla's* cyanobacterial endosymbiont has demonstrated four distinct PCD morphotypes: apoptotic-like, autolytic-like, autophagic-like, and necrosis-like cell death morphotypes. Each morphotype exhibits unique ultrastructural and biochemical alterations, allowing for clear differentiation. Annexin V–green fluorescent protein-based analysis indicates phosphatidylserine externalization, is evident only in autophagic- and apoptotic-like cell death, while DNA damage is a common characteristic of all dying cells (demonstrated by deoxynucleotidyl transferase dUTP nick end-labeling assay) (Zheng et al. 2013). Furthermore, a detailed ultrastructural analysis using transmission electron microscopy (TEM) categorized the distinct death morphotypes. The apoptotic-like cell death featured an intact outer membrane, peptidoglycan layer, and early stage cell shrinkage, thylakoid disintegration along with unique cellular inclusions like carboxysomes and cyanophycean granules. Further, these cells assumed a remarkably irregular dendritic shape instead of disintegrating into apoptotic bodies. The autophagic-like cell death exhibited extensive vacuolation and cytoplasmic disintegration. The peptidoglycan layer undergoes lysis only during the early stages of autolytic- and necrosis-like cell demises. The previous autolytic-like death involves cytoplasmic disintegration, vesicle generation, as well as loss of cyanobacterial cytoskeleton in the later stages, while the later necrotic-like shows rapid dispersion of cellular content following cell wall rupture and disintegration of the actin network. The observed cell death morphotypes were evident among all cyanobacterial cell types, encompassing vegetative cells, heterocytes, and akinetes. Additionally, apoptotic- and autophagic-like cell demises were primarily observed during the early stages of development of the host plant, whereas autolytic- and necrotic-like cell deaths were more predominant in the later developmental stages of the fern. However, the autophagic- and necrotic-like cell death in cyanobacteria bears a resemblance with the plant's autolytic and non-autolytic PCDs, respectively. Nevertheless, apoptotic- and autolytic-like cell deaths exhibit minimum resemblance with any of the known plant PCD morphotypes. The limited experimental evidence and the conserved mutually comparable parameters, attributed to primary differences between the organisms, still leave the origin and propagation of PCD in phototrophs as a subject of debate.

17.3 Significance of PCD in Photoautotrophs

PCD holds paramount significance within the realm of photoautotrophs, as this intricate regulatory framework of orchestrating the mechanism of cell demise underlines from developmental fitness to adaptive responses. This orchestrated demise serves as a strategic conduit, not only ensuring optimal resource allocation by culling damaged or superfluous cells for fostering efficient nutrient allocation but also instrumental in developmental and adaptive modification, enhancing the fitness of organisms.

In higher plants, PCD plays a pivotal role across various aspects of their lifecycle, encompassing reproductive development, senescence, environmental stress responses, and defence mechanisms. Plants employ PCD in tissues undergoing senescence, like leaves and petals, to recycle nutrients and discard unnecessary components. Moreover, PCD is instrumental in processes such as embryogenesis and seed germination, curbing the inheritance of detrimental genetic traits (Consonni et al. 2005). This phenomenon is exemplified by self-incompatibility, a mechanism that triggers PCD in pollen tubes to prevent self-pollination in plants with bisexual flowers. The execution of PCD is highly contextual, relying on the specific triggers that prompt its activation.

Recently the researchers are considerate towards a functional framework that distinguishes between developmental-regulated PCD (dPCD) and environmentally induced PCD (ePCD). dPCD, guided by internal stimuli, is meticulously coordinated in a selected subset of cells at a particular developmental stage. Instances include tracheary element differentiation, endosperm and aleurone cell development, and self-incompatibility responses. While dPCD often involves autolysis, it also integrates autophagic components, as seen in embryogenesis and germination. Conversely, ePCD arises from external stressors like pathogen attacks or abiotic challenges such as heat, salinity, drought, and flooding. The hypersensitive response (HR) exemplifies ePCD, involving cell death specifically occurring at the site of infection to thwart biotrophic pathogens. Notably, HR incorporates autophagic events in neighboring cells to curtail uncontrolled cell death (Liu and Levine 2015). To adapt to adverse conditions, specific plant tissues may trigger PCD, instigating morphological adaptations. For example, under flooding conditions, root cortical cells initiate PCD, forming aerenchyma to enable oxygen movement from shoots to roots. In contrast, PCD is activated when abiotic stress overwhelms a plant's coping mechanisms, leading to metabolic impairment and oxidative stress (Locato and De Gara 2018). Thus, PCD in higher plants occurs in the selective vegetative cells as well as during sexual developments, thus conferring adaptive and evolutionary advantages.

17.3.1 PCD in Vegetative Cells of Higher Plants

17.3.1.1 Autolyis of Pith Cells

The autolytic death of pith parenchyma cells leading to lysigenous aerenchyma formation, characterized by air-filled cavities in stems or roots is observed upon hypoxia, drought, temperature fluctuation, and nutrient deficiency (Beers 1997). Such autolysis in the stem pith correlates with rapid stem elongation and increased sink strength in some species. It was reported that the hypoxia triggers aerenchyma formation in maize and sunflower roots (Ni et al. 2019), where cellulase, polygalacturonase, and α-amylase activities serve as markers for lysigenous cavity formation, hinting at coordinated cell death programs involving enzymatic cascades. Such adaptive measures enhance the resilience of plants in harsh conditions (Fig. 17.1).

17.3.1.2 Vascular Differentiation and Trachery Element Formation

Tracheary elements and fiber-sclereids culminate their development with robust, often lignified cell walls lacking protoplasts. These structures form interconnected conduits for water transport in the xylem, adjacent to the phloem. It was observed that the development of sclereid in Arabidopsis presents avenues for molecular genetic studies on plant PCD in tissue development. Notably, water-storage cell death in xeromorphic leaves of *Sansevieria* and *Oncidium* species exemplifies terminal differentiation yielding altered cell walls without protoplasts. Furthermore, the Zinnia mesophyll cell culture system plays a pivotal role in inducing tracheary element differentiation from parenchyma cells in vitro (Fukuda 1994). Characterized by wall thickenings and semi-synchronous autolysis. While evidence suggests DNA fragmentation during PCD of tracheary elements, its progression deviates from mammalian cell apoptosis, resembling senescent leaf cell death culminating in tonoplast degradation. The precise involvement of hydrolytic enzymes, released post-tonoplast rupture, in PCD as either primary effectors or consequential elements remains a subject of unresolved inquiry.

The genes or proteins responsible for initiating the autolytic phase during tracheary element differentiation are still unknown, but the activity of hydrolytic enzyme surges in later stages, accompanied by endonucleases, RNAases, and proteases

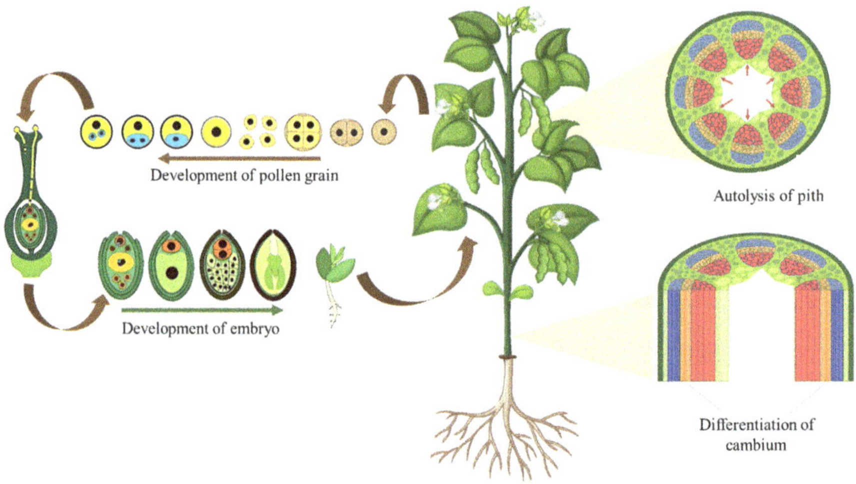

Fig. 17.1 Schematic representation depicting the significance of regulated cell death in plants. Cell death in plants is necessary for differentiation of tracheary elements and pit formation. Furthermore, sexual development also encompasses regulated cell death, involving degeneration of synergids, pollen incompatibility, etc.

were observed. Interestingly, the inhibition of tracheary element differentiation occurred when cysteine and serine protease inhibitors were applied before the initiation of wall thickenings. Insights into cysteine proteases and ubiquitin–protein conjugation offer a partial understanding of their roles in tracheary element differentiation (Woffenden et al. 1998). The presence of protoplasts within certain tracheary elements and fiber-sclereids suggests possible uncoupling of PCD from differentiation, though the precise dependence on differentiation signals remains uncertain. Premature tracheary element death, decoupled from wall modification, could explain cell death in the vascular cavity of pea roots. However, conclusive evidence elicitation and progression of signals for orchestered demise of specific cells for vascular channel formation are still elusive (Fig. 17.1).

17.3.2 PCD during Sexual Developments in Higher Plants

PCD emerges as a pivotal participant in fundamental reproductive events, including sex determination, development of gametes, fertilization, and embryo formation. The scientific community is increasingly drawn to employing established model systems for plant reproduction to unravel the complexities of PCD (Pacini and Dolferus 2016). Studies of several mutants influencing sex determination and embryonic development, along with in vitro experiments using isolated embryo sacs, promise insights into the intricate components of plant PCD pathways. The integration of various disciplines enhances our comprehension of the molecular coordination of PCD within the complex framework of plant reproductive processes (Fig. 17.1).

17.3.2.1 Determination of Sex Types

In the realm of flowering plants, where mostly flowers harbored both male and female sex organs, a fascinating divergence is observed in species like asparagus, maize, and white campion. Here, unisexuality arises through the orchestrated demise of preformed sex organs, an intricate phenomenon examined by Dellaporta and Calderon-Urrea (1994). In maize, the PCD that coordinates the deterioration of both male and female sex organs exhibits unique characteristics, vacuolation, and loss of cell organelles (Cheng et al. 1983). External manifestations reveal sunken cells in the degenerating organs, suggesting tonoplast rupture as a penultimate event in the process of sex organ elimination. Asparagus, in turn, exhibits incomplete vascular development in female sex-organs of male flowers and vice versa, leaving a conundrum—whether this deficiency in vascular maturation propels cell death through phytohormone withholding a consequential byproduct of the cell death cascade remains an enigma (Bracale et al. 1991). Moreover, The genetic foundations of female sex organ abortion elucidate the crucial functions of Ts2 in maize and Gsf1 in *Tripsacum*, which may potentially reveal the biochemical intricacies involved in responding to externally applied feminizing phytohormones (Hansen et al. 1976). The foreseen product of these genes, akin to hydroxysteroid dehydrogenases, holds the promise of catalyzing the conversion of a GA-like precursor into a PCD promote (Fig. 17.1). This dual role implies indirect regulation of yet undiscovered genes or a direct cytotoxic impact, establishing an absorbing facet in the realm of plant biochemistry and sex determination (DeLong et al. 1993).

17.3.2.2 Dehiscence of Anther

Anther dehiscence unfolds as a captivating phenomenon in plant biology, showcasing the orchestrated demise of diverse cell types within a single organ, the anther. This complex mechanism is characterized by the enhanced expression of cysteine protease (Koltunow et al. 1990), leading to PCD. The culmination of anther rupture in tomatoes occurs specifically at the stomium, subsequent to the degeneration of tissues situated between the stomium and the connective tissues in the pollen sacs. In tomato anthers, the orchestration of cell death encompasses the tapetum, epidermal cells near the stomium, endothecium, and connective tissue, with intriguing indications of potential hydrolytic enzyme participation. Notably, cellulase protein levels surge during anther maturation (del Campillo and Lewis 1992). The mRNA of the TA56 thiol proteinase noticeably accumulates in the circular cell cluster positioned between the stomium and the connective tissue. This accumulation is accompanied by an increase in ubiquitin levels within the tissue. (Li et al., 1995). This tightly controlled process suggests the involvement of proteolytic pathways, specifically encompassing both thiol proteases and, notably, ubiquitin conjugation. This occurs prior to the degeneration of specific cell types in anthers.

17.3.2.3 Pollen Incompatibility

The fate of pollen released through anther dehiscence germinates on the epidermis of the stigma cells upon hydration, the pollen tube begins a journey fraught with potential peril, occasionally culminating in its demise. The intricate linkage between the stigma and ovary is facilitated by the long tissue system through the style (Esau 1965), assumes a crucial role. The stigma and style, guardians of reproductive integrity, meticulously distinguish the appropriate and inappropriate pollen grains, leading to the prevention of both interspecific and intraspecific pollen growth. This mechanism, self-incompatibility, an evolutionary safeguard against the constrains of inbreeding, is a subject thoughtfully reviewed by Newbigin et al. (1993), and Nasrallah et al. (1994) (Fig. 17.1).

In members of the Solanaceae, Rosaceae, and Scrophulariaceae, self-incompatibility of the gametophytic type is critically governed by the highly polymorphic S-locus. The pollen tube's fate in a stylar environment bearing identical S-alleles is a deceleration in growth and imminent demise. The self-incompatibility mechanism involves a captivating array of genes encoded by the S-locus, with a particular focus on the stylar ribonucleases referred to as S-RNases. The allelic series of these S-RNases function as allele-specific cytotoxins pivotal for the manifestation of self-incompatibility, unveiling a complex ballet of biochemical interactions. In vitro assays reveal the nonspecific degradation of RNA by S-RNases (McClure et al. 2011). However, the in vivo scenario, as observed during the growth of *Nicotiana alata* styles, intriguingly reveals the selective degradation of ribosomal RNA (rRNA) specifically in pollen tubes exhibiting incompatibility, posing a conundrum. Two proposed models elucidate the enigma—one suggesting the exclusion of S-RNases from the pollen tube, with recognition facilitated by the S-locus product, and the other proposing S-RNase entry into all pollen tubes, with subsequent inactivation or sequestration in compatible ones. The outcome in both models converges on non-autonomous inhibition of protein synthesis, inducing growth arrest and eventual demise. Amid the plethora of hydrolytic markers for autolysis, stylar S-RNase stands as the singular lytic agent indispensable for this intriguing process initiated by the elusive pollen S-locus product, however, a cell death promoter is yet to be unveiled.

17.3.2.4 Maturation of Embryo Sac

After reaching the ovary, the pollen tube autonomously navigates into the embryo sac for fertilization. In angiosperms, the development of a singular embryo sac arises from a viable haploid megaspore, as three out of the four megaspores resulting from meiosis undergo programmed degeneration. Marsilea ferns, for instance, depict megaspore death characterized by chromatin pycnosis, hinting at apoptotic undertones (Bell 1996). Distinctions in pycnotic chromatin morphologies in apoptotic and nonapoptotic PCD guide investigations in angiosperms, urging scrutiny at ultrastructural, biochemical, and molecular levels. This exploration promises fresh insights into PCD intricacies within economically significant seed plants.

PCD occurs within the embryo sac during fertilization, initiated by the entry of the pollen tube through the micropyle, marking the onset of consequential events. At the micropylar end of the embryo sac, two synergids and egg cells are positioned. Upon pollen tube penetration, one synergid typically degenerates, a process postulated to be either necrotic in response to injury or programmed for death, even in the absence of pollination. The termination of pollen tube growth within the synergid triggers an opening in the tube wall, facilitating the release of two sperm cells. One sperm cell engages in zygote formation, while the other plays a role in the initiation of endosperm development. The regulated process of synergid demise is characterized in tobacco and cotton, as well as in pearl millet. Spontaneous changes in calcium distribution precede and accompany synergid degeneration, suggesting that calcium influx may act as an effector in their regulated cell death. Additionally, involvement of phytohormones such as GA and IAA, in promoting synergid degeneration has been observed ye their mechanism remains elusive.

Furthermore, the suspensor, pivotal in early embryo growth promotion, undergoes degeneration during late embryonic stages. Morphological cues, including autophagic vesicles and subsequent autolysis, suggest an autophagic cell death pathway. Notably, chromatin pycnosis in degenerating suspensors aligns with a nonapoptotic trajectory. Abnormalities in embryo growth or embryo destruction often result in heightened suspensor growth, indicative of its plasticity. Mutant analyses, such as the twin mutant in Arabidopsis, reveal altered autolysis programs and underscore the impact of a normally developing embryo on restricting suspensor growth, shedding light on potential initiators of the suspensor's cell death pathway.

17.3.3 PCD in Phytoplanktons

Phytoplanktons, comprised of cyanobacteria and diverse algae, constitute the foundation of marine ecosystems, residing at the ocean's surface. Their abundance and productivity confers them integral to the aquatic food webs (Smetacek 2012). PCD in phytoplankton involves tightly regulated processes leading to self-destruction, contrasting with necrosis. PCD is essential for maintaining ecological balance and population dynamics (Bidle 2016a, b). Besides, PCD in phytoplanktons profoundly influences marine food webs, global oxygen production, and carbon sequestration, thus signified the essentiality of this process beyond the organism or community. Thus, the pertinence of this process of self-elimination can be categorized into adaptive advantages and ecophysiology.

17.3.3.1 Adaptive Significance of PCD in Phytoplanktons

In phytoplanktons, PCD functions as an adaptive mechanism in response to environmental constraints. This adaptive process helps reshape the microenvironment and influences interactions among diverse taxa. (Durand et al., 2016; Durand and Ramsey, 2019). Environmental stresses like nutrient deprivation, high salinity, and constant irradiations trigger PCD among phytoplanktons. PCD in stressful conditions allows elimination of older cells, relocates nutrients in the community, thereby enabling the survival of a subset of cells which divide or germinate in favorable condition to re-establish the colonies of phytoplanktons. Such a mechanism not only maintains the phytoplankton population but also their genetic diversity (Durand et al., 2016). Formation of akinetes in filamentous cyanobacteria under phosphate-starved conditions is one of such example (Fig. 17.2).

Interestingly, the *Dunaliella–Halobacterium* model exemplifies the impact of PCD on interaction between phytoplankton and microbial communities inhabiting together. Organic compounds discharged during the programmed death of *Dunaliella salina* undergo reutilization by viable cells and subsequent re-mineralization orchestrated by *Halobacterium salinarum*, showcasing a nutrient exchange mechanism (Orellana et al., 2013). Similar advantages in nutrient exchange are observed in prokaryotic phytoplankton communities, where organic compounds released during phytoplankton cell death become valuable resources for bacteria (Azam and Malfatti 2007), facilitating adaptation in stressful conditions. Furthermore, Phytoplankton PCD triggers intricate trophic interactions impacting higher trophic levels, thereby influencing the entire marine ecosystem. Grazers and higher predators are intricately linked to phytoplankton dynamics, emphasizing the ecological importance of PCD (Kiorboe 2008). Nevertheless, insights from the *Chlamydomonas* suggest that the advantages of PCD may vary depending on the specific species. The substances liberated during PCD in *Chlamydomonas reinhardtii* CC125 contribute to the fitness of the same species, yet they impede the growth of other *Chlamydomonas* species highlighting the role of PCD in inter-specific competition (Durand et al. 2014).

17.3.3.2 Ecological Significance of PCD in Phytoplanktons

Implications of PCD in phytoplanktons are beyond a single organism or a particular community as these lifeforms are involved in intricate web of microbial interactions and exerting a significant influence on both energy dynamics and biogeochemical processes (Franklin et al. 2006). In the context of *Trichodesmium* bloom, the initiation of PCD prompts the release of exopolysaccharides, initiating a downward flux of organic carbon source. This phenomenon serves as a sophisticated mechanism employed by cyanobacteria to efficiently provide nutrients, particularly carbon and nitrogen, to the benthic sea (Bar-Zeev et al. 2013). Hence, PCD in cyanobacteria

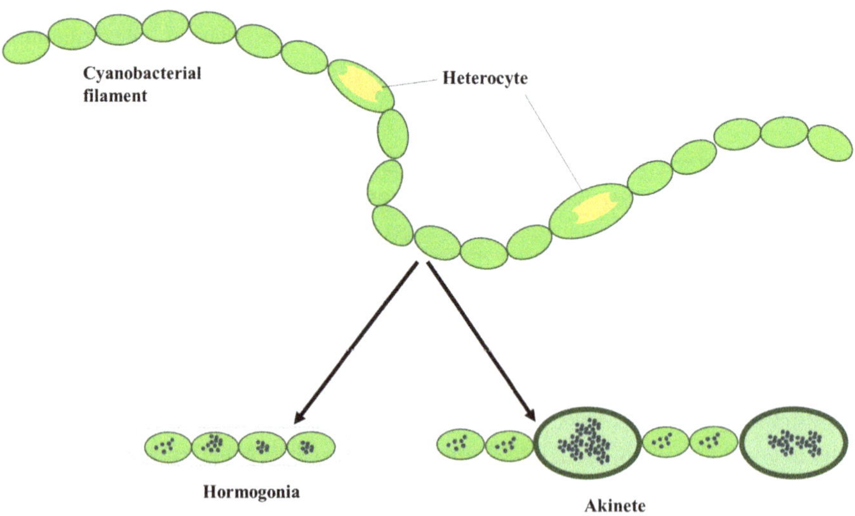

Fig. 17.2 Regulated cell death conferred adaptive advantages in cyanobacteria. Orchestrated demise of cells is important for formation of hormogonia and akinetes

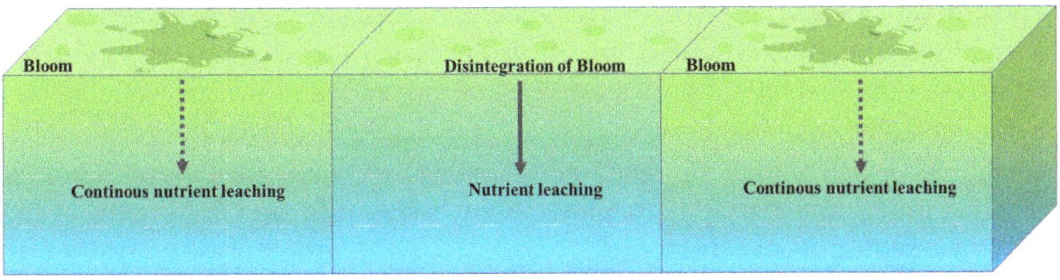

Fig. 17.3 Cyanobacterial bloom dynamics across various seasons. Cyanobacteria are major primary producers in the grazing food chain of aquatic environment. However, in their death releases organic carbon sources like dissolved polysaccharides in the ecosystem. These were used by heterotropic bacteria in the microbial loop, enabling the transfer of nutrient across different niches

serves as a catalyst for the movement and transfer of organic resources within the microbial loop, functioning through an unconventional non-predatory pathway (Scavia and Fahnenstiel, 1987; Franklin et al. 2019) (Fig. 17.3).

Thus, the importance of PCD in *Trichodesmium* transcends the cyanobacterial community, encompassing an ecologically significant and foundational mechanism of energy transfer in the early stages of Earth's oceanic systems., possibly predating the existence of predators. This underscores the pivotal role of cyanobacterial PCD in shaping the energetic and biogeochemical landscapes of marine ecosystems. This provides a comprehensive overview, emphasizing the interconnectedness of phytoplankton PCD with broader ecological and environmental dynamics.

In the realm of phytoplanktons, PCD emerges as a nuanced and vital process with broader ecological implications, from shaping marine food webs to influencing global biogeochemical cycles. As environmental changes continue to unfold, unraveling the necessity of PCD in phytoplanktons becomes not just a scientific endeavor but a key to deciphering the intricate balance of life in our oceans.

17.4 Caspase Homologs in Photoautotrophs

Caspases, recognized as pivotal proteases associated with apoptosis in metazoans, were traditionally considered exclusive to this group. Active caspases exhibit a dimeric configuration, with each monomer possessing p20 and p10 domains. The catalytic histidine-cysteine dyad is housed within the p20 sub-domain, showcasing cleavage specificity of Asp residue at P1 site (Earnshaw et al. 1999). The identification of the caspase-hemoglobinase-like fold in non-metazoan proteins revealed the presence of structural caspase-homologs in these lifeforms, critically challenged the notion conferring exclusive protease activity during death of metazoans (Uren et al. 2000; Koonin and Aravind, 2002). These counterparts, largely known as metacaspases (MCA), orthocaspases (OCA), and paracaspases (PCA), share a conserved p20 domain housing the catalytic HC dyad, yet exhibited cleavage specificity for basic amino acid differentiating it from caspases. Besides, the dimerization of proteins was not required for their catalytic activities. Based on these distinctions, caspase-homologs were subsumed within Peptidase C14B superfamily whereas

Peptidase C14A superfamily harbored archetype caspases, in the MEROPES database. While initially implicated in cell death execution, accumulating evidence underscores the pivotal roles of these caspase homologs in diverse cellular processes, including maintaining homeostasis, responding to stress physiology, influencing cell differentiation, and participating in cell death mechanisms. Photoautotrophs abundantly harbors MCAs, whereas till-date OCAs were exclusively reported from cyanobacteria.

17.4.1 Metacaspases

Metacaspases, constituting the largest sub-family within the broader caspase family, exhibit a remarkable diversity in architectural features, leads to their classification into three distinct types, determined by the structural features of the p20 and p10 domains. Notably, Type I metacaspases introduce an extra layer of complexity, potentially incorporating proline-rich repeats and zinc-finger motifs in the N-terminal prodomain. In contrast, Type II metacaspases, distinguished by the absence of additional N-terminal motifs, feature an elongated linker region positioned between the p20-like and p10-like domains, as elucidated by Vercammen et al. (2004).

The distribution of these metacaspase types across different lineages adds another layer to their intriguing diversity. Type I metacaspases, spanning from Proteobacteria to plants, showcase a broad evolutionary presence. In contrast, Type II metacaspases carve out a more exclusive niche, being confined to the green lineage of plants and algae. A recent discovery has unveiled Type III metacaspases, a unique subclass identified solely in algae emerging after secondary endosymbiosis. Notably, Type III metacaspases, on the other hand, exhibit an atypical rearrangement of domains, where the p10-like domain is positioned at the N-terminal, deviating from its conventional C-terminal location, a departure from the observed pattern in other caspase family members, as reported by Choi and Berges (2013).

Intriguingly, in contrast to their metazoan counterparts, caspases and metacaspases do not depend on dimerization for activation. Instead, the activation mechanism of all three metacaspase types is primarily contingent on the presence of calcium ions. It is noteworthy to mention an exception in the case of Type II metacaspase of *Arabidopsis thaliana*, AtMC9. This metacaspase has exhibited a distinctive calcium-independent activity, a characteristic highlighted by Lam and Zhang (2012). Within the realm of Type II metacaspases, the introduction of CaCl2 in millimolar concentrations induces targeted cleavage in the linker region that connects the p20 and p10 domains. This closely mirrors the activation observed in caspases. This shared activation mechanism suggests a functional parallel, highlighting the intricate interplay of these proteases in cellular processes, as discussed by Lam and Zhang (2012), Piszczek et al. 2012, and Vercammen et al. 2004.

In stark contrast, In Type I metacaspases, autoprocessing between the p20 and p10 domains is absent. For example, TbMC2 from *Trypanosoma brucei* displayed nonspecific in vitro cleavages only at high protein concentrations, unveiling a distinctive activation pattern. Likewise, the N-terminal domain of *Arabidopsis thaliana* metacaspase AtMC1 underwent in vivo cleavage and complete removal, as documented by Coll et al. in 2014 and Coll et al. in 2010. This divergence in activation mechanisms among metacaspase types underscores the nuanced regulation of PCD in different organisms.

It is noteworthy that the absence of the p10 domain in certain caspase family members, such as paracaspases and orthocaspases, results in a unique calcium-independent activation. This highlights the diverse evolutionary trajectories and functional adaptations of metacaspases, shedding light on the nuanced regulation of PCD across different life forms.

In summary, the elaborate structural variations and calcium-dependent activation mechanisms among metacaspase types underscore their diverse roles and evolutionary significance. The exploration of these variations not only contributes to our understanding of PCD but also unravels the intricate regulatory networks that govern cellular processes in various organisms. Further investigations into the functional implications of metacaspases promise to unravel more mysteries, providing a deeper understanding of the evolutionary trajectories of these proteases.

17.4.2 Orthocaspases

Orthocaspases (OCAs) are another class of caspase-homolog abundantly harbored by prokaryotes. The evolutionary trajectory reveals a diverse distribution among bacteria, with cyanobacteria along with α-proteobacteria and δ-proteobacteria depicted rich repertoire of these caspase homologs. Bioinformatics analyses have unveiled a prevalent presence of OCAs in the cyanobacterial phylum. whereas only 4% of the total peptidase-C14 harboring proteins in this diverse group are identified as Type-I Metacaspase-like proteases (MCAs), specifically harbored by *Anabaena, Lyngbya, Moorea, and Nostoc* (Klemenčič et al. 2019). Subsequently, genome-wide exploration involving 132 cyanobacterial strains showcased that approximately 62% of strains harbor putative OCAs, underscoring a rich reservoir within the phyla. Intriguingly, this prevalence is unevenly distributed, with complex cyanobacterial strains, characterized by complicated morphology, broader physiological capabilities, and larger genome sizes—such as those belonging to Nostocales and Oscillatories—displaying a more abundance of OCAs. This striking contrasts with the dearth of OCAs in simpler unicellular cyanobacteria, such as *Prochlorococcus*, *Synechococcales*, and *Synechocystis* (Bhattacharjee et al. 2021). The observation posits a compelling narrative suggesting the evolutionary advantages of OCAs in cyanobacteria, possibly influenced by complex morphology, larger genomes, and the ability of cell differentiation, as seen in Nostocales with heterocyte or akinete development, and formation of hormogonia in both Nostocales and Oscillatoriales. Conversely, the presence and abundance of OCAs in *Microcystis* or *Chroccocus* suggest alternative driving forces beyond morphological and physiological complexities and larger genomes in the acquisition of OCAs (Bhattacharjee and Mishra 2020).

An essential factor influencing the uptake of organic carbon compounds (OCAs) in cyanobacteria corresponds to the crucial role of PCD within their communities. The altruistic response to environmental stresses or harsh ecosystems could favor the presence of OCAs, especially in colonial or multicellular forms, as observed in *Microcystis*. OCAs may instigate regulated cell demise in a portion of the community, thereby facilitating the survival of the remaining population, which could later act as propagules in more favorable conditions. However, the impact of habitat on OCA acquisition remains elusive, as evidenced by observations in *Thermosynechococcus elongatus* BP-1, which lacks OCAs, despite being a mesophilic counterpart. In contrast, thermophilic strains such as *Chroococcidiopsis thermalis* PCC 7203 and halophilic cyanobacterium *Halothece* sp. PCC 7418 harbor an adequate number of these proteases. On the other hand, the absence of OCAs in certain cyanobacterial species, particularly in solitary lifeforms like *Prochlorococcus*, *Synechococcus*, and *Synechocystis*, prompts considerations of genome streamlining as a major factor. This streamlining phenomenon, exemplified by the absence of several DNA-repair genes in *Prochlorococcus* spp., suggests a potential correlation between genome simplification and the loss of OCAs (Dufresne et al. 2005). It raises the argument that solitary lifeforms may not confer altruistic or adaptive advantages upon harboring suicide-responsive genes, thereby opting for a streamlined genome.

Beyond the uneven distribution, an analysis of catalytic motifs reveals a high diversity among cyanobacterial caspase-homologs. While orthocaspases generally exhibit an active site comprised of the archetype His-Cys (HC) dyad, approximately 30% of studied OCAs in cyanobacteria display different motifs such as Tyr-Ser (YS) or Tyr-Asn (YN). These proteases with mutated catalytic motifs are considered as pseudo-variants. Interestingly, the mutation in the active site residues has minimal impact on the overall structure of OCAs. However, the profound effect of substitution lies in the putative loss of peptidase activity (Klemenčič et al. 2019). Nonetheless, these pseudo-variants exhibit cellular functions distinct from proteolysis, adding another layer of complexity to their functional repertoire (Klemenčič et al. 2019; Bhattacharjee et al. 2021). The catalytic motifs vary widely across taxonomical orders, with YS and YN-type mutations predominating. Notably, the YN-type of substitution is more prevalent among Nostocales, suggesting potential advantages conferred by this mutation on a specific lifestyle. Moreover, the exclusive presence of genuine organic carbon compounds (OCAs) in *Gloeobacter* implies an early acquisition of such compounds, which later underwent mutations within the phylum (Bhattacharjee et al. 2021). The global distribution of pseudo-orthocaspases among prokaryotes also highlights the predominance of YS and YN-type mutations (Klemenčič et al. 2019).

The origin of organic carbon compounds (OCAs) in cyanobacteria remains considerably obscure. Preliminary evidence points towards horizontal gene transfer (HGT) events as potential mechanisms underlying the acquisition of OCAs, supported by significant differences in amino acid sequences, indicating a polyphyletic origin. The phylogenetic arrangement of OCAs among cyanobacteria further complicates the understanding, with closely associated species or members of the same orders sparsely placed, and OCAs from the same organisms subsumed into different clades (Bhattacharjee et al. 2021). While not conclusive, these observations suggest multiple HGT events throughout the cyanobacterial life history, contributing to the complex scenario seen today.

Adding complexity to the comprehension of cyanobacterial caspase homologs is the existence of accessory domains, commonly located at the C-terminal region of the peptidase domain, which includes both p20 and p10 subunits. These accessory domains belong to diverse protein superfamilies, presenting a challenge in describing the functionality of caspase homologs. This complexity, not observed in metazoan caspases, underscores the need for a nuanced approach. In-depth in silico studies advocate the multifunctionality of caspase homologs, with accessory domains potentially complementary or unrelated to peptidase activity. For instance, WD40 repeats located at the posterior part of these homologs may be involved in interacting with target proteins, subsequently cleaved by the catalytically active p20 domain. Conversely, domains like formylglycine generating enzyme (FGE)-sulfatase have roles in post-translational modifications unrelated to protease activity. Despite evidence pointing to the potential multifunctionality of these caspase homologs, biochemical studies have predominantly focused on the proteolytic aspect, aiming to relate them to regulated cell demise (Klemencic et al. 2015).

In essence, the intricate exploration of caspase homologs in cyanobacteria unfolds a narrative of structural diversity, evolutionary intricacies, and potential multifunctionality. The nuanced interplay between OCA distribution, catalytic motifs, and accessory domains paints a rich tapestry, opening avenues for further research in deciphering the functional significance of these proteases in diverse microbial ecosystems.

17.5 Future Prospective

Looking into the horizon of PCD in photoautotrophs, the future holds a wealth of opportunities to deepen our understanding and explore new dimensions in this dynamic field. One crucial avenue for further investigation is the comparative analysis of PCD mechanisms across cyanobacteria, microalgae, and plants. While recognizing the significance of PCD in various processes such as altruistic adaptation, hormogonia, and akinete formation, vascular tissue development, and hypersensitive responses, a detailed exploration of these mechanisms at the molecular level is essential. Advanced molecular techniques and comparative genomics could unravel conserved pathways and divergent strategies employed by different photoautotrophs, providing valuable insights into the molecular tapestry of PCD. Another promising frontier is the study of caspase-homologs and their role in managing cell fate decisions in photoautotrophs. Investigating these homologs across different groups could reveal commonalities or unique adaptations in the regulation and execution of PCD. This exploration not only contributes to our fundamental understanding but also holds potential for innovative biotechnological applications. Targeting caspase-like proteins may offer novel strategies to manipulate PCD pathways, influencing growth, stress responses, and developmental processes in photoautotrophs. Beyond fundamental research, the translational applications of PCD knowledge beckon for exploration. Investigating how insights from studying PCD in photoautotrophs can be applied to agricultural improvement, bioenergy production, and environmental management is crucial. Identifying potential targets for crop improvement, enhancing stress tolerance, and developing sustainable practices based on PCD principles are key considerations for addressing global challenges in food security and climate change.

A holistic, multi-omics approach emerges as an imperative for the future of PCD research. Integrating genomics, transcriptomics, proteomics, and metabolomics promises a comprehensive understanding of the intricate networks governing PCD. This systems-level perspective is essential for unraveling the cross-talk between different cellular processes, signaling pathways, and environmental cues that influence PCD dynamics in photoautotrophs. In conclusion, the future of PCD research in photoautotrophs is both thrilling and challenging. By delving into the molecular intricacies, exploring caspase-homologs, understanding evolutionary footprints, bridging gaps for translational applications, and embracing holistic approaches, researchers can contribute to a nuanced and comprehensive understanding of programmed cell death in the vibrant world of photoautotrophs.

Acknowledgements PS is thankful to DST-Inspire for financial assistance in the form of JRF fellowship. AB is thankful to CAS-JRF for financial support. SB is thankful to Council of Scientific and Industrial Research (CSIR) for research grant in the form of SRF.

Conflict of Interest Authors declare that there is no conflict of interest.

References

Aguilera A, Klemenčič M, Sueldo DJ, Rzymski P, Giannuzzi L, Martin MV (2021) Cell death in cyanobacteria: current understanding and recommendations for a consensus on its nomenclature. Front Microbiol 12:631654

Azad AK, Ishikawa T, Ishikawa T, Sawa Y, Shibata H (2008) Intracellular energy depletion triggers programmed cell death during petal senescence in tulip. J Exp Bot 59(8):2085–2095

Azam F, Malfatti F (2007) Microbial structuring of marine ecosystems. Nat Rev Microbiol 5(10):782–791

Balk J, Chew SK, Leaver CJ, McCabe PF (2003) The intermembrane space of plant mitochondria contains a DNase activity that may be involved in programmed cell death. Plant J 34(5):573–583

Balk J, Leaver CJ, McCabe PF (1999) Translocation of cytochrome c from the mitochondria to the cytosol occurs during heat-induced programmed cell death in cucumber plants. FEBS Lett 463(1–2):151–154

Bar-Zeev E, Avishay I, Bidle KD, Berman-Frank I (2013) Programmed cell death in the marine cyanobacterium *Trichodesmium* mediates carbon and nitrogen export. ISME J 7(12):2340–2348

Beers EP (1997) Programmed cell death during plant growth and development. Cell Death Differ 4(8):649–661

Belenghi B, Acconcia F, Trovato M, Perazzolli M, Bocedi A, Polticelli F, Delledonne M (2003) AtCYS1, a cystatin from *Arabidopsis thaliana*, suppresses hypersensitive cell death. Eur J Biochem 270(12):2593–2604

Bell PR (1996) Megaspore abortion: a consequence of selective apoptosis? Int J Plant Sci 157(1):1–7

Bhattacharjee S, Kharwar S, Mishra AK (2021) Insights into the phylogenetic distribution, diversity, structural attributes, and substrate specificity of putative cyanobacterial orthocaspases. Front Microbiol 12:682306

Bhattacharjee S, Mishra AK (2020) The tale of caspase homologues and their evolutionary outlook: deciphering programmed cell death in cyanobacteria. J Exp Bot 71(16):4639–4657

Bidle KD (2015) The molecular ecophysiology of programmed cell death in marine phytoplankton. Annu Rev Mar Sci 7:341–375

Bidle KD (2016a) Programmed cell death in unicellular phytoplankton. Curr Biol 26(13):R594–R607

Bidle KD (2016b) The molecular ecophysiology of programmed cell death in marine phytoplankton. Annu Rev Mar Sci 8:1–28

Borén M, Höglund AS, Bozhkov P, Jansson C (2006) Developmental regulation of a VEIDase caspase-like proteolytic activity in barley caryopsis. J Exp Bot 57(14):3747–3753
Bosch M, Poulter NS, Vatovec S, Franklin-Tong VE (2008) Initiation of programmed cell death in self-incompatibility: role for cytoskeleton modifications and several caspase-like activities. Mol Plant 1(6):879–887
Bozhkov PV, Filonova LH, von Arnold S (2002) A key developmental switch during Norway spruce somatic embryogenesis is induced by withdrawal of growth regulators and is associated with cell death and extracellular acidification. Biotechnol Bioeng 77(6):658–667
Bracale M, Caporali E, Galli MG, Longo C, Marziani-Longo G, Rossi G et al (1991) Sex determination and differentiation in *Asparagus officinalis* L. Plant Sci 80(1–2):67–77
del Campillo E, Lewis LN (1992) Occurrence of 9.5 cellulase and other hydrolases in flower reproductive organs undergoing major cell wall disruption. Plant Physiol 99(3):1015–1020
Cheng PC, Greyson RI, Walden DB (1983) Organ initiation and the development of unisexual flowers in the tassel and ear of *Zea mays*. Am J Bot 70(3):450–462
Choi CJ, Berges JA (2013) New types of metacaspases in phytoplankton reveal diverse origins of cell death proteases. Cell Death Dis 4(2):e490–e490
Coll NS, Smidler A, Puigvert M, Popa C, Valls M, Dangl JL (2014) The plant metacaspase AtMC1 in pathogen-triggered programmed cell death and aging: functional linkage with autophagy. Cell Death & Differentiation 21(9):1399–1408
Coll NS, Vercammen D, Smidler A, Clover C, Van Breusegem F, Dangl JL, Epple P (2010) Arabidopsis type I metacaspases control cell death. Science 330(6009):1393–1397
Consonni G, Gavazzi G, Dolfini S (2005) Genetic analysis as a tool to investigate the molecular mechanisms underlying seed development in maize. Ann Bot 96(3):353–362
Courtois-Moreau CL, Pesquet E, Sjödin A, Muñiz L, Bollhöner B, Kaneda M, Tuominen H (2009) A unique program for cell death in xylem fibers of *Populus* stem. Plant J 58(2):260–274
Curtis MJ, Wolpert TJ (2004) The victorin-induced mitochondrial permeability transition precedes cell shrinkage and biochemical markers of cell death, and shrinkage occurs without loss of membrane integrity. Plant J 38(2):244–259
Danon A, Delorme V, Mailhac N, Gallois P (2000) Plant programmed cell death: a common way to die. Plant Physiol Biochem 38(9):647–655
David KK, Sasaki M, Yu SW, Dawson TM, Dawson VL (2006) EndoG is dispensable in embryogenesis and apoptosis. Cell Death Differ 13(7):1147–1155
DeBono AG, Greenwood JS (2006) Characterization of programmed cell death in the endosperm cells of tomato seed: two distinct death programs. Botany 84(5):791–804
Dellaporta SL, Calderon-Urrea A (1994) The sex determination process in maize. Science 266(5190):1501–1505
DeLong L, Plenefisch JD, Klein RD, Meyer BJ (1993) Feedback control of sex determination by dosage compensation revealed through *Caenorhabditis elegans* sdc-3 mutations. Genetics 133(4):875–896
Doukhanina EV, Chen S, van der Zalm E, Godzik A, Reed J, Dickman MB (2006) Identification and functional characterization of the BAG protein family in *Arabidopsis thaliana*. J Biol Chem 281(27):18793–18801
Dufresne A, Garczarek L, Partensky F (2005) Accelerated evolution associated with genome reduction in a free-living prokaryote. Genome Biol 6:1–10
Durand PM, Ramsey G (2019) The nature of programmed cell death. Biol Theory 14:30–41
Durand PM, Choudhury R, Rashidi A, Michod RE (2014) Programmed death in a unicellular organism has species-specific fitness effects. Biol Lett 10(2):20131088
Durand PM, Sym S, Michod RE (2016) Programmed cell death and complexity in microbial systems. Curr Biol 26(13):R587–R593
Earnshaw WC, Martins LM, Kaufmann SH (1999) Mammalian caspases: structure, activation, substrates, and functions during apoptosis. Annu Rev Biochem 68(1):383–424
Esau K (1965) Anatomy and cytology of Vitis phloem. Hilgardia 37(2):17–72
Fagerstedt KV (2010) Programmed cell death and aerenchyma formation under hypoxia. Waterlogging signalling and tolerance in plants, 99–118

Filonova LH, Bozhkov PV, Brukhin VB, Daniel G, Zhivotovsky B, Arnold SV (2000) Two waves of programmed cell death occur during formation and development of somatic embryos in the gymnosperm, Norway spruce. J Cell Sci 113(24):4399–4411

Franklin DJ, Brussaard CP, Berges JA (2006) What is the role and nature of programmed cell death in phytoplankton ecology? European J Phycol 41(1):1–14

Franklin DJ, Brussaard CP, Berges JA, LaRoche J (2019) Photophysiology of *Phaeocystis Antarctica* at its southern geographic distribution limit. Limnol Oceanogr 64(1):19–33

Fukuda H (1994) Redifferentiation of single mesophyll cells into tracheary elements. Int J Plant Sci 155(3):262–271

Galluzzi L, Vitale I, Aaronson SA, Abrams JM, Adam D, Agostinis P, Turk B (2018) Molecular mechanisms of cell death: recommendations of the nomenclature committee on cell death 2018. Cell Death Differ 25(3):486–541

Galluzzi L, Vitale I, Abrams JM, Alnemri ES, Baehrecke EH, Blagosklonny MV, Kroemer G (2012) Molecular definitions of cell death subroutines: recommendations of the Nomenclature Committee on Cell Death 2012. Cell Death Differ 19(1):107–120

Gechev TS, Van Breusegem F, Stone JM, Denev I, Laloi C (2006) Reactive oxygen species as signals that modulate plant stress responses and programmed cell death. BioEssays 28(11):1091–1101

Greenberg JT, Yao N (2004) The role and regulation of programmed cell death in plant–pathogen interactions. Cell Microbiol 6(3):201–211

Greenwood JS, Helm M, Gietl C (2005) Ricinosomes and endosperm transfer cell structure in programmed cell death of the nucellus during *Ricinus* seed development. Proc Natl Acad Sci 102(6):2238–2243

Gunawardena AH, Pearce DM, Jackson MB, Hawes CR, Evans DE (2001) Characterisation of programmed cell death during aerenchyma formation induced by ethylene or hypoxia in roots of maize (*Zea mays* L.). Planta 212:205–214

Hansen DJ, Bellman SK, Sacher RM (1976) Gibberellic acid-controlled sex expression of corn tassels 1. Crop Sci 16(3):371–374

Hara-Nishimura I, Hatsugai N, Nakaune S, Kuroyanagi M, Nishimura M (2005) Vacuolar processing enzyme: an executor of plant cell death. Curr Opin Plant Biol 8(4):404–408

Hayward AP, Tsao J, Dinesh-Kumar SP (2009) Autophagy and plant innate immunity: defense through degradation. Semin Cell Dev Biol 20(9):1041–1047. Academic Press

Heath MC, Nimchuk ZL, Xu H (1997) Plant nuclear migrations as indicators of critical interactions between resistant or susceptible cowpea epidermal cells and invasion hyphae of the cowpea rust fungus. New Phytol 135(4):689–700

Hegedus DD, Rimmer SR (2005) *Sclerotinia sclerotiorum*: when "to be or not to be" a pathogen? FEMS Microbiol Lett 251(2):177–184

Hoeberichts FA, Woltering EJ (2003) Multiple mediators of plant programmed cell death: interplay of conserved cell death mechanisms and plant-specific regulators. BioEssays 25(1):47–57

Hofius D, Schultz-Larsen T, Joensen J, Tsitsigiannis DI, Petersen NH, Mattsson O et al (2009) Autophagic components contribute to hypersensitive cell death in *Arabidopsis*. Cell 137(4):773–783

Inada N, Sakai A, Kuroiwa H, Kuroiwa T (2000) Senescence in the non-greening region of the rice (*Oryza sativa*) coleoptile. Protoplasma 214:180–193

Kiorboe T (2008) A mechanistic approach to plankton ecology. Princeton University Press

Kladnik A, Chamusco K, Dermastia M, Chourey P (2004) Evidence of programmed cell death in post-phloem transport cells of the maternal pedicel tissue in developing caryopsis of maize. Plant Physiol 136(3):3572–3581

Klemenčič M, Asplund-Samuelsson J, Dolinar M, Funk C (2019) Phylogenetic distribution and diversity of bacterial pseudo-orthocaspases underline their putative role in photosynthesis. Front Plant Sci 10:293

Klemenčič M, Novinec M, Dolinar M (2015) Orthocaspases are proteolytically active prokaryotic caspase homologues: the case of *Microcystis aeruginosa*. Mol Microbiol 98(1):142–150

Koltunow AM, Truettner J, Cox KH, Wallroth M, Goldberg RB (1990) Different temporal and spatial gene expression patterns occur during anther development. Plant Cell 2(12):1201–1224

Koonin EV, Aravind L (2002) Origin and evolution of eukaryotic apoptosis: the bacterial connection. Cell Death & Differentiation 9(4):394–404
Kroemer G, Galluzzi L, Brenner C (2007) Mitochondrial membrane permeabilization in cell death. Physiol Rev 87(1):99–163
Kroemer G, Galluzzi L, Vandenabeele P, Abrams J, Alnemri ES, Baehrecke EH, Melino G (2009) Classification of cell death: recommendations of the Nomenclature Committee on Cell Death 2009. Cell Death Differ 16(1):3–11
Kroemer G, Jäättelä M (2005) Lysosomes and autophagy in cell death control. Nat Rev Cancer 5(11):886–897
Kwon SI, Cho HJ, Jung JH, Yoshimoto K, Shirasu K, Park OK (2010) The Rab GTPase RabG3b functions in autophagy and contributes to tracheary element differentiation in Arabidopsis. Plant J 64(1):151–164
Lam E, Zhang Y (2012) Regulating the reapers: activating metacaspases for programmed cell death. Trends Plant Sci 17(8):487–494
Li YQ, Southworth D, Linskens HF, Mulcahy DL, Cresti M (1995) Localization of ubiquitin in anthers and pistils of Nicotiana. Sex Plant Reprod 8:123–128
Liu Y, Levine B (2015) Autosis and autophagic cell death: the dark side of autophagy. Cell Death Differ 22(3):367–376
Locato V, De Gara L (2018) Programmed cell death in plants: an overview. Plant programmed cell death: methods and protocols, 1–8
Martínez M, Rubio-Somoza I, Carbonero P, Díaz I (2003) A cathepsin B-like cysteine protease gene from Hordeum vulgare (gene CatB) induced by GA in aleurone cells is under circadian control in leaves. J Exp Bot 54(384):951–959
McClure B, Cruz-García F, Romero C (2011) Compatibility and incompatibility in S-RNase-based systems. Ann Bot 108(4):647–658
McLellan H, Gilroy EM, Yun BW, Birch PR, Loake GJ (2009) Functional redundancy in the *Arabidopsis* Cathepsin B gene family contributes to basal defence, the hypersensitive response and senescence. New Phytol 183(2):408–418
Mur LA, Kenton P, Lloyd AJ, Ougham H, Prats E (2008) The hypersensitive response; the centenary is upon us but how much do we know? J Exp Bot 59(3):501–520
Nasrallah JB, Rundle SJ, Nasrallah ME (1994) Genetic evidence for the requirement of the *Brassica* S-locus receptor kinase gene in the self-incompatibility response. Plant J 5(3):373–384
Newbigin E, Anderson MA, Clarke AE (1993) Gametophytic self-incompatibility systems. Plant Cell 5(10):1315
Nguyen HN, Sabelli PA, Larkins BA (2007) Endoreduplication and programmed cell death in the cereal endosperm, *endosperm: developmental and molecular biology*, pp 21–43
Ni XL, Gui MY, Tan LL, Zhu Q, Liu WZ, Li CX (2019) Programmed cell death and aerenchyma formation in water-logged sunflower stems and its promotion by ethylene and ROS. Front Plant Sci 9:1928
Obara K, Kuriyama H, Fukuda H (2001) Direct evidence of active and rapid nuclear degradation triggered by vacuole rupture during programmed cell death in *Zinnia*. Plant Physiol 125(2):615–626
Orellana MV, Pang WL, Durand PM, Whitehead K, Baliga NS (2013) A role for programmed cell death in the microbial loop. PLoS One 8(5):e62595
Pacini E, Dolferus R (2016) The trials and tribulations of the plant male gametophyte—understanding reproductive stage stress tolerance. In: Abiotic and biotic stress in plants-recent advances and future perspectives. IntechOpen
Pak C, Van Doorn WG (2005) Delay of iris flower senescence by protease inhibitors. New Phytol 165(2):473–480
Piszczek E, Dudkiewicz M, Mielecki M (2012) Biochemical and bioinformatic characterization of type II metacaspase protein (TaeMCAII) from wheat. Plant Mol Biol Report 30:1338–1347
Reape TJ, McCabe PF (2008) Apoptotic-like programmed cell death in plants. New Phytol 180(1):13–26
Rojo E, Martın R, Carter C, Zouhar J, Pan S, Plotnikova J, Raikhel NV (2004) VPEγ exhibits a caspase-like activity that contributes to defense against pathogens. Curr Biol 14(21):1897–1906

Scavia D, Fahnenstiel GL (1987) Dynamics of Lake Michigan phytoplankton: mechanisms controlling epilimnetic communities. J Great Lakes Res 13(2):103–120

Senatore A, Trobacher CP, Greenwood JS (2009) Ricinosomes predict programmed cell death leading to anther dehiscence in tomato. Plant Physiol 149(2):775–790

Smertenko AP, Bozhkov PV, Filonova LH, von Arnold S, Hussey PJ (2003) Re-organisation of the cytoskeleton during developmental programmed cell death in *Picea abies* embryos. Plant J 33(5):813–824

Smetacek V (2012) The role of phytoplankton in ocean ecosystems. Naturwissenschaften 99(1):8–13

Sreenivasulu N, Altschmied L, Radchuk V, Gubatz S, Wobus U, Weschke W (2004) Transcript profiles and deduced changes of metabolic pathways in maternal and filial tissues of developing barley grains. Plant J 37(4):539–553

Sundström JF, Vaculova A, Smertenko AP, Savenkov EI, Golovko A, Minina E, Bozhkov PV (2009) Tudor staphylococcal nuclease is an evolutionarily conserved component of the programmed cell death degradome. Nat Cell Biol 11(11):1347–1354

Uren AG, O'Rourke K, Aravind LA, Pisabarro MT, Seshagiri S, Koonin EV, Dixit VM (2000) Identification of paracaspases and metacaspases: two ancient families of caspase-like proteins, one of which plays a key role in MALT lymphoma. Mol Cell 6(4):961–967

Vakifahmetoglu H, Olsson M, Zhivotovsky B (2008) Death through a tragedy: mitotic catastrophe. Cell Death Differ 15(7):1153–1162

Van Doorn WG (2011) Classes of programmed cell death in plants, compared to those in animals. J Exp Bot 62(14):4749–4761

Van Doorn WG, Balk PA, Van Houwelingen AM, Hoeberichts FA, Hall RD, Vorst O, Van Wordragen MF (2003) Gene expression during anthesis and senescence in Iris flowers. Plant Mol Biol 53:845–863

Vandenabeele P, Galluzzi L, Vanden Berghe T, Kroemer G (2010) Molecular mechanisms of necroptosis: an ordered cellular explosion. Nat Rev Mol Cell Biol 11(10):700–714

Vercammen D, Van De Cotte B, De Jaeger G, Eeckhout D, Casteels P, Vandepoele K, Van Breusegem F (2004) Type II metacaspases Atmc4 and Atmc9 of *Arabidopsis thaliana* cleave substrates after arginine and lysine. *J Biol Chem* 279(44):45329–45336

Watanabe N, Lam E (2004) Recent advance in the study of caspase-like proteases and Bax inhibitor-1 in plants: their possible roles as regulator of programmed cell death. Mol Plant Pathol 5(1):65–70

Williams B, Dickman M (2008) Plant programmed cell death: can't live with it; can't live without it. Mol Plant Pathol 9(4):531–544

Woffenden BJ, Freeman TB, Beers EP (1998) Proteasome inhibitors prevent tracheary element differentiation in *Zinnia* mesophyll cell cultures. Plant Physiol 118(2):419–430

Yamada T, Ichimura K, van Doorn WG (2006b) DNA degradation and nuclear degeneration during programmed cell death in petals of *Antirrhinum, Argyranthemum*, and *petunia*. J Exp Bot 57(14):3543–3552

Yamada T, Takatsu Y, Kasumi M, Ichimura K, van Doorn WG (2006a) Nuclear fragmentation and DNA degradation during programmed cell death in petals of morning glory (*Ipomoea* nil). Planta 224:1279–1290

Yamada T, Takatsu Y, Kasumi M, Manabe T, Hayashi M, Marubashi W, Niwa M (2001) Novel evaluation method of flower senescence in freesia (*Freesia hybrida*) based on apoptosis as an indicator. Plant Biotechnol 18(3):215–218

Yamada T, Takatsu Y, Manabe T, Kasumi M, Marubashi W (2003) Suppressive effect of trehalose on apoptotic cell death leading to petal senescence in ethylene-insensitive flowers of gladiolus. Plant Sci 164(2):213–221

Young TE, Gallie DR (2000) Programmed cell death during endosperm development. Programmed cell death in higher plants, 39–57

Zheng W, Rasmussen U, Zheng S, Bao X, Chen B, Gao Y, Bergman B (2013) Multiple modes of cell death discovered in a prokaryotic (cyanobacterial) endosymbiont. PLoS One 8(6):e66147

www.ingramcontent.com/pod-product-compliance
Ingram Content Group UK Ltd.
Pitfield, Milton Keynes, MK11 3LW, UK
UKHW022216051026
15624UKWH00001B/186
* 9 7 8 9 8 1 9 7 1 8 8 2 5 *